Mass Spectrometry:
Techniques and Applications

Mass Spectrometry:

Techniques and Applications

EDITED BY

GEORGE W. A. MILNE

NATIONAL HEART AND LUNG INSTITUTE
NATIONAL INSTITUTES OF HEALTH, BETHESDA, MARYLAND

WILEY-INTERSCIENCE,

A DIVISION OF JOHN WILEY & SONS, INC.
NEW YORK · LONDON · SYDNEY · TORONTO

535.84
M63

Contributors

B. G. Buchanan, Department of Computer Science, Stanford University, California

Maurice M. Bursey, The Venable Chemical Laboratory, University of North Carolina, Chapel Hill, North Carolina

Dominic M. Desiderio, Jr., Department of Biochemistry, Baylor College of Medicine, Houston, Texas

A. M. Duffield, Department of Chemistry, Stanford University, California

Henry M. Fales, National Heart and Lung Institute, National Institutes of Health, Bethesda, Maryland

P. V. Fennessey, Martin Marietta Corporation, Denver, Colorado

Marvin F. Grostic, The Upjohn Company, Kalamazoo, Michigan

Michael K. Hoffman, The Venable Chemical Laboratory, University of North Carolina, Chapel Hill, North Carolina

K. R. Jennings, Department of Chemistry, The University, Sheffield, England

A. A. Kiryushkin, Institute for Chemistry of Natural Products, USSR Academy of Sciences, Moscow, USSR

W. J. McMurray, Section of Physical Sciences, Yale University School of Medicine, New Haven, Connecticut

G. W. A. Milne, National Heart and Lung Institute, National Institutes of Health, Bethesda, Maryland

Yu A. Ovchinnikov, Institute for Chemistry of Natural Products, USSR Academy of Sciences, Moscow, USSR

Kenneth L. Rinehart, Jr., Department of Chemistry, University of Illinois, Urbana, Illinois

A. V. Robertson, Department of Organic Chemistry, University of Sydney, N.S.W., Australia

Ragnar Ryhage, Laboratory of Mass Spectrometry of the Karolinska Institute, Stockholm, Sweden

M. M. Shemyakin, Institute for Chemistry of Natural Products, USSR Academy of Sciences, Moscow, USSR (deceased, June 1970)

R. Venkataraghavan, Department of Chemistry, Cornell University, Ithaca, New York

Sten Wikström, Department of Biophysical Sciences, University of Houston, Houston, Texas

Preface

The affair between organic chemistry and mass spectrometry may be conveniently, if inaccurately, said to have begun in 1960. At about this time equipment suitable for organic mass spectrometry first became commercially available. In compiling this book I have been conscious of the fact that a full decade of work lies behind us and that a correspondingly large amount of data now exists in the literature. The first objective of a work of this sort is to review these papers critically and assess their impact on the development of mass spectrometry. A second aim is to collect essays on the state of the art of the various aspects (or subdisciplines) of mass spectrometry. Thus a record can be established and subsequently consulted by those who are outside the specialized area. Finally, in a book such as this it should be possible to probe a little into the future of mass spectrometry, in order to see what instrumentation may be developed and what new problems tackled at the start, if not in the course, of its second decade.

Neither mass spectrometry nor organic chemistry has been the same since they were introduced to each other by such pioneers as Beynon, Biemann, Djerassi, and McLafferty. Organic chemistry has changed almost beyond recognition during the past ten years, and mass spectrometry has been the biggest single force behind this change. The sensitivity available to the organic chemist, increasing for half a century at the comfortable rate of one order of magnitude per thirty years, suddenly shot up by four orders of magnitude in ten years. New areas opened, long-dormant problems were revived, and completely new interdisciplinary projects became possible. Mass spectrometry, at the same time, was not so much changed by, as modeled around, organic chemistry with the result that today any discussion of resolving power or mass measurement accuracy tends to be centered in such items as the low positive packing fractions of carbon and nitrogen, the high positive packing fraction of hydrogen, and the isotopic compositions of these elements. The organic mass spectroscopist who is suddenly faced with a problem involving, for example, selenium or cobalt will rapidly recognize the narrowness of his viewpoint and may well be excused for reflecting that there are, indeed, more things in heaven and earth than he ever dreamed of.

As editor of a book of this sort, I dealt therefore with several problems. I, of course, had to find contributors who were in command of their subject but not slaves to it. They had to retain sufficient perspective to recognize the importance of a subdiscipline with respect to the science as a whole, and be able to appraise objectively the territory ahead. There are, fortunately, many such people in mass spectrometry and I am fortunate to have been able to persuade 18 of them to collaborate with me in the preparation of this book. A second problem arose because of the potential overlap between many pairs of chapters, while other pairs may allow omissions. Achieving a smooth articulation in both of these cases is an editorial problem and a difficult one. Whatever success can be claimed in this area rightfully belongs to all the contributors, who have communicated with one another in the preparation of their respective chapters. Timing is another editorial problem with which I have had a great deal of assistance from all of the contributors. As a result, all chapters were completed during the first part of 1970, before, it should be noted, the 1970 ASTM-ASMS meeting. Although American authors preponderate, the international character of mass spectrometry is, to some extent, reflected in the list of authors. Five countries are represented and special credit is due the authors of two chapters in particular "Gas Chromatography-Mass Spectrometry" and "Mass Spectrometry in Peptide Chemistry"), who wrote their articles in exemplary English.

The conception of this book was first discussed at the 1968 meeting of the ASTM Committee E-14 in Pittsburgh. One of the ideas brought up at that time was that it would be interesting to seek a substantial contribution for such a book from the younger mass spectroscopists. This idea has been adopted and therefore about half the authors are erstwhile students of (or students of students of) either Biemann, Djerassi, or McLafferty, three noted authors of books on mass spectrometry. Irrespective of age, each author is an acknowledged expert in the area covered by his chapter. The book is thus, in a sense, a research document and should in no way be considered to be an introductory text. Instead, it should be regarded as a collection of critical reviews of the various areas of organic mass spectrometry, to be used by those familiar with general principles.

Finally, I must acknowledge the great deal of assistance I received during the preparation of this book. Most of the chapters have been read by friends and colleagues who have provided me with competent advice on editorial and scientific problems. For help in this area I thank DRS. R. J. Highet, S. Staley, J. Silverton, and S. Markey. I am particularly indebted to Miss V. A. Aandahl, who, in addition to reading manuscripts, provided me with a great deal of advice on computer techniques. I must acknowledge a large debt of gratitude to DR. H. M. Fales, who introduced me to mass spectrometry. Finally, my thanks are due to my wife for her skillful assistance in much of the editorial work and for her patience of the past 18 months.

GEORGE W. A. MILNE

Bethesda, Maryland
February 1971

Contents

AUTOMATIC ACQUISITION AND PROCESSING OF MASS-SPECTRAL DATA 1
By R. Venkataraghavan

PHOTOGRAPHIC TECHNIQUES IN ORGANIC HIGH-RESOLUTION MASS
SPECTROMETRY 11
By Dominic M. Desiderio, Jr.

THE ELECTRICAL RECORDING OF MAGNETICALLY SCANNED HIGH-
RESOLUTION MASS SPECTRA 43
By W. J. McMurray

COMPUTER-ASSISTED INTERPRETATION OF MASS SPECTRA 77
By P. V. Fennessey

GAS CHROMATOGRAPHY—MASS SPECTROMETRY 91
By Ragnar Ryhage and Sten Wikström

AN APPLICATION OF ARTIFICIAL INTELLIGENCE TO THE INTERPRETATION
OF MASS SPECTRA 121
By B. G. Buchanan, A. M. Duffield, and A. V. Robertson

NEWER IONIZATION TECHNIQUES 179
By Henry M. Fales

MASS-SPECTRAL STUDIES EMPLOYING STABLE ISOTOPES IN CHEMISTRY
AND BIOCHEMISTRY 217
By Marvin F. Grostic and Kenneth L. Rinehart, Jr.

MASS SPECTROMETRY IN PEPTIDE CHEMISTRY 289
By M. M. Shemyakin, Yu A. Ovchinnikov, and A. A. Kiryushkin

THE APPLICATION OF MASS SPECTROMETRY TO PROBLEMS IN MEDICINE
AND BIOCHEMISTRY 327
By G. W. A. Milne

MECHANISM STUDIES OF FRAGMENTATION PATHWAYS 373
By Maurice M. Bursey and Michael K. Hoffman

SOME ASPECTS OF METASTABLE TRANSITIONS 419
By K. R. Jennings

AUTHOR INDEX 459

SUBJECT INDEX 473

Mass Spectrometry:
Techniques and Applications

Automatic Acquisition and Processing of Mass Spectral Data

R. VENKATARAGHAVAN

Department of Chemistry, Cornell University, Ithaca, New York

I. Introduction 1
II. Mass-Spectrum Digitizers 3
    A. Semiautomatic Device 3
    B. Fully Automatic Devices 3
        1. Spectro Sadic Digitizer 4
        2. Atlantic Digitizer 5
        3. Galvanometer Digitizer 5
        4. Mascot Digitizer 6
        5. Solid-State Digitizer 6
        6. Digitizer for Gas Chromatography—Mass-Spectrometer System . . . 6
III. Computer-Aided Techniques 7
IV. Conclusion 9
    References 10

I. INTRODUCTION

Compared to other analytical tools the mass spectrometer puts out a wealth of information for a given quantity of a sample. For an effective utilization of this tremendous amount of data, a lot of time and effort is necessary. A survey of the literature shows that many attempts have been made to automate the data collection and processing techniques for a routine analysis of mass-spectral data. Industrial laboratories interested in the analysis of gasoline have done the pioneering work to automate the mass spectrometer. Their attempts have shown that the labor involved can be cut down considerably in the routine analysis of mass-spectral data by making digital data-acquisition systems as an integral part of mass spectrometers. Since 1952, several data-acquisition systems have been described with various degrees of precision and accuracy. Concurrently, several investigators have shown the great potential of the mass spectrometer as an analytical and research tool in a chemical laboratory. With all these developments, it was soon recognized

1

that to utilize all the information available from a mass spectrometer, a data-acquisition system is a necessity rather than a luxury. Initially the cost of such systems was fairly high, but with the advances in technology, the cost has been coming down rather rapidly. Consequently, more laboratories have become interested in the design and development of data-acquisition systems not only for mass spectrometers but also for other analytical instruments.

Before discussing the specific applications of data-acquisition and analysis systems to mass spectrometry, it is worthwhile to examine some of the advantages of such systems and the incentives to design and develop them. The inclusion of a data system in an experiment removes the possibilities for subjective and manual errors that may be introduced in the measurement operations. Thus, it is possible to obtain more precise and accurate results from the experiment. Because of the lack of fatigue a data acquisition system can monitor the control parameters of an experiment and make consistent observations for quite a lengthy period of time. When the system is designed properly, it can use certain criteria to judge the usefulness of the information it is collecting from the experiment or an instrument. Unnecessary information can be discarded to obtain meaningful results. With the incorporation of proper hardware and the usage of necessary software, the system can perform certain routine tasks that the experimenter would normally do. This in essence allows the scientist to concentrate on more difficult and intuitive operations to analyze the data. To obtain optimum performance from a complex research instrument it is essential to understand its characteristics. A data system can aid in the operation by making repeated and quick analysis of the data obtained from the instrument under different conditions. In order to perform a correlation task or to match unknown information with the known data it is essential to keep all the available information in an easily retrievable form. The output device of the data system can record the information in a computer-compatible format, thus making it possible for a rapid access to perform the necessary analysis. Some of the other advantages of the data-acquisition and analysis systems are the ability to have a quick look at the data, and the efficient usage of the experimental facility. It is also possible to have a data-acquisition system perform some of the control operations by incorporating sophisticated hardware and software into the system. In a feedback mode operation a data system can optimize various controls to obtain the best possible data from the instrument. It is thus obvious that the data-acquisition and analysis system can be a very valuable accessory in a laboratory. The more sophisticated a system is, the more tasks it can perform and the more it can relieve the scientist from routine work.

In this article a general review of the data systems and their applications is presented. Section II describes the design and operation of some of the mass spectrum digitizers, and it is followed by the modern approach to data acquisition and analysis in Section III. Specific details of data_acquisition and analysis are discussed elsewhere in the book.

II. MASS-SPECTRUM DIGITIZERS

One of the time-consuming steps in the analysis of mass-spectral information is the measurement of peak heights to determine the relative abundance of all the fragment ions in the spectrum. The measured results are then tabulated as mass number and relative abundance for further evaluation. In specific applications, particularly in the analysis of gasoline fractions, only a few peaks of relevance are used. On the other hand, for a complete analysis of the spectrum, data from all of the fragment ions are required.

A. Semiautomatic Device

As a first step in automation, a contact telereader was devised to read the peak heights and other supplementary information, such as the mass number and the galvanometer sensitivity factors, from the chart paper (1). The contact telereader system has a telereader, teleducer, program unit, and a flexowriter. The teleducer converts the analog voltage measured by the telereader to a digital value. The programmer unit presents these digital values along with the mass number and galvanometer factors to the flexowriter, which can print and/or punch the information. The punched-paper tape containing the mass number and peak heights of all or selected fragment ions is in a computer-compatible format for further analysis of the data. Even though the telereader system took the drudgery out of the measurement operation, it could only be considered a semiautomatic device in the sense that spectra are still recorded on chart paper. To eliminate this intermediate step of recording spectra on chart paper, and then converting to a format acceptable by a computer, fully automatic data-collection systems have been designed and built with varying degrees of sophistication.

B. Fully Automatic Devices

Before discussing the details of individual digitizing equipment, it is better to examine the overall concept involved in the design of such systems. To automate the mass-spectrometer output, it is essential to digitize analog signals corresponding to the mass number and peak amplitude. It is advantageous to sense the presence or absence of a peak before initiating the digitization process since this cuts down the demand on the digitizer. Thus, in general, all the digitization systems have the following components:

1. Peak-sensing device.
2. Digitizer.
3. Output device.

A generalized block diagram of such a system is shown in Figure 1.

Both analog and digital peak-sensing devices have been used to recognize peaks in the spectrum. The analog device in essence is a differentiator, and it senses a peak when there is a change in sign of the slope. In digital systems the same effect is accomplished by incorporation of a digital comparator circuit that compares the currently digitized value against the previous value or values to locate the point of maximum amplitude. Once the peak is

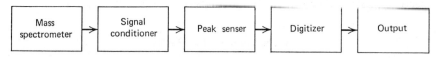

Figure 1. General block diagram of a digitizing system.

sensed, the device triggers the digitizer to convert the analog signal to a digital value. The integral part of the digitizing equipment is the analog-to-digital converter. There are a variety of devices that can be used to convert the analog voltage to a digital value, such as encoders, voltage-to-frequency converters, successive approximation converters, and the like. The output device should produce a record compatible with the computer input format. Some of the output devices that are commonly used are paper-tape punches, teletypewriters, flexowriters, IBM card punches, and magnetic tapes. In addition to showing the essential components, Figure 1 also shows a block labeled as signal conditioning equipment. The purpose of this is to make the output of the mass spectrometer compatible with the rest of the system. Signal conditioning equipment in general incorporates such devices as amplifiers, attenuators, filters, and the like.

1. *Spectro Sadic Digitizer*

The first commercial digitizer, Spectro Sadic, was built by Consolidated Engineering Corporation (2, 3). The system is capable of digitizing the peak amplitudes of preselected peaks at the rate of 8 sec per peak. There is a maximum of 40 peaks that can be preselected for a particular scan in the mass ranges 12 to 122 or 40 to 150. Peak selection is accomplished by incorporation of a special circuit that reads a specific acceleration voltage, compensates for any drift in magnetic current, and brings the corresponding mass into

focus at the collector. The analog voltage corresponding to the peak amplitude of the selected peak is converted to a digital value by the analog-to-
digital converter. When a balance is reached in the A/D converter the
digital value is output to a card reader, a paper-tape punch, or an automatic
typewriter. Some of the disadvantages of Spectro Sadic System are: (1) there
is a limit on the number of peaks that can be digitized, (2) peaks have to be
selected in advance, and (3) ambiguity exists in the analog-to-digital converter
used in the system.

2. *Atlantic Digitizer*

In order to remove any ambiguity that may be present in the conversion
of the analog voltage to a digital value, the Atlantic System substituted an
encoder for the A/D converter (2). To increase the dynamic range of the
digitizer, a logarithmic amplifier was included in the system. The encoder
used in the system employs a cathode-ray technique, has a resolution of one
half unit in 1024 units, and is claimed to be free from ambiguity. The
Atlantic System also included a Miniac digital computer to facilitate both
the data-collection and data-analysis operations. This is one of the early
systems that used a computer on-line for data acquisition.

Both the Spectro Sadic and the Atlantic digitizers are programmed
systems in the sense that only preselected peaks are digitized. For a more
general application it is desirable to digitize all the peaks in a spectrum.
Three such systems are described briefly in the following paragraphs.

3. *Galvanometer Digitizer*

This digitizer includes four galvanometers with different sensitivity, an
optical system including a grating and lens, and a detector system including
a photo cell and an electronic scaler (3, 4). Each galvanometer has its own
optical and detector systems, thus providing four attenuation ranges. The
digitizer also includes a peak-sensing device. During a measurement operation
the galvanometer deflection proportional to the peak amplitude causes a light
beam to go through the optical system. The number of light pulses
corresponding to the galvanometer deflection are counted and scaled
by the detector. The detector output is then fed to any one of the devices
capable of producing a computer-compatible record. The accelerating voltage
of the mass spectrometer is converted to the corresponding mass number by a
mechanical shaft position digitizer. One of the advantages of the galvanometer digitizer is that it can record the amplitude of all peaks with respect
to an independent base line established by the galvanometer. A careful
evaluation of the system showed that the accuracy of the digitized data is
$\pm 1\%$, well above the acceptable limit.

4. *Mascot Digitizer*

This system (3, 5) has been used more extensively than the previously described digitizers because of its precision and reliability in digitizing the mass-spectral data. The input from the peak-sensing device is applied to a comparison circuit, where it is compared with a feedback voltage. When there is a difference in the signals, a multivibrator is triggered and the pulses are counted in a counter circuit. The counter circuit is also provided with a scale-attenuation facility to provide different scale factors. This information, along with the number of pulses, is sent to a readout device to produce a computer-compatible output. Pulse output from the counter circuit is also fed to a series of digital-to-analog converters to supply a feedback voltage to the comparison circuit. Mass number digitization is accomplished in the same way as in the galvanometer digitization system. Further details of the Mascot digitizer can be found in the original references (3, 5). Recently, some modifications to the system have been described that allow the digitizer to be programmed in addition to operating in the normal mode (6). The system speed is roughly $2\frac{1}{2}$ peaks/sec, and the advantages are its precision and the allowed dynamic range on the intensity axis.

5. *Solid-State Digitizer*

The design and operation of a solid-state digitizer has been described for mass spectrometers by Thomason (7). This system employs a digital peak-sensing scheme to recognize the point of maximum amplitude in a peak. To avoid recognizing spurious noise signals as a legitimate peak, several comparisons are made with the previously digitized values. Only when five successive readings follow the prescribed scheme (current value is less the previously digitized value) the data is accepted as a true peak. After obtaining the point of maximum amplitude, the system reads the accelerating voltage and feeds it to a digital voltmeter. The digitized values of peak amplitude and mass number are punched and printed out in the output device. The solid-state digitizer is the first system in which a complete digital logic is used for data acquisition. There are eight distinct logic states governing the operation of the digitizer, and the recognition and response of each state is described by Thomason (7).

6. *Digitizer for Gas Chromatography—Mass-Spectrometer System*

When a gas chromatograph is used as an inlet to the mass spectrometer, it is desirable to take several mass-spectral scans over each of the gas chromatographic peaks. In addition to having the essential components for digitizing the peak amplitudes and mass numbers, this system also has a cycle timer that sweeps the mass spectrometer over the desired range at specified time intervals (8). The digitizer is activated only when a peak is sensed in the gas chromatogram. Thus, several mass spectra can be obtained over one particular gas chromatographic peak. As can be seen, it is essential to scan the mass spectrometer rapidly whenever the components are eluted from the gas chromatograph. To obtain reliable information from this type of system it is necessary to have fast response from the detector of the mass spectrometer. A detailed description of the design criteria for gas-chromatograph/mass-spectrometer systems can be found elsewhere in this book.

The main application of the systems described so far has been in digitizing the mass spectra of a mixture and analyzing the results for the components. Several mathematical techniques have been described for quantitative analysis of the data. Computer programs have been written to read the output from the mass-spectrum digitizer, perform the necessary computations, and print out the composition of the mixture. As mentioned earlier, laboratories interested in the analysis of gasoline fractions have developed a wide variety of techniques for the quantitative analysis of data. The details of these methods are beyond the scope of this chapter.

III. COMPUTER-AIDED TECHNIQUES

Since Beynon's demonstration of the potentialities of high-resolution mass-spectral data for structural identification (9) and Biemann's proposal for an automatic data collection and reduction system (10, 11) several laboratories have reported a wide variety of techniques for data acquisition and analysis. With the availability of low-cost, high-speed computers, there has been an increasing tendency to make the computer an integral part of the data acquisition and reduction system for mass spectrometers. Figure 2 shows a generalized block diagram of the data-acquisition systems used for

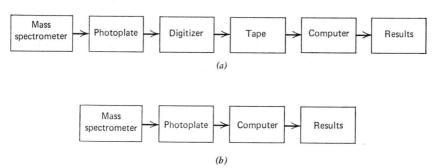

Figure 2. Block diagrams of photoplate data-acquisition system (*a*) off-line, (*b*) on-line.

collecting data from photoplate-recorded high-resolution mass spectra (11–15). Techniques employed for digitizing the electron multiplier output of the mass spectrometer are shown in Figure 3 (16–20). From these block diagrams it can be seen that the system can be either on-line or off-line to a computer. There are several advantages in making a system on-line, the main one being the availability of data within a few minutes after the scan. Incorporation of a digital computer into the system makes it possible to

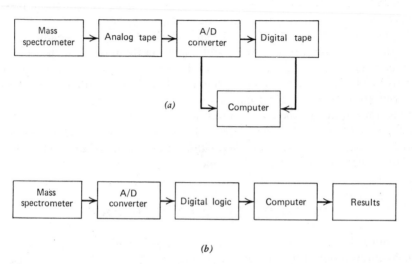

Figure 3. Block diagrams of electrical recording system (a) analog recording and subsequent conversion, (b) direct digital recording.

perform several tasks such as processing the data, correlating information, recording the results, displaying data, and returning control signals to the experiment, besides acquiring data. On-line systems thus add a new dimension in the handling of information with greater flexibility, compared to the digitizers described earlier. Further details of these data-acquisition systems can be found in the following chapters.

Computer-aided methods have been developed to perform sophisticated analysis of the data, and the techniques employed can be broadly subdivided as follows:

1. *Data reduction.* Raw data are reduced to precise mass, relative abundance, and possible elemental compositions.
2. *Preliminary processing.* Reduced data are categorized according to their elemental compositions (element mapping).
3. *Data analysis.* Evaluation of the data is performed, based on previously known fragmentation patterns for structural clues.

4. *Interpretation*. Collection is made of all available information for the prediction of possible structures. Detailed discussion of these techniques can be found in Chapter 4, "Computer-Assisted Interpretation of Mass Spectra."

Sophisticated data-acquisition systems have been designed to analyze the low-resolution mass-spectral data. On-line computers have considerably reduced the amount of labor involved in the interpretation of data and identification of compounds, particularly when the gas chromatograph is used as an inlet system to the mass spectrometer.

IV. CONCLUSION

To utilize effectively the tremendous quantity of data available from a mass spectrum, it is desirable to have a data system capable of doing several meaningful operations. With the present capability it is possible to design and implement a laboratory data system that will operate in an interactive manner, with the experimenter performing several useful and complex calculations in real time. An ideal data system for the mass spectrometers should include a small- or medium-sized computer, a cathode-ray tube (CRT) for visual display, high-speed peripheral units such as discs for auxiliary storage, a printer and a plotter for hard-copy output, and digital logic elements to control and govern the data-acquisition phase. Incorporation of additional sophisticated hardware and software would allow the system to operate in a feedback mode, thus controlling the instrument for optimum performance. The initial cost of such a system will be fairly high, but can be justified on the basis that it will serve a variety of other instruments besides the mass spectrometers and vastly improve the efficiency of the entire operation.

A study of the evolution of digitizing systems for automatic acquisition and analysis, has shown that they are becoming an integral part of the mass spectrometer. The trend recently has been to incorporate a small, low-cost computer into the instrument. By relieving the researcher of the routine tasks of data analysis, the digitizing systems serve a very valuable purpose in a chemical laboratory.

ACKNOWLEDGMENTS

I am indebted to Professor F. W. McLafferty and T. Wachs for a critical review of this manuscript.

REFERENCES

1. B. K. Fritts and C. G. Peattie, *Anal. Chem.*, **28**, 1518 (1956).
2. B. F. Dudenbostel, Jr., and W. Priestley, Jr., *Anal. Chem.*, **26**, 1275 (1954).
3. B. F. Dudenbostel, Jr., and P. J. Klaas, in J. D. Waldron (Ed.), *Advances in Mass Spectrometry*, Vol. I, Pergamon Press, New York, 1959, p. 232.
4. W. H. King, Jr., and A. P. Gifford, *5th Annual Conference on Mass Spectrometry and Allied Topics, ASTM, Committee E-14*, 1957, New York, N.Y.
5. A. P. Gifford, *5th Annual Conference on Mass Spectrometry and Allied Topics, ASTM, Committee E-14*, 1957, New York, N.Y.
6. J. F. Light, *Anal. Chem.*, **37**, 1627 (1965).
7. E. M. Thomason, *Anal. Chem.*, **35**, 2155 (1963).
8. R. F. Klaver and R. M. Teeter, *11th Annual Conference on Mass Spectrometry and Allied Topics, ASTM, Committee E-14*, 1963, San Francisco, California, p. 153.
9. J. H. Beynon, in J. D. Waldron (Ed.), *Advances in Mass Spectrometry*, Vol. I, Pergamon Press, New York, 1959, p. 328.
10. K. Biemann, W. McMurray, and P. Bommer, *12th Annual Conference on Mass Spectrometry and Allied Topics, ASTM, Committee E-14*, 1964, Montreal, Canada, p. 428.
11. D. Desiderio and K. Biemann, *12th Annual Conference on Mass Spectrometry and Allied Topics, ASTM, Committee E-14*, 1964, Montreal, Canada, p. 433.
12. K. Biemann, *Pure Appl. Chem.*, **9**, 95 (1964).
13. R. W. Olsen and A. L. Burlingame, *13th Annual Conference on Mass Spectrometry and Allied Topics, ASTM, Committee E-14*, 1965, St. Louis, Missouri.
14. R. Venkataraghavan, F. W. McLafferty, and J. W. Amy, *Anal. Chem.*, **39**, 178 (1967).
15. D. D. Tunnicliff and P. A. Wadsworth, *Anal. Chem.*, **40**, 1826 (1968).
16. C. Merritt, Jr., P. Issenberg, M. L. Bazinet, B. N. Greene, T. O. Merron, and J. G. Murray, *Anal. Chem.*, **37**, 1037 (1965).
17. W. J. McMurray, B. N. Greene, and S. R. Lipsky, *Anal. Chem.*, **38**, 1194 (1966).
18. H. C. Bowen, E. Clayton, D. J. Shields, and H. M. Stanier, in E. Kendrick (Ed.), *Advances in Mass Spectrometry*, Vol. IV, The Institute of Petroleum, London, 1968.
19. A. L. Burlingame, in E. Kendrick (Ed.), *Advances in Mass Spectrometry*, Vol. IV, The Institute of Petroleum, London, 1968.
20. R. Venkataraghavan, R. J. Klimowski, J. E. Coutant, and F. W. McLafferty, *17th Annual Conference on Mass Spectrometry and Allied Topics, ASTM, Committee E-14* 1969, Dallas, Texas, p. 31.

Photographic Techniques in Organic High-Resolution Mass Spectrometry

DOMINIC M. DESIDERIO, JR.

*Institute for Lipid Research and Department
of Biochemistry, Baylor College of Medicine,
Houston, Texas*

I. Introduction 12
II. Instrumentation 12
   A. Geometry 12
   B. Some Areas of Applicability of Photoplates 14
      1. GC-MS Detection 14
      2. Analysis of Mixtures Without Prior Separation 15
      3. Small Amounts of Sample 15
      4. Metastable-Ion Recording 15
      5. Field Ionization 16
III. The Photographic Plate 16
   A. Glass 16
   B. Gelatin Emulsions 17
   C. The Characteristic Response Curve of Photographic Emulsions . . . 17
   D. Calibration of the Photographic Emulsion 19
IV. Theory of Photographic Processes After Ion Impact 20
   A. The Theory for the Formation of the Latent Image 21
   B. Primary Versus Secondary Processes 22
   C. Mass and Energy Effects on the Photographic Response 23
V. Types of Commercially Available Photoplates 24
VI. Linear and Microdensitometric Measurements 26
   A. Linear 27
   B. Microdensitometry 27
   C. Output 28
VII. Data Reduction 29
   A. Resolution of Overlapping Peaks 29
   B. Peak-Center Determination 31
   C. Peak Area 31
   D. Conversion of Distance to Mass 32
   E. Determination of Elemental Compositions 33
VIII. Forms of Data Output 35
   A. Listing of Elemental Compositions 35
   B. Element Map 35
   C. Heteroatomic Plot 39
   D. Topographical Element Map 39
   E. Ion Types 40
IX. Summary 40
   References 41

I. INTRODUCTION

Mass spectrometry has been used for many years for quantitative and/or qualitative analysis in the fields of petroleum chemistry, inorganic chemistry (spark source for solids), and gas analysis. Recently, analytical and organic chemists, biochemists, and pharmacologists have applied this technique with notable success to their particular structural-determination problems. However, due to the great diversity in these newcomers' backgrounds, all of the basic theories and facts of mass spectrometry have not been conveniently and collectively available to them. In general, this book compiles all of the necessary facts for these newcomers to this field.

The detection of positive ions in a high-resolution mass spectrograph with a photographic plate to produce a permanent record of an experiment, and those techniques necessary to abstract and present the recorded data will be discussed in this chapter. Seven topics will be covered—instrumentation, physical parameters of photoplates, theory of photographic processes occurring after ion impact, types of commercially available photoplates, linear and microdensitometric measurements, data reduction, and various forms of data output.

II. INSTRUMENTATION

A. Geometry

In order to illustrate the need for a photographic plate as an ion detector, a very brief review of the two basic types of geometry that are commerically available will be given here. Historically, the first design of a double-focusing (direction and velocity) mass spectrograph was presented by Mattauch and Herzog (1). As seen in Figure 1, a source (S) provides a very narrow ion

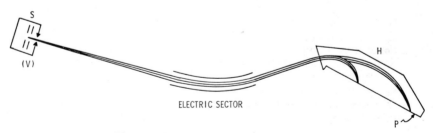

Figure 1. Mattauch-Herzog geometry.

"ribbon" (e.g., the height is a few millimeters, and, for a resolution of 20,000, the width is approximately 0.0002 in., or 5 μ). After traversing a field-free drift region, the ion ribbon enters an electric field region that focuses the energy of the ion beam. The monoenergetic ion beam passes into a magnetic field (H) that separates the masses according to the equation

$$\frac{m}{e} = \text{constant} \times \frac{H^2 r^2}{V} \qquad [1]$$

where m/e is the mass to charge ratio of an ion having a trajectory whose radius of curvature is r in a particular magnetic field (H) and accelerated with

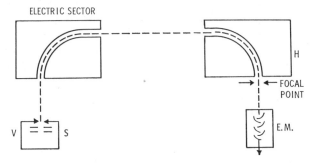

Figure 2. Nier-Johnson geometry.

a potential of V. The ions are focused at a *plane* (P) in which a photographic plate is placed. The ions (m/e) cause a blackening of the photoplate relative to the number of ions and register at varying radii of deflection (r) according to a fixed H and V. In most mass spectrographs, usually 15–50 mass spectra can be recorded per photoplate (dimensions for a Q2 plate 2 × 10.5 × 0.040 in.), each exposure lasting from fractions of a minute up to a few minutes each. In addition, a photomultiplier may be placed at a point in the focal plane, the magnetic field scanned, and a high-resolution spectrum obtained.

The second geometry is that of Nier and Johnson (2) and is schematically represented in Figure 2. The same sequence as above is followed: source, electric field, and magnetic deflection. However, two important differences are noted: the ions are focused at a focal *point*, and therefore, in order to obtain a mass spectrum, it is necessary to scan either H or V. The H is usually scanned, because of the greater mass range provided by the squaring effect.

Chapter 3 will deal more fully with the topic of electrical-scanning mass spectrometers.

B. Some Areas of Applicability of Photoplates

There is not usually a clear-cut choice between the above two types of mass spectrometers to be used in any general type of laboratory. This sort of decision is also beyond the scope of this article. However, five areas of current interest in mass spectrometry will be briefly discussed in order to illustrate those areas in which photoplate detection might have a greater applicability. One should not infer from the following discussion that photoplate techniques are the only, or the best, method of ion detection for every situation. (W. J. McMurray will discuss the merits of electrical scanning of mass spectra in Chapter 3.) The fields are gas-chromatography/mass-spectrometry combination instruments; the analysis of mixtures, usually of biological origin, in which the separation of constituents is either not necessary, or not desirable; extremely small amounts of sample, again usually of biological origin; the recording of metastable ions produced in a double-focusing instrument, with or without the "defocusing" method; and finally, field ionization. In addition, as was mentioned in the introduction, a permanent record of the experiment is obtained, in a convenient, storable form, at a relatively high "density of information" content.

1. *GC-MS Detection*

There are several variables to consider when obtaining the mass spectrum of peaks emerging from a gas chromatograph after the helium has been preferentially removed by a separator. These are the changing concentration of the eluent with time, the variable amount of marker compound necessitated by varying peak heights of eluents, the ability to move very quickly from one exposure to the next whenever two peaks emerge quickly in succession, and the ability to obtain a *complete* high-resolution mass spectrum in seconds whenever a very small peak emerges.

In the first case, the integrability of the photoplate obviates any difficulties experienced with scanning instruments (3–5), such as biasing the mass spectrum whenever a scan is taken for a significant portion of the time that the peak is emerging from the chromatograph.

Whenever the peak heights of the eluents vary widely and the required amount of marker compound must follow the changes approximately, it is possible to place a variable leak between the reservoir and the ion source. This is accomplished with a Varian model 951-5100 in the author's laboratory so that the intensities of both the compound and the marker compound are approximately equal.

If two peaks on the gas chromatogram elute very close to each other, or if resolution of two peaks is not complete, it is often necessary to move the photoplate very rapidly from one exposure to another in order to obtain a "pure" mass spectrum of each compound. On most commercial instruments, this can be accomplished manually within approximately three seconds. However, in my laboratory, it was found convenient to provide a motor drive on the photoplate transport mechanism and actuate the motor by a switch located near the ion beam deflection switch. Changing exposures thus requires fractions of a second (6).

Finally, and in some cases, most importantly, if the elution time, or the peak height, of a peak is very small, it is still possible to obtain a complete mass spectrum within seconds. Naturally, the ion current due to the marker compound must be increased, but a 10^{-10} coulomb charge can be deposited on the photoplate (Q2) within 10 sec and the mass range up to 700 atomic mass units, amu, can be exposed.

2. *Analysis of Mixtures Without Prior Separation*

The detection and structural elucidation of minor bases occurring in crude nucleic acid hydrolysate mixtures are now confronting biochemists. A procedure has been developed whereby unfractionated transfer RNA is degraded by snake venom and bacterial alkaline phosphatase followed by trimethylsilylation. The resultant *mixture* is placed into a direct introduction probe and volatilized. The mass spectrum obtained contains the major nucleosides (adenosine, guanosine, cytosine, and thymine) along with the chemically modified nucleosides. In this case, the long-term integrability of the photoplate permits the detection of the low concentration levels of the minor nucleosides (7).

3. *Small Amounts of Sample*

Frequently, biological systems provide extremely small amounts of material for purposes of structural elucidation. In these cases, as was also noted under Section B-1, the short-term integrability of the photoplate plays a prominent role. An example of this is illustrated by the recent structural elucidation of the hypothalamic hypophysiotropic hormone TRF (thyroid-stimulating-hormone [TSH] releasing factor). After the necessary purification steps (which took a decade to develop), each sheep hypothalamus furnished a fraction of a nanogram of TRF. Therefore it was necessary to extract TRF from millions of sheep. As the sample was extremely valuable, micrograms of material were treated with diazomethane and the structure pyro-Glu-His-Pro-NH$_2$ elucidated (8–10).

4. *Metastable-Ion Recording*

In general, photoplates are not the most common way to detect metastable ions. (For a discussion of metastable transitions, see Chapter 12.)

However, Biemann *et al.* have devised a computer-aided correlation of metastable peaks with the normally obtained high-resolution mass-spectral data (11).

Recently, "defocusing" techniques for a double-focusing instrument have been developed whereby metastable ions that have been formed between the source and the electric sector can be focused onto the detector by varying the accelerating voltage independently of the electric sector (12). McLafferty has developed the photoplate techniques necessary for the detection of the various metastable ions produced at incremental changes in the accelerating voltage (13).

5. *Field Ionization*

In Chapter 7, "Newer Ionization Techniques," ionization methods other than electron bombardment are discussed. One of these—field ionization—produces an ion beam by means of a very high electric field gradient, and the resulting mass spectrum is simplified in that most of the ion current is carried by molecular ions (14). McLafferty *et al.* (15) and Burlingame *et al.* (16) have developed techniques employing a combination electron-impact/ field-ionization ion source. In these cases, the ion-current fluctuations are averaged out by the photoplate integration.

These brief discussions of five areas of current interest in mass spectrometry, in addition to the primary goal of high resolution mass spectrometry— accurate mass measurement—should exemplify the applicability of photographic detection of ions.

III. THE PHOTOGRAPHIC PLATE

The physical parameters (glass, gelatin, and response curve) and the emulsion calibration of the photoplate are discussed in this section to provide a background for a better understanding of events occurring after ion impact, which are discussed later, and to illustrate the advantages and disadvantages inherent in this type of ion detection.

A. Glass

The best way to maintain an emulsion rigidly in the focal plane is to coat a glass plate with an emulsion. The distance between the lines recorded on the plate must be measured with an accuracy of more than $\frac{1}{2}\mu$ to determine masses accurate to within 0.001 mass unit (one millimass unit, mmu). Therefore a photoplate presents the following advantages: very great dimensional stability, no humidity coefficient, and a very small thermal expansion coefficient of $0.2\ \mu/$(in.)(°C), and any dimensional changes due to thermal effects are truly reversible (17). The disadvantages are outgassing of the emulsion in a vacuum; loading photoplates into the spectrograph, necessitating breaking the vacuum; wet chemical development; and scanning of the photoplate to acquire the data.

In order to maintain the emulsion in the focal plane, the glass should be flat, and "waviness" should be minimal. The glass employed for these purposes is very flat—less than $25\ \mu$ variation in depth per linear inch of glass.

B. Gelatin Emulsions

(This discussion is not applicable to evaporated silver bromide photoplates—*vide infra*.)

A photographic plate's emulsion is about $2.5\ \mu$ thick, with silver bromide grains distributed within this volume of gelatin. The average AgBr particle diameter is about $0.8\ \mu$ (18). The gelatin is obtained from collagen— a fibrous protein occurring in the connective tissue of mammals. Gelatin is obtained from lime-processed ox skin and ox bone and contains mostly the amino acids glycine, proline, and alanine, and also extremely low amounts of sulfur-containing amino acids. As we will see later, this sulfur plays a very important role by providing silver sulfide sensitivity specks that increase the energy levels of "traps" that attract migrating electrons and/or silver ions after ion impact.

C. The Characteristic Response Curve of Photographic Emulsions

When an ion strikes an emulsion, a blackening process is initiated at the point of impact. In order to relate the blackening to the number of ions, the curve relating these two parameters must be determined quantitatively. Unfortunately, a simple linear relationship does not hold over the entire response curve from a barely detectable blackening to the depletion of the AgBr grains.

A typical response curve is shown in Figure 3. The curve has a sigmoid shape and consists of four characteristic regions. The curve up to A represents the background of the emulsion and is the sum of the blackening of the glass base, the emulsion, plus any darkening introduced by chemical development. From A to B is the toe region where ions begin to cause grains to become developable. The B to C is a straight-line portion where an increase in the logarithm of the number of ions causes a linear increase in the emulsion blackening. From C to D, the number of available AgBr grains is decreasing and eventually no more blackening occurs with any further ion bombardment.

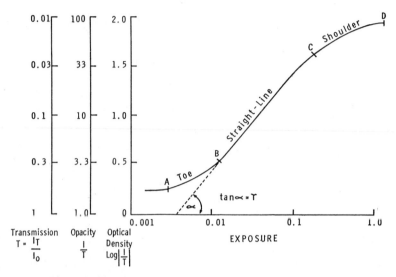

Figure 3. The characteristic curve of a photographic emulsion.

The abscissa of this graph is the exposure level and is the logarithm of the number of bombarding ions. The coordinate is the blackening of the emulsion and is measured by passing a light beam through the blackened regions. The intensity of the transmitted light (I_t) divided by the intensity of the incident light (I_0) is the transmission (T):

$$T = \frac{I_t}{I_0} \tag{2}$$

Opacity (O) is defined as the reciprocal of T:

$$O = \frac{1}{T} = \frac{I_0}{I_t} \tag{3}$$

The most commonly measured variable, optical density (OD), is the logarithm of opacity:

$$\text{OD} = \log |O| = \log \left| \frac{1}{T} \right| = \log \left| \frac{I_0}{I_t} \right| \tag{4}$$

The slope of the straight-line portion of the curve is usually denoted alpha (α), and tan α is usually specified gamma (γ). The intercept of the straight line with the abscissa provides an exposure related to the "speed" of an emulsion.

D. Calibration of the Photographic Emulsion

The blackening of a photographic emulsion in response to ion bombard-ment has been a topic of major importance in the field of inorganic spark-source mass spectroscopy for many years. Much of the work discussed in this chapter is necessarily predicated on this prior work (19–23). Recently, an excellent paper by Hayes (24) discussed the limitations and capabilities in terms of resolution and sensitivity in organic chemical mass spectrography. The effect of instrumental parameters on the minimum line width (2–3 μ) that could be obtained on Q2 photoplates was discussed. Knowing the response curve accurately permits an organic mass spectroscopist to deter-mine relative ion intensities more quantitatively and reproducibly and thus permits deduction of subtle intensity changes from various possible frag-mentation pathways.

In order to produce the characteristic curve shown in Figure 3, a wide dynamic range (ca. 10^3) of exposure levels is necessary. There are two methods at an organic mass spectroscopist's disposal to provide this range of expos-ures. One method is to employ a multiisotopic element and expose a series of exposures, ranging from a barely detectable up to saturation level. Know-ing the ratios among the exposures and measuring the maximum optical density of the lines permits the construction of the response curve.

Another more automatic approach was developed initially to improve the quantitation of the major and minor nucleosides obtained from the enzymatic hydrolysis of transfer RNA. In this method, the complete range of exposure is covered on one exposure, allowing the remainder of the plate to be used for the compounds of interest (7, 25).

The group of ions due to the loss of one chlorine atom from hexachloro-butadiene (HCBD) is employed. An automatic device permits the recording of eight regularly decreasing (usually 20% from the preceding) exposures of HCBD by offsetting the electric sector voltage slightly between exposures (moving each line on the plate ca. 40 μ). Each total ion current is read by an integrator and the total integral printed after each exposure. The maximum optical density of each of the eight exposures for each of the nine usable isotope peaks are measured. These maxima are then plotted against their corresponding exposure levels and a 72-point response curve is obtained (usually, about 10–15 of the weakest lines are not recorded). This method requires only one line on the photoplate and is subjected to the same develop-ment conditions as those exposures on the remainder of the photoplate.

The data points are subjected to an iterative, nonlinear, least-squares computer algorithm that obtains the best-fit curve through the points and determines the parameters in the equation describing the curve (see below, [10]). Usually, the data points have a scatter of 5–7% about the "best-fit"

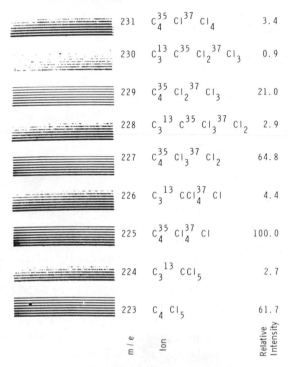

m/e	Ion	Relative Intensity
231	$C_4^{35}Cl^{37}Cl_4$	3.4
230	$C_3^{13}C^{35}Cl_2^{37}Cl_3$	0.9
229	$C_4^{35}Cl_2^{37}Cl_3$	21.0
228	$C_3^{13}C^{35}Cl_3^{37}Cl_2$	2.9
227	$C_4^{35}Cl_3^{37}Cl_2$	64.8
226	$C_3^{13}CCl_4^{37}Cl$	4.4
225	$C_4^{35}Cl_4^{37}Cl$	100.0
224	$C_3^{13}CCl_5$	2.7
223	C_4Cl_5	61.7

Figure 4. (M-Cl)$^+$ region of hexachlorobutadiene (SWR).

curve. A photograph of the (M-Cl)$^+$ region of HCBD on an SWR photoplate is given in Figure 4 to illustrate the eight superimposed spectra.

IV. THEORY OF PHOTOGRAPHIC PROCESSES AFTER ION IMPACT

An organic mass spectroscopist should at least be aware of the events occurring on the atomic level after an ion has bombarded a silver bromide crystal and what steps then render that crystal developable. This discussion will not presume to go to the greatest depths of what is known about these processes. An excellent book by Mees and James (26) is available to those interested in more extensive and/or intensive detail. This chapter is written from the viewpoint of what an organic mass spectroscopist should know in order to maximize mass accuracy, resolution, and ionic-abundance calculations. Latent image formation, primary and secondary blackening processes, and mass and energy effects on the photographic response will be discussed. The current controversy about latent image formation will not be resolved here, but merely illuminated for the sake of completeness for this discussion.

A. The Theory for the Formation of the Latent Image

Historically, the latent image was described as the change produced by exposure to light and which then could subsequently be rendered visible by development. As we shall see below, the latent-image formation appears to be a two-stage process (27):

1. The formation of stable developable silver atom specks.
2. The clustering of the specks into larger specks that could then be developed.

There are two current theories attempting to explain the latent-image formation. One was developed by R. W. Gurney and N. F. Mott (28), and the other by J. W. Mitchell (29–31). The two theories differ mainly in the nucleation stage. The Gurney-Mott theory assumes that an electron is initially trapped, whereas the Mitchell theory assumes that a silver ion is initially trapped. Conduction electrons and positive holes (Br^0) are freed in the silver bromide crystal whenever a bombarding ion transfers sufficient energy to the crystal. In this discussion a trap is an imperfection or fault in the crystal structure. The proximity of the sulfide ion mentioned previously increases the energy or depth of these traps and thus can influence the basic sensitivity of an emulsion. As is represented in Figure 5, the Gurney-Mott

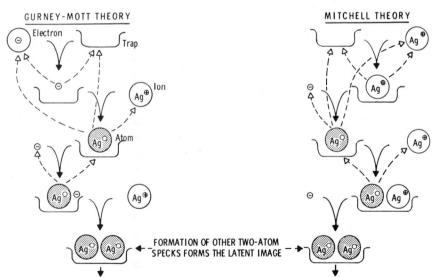

Figure 5. Schematic representation of latent image formation (adapted from Mees and James, p. 104).

theory assumes that an electron is trapped and that a silver ion migrates to this trap. This process of forming two-atom specks occurs at all of the available trapping sites. All of the above steps can be reversible except the very last step, the formation of the two-atom speck. It has been found that these specks are quite stable and last at room temperature for days (32). The collection of all of the formed two-atom specks then is known as the "latent image."

The Mitchell theory is quite similar except that it is assumed that a silver ion is initially trapped and that an electron migrates to the trap and neutralizes the silver atom. This process is repeated, and once again the stable two-atom speck is formed.

B. Primary Versus Secondary Processes

A positive ion bombarding a silver bromide grain can render that grain developable in two different ways: either by directly striking that grain or by causing light to be emitted in the gelatin by the energy of the impact and releasing a photon that travels through the emulsion and strikes a grain. The differences between these two blackening processes are illustrated by the following schematic, Figure 6, and consideration of the following three facts:

1. Emulsion thickness (approximately 2.5 μ).
2. The average silver bromide grain size (approximately 0.8 μ).
3. the depth of penetration of 10–20 keV ions (approximately 0.2 μ) (33).

Thus, as can be seen in Figure 6, a primary blackening process is one in which an ion strikes a grain located within 0.2 μ of the surface, rendering that grain

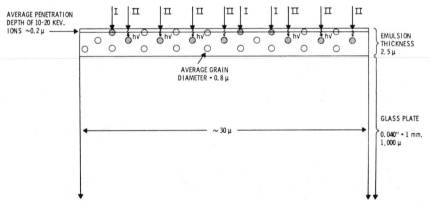

Figure 6. Schematic representation of primary and secondary photographic processes.

immediately developable. The secondary process occurs whenever an ion strikes the gelatin and there are no grains in the neighborhood of the point of impact. P. Brix concluded that the ion impact causes luminescence of the gelatin, and the energy (photon) released renders a grain farther into the emulsion developable (34). This secondary process can act at a greater distance than that of the primary process because of the transparency of gelatin towards light.

C. Mass and Energy Effects of the Photographic Response

E. B. Owens and N. A. Giardino of Lincoln Laboratories (MIT) have determined the emulsion sensitivity of the photographic plate versus ion

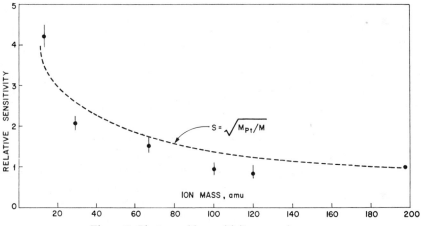

Figure 7. Photographic sensitivity versus ion mass.

mass and versus ion energy (35). The results of their experiments with ion mass versus ion sensitivity are shown in Figure 7. The abundances of the various responses are all related to the molecular ion of platinum in order to remove plate-to-plate variations. The data display a curve that decreases with the square root of the mass. There is an abrupt decrease in sensitivity up to approximately mass 50 and a very slow decrease in relative intensity up to mass 200. Figure 8 shows the photographic sensitivity versus ion energy for the ions silicon-29 and molybdenum-100 over the range of ion energies 5–50 kV obtained by including multiply-charged ions (ion energy = ion charge × accelerating potential). It is seen from this graph that the photographic sensitivity is proportional to the ion-acceleration voltage. The difference between the apparent yields of [29]Si and [100]Mo follows directly

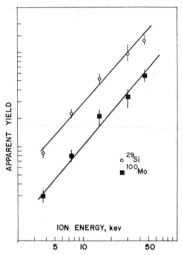

Figure 8. Photographic sensitivity versus ion energy.

from sensitivity $= f(\sqrt{mass})$. This lower sensitivity is noticed when one wishes to work at a higher mass range. One must decrease the ion accelerating voltage to reach this higher mass [1], thus decreasing the sensitivity of ion detection.

Hayes has plotted the charge density required to produce an optical density $= 0.30$ (equivalent to transmission $= 50\%$) for various perfluoro-kerosene ions versus m/e 181 as the point of normalization (24). Plotting data in this manner shows that relative sensitivity varies with $(m/e)^{-1.5}$. To ensure that this difference in the power is not due to a difference in the definition of sensitivity, Hayes adapted the method of Owens and Giardino (35) to recalculate sensitivities. He found the same dependence on $(m/e)^{-1.5}$.

V. TYPES OF COMMERCIALLY AVAILABLE PHOTOPLATES

There are three more common types of commerically available photo-plates suitable for use as ion detectors in high-resolution mass spectrographs. The most commonly employed photoplate is the Q2 photoplate from Ilford, Ltd. (Ilford, Essex, England). Kodak (Rochester, N.Y.) produces an SWR (short-wave radiation) photoplate, and Technical Operations (Tech-Ops, Burlington, Mass.) has recently introduced a photoplate of silver bromide evaporated directly onto a glass base (36, 20).

A computer plot of the characteristic curves of these three photoplates on one graph is given in Figure 9. The data were experimentally obtained by

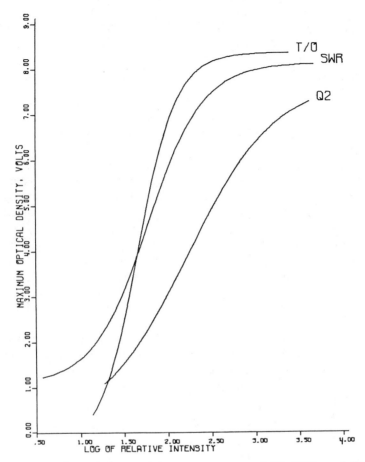

Figure 9. Comparison of characteristic response curves of Q2, SWR, and Tech/Ops emulsions.

the procedure outlined above in Section III-D. Optical density is related to the given voltage reading by:

$$OD = \log \left| \frac{V_1}{V_1 - V_2} \right|$$

where V_1 is the voltage at zero optical density (10 V), and V_2 is the voltage at the peak maximum. The following critical factors are noticed by the comparisons in Figure 9:

1. SWR has the highest sensitivity in terms of dynamic range of relative exposure.

2. Tech-Ops has the lowest limit of detectability and the lowest noise level (thus potentially the highest signal-to-noise ratio), the highest saturation level, the longest linear portion, and the highest contrast (gamma).

3. Q2 is a "slower" (ca. 4 ×) emulsion more limited in threshold and saturation levels.

The type of experiment to be performed dictates which emulsion should be employed: highest accuracy for quantitation—highest gamma—T/0 (Ref. 19, p. 77): extremely low level of exposure—SWR or T/0: general use—Q2.

The Q2 emulsion is usually developed in Kodak Microdol X and the SWR emulsion in Kodak HRP. After development, the plate is stopped (acetic acid), fixed (hypo), and washed according to normal development procedures. The Tech-Ops plate is developed according to the manufacturer's instructions with chemicals provided.

The development times and/or concentrations of the solutions can be varied to accommodate various laboratory conditions or requirements. However, the agitation and temperature of the solutions are critical. A homogeneous solution must be maintained to reduce rebromination effects by periodically bursting or continuously bubbling the solutions with an inert gas, usually nitrogen. Finally, to insure proper and reproducible developing times, the temperature of the solutions should be regulated to within ±1–2°C.

After processing, extreme care should be taken to avoid scratching or warping the emulsion. The recommended storage climate is 68–75°F and 40–60% relative humidity, preferably in enclosed solid boxes.

VI. LINEAR AND MICRODENSITOMETRIC MEASUREMENTS

The most important data to be obtained from a high-resolution mass spectrum recorded on a photoplate are the line positions. Very accurate knowledge of the line positions permits very accurate calculation of the masses which, in turn, will define the elemental compositions of the ions. If these line positions are measured with a high degree of precision, then the number of possible elemental compositions will be low. For example, if one could measure a line position within ±0.5 μ, the mass will be known to within approximately ±0.5 mmu. Whenever a compound contains a reasonable number of only C, H, N, and O, and the mass of the ion is a few hundred mass units, then perhaps two or three elemental compositions will be found to fit within this error limit.

Of secondary importance to an organic mass spectroscopist is the relative intensity of an ion. This priority is different from inorganic work, where the abundances of the ions are extremely important for quantitative analysis or for lower limits (ppb) of detection.

The device to measure distance and density data from a photoplate is called a microdensitometer—linear comparator. The distance that the linear comparator has traveled is divided into accurate fractional distances by an encoder, and a microdensitometric system measures the optical-density profile of each line. These two measurements are discussed separately in this section.

A. Linear

In order to obtain the mass of an ion to more than $\frac{1}{2}$ mmu accuracy, the position of that line must be measured with at least $\frac{1}{2}$-μ accuracy. Therefore the position of the photoplate carriage must be known with this accuracy. In the instruments that are commercially available, the plate carriage is moved by a very finely threaded, hardened steel screw. Each single rotation is accurately machined (e.g., with a 1.0000-mm pitch) so that one rotation of the screw corresponds to, for example, precisely 1.0000 mm of plate travel. The motor that automatically rotates the screw is continually variable—usually providing a carriage speed from 0 to about 100 mm/min. Theoretically then, a 250-mm plate can be scanned in less than 3 min. However, due to the limitations imposed by the data-acquisition system, the speed is usually less than this upper limit. An encoder is the device that divides each rotation into a thousand or more bits of fractional rotations of plate travel. To provide one micron accuracy, each mm of plate travel is divided into 1000 units (To provide $\frac{1}{2}\mu$ accuracy, 2000 parts; and $\frac{1}{4}$-μ accuracy, 4000 parts.) In this manner, the optical-density measurement, as described below, is known precisely up to each and every $\frac{1}{4}$-μ increment of photoplate travel.

B. Microdensitometry

The following microdensitometric system produces the optical density profile of each line on the plate:

1. *Light source.* Usually a tungsten bulb provides the intensity (I_0, [2]) needed to measure the blackening. Quartz-iodine lamps provide a more coherent, parallel light beam.

2. *Reflecting mirror.* The light source should be mounted directly under the photoplate to reduce any scattering and/or divergence of the light beam. In those cases where this is mechanically difficult, the substage illuminating beam must usually be reflected by a mirror. This should be a high-quality, first-surface mirror to reduce multiple internal light reflections.

3. *Substage collimating lenses.* Whenever a light source cannot provide

a coherent parallel light beam, a good-quality lens system is necessary to focus the light beam properly.

3a. *Preslit.* In order to transmit light of only the dimensions of the line, and reduce scattered light reflecting into the photomultiplier, a slit defining the light beam and of dimensions slightly larger than a line should be placed before the photoplate (24, p. 1967).

4. *Platen.* A glass platen usually provides support for the photoplate. However, it is better to invert the plate, emulsion down; support the plate at the edges or ends; and scan the plate in this manner. This procedure would further reduce any reflections and dispersions of the light beam.

5. *Lens.* After passing through the photoplate, the light beam is magnified. This is usually a low-power (e.g., 5 ×) magnification. For greatest flexibility, a zoom lens would be helpful.

6. *Beam splitter.* A half-silvered prism system divides the light beam into two halves—one half continues straight and the other half is deflected, usually 90°.

7. *Display screen.* After the beam splitter, another lens magnifies the image (e.g., 10 ×) and presents the image on a ground glass screen for visual observation.

8. *Slit.* The deflected half of the split beam travels through a closable (e.g., 0–20 μ) and rotatable (e.g., $\sim \pm 5°$) slit assembly. The effective slit width is actually the physical width of the slit divided by the first magnifying lens (e.g., $15 \mu \div 5 = 3 \mu$ in the plane of the photoplate). The maximum obtainable resolution from the comparator is thus defined by this slit.

9. *Photomultiplier tube.* A stabilized-voltage, high quality, photomultiplier tube finally converts the resultant light beam into an analog voltage corresponding to the optical density profile on the photoplate. Typically, the PM tube output is from 0 to -10 V, and is digitized by an analog-to-digital (A/D) converter.

C. Output

The encoder output of the distance traversed by the photoplate is coupled with the simultaneously digitized optical-density output and presented to the data-recording system (punched paper tape, punched cards, magnetic tape, or computer) (37–40).

Figure 10 shows a schematic representation of a microdensitometer-linear comparator.

Gaertner Scientific Company (Chicago, Ill.), Grant Instruments (Berkeley, Calif.), Mann Company (Burlington, Mass.), Jarrell-Ash (Waltham, Mass.), and Joyce-Loebl (Gateshead, England) manufacture commercial instruments embodying various combinations of the above components. Encoders are produced by Datex (Duarte, Calif.) and Gurley (Troy, N.Y.).

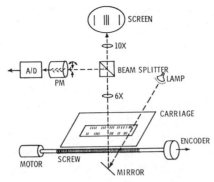

Figure 10. Schematic of microdensitometer-linear comparator.

VII. DATA REDUCTION

A. Resolution of Overlapping Peaks

In any given mass spectrum, there are probably doublets that cannot be resolved to a 10% valley (the valley commonly employed when defining resolution), no matter what the resolution of the mass spectrometer was. [The various instrumental parameters and techniques of ion exposure that relate to the attainable resolution are discussed by Hayes (24).] Thus computer techniques must be available to separate overlapping ion profiles. It can therefore be seen that the amount of work that a computer program must do to resolve overlapping peaks is inversely proportional to the resolution of the mass spectrometer.

It is known that the normal Gaussian curve approximates the ion profiles well (41). Therefore, we may describe an ion profile as follows:

$$y = \frac{1}{\sqrt{2\pi}\sigma} \exp\left[-\tfrac{1}{2}\left(\frac{x-\mu}{\sigma}\right)^2\right] \qquad [5]$$

where
y = the optical density
x = distance reading along the ion profile
μ = mean distance (center of peak)
σ = distance from μ to inflection point (at inflection point, dy/dx is maximum)

Once a mathematical model has been established, computer techniques can be employed to resolve overlapping profiles. For instance, we may generally represent [5] as a function of x and various parameters P:

$$y = f(x, P_1, \ldots, P_n) \qquad [6]$$

The best fit to the theoretical equation is obtained whenever the parameters are chosen to minimize the following function,

$$\varphi = \sum_{i=1}^{n} W_i (Y_i - \hat{Y}_i)^2 \tag{7}$$

where W is a weighting function for the theoretical value Y and the evaluated function $\hat{Y}$. Such an iterative function is quite laborious and difficult, if not impossible in some cases, to evaluate manually. The whole procedure can be programmed for a digital computer quite easily.

As an example, Figure 11 shows a computer plot of four overlapping peaks. The amount of overlap is 80, 72, and 50%. The equation describing this complex profile is

$$Y = \sum_{i=1}^{4} \frac{P_{3i-2}}{\sqrt{2\pi}\, P_{3i-1}} \exp\left[-\tfrac{1}{2} \left(\frac{X - P_{3i}}{P_{3i-1}} \right)^2 \right] \tag{8}$$

where P_i are the parameters to be evaluated. After iteration (about 20 iterations) the four indicated peaks were found. This procedure can be executed in fractions of a minute by a 7094 computer.

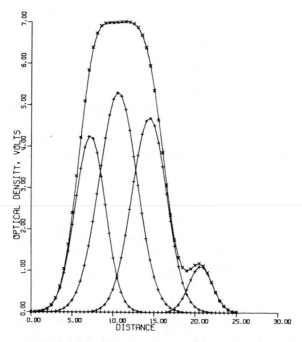

Figure 11. Resolution of overlapping ion-density profiles: P_1 through $P_{12} = 39.1, 1.8, 7.2;$ 61.4, 2.3, 10.6; 49.8, 2.1, 14.4; and 7.9, 1.4, 20.7.

B. Peak-Center Determination

For multiplets resolved by the preceding procedure, the peak center is P_{3_i} in [8]. (In a Gaussian formula, this is equivalent to μ.) For singlets, any one of three different methods could be employed to determine the peak center of an ion profile:

1. Fit a Gaussian curve through the data and obtain μ.
2. Calculate the first moment (M_1) about the mean of the profile:

$$M_1 = \frac{1}{n} \sum_{i=1}^{n} (X_i - \bar{X})^2 \qquad [9]$$

3. Fit a parabola through the three highest points, differentiate the equation, and calculate where $dy/dx = 0$.

In my laboratory, a test employed on the ion profiles to determine whether a given peak is a multiplet is to calculate both 2 and 3 (42). If they are significantly different (for example, $\gtrless 0.75\ \mu$), that profile is probably a multiplet and can be treated appropriately.

C. Peak Area

In order to estimate the number of ions that had bombarded the photo-plate to produce a given line, one must measure the area under the density-distance profile. Because each point on this curve is in optical-density units (a dimensionless logarithmic quantity), each density unit must be converted to its antilogarithmic counterpart, or intensity value. The sigmoid curve obtained above (Figure 3) gives the relationship between density and intensity.

The general equation for this type of curve is given by

$$Y = A_1 + \frac{1}{A_2 + A_3 e^{-A_4 x}} \qquad [10]$$

Typical values obtained for the parameters from a Q2 plate are

$$A_1 = 0.812 \qquad A_3 = 8.468$$
$$A_2 = 0.134 \qquad A_4 = 2.68$$

Once the equation for a given photoplate is known, density is converted to intensity by either evaluating the equation for each density value or using a table of values for a preevaluated equation.

In either event, the area under the density-distance profile is given by

$$\text{area} = \sum_{i=1}^{n} (\text{int})_i \qquad [11]$$

For accurate calculation of ion abundances, two correction factors should be taken into account, (1) the decreasing sensitivity of a photographic emulsion towards increasing mass, and (2) the spreading of the ion beam along its vertical axis (24). The vertical spread on the vertical axis is easily corrected for by letting h = the height of the beam at a distance L from the ion source; $h + \Delta h$ = the height of the beam at a distance $L + \Delta L$, and then,

$$\frac{h}{h + \Delta h} = \frac{L}{L + \Delta L} \qquad [12]$$

This correction factor corresponds to approximately 1 %/10 mm of plate.

Usually, the largest value in a spectrum is assigned 1000 units, and all other intensities are normalized to this value.

D. Conversion of Distance to Mass

According to equation one, it can be inferred that m/e can be determined by accurately measuring H, r, and V. This is true theoretically, but in practice it is very difficult to measure these values with the seven-significant-digit accuracy necessary for unambiguous mass determination. Thus, this absolute mass-determination method is impractical, and a relative method must be employed. Because of the high resolution attainable (usually one part in 20,000 or higher), a compound containing elements whose accurate fractional masses are quite different from those elements commonly encountered can be used an internal "mass-marking standard." As carbon, hydrogen, nitrogen, and oxygen are most commonly encountered in organic problems, a perfluorinated hydrocarbon serves well as a mass marker, as seen from the accurate masses of these elements (Table 1) and considering various common elemental compositions. (For example, at nominal mass 131, $C_7H_{15}O_2$ = 131.1072, whereas C_3F_5 = 130.9920.)

A very useful consequence of using a perfluorocarbon as a mass-marking compound is its regular fragmentation pattern. Beginning with the

TABLE 1

Element	Atomic Weight
^{12}C	12.000000
^1H	1.007825
^{14}N	14.003074
^{16}O	15.994915
^{19}F	18.998405

most abundant ion, 69 (CF_3), the most intense ions are 81, 93, 105, 119, 131, 143, 155, 169, 181, 193, and so on, the same masses recurring every hundred mass units. This fact—regular spacing—plus the fact that, at any given nominal mass, the fluorocarbon mass is usually the lowest accurate mass, makes the mass-calculation procedure amenable to automatic computer calculations. The computer can find (or be told) where the first two marker lines are located and then linearly extrapolate the square roots of the masses to find the third marker line, and so on, until all of the intense marker lines are found. The weaker marker lines and the lines due to the compound of interest are calculated by linear interpolation between the regularly spaced intense marker lines. At this point, the mass-marker compound has served its purpose, and all of its masses are printed, along with the deviations between the found and the theoretical masses, to provide an indication of the accuracy with which the compound masses were calculated (43).

Compounds other than perfluorocarbons have been employed recently as mass markers. Tris (perfluoroethyl, -propyl or -heptyl)-s-triazines (44) are commercially available (45) and have provided accurate mass measurements up to m/e 1185. A recent publication by Biemann described routine and accurate mass measurement up to m/e 1700 on evaporated silver bromide plates (46). Perfluoroalkyl phosphonitrilates have provided intense peaks up to m/e 3628 (47). However, both the triazines and the phosphonitrilates suffer from the same disadvantage—they do not have a regularly spaced series of intense ions across the entire mass range. This deficiency may possibly be decreased but not necessarily removed by synthesizing a mixture of mixed lengths of the alkyl side chains.

Another type of curve fitting is employed by some laboratories (41). Rather than linear extrapolation and interpolation, the mass versus distance relationship is determined by fitting all the data points to a higher power polynomial, such as

$$\text{mass} = A_n X^n + A_{n-1} X^{n-1} + \cdots + A_1 X + A_0 \qquad [13]$$

where n is usually 5. This curve can be determined easily by a computer using standard least-squares iterative procedures.

After calculating the masses of the compound lines, with either method, the computer then has a list of masses that must be converted to a list of corresponding elemental compositions.

E. Determination of Elemental Compositions

There are two ways a computer could calculate all of the possible elemental compositions within a few millimass units of an experimentally found mass. One way is to have a computer precalculate a list of all possible combinations of all of the elements that could be encountered in a given laboratory employing the maximum expected number of each element. That list would easily reach millions of possibilities and could fit only on reels of magnetic tape. For each determined mass, a computer could search the tapes for all masses fitting within the experimental error, and then print out the found elemental compositions.

The second way is to have the computer add, in all of the chemically possible combinations, the atomic weights of the particular elements that could be present in the specific compound being analyzed. The masses would be built up from zero each and every time, and whenever a theoretical mass was found within the experimental error, it would be printed (37). This method is quite fast, which might be surprising because of the amount of work that must be done for each mass. But, it must be remembered that all the computer does is what it can do most quickly—add. A subtle variation in the second method is to work on the accurate fractional mass, truncating the nominal mass (48). This is possible because the unique part of an atomic weight is its mantissa, not the characteristic.

Thus, the algorithm for calculating elemental compositions is as follows. The mass is presumed to be a saturated hydrocarbon and the maximum number of carbons and hydrogens is calculated. If this mass is within the

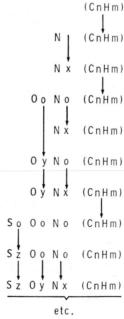

etc.

Figure 12. Computer alogrithm for calculating all possible elemental compositions.

experimental error, its composition is printed. If not, a heteroatom is sub-tracted and all of the C–H combinations are searched. If no composition is found yet, two of the same heteroatoms are subtracted and the C–H se-quence performed. This sequence is repeated until the maximum number of the first heteroatom is reached. Then a second heteroatom is subtracted in the same manner. The algorithm is repeated until all possible combinations of the maximum number for each element have been investigated.

A diagrammatic representation of this procedure is in Figure 12.

VIII. FORMS OF DATA OUTPUT

A. Listing of Elemental Compositions

In the evolutionary process of obtaining and using high-resolution mass-spectral data, various forms of data presentation were developed. At first, the elemental compositions found to fit each succeeding mass were printed out. These outputs consisted of (1) a numerical value of the intensity, (2) the mass determined from its line position, (3) the theoretical mass of the combination, (4) the difference (in mmu) between the theoretical and deter-mined masses, and (5) the numbers of each of the elements of the combination found to lie within the given range. These data are represented in Figure 13 (a partial listing) and are labeled INTENSITY, DETM, CALC, ERROR, and C, H, N, O, respectively. Even though such listings do contain all of the data, the format does not lend itself to facile interpretation (37, 43).

B. Element Map

As the reduced data (elemental compositions) are stored in a computer at one particular point in the data reduction, it is an easy matter to rearrange the data in a more meaningful way. For molecules that do not contain many heteroatoms or many types of heteroatoms, the best arrangement was found to be in terms of increasing heteroatomic content. As the data are inherently ordered in increasing mass, this procedure automatically places related ions next to each other. The data now appears in a much more condensed and meaningful arrangement and also can be looked at in one glance, rather than necessitating leafing through pages of computer output. Such a representation was termed an "element map" for reasons that will become apparent in the following discussion (49, 50).

In order to arrange the ions according to their heteroatomic content, they are sorted into groups—those ions containing only carbon and hydrogen,

INTENSITY	DETM	CALC	ERROR	C	H	N	O
178	73.0292	73.0290	.23	3	5	0	2
17	73.0658	73.0653	.43	4	9	0	1
28	74.0361	74.0368	-.67	3	6	0	2
7	75.0452	75.0446	.63	3	7	0	2
5	77.0392	77.0391	.12	6	5	0	0
6	78.0459	78.0469	-1.06	6	6	0	0
9	79.0546	79.0548	-.13	6	7	0	0
6	80.0626	80.0626	-.01	6	8	0	0
23	81.0702	81.0704	-.21	6	9	0	0
11	82.0780	82.0782	-.21	6	10	0	0
22	83.0498	83.0497	.09	5	7	0	1
62	83.0864	83.0861	.34	6	11	0	0
45	84.0582	84.0575	.72	5	8	0	1
12	84.0936	84.0939	-.34	6	12	0	0
5	85.0296	85.0290	.69	4	5	0	2
9	85.0652	85.0653	-.16	5	9	0	1
54	85.1027	85.1017	.93	6	13	0	0
5	86.0373	86.0368	.56	4	6	0	2
29	87.0451	87.0446	.48	4	7	0	2
5	87.0821	87.0810	1.10	5	11	0	1
998	88.0524	88.0524	-.06	4	8	0	2
220	89.0587	89.0603	-1.58	4	9	0	2
5	91.0554	91.0548	.58	7	7	0	0
6	93.0711	93.0704	.70	7	9	0	0
6	94.0775	94.0782	-.71	7	10	0	0
26	95.0856	95.0861	-.43	7	11	0	0
7	96.0562	96.0575	-1.31	6	8	0	1
7	96.0933	96.0939	-.61	7	12	0	0
33	97.0659	97.0653	.51	6	9	0	1
21	97.1021	97.1017	.37	7	13	0	0
39	98.0739	98.0732	.77	6	10	0	1
6	98.1102	98.1095	.64	7	14	0	0
6	99.0467	99.0446	2.09	5	7	0	2
7	99.0788	99.0810	-2.23	6	11	0	1
6	99.1176	99.1174	.27	7	15	0	0
6	100.0529	100.0524	.46	5	8	0	2
535	101.0614	101.0603	1.10	5	9	0	2
6	101.0973	101.0966	.63	6	13	0	1
117	102.0669	102.0681	-1.17	5	10	0	2
5	107.0870	107.0861	.89	8	11	0	0
10	109.1019	109.1017	.17	8	13	0	0
12	111.0818	111.0810	.79	7	11	0	1
5	111.1182	111.1174	.79	8	15	0	0
7	112.0893	112.0888	.53	7	12	0	1
5	112.1261	112.1252	.87	8	16	0	0
5	114.0682	114.0681	.16	6	10	0	2
100	115.0771	115.0759	1.19	6	11	0	2
5	121.1023	121.1017	.58	9	13	0	0

Figure 13. Partial listing of elemental compositions.

those containing in addition one oxygen, two oxygens, and so on, those containing one nitrogen, and so on, those containing one nitrogen and one oxygen, and so on, until all possible combinations of all of those heteroatoms present have been printed. These compositions are listed in vertical columns next to each other but in such a way that compositions of the same nominal mass fall on the same line. As an example of this graphical output, an element map of ethyl undecanoate is given in Figure 14. At the left is the nominal masses (to save space, nominal masses where no ions are found are omitted). The next column lists those ions containing only carbon and hydrogen; the number of each is printed, separated by a slash. The following number is the difference (in truncated mmu) between the theoretical and determined masses. Finally, the intensity of the ion is represented by a number of asterisks. The number of asterisks is calculated by 2.7 times the logarithm of the relative intensity (based on 1000 units for the most intense ion). The factor 2.7 was chosen merely because it permits displaying the intensity range 1–1000 using 1–8 asterisks. The maximum number eight was dictated by the space available when placing eight output columns on a line printer having 132 spaces.

In the next column to the right, all ions containing an additional oxygen are listed in a similar manner. It should be noted that entering only the number of carbons and hydrogens suffices to define the elemental composition of the ion because the heteroatoms are given in the column heading. Thus the line

$$87 \qquad 5/11 \ 0* \qquad 4/7 \ 0***$$

shows that two compound ions are present at the nominal mass 87 in the mass spectrum of ethyl undecanoate. One of them has a composition of $C_5H_{11}O$ while the other corresponds to $C_4H_7O_2$. Both measurements were made with an error of 0 mmu (which, due to truncation, could have been as large as ± 0.99 mmu).

Thus the number of carbon and hydrogen atoms increases from top to bottom and the heteroatoms from left to right. The elemental composition of the molecular ion (if sufficiently abundant to be detected) is shown in the lower right-hand corner $(C_{13}H_{26}O_2)$, and above it are those ions that had lost only hydrocarbon fragments. In Figure 14, all but three carbons and five hydrogens can be lost without losing oxygen (smallest ion $C_3H_5O_2$). This implies that this is the smallest O_2-containing unit in the spectrum above m/e 72.

The hydrocarbon ions (first column) run up to $C_{11}H_{19}$, indicating the presence of a long hydrocarbon chain. The O_1-column goes up to $C_{11}H_{21}O$, which is C_2H_5O less than the molecular ion. These facts demonstrate the presence of an ethyl ester (C_2H_5O) of a saturated, unbranched, long-chain $C_{11}H_{21}O$ acid, that is, undecanoic acid.

	CH	CHO	CHO$_2$	CHO$_3$	CHO$_4$
73		4/ 9 0***	3/ 5 0*****		
74			3/ 6 0***		
75			3/ 7 0**		
77	6/ 5 0*				
78	6/ 6-1**				
79	6/ 7 0**				
80	6/ 8 0**				
81	6/ 9 0***				
82	6/10 0**				
83	6/11 0****	5/ 7 0***			
84	6/12 0**	5/ 8 0****			
85	6/13 0****	5/ 9 0**	4/ 5 0*		
86			4/ 6 0*		
87		5/11 0*	4/ 7 0***		
88			4/ 8-1*********		
89			4/ 9-2*******		
91	7/ 7 0*				
93	7/ 9 0**				
94	7/10 0**				
95	7/11 0***				
96	7/12-1**	6/ 8-1**			
97	7/13 0***	6/ 9 0*****			
98	7/14 0**	6/10 0****			
99	7/15 0**		5/ 7 1**		
100			5/ 8 0**		
101		6/13 0**	5/ 9 0********		
102			5/10-1*****		
107	8/11 0*				
109	8/13 0**				
111	8/15 0*	7/11 0**			
112	8/16 1*	7/12 0**			
114			6/10 0*		
115			6/11 1*****		
121	9/13 0*				
122	9/14-1**				
123	9/15 1*				
124	9/16 0**				
125	9/17 0*	8/13 0**			
126	9/18 0**				
129			7/13 0****		
130			7/14-1**		
135	10/15 1**				
139		9/15 0**			
143			8/15-1*****		
149	11/17 0**				
150	11/18-1**				
151	11/19 0**				
155		10/19 0**			
157			9/17 1*****		
166		11/18 1*			
167		11/19 0**			
168		11/20-1*			
169		11/21 0******			
171			10/19 0*****		
185			11/21 0***		
186			11/22-1*		
187			11/23-1*		
214			13/26-1****		
	CH	CHO	CHO$_2$	CHO$_3$	CHO$_4$

Figure 14. Element map of ethyl undecanoate.

38

This detailed description of a relatively simple molecule whose structure could have been easily deduced from a conventional low-resolution mass spectrum was presented only as an introduction to the technique of element mapping and also to present the fact that the use of the actual elemental compositions, rather than nominal masses, adds an entire new dimension and greater facility to the interpretation of mass spectra.

With this detailed discussion of element mapping and with the attendant introduction to working with elemental compositions, three modifications to the basic output format and data interpretation will now be presented: (1) heteroatomic plots, (2) topographical element maps, and (3) ion types.

C. Heteroatomic Plot

In the heteroatomic plot (51) one obtains a plot of relative intensity versus the carbon/hydrogen content. In terms of structural elucidation, this form of output is manageable only for compounds containing very few heteroatoms; otherwise, one must simultaneously search many separate such plots.

D. Topographical Element Map

A topographical element map presents a three-dimensional output relating the relative intensity of an ion, the number of carbon atoms, and the heteroatomic content (see Figure 15). One must then calculate the degree of saturation from the position of the relative intensity arrow on the carbon-hydrogen axis in order to obtain the elemental composition (52).

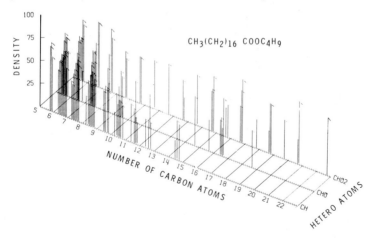

Figure 15. Topographical element map of butyl stearate.

E. Ion Types

If the data are arranged so that all of the ions representing the various homologous series (for instance, alkyl, acyl, alkyl amine, and the like) are gathered together, an "ion type" output results. The data are again subdivided according to the heteroatomic content of the ion, and further grouped according to their degree of saturation. For instance, in the ion type of the series $C_nH_{2n-1}O_2$ of the methyl ester of a C_{20} isoprenoid fatty acid, intensity maxima are shown at carbon numbers 3, 5, 8, 10, 13, 15, 18, and 20, thus making the branching points conspicuous (53, 54).

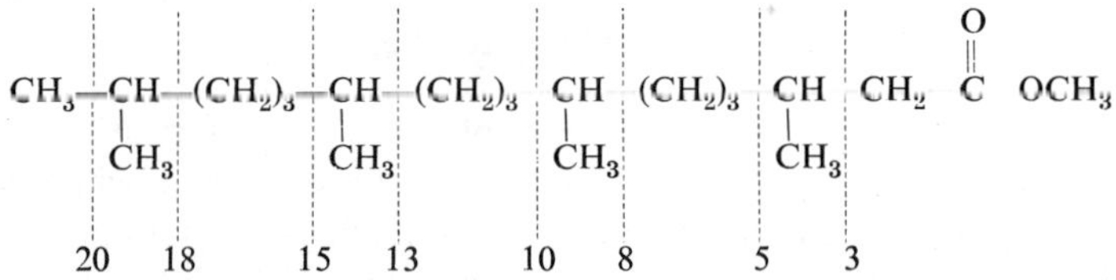

$$
\begin{array}{c}
\underset{20}{CH_3}\text{---}\underset{18}{\underset{|}{CH}}\text{---}(CH_2)_3\text{---}\underset{15}{\underset{|}{CH}}\text{---}(CH_2)_3\ \underset{10}{\underset{|}{CH}}\ (CH_2)_3\ \underset{5}{\underset{|}{CH}}\ CH_2\ \overset{O}{\overset{\|}{C}}\ \underset{3}{OCH_3} \\
\underset{}{CH_3}\qquad\ \underset{13}{CH_3}\qquad\ \underset{8}{CH_3}\qquad\ CH_3
\end{array}
$$

This discussion serves only as an introduction to the various methods available for presenting mass spectral data. Each individual chemist will have specific needs and desires that will dictate the necessary format for his particular data.

IX. SUMMARY

One of the most common methods of obtaining a complete high-resolution mass spectrum in permanent form is photoplate recording. This chapter discusses the advantages, disadvantages, and applicability of this form of ion detection. The physical aspects and the underlying atomic phenomena caused by ion impact are discussed. The apparatus to acquire the information from the photoplate, the computer algorithms to reduce the data to elemental compositions, and the various ways to display the data in order to facilitate interpretation are examined.

ACKNOWLEDGMENTS

I express my gratitude to Drs. J. A. McCloskey and P. Vouros for their helpful comments concerning the manuscript; to Miss C. Weise for assistance in preparing the computer plots; to Drs. F. McLafferty and E. Owens for permission to use their data; to my wife for helping with the editing; and to C. Renaud and C. Howell for their typing assistance.

Financial support from NIH (GM-13901) is gratefully acknowledged.

REFERENCES

1. J. Mattauch and R. Herzog, *Z. Physik*, **89**, 786 (1934).
2. E. G. Johnson and A. O. Nier, *Phys. Rev.*, **91**, 10 (1953).
3. F. A. J. M. Leemans and J. A. McCloskey, *J. Amer. Oil Chem. Soc.*, **44**, 11 (1967).
4. J. T. Watson in L. S. Ettre and W. H. McFadden (Eds.), *Ancillary Techniques in Gas Chromatography*, Interscience Division of Wiley, 1969.
5. J. T. Watson and K. Biemann, *Anal. Chem.*, **36**, 1135 (1964).
6. J. Leferink, P. Krueger, and J. A. McCloskey, personal communication.
7. D. M. Desiderio, N. R. Earle, P. M. Krueger, A. M. Lawson, L. Smith, R. N. Stillwell, K. Tsuboyama, J. Wijtvliet, and J. A. McCloskey, *16th Annual Conference on Mass Spectrometry and Allied Topics, ASTM, Committee E-14*, 1968, Pittsburgh, Pennsylvania, p. 228.
8. R. Burgus, T. F. Dunn, D. M. Desiderio, and R. Guillemin, *C. R. Acad. Sci. (Paris)*, **269**, 1870 (1969).
9. R. Burgus, T. F. Dunn, D. M. Desiderio, D. N. Ward, W. Vale, and R. Guillemin, *Nature*, **226**, 321 (1970).
10. D. M. Desiderio, R. Burgus, T. F. Dunn, D. N. Ward, W. Vale, and R. Guillemin, *18th Annual Conference on Mass Spectrometry and Allied Topics, ASTM, Committee E-14*, 1970, San Francisco, California, p. B67.
11. N. R. Mancuso, S. Tsunakawa, and K. Biemann, *Anal. Chem.*, **38**, 1775 (1966).
12. K. R. Jennings, *J. Chem. Phys.*, **43**, 4176 (1965).
13. T. W. Shannon, T. E. Mead, C. G. Warner, and F. W. McLafferty, *Anal. Chem.*, **39**, 1748 (1967).
14. H. D. Beckey, H. Knöppel, G. Metzinger, and P. Schulze, in W. L. Mead (Ed.), *Advances in Mass Spectrometry*, Vol. 3, Institute for Petroleum, London, England, 1966.
15. E. M. Chait, T. W. Shannon, J. W. Amy, and F. W. McLafferty, *Anal. Chem.*, **40**, 835 (1968).
16. P. Schulze, B. R. Simoneit, and A. L. Burlingame, *Int. J. of Mass Spectr. and Ion Phys.*, **2**, 181 (1969).
17. Kodak Pamphlet No. Q-35.
18. J. M. McCrea, *12th Annual Conference on Mass Spectrometry and Allied Topics, ASTM, Committee E-14*, 1964, Montreal, Canada, p. 541.
19. E. B. Owens in A. J. Ahearn (Ed.), *Mass Spectrometric Analysis of Solids*, Elsevier, New York, 1966.
20. R. E. Honig, J. R. Woolston, and D. A. Kramer, *Rev. Sci. Instrum.*, **38**, 1703 (1967).
21. A. J. Ahearn in A. J. Ahearn (Ed.), *Mass Spectrometric Analysis of Solids*, Elsevier, New York, 1966.
22. J. M. McCrea, *Appl. Spectrosc.*, **21**, 305 (1967).
23. P. R. Kennicott, *15th Annual Conference on Mass Spectrometry and Allied Topics, ASTM, Committee E-14*, 1967, Denver, Colorado, p. 145.
24. J. M. Hayes, *Anal. Chem.*, **41**, 1966 (1969).
25. P. Vouros, D. M. Desiderio, J. V. M. Leferink, and J. A. McCloskey, *18th Annual*

Conference on Mass Spectrometry and Allied Topics, ASTM, Committee E-14, 1970, San Francisco, California, p. B372.

26. C. E. K. Mees and T. H. James (Eds.), *The Theory of the Photographic Process*, 3rd ed., Macmillan, New York, 1967, Chapter 3.

27. Ibid., Chapter 5, p. 103.

28. R. W. Gurney and N. F. Mott. *Proc. Roy. Soc. (London)*, *Ser. A*, **164**, 151 (1938).

29. J. W. Mitchell, *Rept. Progr. Phys.*, **20**, 433 (1957).

30. J. W. Mitchell, *J. Photo. Sci.*, **6**, 57 (1958).

31. Ibid., **9**, 328 (1961).

32. E. B. Owens, in A. J. Ahearn (Ed.), *Mass Spectrometric Analysis of Solids*, Elsevier, New York, 1966, p. 60.

33. Ibid., p. 63.

34. P. Brix, *Z. Physik*, **126**, 35 (1949).

35. E. B. Owens and N. A. Giardino, *Anal. Chem.*, **35**, 1172 (1963); and also references in Ref. 20.

36. M. H. Hunt, *Anal. Chem.*, **38**, 620 (1966).

37. D. M. Desiderio, Ph.D. thesis, M.I.T., 1965.

38. R. Venkataraghavan, F. W. McLafferty, and J. W. Amy, *Anal. Chem.*, **39**, 178 (1967).

39. D. M. Desiderio and T. E. Mead, *Anal. Chem.*, **40**, 2090 (1968).

40. D. M. Desiderio and T. E. Mead, *16th Annual Conference on Mass Spectrometry and Allied Topics*, ASTM, *Committee E-14*, 1968, Pittsburgh, Pennsylvania, p. 125.

41. D. D. Tunnicliff and P. A. Wadsworth, *Anal. Chem.*, **40**, 1826 (1968).

42. C. Cone, P. Fennessey, R. Hites, N. Mancuso, and K. Biemann,*15th Annual Conference on Mass Spectrometry and Allied Topics*, ASTM, *Committee E-14*, 1967, Denver, Colorado, p. 114.

43. D. M. Desiderio and K. Biemann, *12th Annual Conference on Mass Spectrometry and Allied Topics*, ASTM, *Committee E-14*, 1964, Montreal, Canada, p. 433.

44. T. Aczel, *Anal. Chem.*, **40**, 1917 (1968).

45. "Mass Spec Kit," Peninsular Chemresearch, Inc., Gainesville, Fla.

46. C. Hignite and K. Biemann, *Org. Mass Spectrom.*, **2**, 1215 (1969).

47. H. M. Fales, *Anal. Chem.*, **38**, 1058 (1966).

48. J. Lederberg, "Computation of Molecular Formulas for Mass Spectrometry" Holden-Day, San Francisco, 1964.

49. K. Biemann, P. Bommer, and D. M. Desiderio, *Tetrahedron Lett.*, **26**, 1725 (1964).

50. K. Biemann, P. Bommer, D. M. Desiderio, and W. J. McMurray, in W. L. Mead (Ed.), *Advances in Mass Spectrometry*, Vol. III, Elsevier, Amsterdam, 1966.

51. A. L. Burlingame, *13th Annual Conference on Mass Spectrometry and Allied Topics*, ASTM, *Committee E-14*, 1965, St. Louis, Missouri.

52. R. Venkataraghavan and F. W. McLafferty, *Anal. Chem.*, **39**, 278 (1967).

53. K. Biemann, W. McMurray, and P. V. Fennessey, *Tetrahedron Lett.*, **33**, 3997 (1966).

54. P. V. Fennessey, Ph.D. thesis, M.I.T., July 1968.

The Electrical Recording of Magnetically Scanned High-Resolution Mass Spectra

W. J. McMURRAY

Section of Physical Sciences, Yale University School of Medicine, New Haven, Connecticut.

I. Introduction 43
II. Mass-Spectrometric System 45
   A. Scanning System 45
   B. Detection System 48
      1. Theoretical Considerations 48
   C. Ion Statistics 51
   D. Operational Technique 55
      1. Resolution 55
III. Data Acquisition 57
   A. Data-Acquisition System 57
   B. Analog-Digital Converter 57
   C. Multiplexers 58
   D. Threshold Comparators 59
   E. Digital Recording Devices 60
   F. Analog Recording Devices 61
   G. Determination of Peak Maxima 62
   H. Selection of the Analog-to-Digital Converter 62
IV. Data Processing 67
   A. Peak-Profile Analysis 69
   B. Threshold Determination 69
   C. Deconvolution 70
   D. Peak-Position Determination 70
   E. Peak Identification 71
   F. Time-to-Mass Conversion 73
   G. Elemental Composition Determination 74
   References 75

I. INTRODUCTION

The development of the technique of electrical recording of magnetically scanned high-resolution mass spectra began in late 1964. Earlier, Biemann, Bommer, and Desiderio (1) had shown that the full utilization of all information present in a high-resolution mass spectrum required the exact mass

43

measurement of all the ion peaks in the mass spectrum. Furthermore, Watson and Biemann (2) had demonstrated the recording of high-resolution mass spectra of gas chromatographic effluents with a double-focusing mass spectrometer that utilized a photographic plate to record the mass spectra. However, a photographic recording system is possible only with mass spectrometers that contain a focal plane, that is, a geometry that simultaneously focuses ions of all mass-to-charge ratios. The question at this time was, could high resolution mass spectra be obtained from gas chromatographic effluents by the electrical recording of the analog output from a fast magnetically scanned high-resolution mass spectrometer? At the start, conflicting conclusions had cast doubt on the feasibility of obtaining complete high-resolution mass spectra by this technique. McFadden and Day (3), after examining the relationship between scan time, sensitivity, and resolution on an instrument of low resolving power ($M/\Delta M \sim 400$), extrapolated their results to higher resolution ($M/\Delta M \sim 10,000$) and concluded that fast scanning and electrical recording of mass spectra at this resolving power was impractical because of ion statistical considerations. Later, Merritt et al. (4) showed that the fast-scan technique was feasible by recording mass spectra at a resolution of 9000 and a scan speed of 13 seconds per decade. For these experiments, the spectra were recorded on analog tape and the mass measurements were obtained by measuring the distance between the reference peaks and the sample peaks on the oscillographic recordings obtained on slow playback of the analog tape into the oscillographic recorder. The reference peaks are obtained by introducing simultaneously into the ion source with the unknown sample a compound for which the accurate mass for each of its fragments is known. These first mass measurements showed errors larger than those routinely obtained with the photographic recording technique. Green, Merren, and Murray subsequently reported (5) improved mass measurements on selected peaks utilizing the centroid instead of half widths as the measure of peak location. By utilizing automated data-acquisition techniques and digital computers, McMurray, Green, and Lipsky (6) demonstrated that mass-measurement accuracies in the order of 10 ppm on all the peaks in the spectrum could be routinely obtained. Their spectra were also recorded on analog tape. The analog tape was played back through an analog-to-digital converter to produce a digital mass spectrum. The digital spectrum, recorded on digital magnetic tape, was then processed by a computer that converted all the peaks in the spectrum to exact masses and elemental compositions. Not only did the computer permit the rapid conversion of peak position to accurate mass, but the program also incorporated a correction for a systematic error in the exponential scan. The spectra were obtained at $M/\Delta M \sim 10,000$ and at scan speeds of 10 and 72 sec/decade. It was now apparent that the electrical recording of high-resolution mass spectra by rapidly scanning the magnetic field was indeed practical. It was also clear that high-speed, sophisticated data-acquisition systems, combined with computer processing, were required if the full capability of the method was to be realized.

The intervening years have witnessed a rapid development in data acquisition systems. This has resulted both from the rapid technological advances in this area and the ingenuity of the various investigators. Indeed, the accuracy of the results obtained with most of the systems now in use are limited by the mass-spectrometer instrumentation (ion statistics, signal-to-noise, and the like) and not by limitations in the data-acquisition devices. It is impossible in a chapter such as this to acknowledge the contributions of each individual to the developments that have made this technique for acquiring mass spectra so widely accepted. However, I have tried to include in this review the lead references from which a detailed study of the various topics relevant to this technique may be initiated.

II. MASS-SPECTROMETRIC SYSTEM

This section will review that portion of the technique of electrical recording of high-resolution mass spectra produced by magnetic scanning that deals exclusively with the mass spectrometer. The discussion will stress the instrumental specifications required if fast-scan high-resolution mass spectra are to be recorded. The technique of recording high-resolution mass spectra should be and is independent of the "front end" of the mass spectrometer; that is, no special inlet systems, ion sources, or sample introduction methods are required. Hence the portions of the mass spectrometer that are similar in all types of instruments need not be described in this section. Our discussion will deal with the scanning system and recording system of the mass spectrometer and statistical considerations. Finally, this section will include a brief description of the operational technique for recording high-resolution mass spectra with the fast-scan technique.

A. Scanning System

The fundamental equation relating mass (m/e, that is, the mass-to-charge ratio) to the accelerating voltage (V) and to the magnetic field strength (H) for magnetic deflection mass spectrometers is

$$\frac{m}{e} = \frac{R^2H^2}{2V} \qquad [1]$$

where R is the radius of the magnetic field. It is readily apparent from [1] that ions of differing m/e may be brought to focus on the collector of those mass spectrometers possessing a focal point (in contrast to a focal plane where all ions are simultaneously brought to focus) by either altering the magnetic field strength (H) or the accelerating voltage (V). Commercially available instruments have utilized both methods to scan low-resolution $(M/\Delta M \sim 1000)$ mass spectra, and some mass spectrometers permit the selection of either method to scan a spectrum.

Uniformly changing the accelerating voltage as a method of scanning high-resolution mass spectra has some serious drawbacks. Since the sensitivity of the mass spectrometer is directly related to the magnitude of the accelerating voltage, it can be seen from [1] that the sensitivity will decrease as the scan proceeds from low to high mass, the region in which maximum sensitivity is desired. The second obstacle to voltage-scanning high-resolution mass spectra is that high-resolution mass spectrometers are double-focusing instruments that, in general, utilize a combination of magnetic and electrostatic analyzers in the appropriate geometric relation to achieve their high resolution. The ratio between the accelerating voltage and the electrostatic analyzer voltage must be maintained constant. The difficulty in maintaining a constant ratio in these voltages during a scan prevents altering the accelerating voltage as a method of scanning mass spectra. As a result of these obstacles, high-resolution mass spectra have only been scanned by altering the strength of the magnetic field.

Since it is incorporated in the commercially produced Nier-Johnson type mass spectrometers, the exponential scan, that is, an exponential dependence of time (t) with mass $(m/e \propto e^t)$, is the most widely employed scan function. The advantage of the exponential scan function is that assuming constant resolution, the time spent on each peak is constant and is independent of the mass range; that is, the width of the peak at some percent of the peak height is the same over the entire scan. The time per peak (t_p) between the true 5% intensity points can be calculated and is given (7, 8) by [2]:

$$t_p = \frac{0.43 * t_{10}}{(\text{R.P.})_S} \qquad [2]$$

where t_{10} is the time to scan a decade in mass (i.e., from m/e to 0.1 m/e) and $(\text{R.P.})_S$ is the static resolving power for which the mass spectrometer has been set. Equation 2 shows that the time per peak depends only on the scan speed and the resolution of the mass spectrometer. The time per mass unit $(t_{m,m-1})$ can easily be calculated from [2] by substituting $M/\Delta M$ for $(\text{R.P.})_S$ and setting $\Delta M = 1$. These substitutions yield [3]:

$$t_{m,m-1} = \frac{0.43 * t_{10}}{M} \qquad [3]$$

The exponential scan offers other advantages. Since the peak width is a measure of the resolution, the constancy of the peak width over the entire mass range permits a rapid check of the dynamic resolving power [$(R.P.)_D$], that is, the resolving power of the mass spectrometer during the scan. As the later discussion on the operational technique will show, this characteristic simplifies the procedure to maximize the focus of the instrument. It has also been shown that for exponential scanning the ion statistics are independent of mass, and for a given sample size consumed during a specified mass range, independent of the scan rate. Furthermore, either the peak areas or peak amplitude can be used without correction for the measure of relative abundance.

As pointed out by Burlingame et al. (9, 10), the exponential dependence of mass with time has one drawback. In a computer, an exponential is evaluated by calculating a power series until it converges to the required precision. These authors proposed that more simplified (in terms of computability) scan functions be utilized for scanning high-resolution mass spectra. They built a scan-function generator that changed the magnetic field linearly with respect to mass. This means that mass is proportional to time squared ($m/e \propto t^2$). Two advantages are claimed for this function. They point out that a quadratic scan function has an advantage in that the high mass region, where mass measurement accuracy is most important, is traversed more slowly than in an exponential scan. As a result, more ions are collected in a peak, and the time interval between two data points at constant digitizing rate will represent a smaller mass difference. The second advantage of a quadratic scan function is that the data reduction is simplified because an equation of $m/e \propto t^2$ can be approximated with a four- or five-terminal polynomial. The apparent disadvantages to the quadratic scan are that the peak width changes with mass and that the spectrum must be normalized to represent the true relative abundances over the entire mass range. These authors also suggested that a linear mass scan, that is, $m = a + bt$, which could be accomplished by changing the magnetic field as $t^{1/2}$, might be advantageous. Beyond this suggestion, no results with this scan function were reported.

In practice, the scan functions exhibit some deviations from the ideal mathematical function. The hysteresis and inductance of the magnet and the stability of the magnet scan circuits are among the factors that produce deviations in the scan function. Since the deviations from the ideal functions are usually systematic, the ideal mathematical functions can be modified to take into account the experimental deviations during the computer processing of the data. The details for this correction procedure will be discussed in the section on data processing.

B. Detection System

1. *Theoretical Considerations*

The bandwidth of the detection system required to preserve the resolution and sensitivity of the mass spectrometer depends on the speed with which the spectrum is scanned and the resolution at which the mass spectrometer is operated. If the bandwidth of the recording system is overly limited, the resulting mass spectral peaks will be broadened and their amplitudes reduced. In general, as scanning speeds are increased, the limited bandwidth of the recording system will reduce both the resolving power and the sensitivity to values lower than static values. Banner (7, 8) has examined the effect of scan time and of the time constant of the recording system on the sensitivity and dynamic resolution at various values of static resolution. More recently, the distortion of a Gaussian curve due to the effect of a finite instrument time constant has been considered (11).

Using as a model for a mass spectral peak a triangular peak shape and

VARIATION OF RESOLVING POWER.

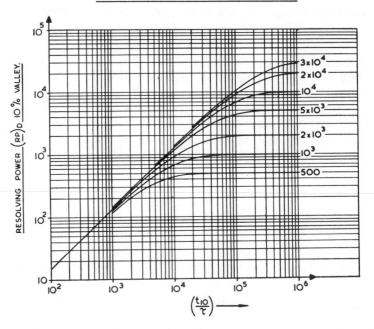

Figure 1. Dynamic resolving power versus (t_{10}/τ), for various values of static resolving power.

VARIATION OF APPARENT SENSITIVITY.

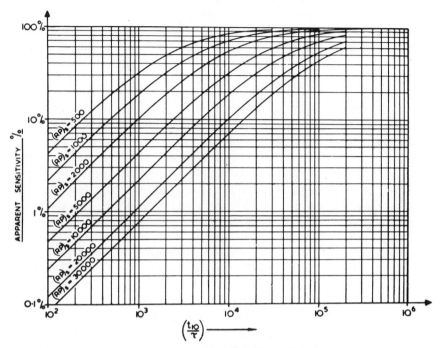

Figure 2. Apparent sensitivity as a percentage of static sensitivity versus (t_{10}/τ), for various values of static resolving power.

assuming that the amplifier and recording system of the mass spectrometer may be represented by a simple resistance (R) and capacitance (C) filter circuit, Banner predicted the effect of the time constant (τ) of the system and the scan speed on the sensitivity and resolving power $[\tau = R \cdot C = 1/(2\pi f)$ where f is the upper frequency response of the recording system]. The predicted results from the model are summarized in Figures 1 and 2, in which t_{10} is the time to scan a decade in mass.

Figure 1 is a plot of the dynamic resolving power versus (t_{10}/τ) for various values of static resolving power. It shows that when the scan time is reduced or the time constant of the recording system is increased, the dynamic resolving power is most severely reduced for high static resolving power.

Figure 2 is a plot of the apparent sensitivity as a percentage of the static sensitivity plotted against (t_{10}/τ) for various values of static resolving power. It shows, as expected, that the greatest peak clipping occurs at the high static resolving powers.

To check his model, Banner compared his predicted results with experimental results and observed that, under slow scan conditions, there was excellent agreement between theory and practice for (R.P.)$_D$ at (R.P.)$_S$ equal to 590 and 8000. For fast scans, the experimental results for (R.P.)$_D$ were 11 and 15% greater than the theory predicted. For sensitivity, the predicted values for fast scans are 15 and 31% lower than the experimental values.

From Banner's model, it can be seen that the time constant of the recording system coupled with the acceptable loss in resolution and sensitivity from the static values determines the maximum scan speed at various resolutions. For example, in a ten-second scan of a decade in mass at a resolution of 10,000 (10% valley definition) and with a recording system having a bandwidth of 10 kHz, distortion of the peak would result only in a 3% loss in the resolving power and less than 7% loss in peak height.

To obtain the required broad frequency range and high sensitivity, the mass spectrometer must be equipped with an electron multiplier. With an electron multiplier gain of 10^6, the signal-to-noise ratio for a single ion can be as large as 50 to 1 (5). While a discussion of electron multipliers is beyond the scope of this chapter, it should be noted that the gain will depend on the mass, charge, energy, chemical nature, and physical size of the bombarding particle (12). With the high-bandwidth, fast-scanning conditions, output signal-to-noise ratios of at least 2500 to 1 have been obtained. Thus, allowing a minimum useful ratio of 2.5 to 1 for the weakest measurable ion peaks, this type of detection system will permit a useful peak intensity range of 1000 to 1 (5).

The selection of the appropriate bandwidth for the particular scan speed and resolution can be approximated from the empirical equation [4]:

$$B \simeq \frac{7.3*R}{t_{10}} \qquad [4]$$

where B is the bandwidth, R is the resolution, and t_{10} is the time to scan a decade in mass. Thus, if $t_{10} = 8$ sec and $R = 10^4$, then $B \simeq 10$ kHz; whereas if $t_{10} = 80$ sec and $R = 10^4$, then only a bandwidth of 1 kHz is required to keep the loss in peak amplitude below 10%.

Recently, Klimowski et al. (13) reported an equation that summarizes the relationship between the interval, location, resolution, signal amplitudes, and bandwidth of a peak. Using a triangular wave as a model for a mass spectral peak, they derived [5]:

$$\frac{dm}{m} = \pm \frac{1}{2R}\left[\frac{di}{I_0}\right] = \pm \frac{1}{2R}*\sqrt{\frac{8edf}{I_0}} \qquad [5]$$

where e = coulombic charge
     I_0 = maximum signal amplitude (amperes)
    df = bandwidth (hertz)
    R = resolution
    m = mass
    di = differential of amplitude
    dm = differential of mass

Equation 5 states that, at the 95% confidence level, the peak center has a probability of occurrence within an interval, dm, that is inversely related to resolution and the square root of the amplitude and directly related to the square root of noise interference. Equation 5 is also a statement of the signal-to-noise (S/N) ratio, since $I_0/di = S/N$.

This qualitative relation can be used to predict the effect of different instrument operating conditions on peak locations (mass measurement), resolution, and sensitivity [maximum signal amplitude (I_0) for a constant sample flow]. If one assumes that the principle source of noise is ion statistics and that the instrument is operated at constant bandwidth, [6] can be derived comparing two sets of operating conditions:

$$\frac{\left[\dfrac{dm}{m}\right]_1}{\left[\dfrac{dm}{m}\right]_2} = \frac{R_2}{R_1}\sqrt{\frac{I_{02}}{I_{01}}} \qquad [6]$$

From [6] one can predict:

1. If the sensitivity (of the mass spectrometer) varies inversely as the square of resolution, constant mass measuring accuracy is predicted at all resolutions.

2. If the sensitivity varies less than the inverse square of resolution, operation at higher resolution is favored because improved mass measuring accuracy is predicted.

3. If the sensitivity varies more than the inverse square of resolution, going to lower resolution is favored.

4. If either 1 or 3 is generally true, in the presence of other than statistical noise, operating mass spectrometer at lower resolution is favored.

C. Ion Statistics

Plots of the peak profiles of peaks with a wide range of intensities leave little doubt that ion statistics cause distortion of the peak profiles, particularly if only a small number of ions are present in the profile. Campbell and Halliday (14) examined the theoretical limit on mass measurement set by ion statistics. For this analysis, they employed an isosceles triangle as a model for the mass spectral peak and assumed a peak width of 100 ppm. Their results are presented in Table 1. According to these calculations, one should obtain an accuracy of mass measurement of 10 ppm with a peak containing between 4 and 30 ions, depending on the confidence level required in the result. However, in practice, it takes from 5 to 10 ions to establish an

TABLE 1
PPM Errors Expected in Mass Measurement of 100 ppm Peak[a]

Confidence level/N_p	$\pm 64\%$	$\pm 95\%$	$\pm 99\%$
1	20.4	38.8	45.0
2	14.4	28.0	35.3
3	11.8	22.9	29.4
4	10.2	19.9	25.7
10	6.45	12.7	16.5
30	3.73	7.30	9.57
100	2.04	4.00	5.25
300	1.18	2.31	3.03
1000	0.645	1.27	1.66
$N_p \geq 300$	$20.4/\sqrt{N_p}$	$40.0/\sqrt{N_p}$	$52.6/\sqrt{N_p}$

[a] N_p is the average number of ions in a peak.

undoubted peak shape (5). Hence these predictions represent the upper limits of mass-measurement accuracy. In practice, it probably takes 10 to 15 ions to achieve a single standard deviation (14).

The measurements of peak areas show a good correlation between the errors observed and the errors expected theoretically from ion statistical considerations. These results (5, 15) are summarized in Table 2. Other measurements have shown that the reproducibility of intensity measurements are limited by ion statistics (16). These results showed that either peak height or peak area can be used as a measure of intensity with almost the same reliability.

Campbell and Halliday extended their study on ion statistics to examine the interaction between sample size, mass measurement, resolution, and intensity measurement. The intensity range that can be mass measured depends on the amount of sample consumed during the scan period and on the transmission of the mass spectrometer. As the resolution is increased, a peak will contain fewer ions from a given sample size, but there should also be fewer stray ions. In addition, the theoretical relation $\sigma = 10^6/(R^*\sqrt{24N_p})$ ppm indicates that fewer ions are required to define a peak position (N_p is the average number of ions in a peak). They contend that the amount of sample required during the total scan period should be inversely proportional to $S^*R^*\sigma^2$ where S is the sensitivity, R is the resolution, and σ is a standard

TABLE 2

Ion Statistical Errors on Areas of Peaks Obtained with a
Mixture of Tribenzylamine and Perfluorokerosene.[a]

Peak	N_p	$100/\sqrt{N_p}\%$	$\phi\%$
C_7F_{11}	140	8.5	10.7
$^{19}CC_{20}H_{21}N$	50	14.1	15.3
$C_{21}H_{21}N$	230	6.6	8.3
$C_{21}H_{20}N$	50	14.1	12.1
C_8F_{10}	15	25.8	25.0
$^{13}CC_5F_{11}$	15	25.8	38.5
C_6F_{11}	230	6.6	3.7

[a] N_p is the average number of ions in a peak; the theoretical percentage standard deviation is, therefore, $100/\sqrt{N_p}$. The experimental value is σ.

deviation or error. Figure 3, curve 1 shows the predicted variation of sample size for an intensity range in the order of 50 to 1 for a constant mass measurement accuracy ($\sigma = 10$ ppm), assuming one microgram is needed at $R = 10{,}000$. The sample size increases rapidly at the higher resolutions due to the increasing effect of electronic instabilities and ion optical aberrations on the resolution. At lower resolutions, where R is inversely proportional to S, there is little variation in sample size. Curve 2 in Figure 3 is constructed assuming that $R^* \sigma$ is constant and is, in effect, a curve of constant N_p. It shows that a 0.01 μg of sample should yield a useful intensity range of 50 to 1 at $R = 1000$.

The authors extended their argument to examine the variation of measurable intensity range (IR) with resolution. These conclusions are presented in Figure 4. Curve 1 shows the constant 50 to 1 intensity ratio that would be obtained at mass measurement accuracy of 10 ppm and a sample size indicated by curve 3 (or curve 1, Figure 3). If the same sample

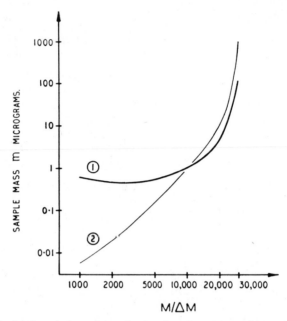

Figure 3. Calculated variation of sample size, in micrograms, with resolution, $M/\Delta M$, assuming mass measurement required to (1) 10 ppm and (2) $10^{-5}\,\Delta M/M$ ppm.

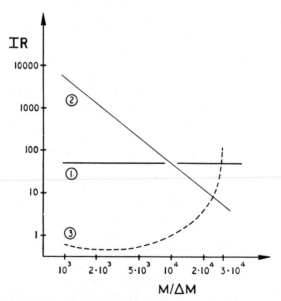

Figure 4. Variation of measurable intensity range, IR, with resolution, $M/\Delta M$, for sample size in micrograms, assuming mass measurement required to (1) 10 ppm and (2) $10^{-5}\,\Delta M/M$ ppm.

size is used, the variation of the intensity ratio when $R*\sigma$ is constant is shown in curve 2, Figure 4.

D. Operational Technique

Acquiring good high-resolution mass spectra routinely requires that the instrumental parameters be properly set for the mass-spectrometric conditions to be employed. While each operator has his own sequence of steps that he performs to "set up" the instrument, the end result is (hopefully) the same in each case. The discussion of the operational technique will be oriented toward my experience using the MS-902 (Associated Electrical Industries, Manchester, England), but I believe that the techniques are applicable to all instruments employing exponential scans.

1. *Resolution*

The selection of the resolution for which the mass spectrometer is set is determined by the problem to be solved and the capabilities of the mass spectrometer. Some guidelines based on instrumental parameters for the selection of resolution were discussed in a previous section.

A static or peak matching mode of operation is used to set the resolution. To begin, the source conditions are adjusted to maximize the ion current on the beam monitor and the magnet is focused on a convenient peak in the m/e 200–300 mass range, for example, the m/e 219 fragment of perfluoro-tributylamine or perfluorokerosene. The source and collector slits are adjusted to give the desired resolution with the maximum sensitivity. The slit widths and the technique for obtaining these conditions are documented in the instrument-user manuals. Prior to and during the course of the slit manipulations, the source-focusing conditions are continually being optimized. During the final adjustments, the magnet is moved horizontally if necessary, until the maximum ion current for the particular peak used in the setting up (m/e 219 in our example) is observed on the collector meter. The resolution in this static mode is measured using the peak matching feature, that is, by measuring the number of parts per million the peak must be displaced from itself for a 10% valley to be observed. If the 10% valley is observed after 100 ppm displacement, the resolution is equal to 10,000. This is defined as the static resolving power $(R.P.)_S$.

Now, with the source controls maximized and slits adjusted for the required static resolution, the focusing conditions for the desired scan speed must be determined. It has been observed that the position of optimum focus of the magnet differs from the static positions by increasing distances as the scan speeds are increased. Generally, scan speeds faster than 2 or 3 mins/ decade require some adjustment to the horizontal position of the magnet if uniform resolution is to be maintained over the entire mass range.

The determination of the magnet position that maximizes the dynamic resolving power $(R.P.)_D$ is a trial-and-error procedure. Generally, from experience the approximate distance the magnet must be moved from the static position for a particular scan speed is known after this initial positioning of the magnet. After each scan a small horizontal adjustment to the magnet position is made. The direction of the move relative to the static position is determined by the increase or decrease in resolution and by its uniformity over the entire mass range. The latter quality is very important, since changes in the magnet position may improve resolution at one end of the mass range but degrade the resolution at the opposite end of the mass scale. The magnet is adjusted until the optimum conditions are obtained.

To determine the resolution under these dynamic conditions, one utilizes the property of exponential scans that the peak width is constant and independent of mass. Hence the magnet position can be maximized by comparing the peak widths over the entire mass range from scan to scan. This is most easily accomplished visually using a storage oscilloscope and allowing the occurrence of a peak to trigger the horizontal sweep of the oscilloscope. In this way, the peak profiles can be displayed for the entire mass range. If perfluorokerosene is used in the application, it aids the visual observation if the input voltage to the oscilloscope is manually attenuated to keep all the peaks on the same amplitude scale. Then the comparison is easily made, since one is measuring the widths with peaks of nearly uniform height. The shape of the peaks is also helpful, since the appearance of shoulders or tails on the peaks or excess noise is indicative of some faults in the mass spectrometer or improper instrument parameters—for example, too limited a bandwidth in the detection systems.

While our system connects the analog output from the mass spectrometer directly to the oscilloscope, those groups that have the oscilloscope receiving its input from a computer-centered data-acquisition system (9, 10) can display selected peaks over the entire mass range. Further, the data can be automatically scaled and compared by computer. While computer systems for automatically focusing the mass spectrometer have not as yet been reported, this function is certainly within the realm of the computer. As an example of how this is done, we have used the following technique in our laboratory. In this case, the computer determines the widths of a selected group of peaks over the entire mass range and then fits these widths to a straight line. The magnitude of the slope of the line indicates whether the magnet must be moved, and the sign of slope indicates the direction in which it is to be moved. We have not as yet determined whether the magnitude of the slope can be correlated with the magnitude in the distance the magnet must be moved. Presumably, if it is observed that the widths are too large when the conditions are maximized, the computer could conclude that some improper conditions are present and signal the appropriate warning, although this latter situation has not been tested.

Our experience has been that once the focus conditions have been maximized, they do not need to be adjusted during the course of the working day. However, in practice we generally monitor the widths of the peaks on the oscilloscope as a check on the instrument's operation.

At this point, the instrument is ready to record a spectrum. Generally, a complete spectrum of the reference compound is recorded and processed on the computer. This is done to check the entire mass-spectrometer/data-acquisition/data-processing system. If this operation is successfully completed, the sample to be measured is introduced to the instrument.

III. DATA ACQUISITION

A. Data-Acquisition System

Data acquisition has been one of the rapidly developing areas of mass spectrometry. This can be observed from the number of papers on this topic that have been presented at recent meetings on mass spectrometry. Some of these papers describe the mass-spectral data-acquisition systems that have been developed in various laboratories. Most of the data-acquisition systems in operation are unique but are based on the same principle. This results in part from the broad price range of such systems and the desire or necessity to utilize equipment or other computing facilities already present in the laboratory. Further, the rapidly changing technology in electronics results in rapid obsolescence of specific pieces of hardware. These technological advances often permit the features or performances of the older systems to be duplicated at a fraction of the cost. As a result, instead of discussing specific data-acquisition systems and specific pieces of hardware, we shall outline a general data-acquisition system (which will incorporate the features of most of the newer systems) and discuss the basic performance requirements of the various pieces of hardware. The scheme for the generalized data-acquisition system is illustrated in Figure 5.

B. Analog-digital Converters

The essential component of any data-acquisition system is the analog-to-digital converter (ADC), the device that converts the analog output of the

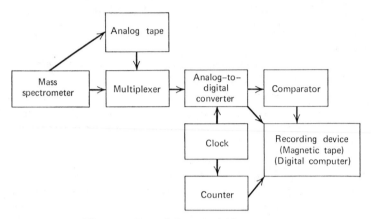

Figure 5. General data-acquisition system.

mass spectrometer into the digital values required as input for digital computers. Analog-to-digital converters can be characterized in terms of their conversion times and their resolution. Conversion times as high at 250 kHz and resolution or bit capacity ranging from eight to fourteen bits plus sign are typical [a bit is a single binary numeral in a binary word (17)]. With computer controlled ADC's, the resolution can be program-selectable, but the conversion time will generally decrease as the resolution is increased.

It should be mentioned that there are several different methods for converting analog signals to digital data. In a recent book, Malmstadt and Enke (17) have reviewed the various methods and have discussed the attributes and qualifications of each technique. These authors also described various sampling techniques such as the use of sample-and-hold amplifiers. This device used in conjunction with the ADC samples the varying analog signal at a particular instant in time and then holds the signal long enough to allow an analog-to-digital conversion of the sampled value. Their book serves as a useful introduction to the techniques and terminology of analog conversion for those unfamiliar with this topic. In a later section of this chapter, the factors to be considered in the selection of an ADC suitable for recording high resolution mass spectra will be examined.

C. Multiplexers

Most data-acquisition systems include a multiplexer, a device that permits multiple analog-input signals to be sampled by a single ADC. In effect, the multiplexer is a high-speed switch. The maximum number of channels, input signals, or addresses available will vary with the individual multiplexer. Also, the mode of switching will vary with each multiplexer from the relatively simple sampling of one or more channels in sequence to the relatively complex random sampling of each channel. The latter mode operates when a computer that can control the channel to be sampled, performs an analysis on the previous data value and determines from the result the next channel to be sampled. The computer then directs the switching of the multiplexer to the appropriate channel in order to obtain the desired data.

There are some readily apparent applications of multiplexers to mass-spectrometric data acquisition. This device has been used to permit the sampling of the output from the total ion current monitor while sampling the spectrum. Obviously, a programmable device is required, since the sampling rate of the total ion-current monitor is at least a factor of 10 less than the sampling rate for the spectrum. In an alternative approach, switching may be initiated when a peak is sensed by peak detecting hardware.

The multiplexer also permits the multiple channels of data required to record a mass spectrum on analog tape to be digitized in a single playback of the tape. In this case, a sequential mode of switching from channel to channel could be used. However, if a faster data rate is desired, the random mode of channel selection would permit the magnitude of the sample to be examined, and if a less sensitive data channel is required, to switch to the appropriate channel. Another application occurs in the multiinstrument environment in which the multiplexer permits several mass spectrometers or a combination of other instruments to utilize the same ADC.

D. Threshold Comparators

Many data-acquisition systems utilize a digital device called a comparator. This device permits the value developed by the ADC to be compared to some preset digital value. This preset value may be variable and may be continually updated if a computer analyzing the incoming data can communicate a new value to the comparator. Otherwise a constant preset value is used during a data-acquisition sequence. The primary function of the comparator is obvious when one realizes that when scanning a high-resolution mass spectrum, the mass spectrometer is producing peaks less than 5% of the time. For example, at a resolution of 10,000 it would require in the order of 23,000 contiguous peaks to completely occupy the total scan time from mass M to $0.1M$ (15). In the average high-resolution mass spectrum, one generally observes less than 500 peaks. The comparator, functioning as a threshold device, avoids the transmission of the very high proportion of empty information (baseline) in the spectrum to the computer or other recording device. The preset value is set equal to the threshold value, and the comparator compares the values developed by the ADC with the threshold value. Only those values that exceed this threshold are transmitted to the recording device. If this threshold mode of operation is employed, the number of samples eliminated must be "counted" if the time of occurrence of a peak from either the last peak or from the start of the scan is to be known. This is most easily accomplished by incorporating a counter into the data-acquisition scheme. This counter is incremented each time the clock, which controls the rate at which the ADC samples the signal, pulses the ADC. The sample count is then read by the data-acquisition system, generally when the transmission of the peak profile has been concluded. At this point, the counter can be reset to zero or continue to maintain the total count as the scan proceeds, depending usually on the number of bits available to store the count. In order to reduce the problem of assigning a peak to occasional noise spikes, some comparators contain the necessary logic to require a certain number of consecutive samples, usually about four, to exceed the threshold value before the recording devices are activated.

The use of the comparator limits the amount of data recorded, and as a result substantially lowers the time to process the resulting spectrum on the computer. However, if the threshold is set too high, many of the low-intensity peaks will be overlooked, or many of the samples in the peak profile will be eliminated. Therefore, with certain valuable samples, it may be advantageous to lower the threshold and record the entire spectrum, baseline, and peaks, and then perform the thresholding during the subsequent computer processing of the recorded spectrum. With this method, one can experiment with several values for the threshold until the ideal value for processing is determined. However, this mode of operation requires a bulk storage device be available to store all the raw data.

It is possible to accomplish most of the function of the comparator by computer software. When a computer is the recording mode, a portion of the core of the computer is divided into logical blocks of storage or buffers into which the data is continuously deposited. If the data rate is sufficiently slow, the data values in these buffers can be compared to a threshold value prior to recording on tape or disk. The empty buffers—that is, those containing no peaks—are not recorded, but they are counted in order that the time of the data value be known. This software approach to thresholding consumes computer time and, as a result, functions at lower data rates and slower scan speeds than the hardware method.

E. Digital Recording Devices

The recording devices currently used are digital magnetic tape, disk, or the core storage of a digital computer. The small inexpensive computers that perform the data acquisition and processing and that are also capable of controlling some of the external devices, now predominate. Because of limited available storage, the thresholding and reduction of the peak profile to peak position and peak abundance must be performed during the acquisition sequence. Magnetic tape is used when the thresholded or nonthresholded spectrum is recorded but processed off-line or on a remote computer. If magnetic tape is the recording medium, it is advantageous to have sufficient logic and buffering available so that the data is recorded with the appropriate interrecord gaps between the blocks of digital data. This greatly simplifies reading the data from the magnetic tape into the digital computer.

F. Analog Recording Devices

The scheme illustrated in Figure 5 shows an analog FM tape recorder as an off-line recording device. The advantages and limitations of analog tape have been clearly presented (18), and its use in mass spectrometry demonstrated (4, 6). The principal advantage is the capacity of analog tape to expand or compress the time scale of the actual data. The recorded signal may be reproduced at a faster, a slower or the same tape speed as the original recording. For high resolution mass spectrometry, the spectra are recorded at high tape speed, in order to attain the required wide frequency range, and reproduced at slower speed, which in effect permits higher sampling rates of the spectrum than would be possible in real time. Earlier data-acquisition systems made effective use of analog tape (6, 19). For specific operations, such as gas-chromatography-high-resolution mass spectrometry, where very high sampling rates are necessary, analog tape recording may be superior to even the newer data-acquisition systems.

The analog tape does have some limitations that make its use less than ideal. The biggest drawback is that the signal-to-noise ratio per channel is a low 100:1 (18). This means that to obtain the necessary dynamic range, which is inherent in a mass spectrum, one must record either multiple channels— with each channel recording a signal at different attenuation in a manner similar to that employed with oscillographic recorders—or, as an alternative, a single channel recording the analog signal in a logarithmic form. Either technique would increase the subsequent time for data processing, since the peak profiles would have to be reconstructed from the recorded data. The multiple-channel method would also increase the data-acquisition time by a factor equivalent to the number of channels to be digitized. No comparison of these two analog-recording techniques, in terms of accuracy and of mass measurement and intensity, has been made. The accuracies in both measurements would be dependent on the number of channels and the attenuation factor between channels. These factors also determine the maximum signal to noise (dynamic range) permitted by the recording system. Other sources of error in the use of analog tape are fluctuation in tape speed and tape flutter (short-term tape speed deviation), although use of an internal reference can minimize errors due to variations in tape speed. Even without this standard, mass measurements with an average error of 10 ppm have been reported (6) for spectra recorded in the logarithmic form.

G. Determination of Peak Maxima

Most of the data-acquisition systems use the principle of making measurements of the magnitude of the voltage signal from the spectrometer at uniform intervals of time. A general scheme for such an operation is shown in Figure 5. However, Carrick (20) has proposed that the complete profile of the peak is not required to determine its position. Instead, if well-shaped, statistically good peaks are present, the location of the peak can be specified by the exact time of the incidence of the peak maximum. He has constructed a data-acquisition system utilizing this principle to record high-resolution mass spectra. This device uses the first derivative to locate the peak maximum. At this point, the time elapsed from the start of the scan is read off a master counter into a register prior to punching on paper. A separate set of circuits is required to record the peak height. The prototype system is capable of handling 20 (unevenly spaced) peaks per second. The system produced mass measurements better than ± 5 ppm from 104 secs/decade scans at a resolution of 8000 using a 10 kHz clock. Improved results were predicted if faster clock rates were utilized.

He also considered the situation with this technique if poor ion statistics or multiplets were encountered in the scan. For peaks containing few ions, there would be several maxima across a peak width, and the differentiator would give a time reading for each maximum. These time values, combined with the intensity values, could then be treated as a conventional peak profile. Multiplets could only be detected by considerations in the errors of mass measurements. Since this system proposed by Carrick represents a departure from the conventional data-acquisition systems, it will be interesting to follow its future developments.

H. Selection of the Analog-to-Digital Converter

The selection of the appropriate ADC for the conventional systems is one of the most important considerations in the design of the data-acquisition system. This selection must be based on the capacity of the ADC to permit accurate measurement of peak position and peak abundance. The factors that will affect these measurements are the rate at which the analog output of the mass spectrometer is to be sampled and the precision with which each sample is converted to digital form. The effect of these factors on the measurement of peak position and peak abundance has been analyzed by Halliday (15), and the discussion that follows is a summary of his conclusions.

Before proceeding to his conclusions, some general comments concerning the digitizing of mass-spectral peaks are in order. It should be apparent that although peaks have equal width at a given fraction of their own height (assuming an exponential scan at constant resolving power), the width of a peak at a given voltage level will vary because of the variation in the maximum intensity. As a result, the number of digital conversions present in a peak profile will be different for peaks of different height, increasing as the intensity of the peak increases. A second observation is that, although a peak is quite symmetrical (assuming that the instrument is properly focused), the heights of the digital samples are not symmetrical. This arises from the fact that the peaks arrive at irregular times relative to the uniform time interval between samples; that is, a peak may arrive just prior to the next sample or just after the last sample was determined.

These observations indicate that it is important to consider the number of digital samples that should be obtained between the 5% intensity points of an average peak and the number of bits of resolution of the ADC, that is, the number of voltage levels that are available for each sample conversion. For economic reasons, the sampling rate selected should be limited to that consistent with accurate representation of the peak position and peak abundance.

Halliday described the variation in the peak area and peak centroid produced by using a limited number of samples to describe the peak as the "ratchet effect." The errors were estimated from a model system based on a triangular peak containing two or three samples, where the number of samples depended on the time position of the apex of the triangle relative to the sampling time. This model predicts that when there are three conversions there is continuous variation of 7% in the peak area. This variation is due to the asymmetry in the three voltage samples as they move across the peak profile. If the peak apex occurs in the central part of the time interval between samples, a large drop (30%) in peak area is predicted, due to the presence of only two samples in the peak profile. With three samples, the estimated centroid can be in error by as much as ±5.4 ppm with the maximum error between two such peaks being as large as 11 ppm. The standard deviation expected on a large number of peaks is about ±3 ppm. However, if the peak position was estimated from the time of the largest digital sample, the maximum error would be ±9.2 ppm.

It is obvious that these variations in peak area and peak position would be reduced by increasing the number of samples to describe a peak. In an attempt to validate the model and to obtain some experimental basis for a desirable number of samples to describe a peak profile, the following experiment was conducted. The masses from a spectrum in which about 33 samples were taken between the 5% intensity points (if these points were above the threshold) were calculated. The calculation was repeated twice, in one case

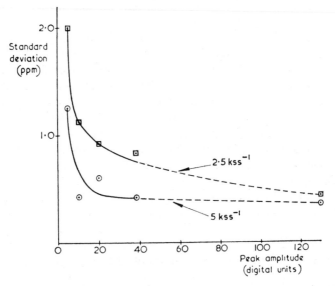

Figure 6. Error relative to 10 kHz sampling rate versus peak amplitude.

using only the odd-numbered samples and, in the second case, using only the even-numbered samples. This calculation in effect lowered the sampling rate by a factor of two. All the mass measurements thus obtained were subtracted from those masses originally determined when all the samples were used. The same calculation was then repeated four times, using every fourth sample; starting with the first, then the second, third, and fourth. This calculation simulated a sampling rate, a factor of four below the original 10 kHz rate, and yielded approximately eight samples in a peak profile. The results of these experiments are plotted in Figure 6. The plot shows that the error increases when the sampling rate is reduced, with the greatest increase, 2 ppm, observed with the lower peak intensities, where the number of samples in the peak was about five. On the basis of these results, it was concluded that a useful number of samples to obtain per peak was about 20.

In Table 3 are listed two equations. The first equation permits the calculation of the sampling rate in kilosamples per second (ks/sec) required to yield N samples in a peak during an exponential scan in terms of the resolution R and the decade scan time t_{10}. The second equation, a variation of [4], shows the dependence of the bandwidth b on the ratio R/t_{10} and f, the fractional peak distortion, that is, the loss of peak height and the increase in peak width. The time-dilation factor, ϕ, is equal to the ratio of the recording speed over the playback speed and is applicable only if the analog tape recorder is used.

TABLE 3

$$\phi S(ks/sec) = 2.3 \times 10^{-3} N\, R/t_{10}$$

$$b(kHz) > \frac{5 \times 10^{-4}\, R/t_{10}}{f}$$

R = resolution
t_{10} = decade scan time (sec)
N = number of digital samples between 5% intensity points
f = fractional peak distortion (loss of peak height, increase in peak width)
S = A/D conversion rate
ϕ = off-line time dilation factor
b = detector system band width $0.1 < f < 0.3$

The number of bits available for conversion in the ADC must match to any errors expected in the ion-peak intensities and their profiles. The source of these errors are ion statistics and electronic noise. As noted in a previous section, the errors observed in the measurement of peak areas correlated well with the errors predicted from ion-statistical considerations. Since ion statistics seem to be the main cause of errors observed in the measurement of peak intensities, it is possible to predict theoretically the errors to be expected when measuring the ratio of two peak intensities. These predicted errors are illustrated in Figure 7. Along the abscissa in Figure 7 is plotted the number of ions in the major (base) peak of the mass spectrum; and along the ordinate, the ratio (R) between a weaker peak and the major peak in the spectrum. The curves on this graph show lines of constant error ε_R to be expected from ion-statistical effects. From the figure, it can be seen that the predicted error decreases as the number of ions in the base peak increases and as the ratio between the base peak and the other peak approaches one. The right side of Figure 7 shows the errors ε_e that would be expected if the only uncontrolled fluctuations in the mass spectrometer signal were due to electronic noise. These predicted errors refer to the total peak area, but Halliday used these errors to estimate the average percentage error (x)

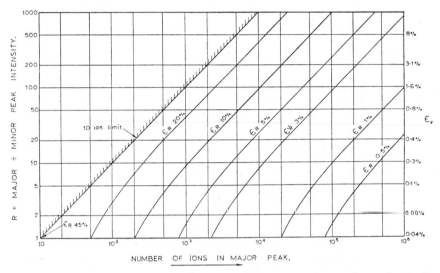

Figure 7. Ion statistical error ϵ_R and electronic noise error ϵ_e on measurement of ratio R.

allowable per digital conversion. The value of x was determined from the expression $x = \sqrt{N_w(\varepsilon_R^2 + \varepsilon_e^2)}$, where N_w is the number of samples in the weaker peak when 22 samples are expected between the 5% intensity points. This error, to be expected from ion statistics and electronics, can be compared to the error expected from the limited bit capacity of the ADC. The results of this analysis are summarized in Figure 8. The analog curves were drawn assuming $N_p = 22$, $S/N = 2500$, and S/T(threshold) = 1000 for an MS-9 instrument. The graph contains four curves representing $x\%$ for spectra containing different ion abundances in the major peak and assuming that the base peak yields the maximum analog signal. It shows that the average error that may be allowed in each digital conversion in a spectrum in which the base peak contains 10^4 ions can be as great as 13% for peaks of highest intensity and can increase to 80% for a peak 400 times weaker than the base peak. The three hatched lines show the error to be expected from digitizing the linear voltage output from the mass spectrometer with a 7, 9, and 11-bit digitizer. With a 7-bit digitizer, a one-bit error on a 100 volt signal (MS-9 maximum analog output) corresponds to an error of 0.8%. Comparing this curve with the analog curve derived for 10^4 ions in the base peak shows that as long as the ratio between the two peaks is less than 40 to 1, the errors introduced by the 7-bit digitizer will be less than the errors introduced by ion statistics and electronic noise. If 10^5 ions are present in the base peak, a 9-bit converter would be satisfactory as long as the ratio between the major and minor peaks is less than 60 to 1. Finally, the plot shows that if 11 bits are

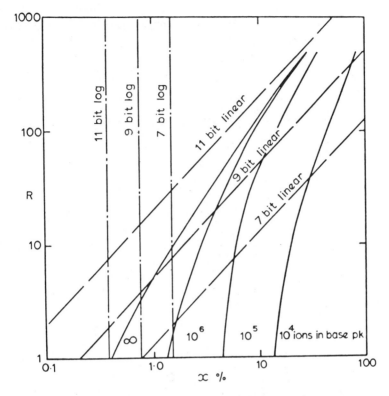

Figure 8. Average error for each digital conversion.

available for conversion, there should be no degradation of mass-spectral information. If an output of the mass spectrometer is converted from linear to logarithmic form, then, little or no degradation of the signal occurs even with a 7-bit ADC. The conversion of the output voltage from linear to logarithmic form results in increased data-processing time, since the logarithmic values must be reconverted to linear values during the profile analysis.

IV. DATA PROCESSING

The high-resolution mass spectrum can be extracted from the data-acquisition system only with the aid of the digital computer. The computers used for this purpose span the entire range from the small, relatively inexpensive, small-core, small-word size machines to the biggest, fastest computers currently available. Because of the diversity in computers and data-acquisition systems utilized, the software packages used to process the data are unique for each combination. Fortunately, the data processing of a mass spectrum from the raw data can be broken down into a set of well-defined sequential operations. The sequence of steps is illustrated in Figure 9. It assumes that no data reduction has occurred during the data acquisition. In a large computer, such as that used for off-line processing, all of these

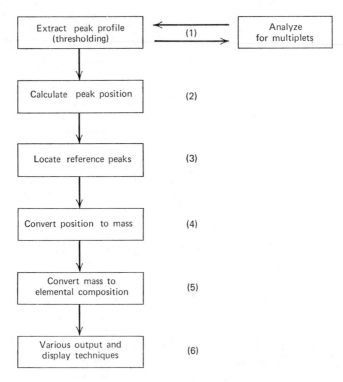

Figure 9. General software scheme for processing high-resolution mass spectra.

blocks or steps could be incorporated into one large program that is resident in the computer at all times; whereas, on the smaller computers, each block could, in essence, be a separate program that only needs access to the data processed by the previous program, but is otherwise independent of the previous program. Since some data-acquisition systems also do preliminary data reduction on line with the data acquisition, the resulting input data from some systems may be in a more advanced state of processing; for example, the peak profile may have already been reduced to the peak position. Therefore, in such a situation, the processing programs would begin at step 3.

A. Peak-Profile Analysis

Let us examine in more detail the programs depicted by the boxes in Figure 9. Prior to the start of the data processing, the scan conditions, the data, the resolution, and the data-acquisition conditions, the title and any comments are read into the computer. The first step in the data reduction is to extract the peak profiles from the raw data. If the thresholding to eliminate the baseline has been done by the data-acquisition hardware, then all that is required is to separate the stored records into the peak profiles along with their corresponding time reference. If, however, the thresholding is to be done by the software, then the records of raw data are read in from the bulk storage devices, usually magnetic tape. To eliminate those values derived from the baseline, the raw data values in these records are compared to the thresholding value. Since the digitizing rate is selected so that at least 20 data values are present between the two points in the peak profiles that are at 5% of the peak height, it is not necessary to check each raw value to detect a peak, but only every second or fourth value in the record. Since only 3–5% of data represent peaks, this technique speeds up the processing of the raw data. If one of the raw data values is found to exceed the threshold, then the entire record is examined value by value until the entire peak profile has been extracted.

B. Threshold Determination

Before we continue with the peak-profile analysis, we should discuss how the threshold value is determined. The first requirement is to calculate the digital values for the baseline and for the peak-to-peak noise. To determine these values, a section of the spectrum that is devoid of peaks, usually the highest mass region, is selected. This region is especially convenient for instruments that scan from high mass to low mass, since it is contained in the first few records of recorded data. The mean and standard deviation of 1000 data points from the selected mass region are determined. The mean is used as the measure of the baseline, and the standard deviation is used as a measure of the peak-to-peak noise. These two values are now used to estimate the threshold value. We generally compute the threshold value by adding the product of one or two times the standard deviation to the baseline value. The multiple of the standard deviation depends on the intensity of the spectrum; the higher value is used if the spectrum is very intense. Once a value for the threshold has been determined, it is constant as long as the recording system (multiplier, etc.) is not changed and the spectra are of comparable intensity.

C. Deconvolution

Having extracted the peak profile, it may be desirable to analyze whether the profile represents a single peak or is produced by the overlapping of two or more peaks. Several techniques (21, 22) specific to mass spectrometry have been reported to detect this. If the resolution is constant over the entire mass scale, a comparison of the width of a peak, at various percentages of the peak height, can be used for this purpose (21). Burlingame has also used the magnitude of mass-measurement errors to detect multiplets that are then reexamined on a second pass of the data (22). Other techniques for detecting multiplets are presented in Chapter 2, "Photographic Techniques in Organic High-Resolution Mass Spectrometry." Once a multiplet has been detected, it can be flagged to be deconvoluted later or it can be deconvoluted immediately. We utilize the off-line technique, preferring to look at the results to determine whether deconvolution is necessary or whether the results of the deconvolution are predictable. The reason for this approach is that the deconvolution takes considerable computer time and is utilized only when absolutely necessary. The deconvolution utilized (21) is an iterative method described by Lusibrink (23). For peak models, rather than using standard mathematical functions, this method utilizes a function determined from curve-fitting peak profiles known to be singlets, generally PFK peaks, and ideally these peaks are from the same mass range and are of similar intensity. A least-square criterion determines the best fit.

D. Peak Position Determination

With the peak profile in hand, the next step is to determine the peak position. Using a statistical approach, Campbell and Halliday (14) investigated three ways of estimating peak position. The three methods were (1) the center of maximum intensity, (2) the center of the peak area, that is, the "median," and (3) the weighted mean of the peak area, that is, the "mean" or centroid. They concluded that the mean provided an unbiased estimate of peak position. The centroid is calculated according to the equation:

$$y = \frac{\sum\limits_{n=1}^{k} n*A_n}{\sum\limits_{n=1}^{1} A_n}$$

where y = the peak position, k = number of samples in the peak profile, and A_n is the amplitude of each point in the profile. The peak position and peak intensity can be calculated as follows:

$$\text{peak position} = \text{Count} + y$$

$$\text{peak intensity} = \sum_{n=1}^{k} A_n$$

where count is the number of data points from start of the scan or from the previous peak. Burlingame (24) has reported that the centers of gravity yielded more precise results for peak position than those determined from peak tops. The center of gravity calculates the weighted mean the same as the centroid but is a more attractive algorithm since it utilizes only addition operations and not multiplications. The center of gravity algorithm is as follows:

$$C.G. = \frac{\displaystyle\sum_{j=1}^{k} \sum_{n=1}^{i} A_n}{\displaystyle\sum_{n=1}^{k} A_n}$$

$$\text{peak position} = \text{count} - C.G.$$

Bowen et al. (25) investigated the possibility of improving the centroid method by eliminating the noisy portions of the peak, but found no improvement over the straightforward method. Recently, Klimowski *et al.* (13) extended the statistical analysis of the centroid method by replacing the triangular-peak-shape model with a Gaussian-peak-shape model, which resembles more closely the shape of an experimental peak. From their analysis, they also concluded that the centroid method of calculating peak centers is fundamentally correct.

The question of whether to incorporate digital smoothing prior to the determination of peak position and peak area probably depends on transmission noise between spectrometer and data-acquisition device, on the analog-to-digital converter, and on the amount of filtering permissible from the spectrometer. Klimowski *et al.* reported (13) that their data-processing sequences include a five-point smoothing of the peak profile. In contrast, we have observed that incorporation of a nine-point smoothing operation involved at most only a marginal improvement in mass-measurement accuracy and was not worth the increased computer time that the step involved. Therefore, a comparison of the results of mass measurements utilizing smoothed and unsmoothed data should be examined before this step is incorporated.

E. Peak Identification

The next operation after determining all the peak positions is to locate those peaks derived from the reference compound that are to be utilized in the time-to-mass conversion. There are two basic techniques for accomplishing this step. One technique involves locating the base peak (m/e 69) of PFK and its ^{13}C satellite at m/e 70 (26). With these peaks one can then predict the position for the remaining reference masses. An alternate approach is to locate a reference mass at the high-mass region of the spectrum. While this appears to be a less desirable method, it is facilitated by the high reproducibility of the scans (9) when the scans are initiated from the same magnetic field setting. Hence a combination of time and intensity of the reference peaks from previous spectra makes the identification of high mass peaks very simple. All these techniques are easily automated on a computer, if the necessary storage devices, such as disks, are available. After locating the reference peaks in the spectrum, the location of the remaining peaks can easily be predicted. We have employed a modified exponential function to predict the position of the remaining reference peaks. The modification to the ideal exponential function is required because the actual scan shows a small deviation from the true exponential.

The mathematical relation is

$$M_4 = \frac{M_3 \exp (t_3 - t_4)}{T_{3-4}} \qquad [7]$$

where T_{3-4} is predicted from the relation:

$$T_{3-4} = T_{2-3} - (T_{1-2} - T_{2-3}) \frac{(M_2 - M_4)}{(M_1 - M_3)} \qquad [8]$$

where M_1, M_2, M_3, and M_4 ($M_1 > M_2 > M_3 > M_4$) are consecutive reference masses that occur at time t_1, t_2, t_3, and t_4, ($t_4 > t_3 > t_2 > t_1$). T_{1-2} and T_{2-3} are given by

$$M_2 = \frac{M_1 \exp (t_1 - t_2)}{T_{1-2}} \qquad [9]$$

$$M_3 = \frac{M_2 \exp (t_2 - t_3)}{T_{2-3}} \qquad [10]$$

Given the initially identified reference times (t_1, t_2, t_3) the predicted time (t_4) corresponding to reference mass M_4 is calculated.

This time is then compared to the array of experimental times from the spectrum in order to locate the time that is the closest match to t_4. The match must fit within a specified time interval, usually the time corresponding to a peak width or less, and the peak must also fit a minimal intensity criterion. Once the experimental time, that is, the actual time in the spectrum has been located, its value is substituted for t_4 and the corrected value for T_{3-4} is calculated. These corrected values are then used to predict the next reference time, and the sequence of predictor-corrector operations is continued until all the reference masses have been identified. If one reference time is not located, it is assumed that the peak, due to its low abundance, is absent from the spectrum. The time predicted for the position of this reference peak is utilized for the peak position in the next predictor-corrector sequence. If a second reference time cannot be located, the search for the remaining reference peaks is terminated. However, if the new reference peak is located, the search for the remaining reference peaks is continued after indicating in the listing of the reference peaks and their positions which reference peak was absent from the spectrum.

The above discussion assumed that the extrapolation sequence proceeds from high mass to low mass. The extrapolation conducted from low mass to high mass would proceed by the same sequence of operations after locating the initial reference peaks, except that the order of increasing magnitude of M_1 to M_4 and t_1 to t_4 would be reversed. It should be obvious that the mathematical function used to predict the time position of the reference masses depends primarily on scan function. For extrapolation, Burlingame et al. (9, 10) use a quadratic function of the form $M = A + B + Ct^2$, which coincides with the quadratic scan function employed on their mass spectrometer. Ideally, the simplest function, that is, the most rapidly computable function that can predict the next reference position with the desired accuracy (less than a peak width), is the most desirable function for this application.

F. Time-to-Mass Conversion

The next step in the data-processing sequence after locating the position of the reference peaks is to convert the remaining time positions in the spectrum to masses. Several mathematical functions have been utilized to accomplish this conversion. Several groups have reported the use of higher term polynomials, of the form $M = A + Bt^1 \cdots Nt^n$, the number of terms in the function varying from system to system. Klimowski has reported (13) satisfactory results utilizing a third-degree Lagrange polynomial for the time-to-mass conversion. Modified exponential functions have been used extensively for this purpose.

Bowen et al. (25) evaluated two general methods of interpolation, utilizing modified exponential functions. In the first case, equations were generated that precisely fitted one or more reference peaks on either side of the peak to be interpolated. They found that the best results were obtained with this precise-fit method, using four reference masses with equations of the form:

$$M = M_0 \exp [a(t - t_0) + b(t - t_0)^2 + c(t - t_0)^3].$$ [11]

The second method utilized equations generated by a linear-regression method. In this case, the fitting equation was not constrained to pass through any of the points used as reference, but was the best least-squares fit to these points. Results superior to the precise-fit method were observed with an equation of the form:

$$M = A \exp (at + bt^2 + ct^3) \qquad [12]$$

These authors (25) suggested that this improvement is due to the fact that the least-squares method does not require a precise fit to each data point, each of which may be subject to statistical inaccuracies.

We have utilized an exponential function that incorporates a first-order correction. The equation is of the form:

$$M_u = \frac{M_2 \exp (t_2 - t_u)}{T_u} \qquad [13]$$

where

$$T_u = T_{2-3} + (T_{1-2} - T_{2-3}) \left(\frac{t_3 - t_u}{t_3 - t_1} \right) \qquad [14]$$

in which T_{1-2} and T_{2-3} are the same time constants calculated in the extrapolation sequence. Other groups have reported (9, 20) improved accuracies in mass measurements by replacing the above precise-fit time-time constant relation by one utilizing least-squares curve-fitting techniques on the time constant versus time curve. However, in one case (9), the improvement depended markedly on the number of reference masses and the number of coefficients used. These combined results suggest that the linear-regression least-squares method is more suitable for time-to-mass conversion than the precise-fit methods.

It is not certain at this time whether the same mathematical function will yield the best results on all mass spectrometers of the same make or design. The success of the commercial data-processing package for A.E.I.—MS-9 instruments indicates that there probably is at most little difference between the same commercial instruments.

It is apparent that several types of functions have been utilized for the time-to-mass conversion. Whether any of them represents the ideal function is uncertain. They do yield acceptable results. While improvements in the accuracy of mass measurement may be obtained by mathematical techniques, it seems at this time that the major sources of error are derived primarily from other sources, such as ion statistics, noise, and the like.

G. Elemental Composition Determination

The final step in the data-processing sequence is to convert the masses to elemental compositions. The technique we use is identical to that described in Chapter 2, "Photographic Techniques in Organic High-Resolution Mass Spectrometry." Several algorithms have been published that describe the conversion of mass to elemental composition. Burlingame *et al.* (10) have described their technique, which is similar to that reported in the above-mentioned chapter on photographic techniques. Different algorithms have been reported by Lederberg (27, 28) and by Kendrick (29).

Before concluding this section, mention should be made of the technique of averaging the mass measurements of several individual scans. Burlingame *et al.* (22) observed that the mean mass gave an improvement in precision approaching the factor predicted from the square root of the number of scans averaged, that is, the factor of two improvements by averaging four scans. Besides improving the accuracy, this result also proves that the errors on each mass measurement are random and not due to some systematic error in the data-acquisition or data-reduction system. This technique should be valuable for those compounds that yield several elemental compositions for each mass if broad error ranges are permitted.

After completion of the conversion of the masses to elemental composition, the presentation of the results can proceed by one of the methods described in Chapter 4, "Computer-Assisted Interpretation of Mass Spectra."

REFERENCES

1. K. Biemann, P. Bommer, and D. M. Desiderio, *Tetrahedron Lett.*, **1964**, 1725.
2. J. T. Watson and K. Biemann, *Anal. Chem.*, **36**, 1135 (1964).
3. W. H. McFadden and E. A. Day, *Anal. Chem.*, **36**, 2362 (1964).
4. C. Merritt, Jr., P. Issenberg, M. L. Bazinet, B. N. Green, T. O. Merren, and J. G. Murray, *Anal. Chem.* **37**, 1037 (1965).
5. B. N. Green, T. O. Merren, and J. G. Murray, *13th Annual Conference on Mass Spectrometry and Allied Topics*, ASTM, Committee E-14, 1965, St. Louis, Missouri, p. 204.
6. W. J. McMurray, B. N. Green, and S. R. Lipsky, *Anal. Chem.*, **38**, 1194 (1966).
7. A. E. Banner, *J. Sci. Instrum.*, **43**, 138 (1966).
8. A. E. Banner, *13th Annual Conference on Mass Spectrometry and Allied Topics*, ASTM, Committee E-14, 1965, St. Louis, Missouri, p. 193.
9. A. L. Burlingame, D. H. Smith, and R. W. Olsen, *Anal. Chem.*, **40**, 13 (1968).
10. A. L. Burlingame, in E. Kendrick (Ed.), *Advances in Mass Spectrometry*, Vol. 4, Institute of Petroleum, London, 1967, p, 15.
11. I. G. McWilliams and H. C. Bolton, *Anal. Chem.*, **41**, 1755 (1969).
12. J. H. Beynon, *Mass Spectrometry and its Application to Organic Chemistry*, Elsevier Publishing Co., Amsterdam, 1960.
13. R. J. Klimowski, R. Venkataraghavan, F. W. McLafferty, and E. B. Delany, *Org. Mass Spectrom.*, **4**, 17 (1970).
14. A. J. Campbell and J. S. Halliday, *13th Annual Conference on Mass Spectrometry and Allied Topics*, ASTM, Committee E-14, 1965, St. Louis, Missouri, p. 200.

15. J. S. Halliday, in E. Kendrick (Ed.), *Advances in Mass Spectrometry*, Vol. 4, Institute of Petroleum, London, England, 1967, p. 239.
16. W. J. McMurray, S. R. Lipsky, and B. N. Green, in E. Kendrick (Ed.), *Advances in Mass Spectrometry*, Vol. 4, Institute of Petroleum, London, England, 1967, p. 77
17. H. V. Malmstadt and C. G. Enke, *Digital Electronics for Scientists*, Benjamin, New York, 1969.
18. P. Issenberg, M. L. Bazinet, and C. Merritt, Jr., *Anal. Chem.*, **37**, 1074 (1965).
19. H. G. Boettger, *15th Annual Conference on Mass Spectrometry and Allied Topics*, *ASTM, Committee E-14*, 1967, Denver, Colorado, p. 90.
20. A. Carrick, *Int. J. of Mass Spectrom. and Ion Phys.* **2**, 333 (1969).
21. M. Barber, B. N. Green, W. J. McMurray, and S. R. Lipsky, *16th Annual Conference on Mass Spectrometry and Allied Topics*, *ASTM, Committee E-14*, 1968, Pittsburgh, Penna., p. 91.
22. A. L. Burlingame, D. H. Smith, T. O. Merren, and R. W. Olsen, *16th Annual Conference on Mass Spectrometry and Allied Topics*, *ASTM, Committee E-14*, 1968, Pittsburgh, Penna., p. 109.
23. T. R. Lusibrink, Ph.D. thesis, University of California, August, 1965.
24. D. H. Smith, R. W. Olsen, and A. L. Burlingame, *16th Annual Conference on Mass Spectrometry and Allied Topics*, *ASTM, Committee E-14*, 1968, Pittsburgh, Penna., p. 101.
25. H. C. Bowen, E. Clayton, D. J. Shields, and H. M. Stanier, in E. Kendrick (Ed.), *Advances in Mass Spectrometry*, Vol. 4, Institute of Petroleum, London, England, 1967, p. 257.
26. M. B. Neher, R. L. Foltz, M. E. Hasofurther, and R. B. Randall, *16th Annual Conference on Mass Spectrometry and Allied Topics*, *ASTM, Committee E-14*, 1968, Pittsburgh, Penna., p. 95.
27. J. Lederberg, *Computation of Molecular Formulas for Mass Spectrometry*, Holden Day, San Francisco, Calif., 1964.
28. H. Budzikiewicz, C. Djerassi, and D. H. Williams, *Structure Elucidation of Natural Products by Mass Spectrometry*, Vol. 2, Holden Day, San Francisco, Calif., 1964.
29. E. Kendrick, *Anal. Chem.*, **35**, 2146 (1963).

Computer Assisted Interpretation of Mass Spectra

P. V. FENNESSEY

Martin Marietta Corporation, Denver, Colorado

I. Introduction 77
II. History 78
III. Computer Interpretation 80
    A. *de novo* Searching of Mass-Spectral Data 81
    B. Library Searching 82
    C. Compound-Type Identification 84
IV. Partial Interpretation of Mass-Spectral Data 85
V. Conclusions 88
    References 88

I. INTRODUCTION

During the past few years mass spectroscopists have witnessed a steady increase in the use of logic devices, including computers, in their routine work. Terms such as "flip-flop" and "or gate" are becoming more commonplace both in the literature and in private discussions. In fact, few people today would even consider the purchase of a high-resolution mass spectrometer without first having some assurance that adequate computer facilities are available. What has caused this dependence? What are the effects on the laboratories and on their capability to solve chemical problems? These and other questions are being asked, notably by research directors and government agencies who must supply the capital necessary to purchase the equipment.

However, for all of us it is important both to understand the history of this computer revolution in our field and to get an idea of what the future may hold. In an attempt to answer some of these questions, this chapter will cover a brief history of the role of the computer in mass spectrometry and then address more specific examples of the use of a computer-mass spectrometer combination in the solution of real chemical problems in the laboratory.

77

II. HISTORY

The study of positively charged particles in electric and magnetic fields that was carried out by J. J. Thomson in 1912 (1) and expanded by A. J. Dempster (2) marked the beginning of mass spectrometry as a field of research. The mass spectrometer, however, was left almost exclusively in the hands of the physicists and physical chemists until the 1950s; since then, it has steadily gained acceptance as an important instrument in the solution of organic chemical problems. This widespread acceptance combined with the availability of excellent commercial instruments has led to the addition of the mass spectrometer as a necessary tool in a modern chemical laboratory. The interest has been stimulated by the appearance of several books (3–10) dealing with the application of mass spectrometry to organic chemical problems and, even more important, by the ever-increasing number of articles in the chemical, biochemical, and medical literature demonstrating the use of this technique in solving structural problems.

Equally important have been those technical advancements that led to the development of high-resolution mass spectrometry (11) and that have made these instruments sufficiently reliable to be used on a daily basis in an organic chemistry laboratory. Using these, one is able to resolve two ions differing in mass by as little as 0.004 atomic mass units at mass 200. More important, one can measure the mass of ions with an accuracy of a milli-atomic mass unit and deduce their elemental compositions with little or no ambiguity. As the technology in both instrument design and measuring systems improves, there should be a corresponding increase in this resolution.

Interestingly enough, it is this measuring system for high-resolution spectra that acted as the impetus for the mating of the computer and mass spectrometer. The sheer volume of data that can be acquired from each complete mass spectrum (i.e., hundreds of thousands of points) could only be handled by a machine. Thus a link of communication (i.e., cards, tape, hard wire) was set up between the spectrometer and the computer, and mass spectroscopists began to include the reading and understanding of computer manuals in their daily routine.

A new dimension was added to both low- and high-resolution mass spectrometry by the development of techniques that allow one to directly couple a gas chromatograph to the mass spectrometer (12). This provides a combination that gives one all of the advantages of a gas-chromatographic system and adds the ability to obtain the complete mass spectrum of each of the components separated on the column. This system has opened broad new horizons both to the natural-product chemists and to the synthetic chemists, for now both the analysis of complex mixtures of closely related substances and the monitoring of a single reaction are possible without the need of elaborate separation steps that are not only time-consuming but also a source of contamination, side reactions, and substantial loss in the quantity of material.

The use of a gas chromatograph mass spectrometer combination produces an inordinate volume of data. In order to make this method more useful and more general, ways were developed that aided the chemist in reaching conclusions as to the nature of the compounds giving rise to each spectrum. Central to this development were the logic devices that are used to reduce the large data volume to something that can be conveniently handled by the chemist. This dependence on computers then spread to low-resolution investigations.

Before 1968 there were a few commercial instruments to help in the acquisition of data from low-resolution mass spectrometers (13). Some of these devices, termed "mass markers," monitor the change of either accelerating potential or magnetic field during a spectral scan and produce a mark on the spectrum at selected mass-unit intervals. Other devices, termed "digitizers," control the scanning circuitry and produce an encoded spectrum (i.e., list of masses and associated intensities). These early approaches suffered from at least one or a combination of the following problems with respect to a gas-chromatograph/mass-spectrometer system: (a) very slow speed, (b) restricted mass range, and (c) frequent and complex calibration. All early approaches neglected the important human factors involved in the interpretation of a large number of spectra and the physical problem of recording and tracking all the data.

A programmable digital computer presents a solution to all the problems mentioned above. One gains the additional advantage of having the computer compare each spectrum to a library of spectra and interpreting (i.e., identifying) many of the spectra. As a result, complex gas chromatograms containing more than 100 different compounds can be analyzed by a chemist on a more or less routine basis.

There was one additional factor that influenced the quick spread of computers into the laboratory, and this was the computer itself. A revolution has swept the computer world and forced the movement away from computer centers with a few highly skilled operators to either a large number of sub-stations or individual small computers. This transition away from working in the sterile surrounding of a typical computer room to one's own laboratory is a major event in the evolution of the computer. Coupled with the availability of computers has been a reduction in their cost. As a result, one can purchase a complete and versatile system for a mass spectrometry application for less than $40,000.

The result of all these changes can best be seen in the capability of the organic chemist or biochemist to tackle complex problems. The computer has taken the jobs that require either a large volume of data or a repetition of the same process many times. Tasks such as identifying the mass and intensity of peaks in the spectrum, producing permanent numerical records of each spectrum, or maintaining large reference files can all be easily accomplished by a computer. This has a direct bearing on the number of spectra one can run in one day. For example, by using a computer-based system, one can handle a problem in which there are 100 to as many as 200 components in a single chromatogram (14). Another way of saying this is: with complex gas chromatographs, physically obtaining good mass spectra was, at one time, the slowest task, and now the interpretation of the data is the limiting task.

A simple extension of the computer's role in mass spectrometry led to a search comparing each unknown spectrum to a library of spectra. This approach has the obvious advantage of focusing the chemist's attention on the real unknown compounds. For example, in a complex chromatogram, one often has a number of simple compounds interspersed with true unknowns. A computer-library comparison can usually identify these simple compounds and leave only a list of new or unknown compounds. In addition, one normally gets a good indication of the structural type of the unknown compounds even when a positive identification cannot be made (i.e., a benzene derivative has many ions at the same mass as other compounds with a similar basic structure). Yet, as is shown in the next section, the computer can do much more in the actual interpretation of mass spectra.

III. COMPUTER INTERPRETATION

The ultimate goal in computer-aided interpretation is to be able to produce a complete compound structure identification starting with only the mass-spectral data. But the problem is difficult. The lack of a detailed knowledge of the fragmentation pathways of charged ionic species forces one to rely on probabilities that can become so complex and interrelated that a complete structure elucidation by a batch computer program is, at present, impossible. There have been some significant advances during the past few years in those cases in which the complexity of the compounds studied is limited or a considerable amount of compound information is put into the computer as part of the basic data.

In dealing with low-resolution mass-spectral data, two directions in the applications of computers to interpretation have been followed. One is centered about an approach that identifies the compound class and then uses highly specialized programs to interpret the spectrum. The other approach is to build up a library of thousands of spectra and identify the unknown by a comparison technique.

A. *de novo* Searching of Mass-Spectral Data

The work of Pettersson and Ryhage (15, 16) serves as an example of a method for the interpretation of specific types of compounds. In their approach, the unknown spectrum is first examined by a rectangular array (17) to determine whether a program exists for the unknown compound. A rectangular array is simply an arrangement of the data resulting from the sums of the intensities of masses differing by 14 mass units. The distribution of the intensity sum for each column is then used as an indication of the compound type.

Specific compound identification is accomplished by programs that closely simulate the interpretive patterns that are used by the chemist. The approach does utilize *all* the peaks in the spectrum, and begins first by identifying the molecular ion. Once a molecular ion has been established the various peaks are checked for structural branches (i.e., methyl group). To date, few compound types have been treated in this manner, and those that have are simple, monofunctional compounds.

While these programs represent an interesting start along lines that may lead to generally useful techniques, they point out some of the complex pitfalls facing any completely automatic, interpretative scheme. The first is the determination of the compound type—a challenging task to the computer for all simple monofunctional compounds and an extremely difficult one for polyfunctional or highly branched chemical structures. Other problems arise when one tries to program a computer to find the molecular ion in a completely unknown spectrum. Here such things as impurities from the isolation or preparation phases or homologues [i.e., compounds with the same general structure but differing by methylene units $(CH_2)_n$] can lead to confusing results in a completely automatic interpretation approach.

High-resolution data do resolve some of the problems, and one of the first attempts in this area dealt with the problem of determining the elemental composition of the molecule in the spectrum of a compound of unknown structure from the high-resolution data without requiring additional information that very often is not available (e.g., which heteroatomic elements are definitely present or definitely absent). The plan was to avoid as many restrictions as possible so that the program could handle any type of organic compound. Teaching the computer just what could or could not be a molecular ion became a very complex and exasperating job, for it required the programmer to be able to accurately describe the properties of a molecular ion in simple mathematical form. The initial results showed great promise (18), but as more and more of the decision steps were turned over to the computer, the problems became very complex.

Many compounds do not display a molecular ion in their mass spectrum because of some energetically very favorable process leading to a fragment ion, and these too must be considered. Furthermore, because of the time lag between submission of the data and receipt of the final results, in most cases it was more profitable to do the search by hand and depend on the batch-processing computer output only for those compounds where it was very difficult if not impossible to do such a manual search.

B. Library Searching

Another interpretation approach has been the library search. In a broad sense this, too, is computer interpretation. For a large volume of the more routine "unknowns" are compounds of known structure and recorded spectra. In these cases the manual interpretation involves a tentative assignment of the structure and a search through the literature to confirm the results. The obvious advantage in automating this approach is in the utilization of manpower. Those compounds that fall into the class described above can be identified by the computer, leaving the real unknowns to challenge the experienced chemists. A considerable amount of effort is being directed toward the solutions of the problems associated with a library search interpretation (15, 16, 19, 20–23).

To date, several thousand mass spectra have been punched on cards in a computer-compatible format. Each of these spectra include all mass and intensity values (including metastable and doubly charged peaks), the name of the compound, the nominal molecular weight, the serial number (i.e., API 191), and the source. Various instrumental parameters such as the temperature of the ion source and inlet system, the ionization energy, and the instrument type and model were also included when these were known.

Because of the large volume of data, the bulk of research has centered around methods of storing and searching this complex file of spectra. If, for example, one has a file for 8000 mass spectra and for each spectrum maintains a mass range of 0–500 (9 bits of computer storage) and a dynamic intensity range of 0–500 (9 bits), the total storage required for all these spectra is $9 \times 9 \times 8000 = 648{,}000$ bits. If, at the same time, it required 1 min to search the file and one acquired mass spectra at the rate of one every 3 sec, it would require 10 hrs to search the data from a 30-min gas chromatograph.

Some investigators have reduced the search-time problem by abstracting the N most intense peaks from each spectrum where N is variable (15, 20). Using this approach, one completely searches only a small percentage of each of the spectra in the library. The time required to complete a search for each unknown compound is reduced by this method. But a new storage file must be created for the abstracted data, and only part of the problem is solved.

Additional work has pointed to the need for an index that indicates how well two spectra compare (19, 21, 22). By inspection of a table of comparison indices one should be able to tell if the compounds are identical, are similar, or are different. The need for a similarity indication comes from the fact that different mass spectrometers using different sample inlet methods (i.e., probe, molecular leak, gas chromatograph) produce, for the same compound, spectra with obvious intensity differences. These differences almost always preclude a one for one spectral correspondence for a compound and its library equivalent. This means a 70–90 % similarity is normally indicative of a true identification. Moreover, a similarity index can lead the chemist onto the right track of an identification by pointing to a compound type or class into which the unknown is shown to fall.

The best comparison of two mass spectra can be accomplished by matching each mass and intensity value for the complete unknown mass spectrum to each entry in the library, and would be the method of choice if one could have unlimited computer storage combined with a 10- to 100-fold increase in the speed of the machine steps needed for each comparison. When one is dealing with small- or medium-sized computers of present-day speed, the only practical approach seems to be the abbreviated spectra. This can both accelerate the search process and reduce the amount of storage space necessary for a large complex library of spectra.

A considerable amount of work has gone into the solution of condensed or abbreviated mass-spectral files. One of the most promising methods to date uses the two most intense peaks in each 14-mass unit (22). The major advantages of this approach are that it is systematic and complete, and utilizes the interpretatively significant peaks. The importance of a systematic search is apparent in that all spectra are compared on an equal basis. If one condenses library spectra by choosing the two most intense peaks between masses 6–19, 20–33, and the like, each unknown spectrum is abbreviated in exactly the same manner before it is compared to the library data.

Further, this method utilizes all significant peaks in a spectrum by giving the same weight to the high mass peaks as to those generally more intense low mass peaks. The molecular ions will, therefore, be among the masses compared in each library search. The separation of each section by fourteen mass units is extremely significant because of the fact that in the majority of cases one deals with carbon compounds with a basic 14 amu methylene unit (CH_2). The origin of the selection of this technique was based on the steps a chemist considers in the interpretation of a mass spectrum. One primarily utilizes the peak clusters rather than the most abundant peaks in an interpretation, and this method of abbreviation assures one that significant peaks in these clusters are selected for comparison.

In conclusion, a library search resulting in an interpretation (either absolute identification or indication of compound types) can best be carried out by comparing an abbreviated unknown spectrum to a library of abbreviated spectra; by having a similarity indication for each of the spectra searched; and by an automated process that gives the chemist an opportunity to deal with only the more intellectually challenging spectra. There are a few examples of an interpretation without the use of a library.

C. Compound-Type Identification

The ability of a computer program to pick out automatically a specific functionality can be a big timesaver, and the case of simple compounds this can lead to a complete identification. The feasibility of automatically determining a structure or a limited list of structures has been shown using high-resolution data (24). Once the compound type was identified by the computer, a specific program was called for its interpretation. Using only a limited number of the ions in the spectrum to reach their conclusions, the programs are able to produce the correct structure for a large variety of each of the types of compounds tried.

The restrictions placed on the type of compound amenable to this approach severely limit its general application. As in the case of the programs using only low-resolution data (15, 16), the types of compounds are limited to rather simple, monofunctional ones. The type of functional group must be known or determined before the interpretative program is executed, and a specific program must be available for each compound type. Work has only just begun in this area, and continued effort may lead to some real advances in a complete interpretative computer system.

Most notable among the automatic interpretative systems is the polypeptide sequencing programs that have been developed independently in a few laboratories (25–27). In this case the computer is given a set of important fixed parameters that reduce the vast number of possible interactions and therefore the possible structures to a finite number. The fact that only 22 amino acids are known to exist in naturally occurring peptides and that the peptide chain consists of repeating units ($-NH-CHR-CO-)_n$, the ability to label the N-terminal and the carboxyl ends of the compound with known blocking groups, and a knowledge of the fragmentation processes in this type of molecule all aid in making a computer interpretation feasible. But the single most important factor is that the program does not attempt to explain during its interpretation process all of the ions in the mass spectrum. Instead, only a few lines (in most cases less than 10%) are used to reach the final sequence or sequences. Considering the general structure of a peptide with substituents (R-groups) on the carbon adjacent to the carbonyl, it is possible to see that the final computational problem is only one of searching for all possible fragments and substituting the list of known amino acids for the different R groupings. Finally, a possible structure on the basis of the summed intensities of the fragments in each calculated sequence can be proposed.

The polypeptide sequencing programs have been shown to be highly successful but are also rather limited in the same manner that the programs producing a complete interpretation of a single compound are; namely, to a single class of organic compounds whose chemistry is known. In fact, all of the work dealing with computer interpretation leads one to the conclusion that a single and completely automatic system for the general interpretation of mass spectral data is outside the reach of present technology. The successful work, where one limits the scope of the computer's role to relatively simple compounds, does show the great potential that a computer offers to practical problems. An approach to the solution of some of these limitations is the use of a dynamic man-computer interaction to bridge this apparent technological gap.

IV. PARTIAL INTERPRETATION OF MASS-SPECTRAL DATA

There are a number of advantages to be gained from the use of a digital computer as an interpretative tool. Although one usually attributes speed and efficiency to methods employing these machines, it is the thoroughness of the algorithms that is the more important factor in dealing with mass-spectral data. In the case of an unaided (manual) interpretation one is very likely to become fatigued after considering only the first or second solution compatible with the data, and look no further. The computer, on the other hand, will pursue a given course until it has considered all possibilities, which may correspond to many hours or even days of a manual process. Moreover, the combination of these two factors, speed and thoroughness, opens new avenues of approach to the problems of interpretation.

Programs have been developed using large high-speed computers to aid in the interpretation process and one of these, called "ion type," has evolved into the backbone of a more elaborate interpretation scheme (28). The ion type consists of an arrangement of the data so that all of the ions representing a homologous series (alkyl, acyl, or alkyl amine ions, etc.) are grouped together. This type of presentation has the advantage that the masses, in addition to being separated according to their heteroatom content, are further divided by their degree of saturation. Listing the intensities according to carbon number within each such group leads to a presentation of homologous series of ions. This makes it easy to scan the data and see, for example, the minimum chain length and branching points of the alkyl ions (i.e., at the carbon number corresponding to a branch, the intensity is normally larger than that of the neighboring ions) as well as being able to detect clusters of ions at lower saturation and see how these change with increasing heteroatom content. Although this method does not furnish a system whereby one can determine all of the functionalities in a complex molecule, it does allow one to readily detect at least one and in many cases two or three of these. Because of the apparent value of this type of data organization, it is felt that the approach could play an important role in any interpretative program.

With the advent of time-sharing in the computer industry (29, 30), the chemist is able to have access to a large, complex computer through a typewriter in his own laboratory. Using this approach, one is able to use the computer as an intellectual assistant in the interpretation of simple or complex unknown compounds (31).

This type of system offers many advantages as a method of solving problems in interpretation of mass-spectral data; foremost among these is the relief from not having to foresee or predict all of the decision steps, which was a necessary process in the automatic programs (those programs that do not require human intervention). A time-shared approach not only makes each session completely individual in nature but also, because of the vast amount of storage and the open-ended nature of the programs, the experience gained during the course of a dialogue with the computer can be saved. This discourse can be used either as an aid to those users who are less familiar with the system or as a reminder of a particular difficulty during a later session. Since the main source of communication with the computer is from a modified electric typewriter, one always has a complete record of the user-computer dialogue, which can be used for later reference. After the advantages of a time-sharing system were realized, work was reported on the development of programs that would combine the intuition and experience of the chemist with the high speed and accuracy of a digital computer (32).

The major research effort along these lines has been in two laboratories: Stanford and MIT. The former is discussed in chapter 6, "An Application of Artificial Intelligence to the Interpretation of Mass Spectra," and had its origins centered around low-resolution mass-spectral problems. The MIT work that is described in this section began using only high-resolution data, and both systems have progressed so as to utilize a man-computer interaction approach.

In dealing with simple unknowns or with complex unknown mixtures, the first step usually involves acquiring low-resolution mass spectra. These data are then interpreted by a computer library search and, if necessary, manually. High-resolution spectra, on the other hand, are obtained mainly on compounds that are more complex and are not expected to have been investigated previously by mass spectrometry. This type of data, therefore, eliminates any spectral library comparisons and requires interpretation formulated on basic principles. The power of a combination of these two methods, namely acquiring low-resolution mass spectra from the effluent of a gas chromatograph to determine the simple compounds and a high-resolution run to obtain information on the more complex data, has been demonstrated (33).

Because of the apparent value of this approach, a program aimed toward the interpretation of complex mass spectra should deal primarily with high-resolution information. Further reasons for this choice rest with the facts that high-resolution mass spectra can be obtained on a more or less routine basis and that the accuracy of the data (i.e., 106.0656 ± 0.003 amu) allows one to calculate the elemental composition of all of the ions in the spectra (e.g., $106.0656 = C7 \ H8 \ N1$).

The programs developed for the interpretation of this data had two very important basic principles built into them, namely convenience and expandability. Convenience means ease of operation and understanding of just how to make the computer work for the chemist. The commands are simple and versatile. In this light it is also important that the chemist's control terminal be as close to his desk as is possible, because in this manner the chemist has easy access to notes, books, and the other reference material that play an important part in the actual interpretative process.

Every time one works with the computer to solve a problem, a learning process occurs. Those parts that affect the chemist are recorded and can be utilized in future sessions, while those facts that involve the computer programs must bring about a change in those programs. Therefore, one of the important considerations in an experimental computer program must be expandability. For example, it must be easy to add a new data-manipulation command or to change the internal parts of a function independently. Using this approach of expandability, one develops a logical arrangement of the computer program such that there is no predefined beginning or end and that the different areas of the computer routines are virtually independent of one another.

A set of programs using the general principles outlined above has been developed and used for a man-computer interpretation of complex chemical compounds (34). The present set of commands consists of twelve different operations, of which seven are interpretative aids and five are housekeeping controls.

The actual commands for interpretational aid range in complexity from asking for a certain section of the raw spectral data to be printed to a detailed search of every mass in the spectrum for a broad range of heteroatoms. This latter command (test for elements) uses a combination of approaches to help the chemist select the correct heteroatoms in the unknown. For example, by looking for pairs of masses whose exact mass differ by 1.998 amu and then calculating an abundance ratio using the intensity of the two masses, the computer can obtain an indication as to whether chlorine, bromine, or sulfur are present in the unknown. These and many other commands using similar techniques combining the speed, thoroughness, and accuracy of a computer with the experience and intuition of a chemist have been shown to be invaluable when one is faced with the interpretation of spectra of complex unknown compounds (35).

V. CONCLUSIONS

The exact role of the computer has not yet been completely defined. There is no question that machines possessing "artificial intelligence" will play an increasingly important part in our future. This will certainly be true in the mass spectrometry laboratories around this country and the world.

The initial trend of this is apparent today when one studies the mass-spectral literature. Furthermore, it is the low-resolution spectral library searches and computer files that are being pursued most diligently. This is quite natural considering that the bulk of laboratory and research work is done using a low-resolution mass spectrometer and that a wealth of data is available for these libraries. There is a clear importance that all of the work takes on in the formulation of a library of spectra. The task has already begun but will require support from all laboratories before becoming really effective.

As more and more of the high-resolution mass spectrometers are used for the elucidation of the nature of compounds of unknown structure and composition, an increased dependence on a man-computer interaction will be seen. This is partly because the data are voluminous and as such are difficult to deal with, and partly because one knows so little about the detailed fragmentation pathways available to a molecule of moderate complexity. The symbiosis between a man and a computer provides a method for the chemist to examine speedily and efficiently all data produced in a complete high-resolution mass spectrum.

REFERENCES

1. J. J. Thomson, *Rays of Positive Electricity and Their Application to Chemical Analysis*, Longmans, Green and Co., Ltd., London, 1913.

2. A. J. Dempster, *Science*, Dec. 10, 1920.

3. J. H. Beynon, *Mass Spectrometry and its Application to Organic Chemistry*, Elsevier, Amsterdam, 1960.

4. K. Biemann, *Mass Spectrometry, Applications to Organic Chemistry*, McGraw-Hill, New York, 1962.

5. F. W. McLafferty (Ed.), *Mass Spectrometry of Organic Ions*, Academic Press, New York, 1963.

6. H. Budzikiewicz, C. Djerassi, and D. H. Williams, *Interpretation of Mass Spectra of Organic Compounds*, Holden-Day, San Francisco, Calif., 1964.

7. H. Budzikiewicz, C. Djerassi, and D. H. Williams, *Structure Elucidation of Natural Products by Mass Spectrometry, Vol. I: Alkaloids*, Holden-Day, San Francisco, Calif., 1964.

8. H. Budzikiewicz, C. Djerassi, and D. H. Williams, *Structure Elucidation of Natural Products by Mass Spectrometry, Vol. II: Steroids, Terpenoids, Sugars, and Miscellaneous Classes*, Holden-Day, San Francisco, Calif., 1964.

9. F. W. McLafferty, *Interpretation of Mass Spectra*, Benjamin, New York, 1966.

10. H. Budzikiewicz, C. Djerassi, and D. H. Williams, *Mass Spectrometry of Organic Compounds*, Holden-Day, San Francisco, Calif., 1967.

11. J. H. Beynon in J. D. Waldron (Ed.), *Advances in Mass Spectrometry*, Pergamon Press, Oxford, 1959, p. 328.

12. J. T. Watson and K. Biemann, *Anal. Chem.*, **37,** 844 (1965) and references cited therein.

13. See the discussions in Chapters 1 and 2 of Ref. 4.

14. J. A. Vollmin, I. Omura, J. Seibl, K. Grob, and W. Simon, *Helv. Chim. Acta*, **49,** 1768 (1966).

15. B. Pettersson and R. Ryhage, *Ark. Kemi*, **26,** 293 (1967).

16. B. Pettersson and R. Ryhage, *Anal. Chem.*, **39,** 790 (1967).

17. M. C. Hamming and R. D. Grigsby, *15th Annual Conference on Mass Spectrometry and Allied Topics*, Denver, Colo., May 1967, p. 107.

18. K. Biemann and W. McMurray, *Tetrahedron Lett.*, 647 (1965).

19. S. Abrahamsson, S. Stenhagen-Stallberg, and E. Stenhagen, *Biochem. J.* **92,** 2PB (1964); S. Abrahamsson, G. Haggstrom, and E. Stenhagen, *14th Annual Conference on Mass Spectrometry and Allied Topics*, Dallas, Tex., May 1966, p. 522.

20. I. C. Smith, W. Kelly, A. Brickstock, and J. G. Ridley, *15th Annual Conference on Mass Spectrometry and Allied Topics*, Denver, Colo., May 1967, p. 102.

21. L. R. Crawford and J. D. Morrison, *Anal. Chem.*, **40,** 1464 (1968).

22. R. A. Hites, Ph.D. thesis, Massachusetts Institute of Technology, Cambridge, Mass., 1968.

23. R. A. Hites and K. Biemann, in E. Kendrick (Ed.), *Advances in Mass Spectrometry*, Vol. 4, The Institute of Petroleum, 1968, p. 37.

24. A. Mandelbaum, P. Fennessey, and K. Biemann, *15th Annual Conference on Mass Spectrometry and Allied Topics*, Denver, Colo., May 1967, p. 111.

25. M. Senn, R. Venkataraghavan, and F. W. McLafferty, *J. Amer. Chem. Soc.*, **88,** 5593 (1966).

26. K. Biemann, C. Cone, B. R. Webster, and G. P. Arsenault, *J. Amer. Chem. Soc.*, **88,** 5598 (1966).

27. M. Barber, P. Powers, M. J. Wallington, and W. A. Wolstenholme, *Nature*, **212,** 784 (1966).

28. K. Biemann, W. McMurray, and P. V. Fennessey, *Tetrahedron Lett.*, 3997 (1966).

29. C. Strachey, paper presented at the UNESCO Information Processing Conference, June 1959.

30. F. J. Corbato, M. M. Daggett, and R. C. Daley, *Proc. 1962 Spring Joint Computer Conferences.*, AFIPS, May 1962, pp. 335–344.

31. F. J. Corbato and R. M. Fano, *Scientific American*, Sept. 1966, p. 129.

32. K. Biemann and P. V. Fennessey, *Chimia*, **21,** 266 (1967).

33. S. P. Markey, Ph.D. thesis, Massachusetts Institute of Technology, Cambridge, Mass., 1968.

34. P. V. Fennessey, Ph.D. thesis, Massachusetts Institute of Technology, Cambridge, Mass., 1968.

35. J. M. Hayes, Ph.D. thesis, Massachusetts Institute of Technology, Cambridge, Mass., 1966.

Gas Chromatography—Mass Spectrometry

RAGNAR RYHAGE AND STEN WIKSTRÖM

Laboratory for Mass Spectrometry, Karolinska Institute,
Stockholm, Sweden

I. Introduction 91
II. Gas Chromatography 93
III. Interface in GC-MS Combination 95
   A. The Jet Separator 95
   B. The Porous Glass Separator 98
   C. The Teflon Capillary Separator 99
   D. The Porous Silver Membrane I 99
   E. The Porous Silver Membrane II 100
   F. The Porous Stainless-Steel Separator 101
   G. Silicon Rubber Membrane 102
   H. Silicon-Coated Silver Membrane 103
   I. Combined Membrane and Porous Silver Separator 103
   J. Variable Conductance Separator (Slit Separator) 104
   K. General Discussion 104
IV. Mass Spectrometers Used in the GC-MS Combination 108
V. GC-MS Applications 110
VI. Specific Detectors 114
VII. Computers in Combination with GC-MS 114
VIII. Summary 116
   References 117

I. INTRODUCTION

Separately, the mass spectrometer and the gas chromatograph are two powerful instruments in the analysis of organic compounds. The mass spectrometer, from its beginning in 1910 up until about 1950, was used mainly by physicists, but afterwards it became more and more useful in solving problems in analytical chemistry. Conversely, the gas chromatograph, since its development in 1952, was an instrument used solely by analytical chemists. The combination of these techniques adds a new dimension to the uses of these instruments. The main problem in combining these techniques involves the transferring of the eluate at atmospheric pressure from the gas

91

chromatograph to the mass spectrometer without exceeding the vacuum limits of the latter. This was initially accomplished by condensing the eluate emerging from the GLC column on the inner surface of a glass tube, which then was transferred to the inlet system of the mass spectrometer. This method was rather time-consuming, and the idea of directly combining these instruments through some kind of pressure-reduction system arose. The earliest attempt at direct combination was made by Holmes and Morrell in 1957 (1). They split the effluent of a packed gas chromatographic column, and analyzed a portion of this effluent with a mass spectrometer scanning a preselected mass range. Thus they successfully analyzed low-molecular-weight components of city gas. In 1959, Gohlke, using this technique, extended the range of compounds to include acetone, benzene, toluene, ethylbenzene, and styrene (2). A time of flight (TOF) mass spectrometer was used and equipped with an oscilloscope for the display of the mass spectrum. The high scan rate of this type of instrument makes it very suitable also for capillary column gas chromatography combinations (3). In 1960 Lindeman et al. made further improvements on the method by recording the entire mass spectra on an oscillograph recorder (4).

In 1963 McFadden et al. successfully used capillary columns connected via a splitter to a mass spectrometer for the analysis of flavors (5). Dorsey et al. showed that the C_9-paraffins could be analyzed by a capillary column connected directly to the mass-spectrometer ion source (6). Aside from splitting or direct coupling, an in-line cold trap combination reported by Ebert in 1961 allowed for the necessary pressure reduction in the combined technique (7).

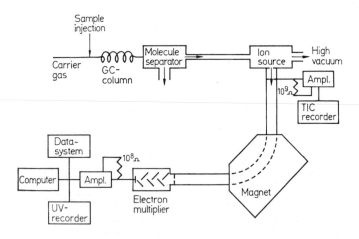

Figure 1. A general block diagram of a GC-MS instrument.

Our laboratory became involved in the combination of gas chromatography-mass spectrometry in 1959, when we modified a magnetic sector instrument to obtain a fast scan speed (8). The modified instrument could scan m/e 12–500 in 1 sec, allowing the analysis of organic compounds in the microgram range. The combination of a gas chromatograph and the above-mentioned mass spectrometer by the use of a jet molecular separator was effected in 1963 (9). Almost simultaneously a description of the porous glass separator used in combination with a high-resolution mass spectrometer was published (10). Today all low-resolution magnetic mass spectrometers for organic work can in some way be connected to a gas chromatograph via a molecular separator, and analysis of components in a mixture of organic compounds can now in general be analyzed by the combined instrument.

The relatively fast development of computerization in the gas-chromatography and the mass-spectrometry field will soon make it possible to integrate a small computer as a standard part of the GC-MS instrument. Figure 1 shows a block diagram of such an instrument, which is discussed later in this chapter.

II. GAS CHROMATOGRAPHY

The main difference in the operation of an ordinary gas chromatograph, as compared to the gas chromatograph connected to a mass spectrometer via a molecular-separator interface, is that the combined instrument is more sensitive to the carrier-gas flow rate and to the liquid phase bleeding effect. For a single GC the flow rate can more easily be optimized for each column than when interfaced to the MS, and different types of carrier gas can be used. For an interface that works on the principle of diffusion, helium or hydrogen should be used for a good separation effect. When a part of the total ion current in the mass spectrometer is used for GC detection, the energy of the bombarding electrons should be kept below 24.8 eV for helium, 18.0 eV for hydrogen, and 15.5 eV for nitrogen in order to exclude these ions from the beam. However, normally for a mass spectrum it is desirable to use a higher electron energy, since the ionization curve consists of a rather steep slope up to about 25 eV and then increases slowly up to 100 eV. For this reason 20 to 25 eV is used for the gas chromatogram, if helium is used as a carrier gas, except when mass spectra are taken when a standard energy of 70 eV is used.

The inlet pressures for a combined GC-MS are usually applied through a flow controller, which works at constant pressure of 3 kg/cm² or 35 psi. The flow rate is controlled and kept relatively constant even if the column is temperature-programmed, and this will have the advantage of giving the interface stable working conditions. The behavior of the column will change somewhat when the column exit is working below atmospheric pressure. Some investigations suggest that these working conditions give better column efficiency than when working at atmospheric pressure at the column exit (11). Nevertheless, some interfaces between the GC and MS, such as the rubber membrane interface, do not require a low input pressure. These interfaces have the advantage of allowing one to feed the remainder of the effluent, after passing the input of the interface, to a standard GC detector and receive a normal GC record. A GC detector can also be connected to an effluent splitter with the entrance to the interface at atmospheric pressure.

The choice of the proper liquid phase is essential in GC-MS. The phase must effect the desired separation while producing a minimum of column bleed. Many peaks arise from the column bleed and, together with the separated compound, give a mass spectrum that is difficult to interpret. This problem is aggravated when small amounts of sample are studied. Although it is possible to subtract background spectra from sample spectra, it remains desirable to minimize column bleeding. A thermostable liquid phase should be chosen. The amount of bleeding is proportional to the surface area of the coated support, and will increase with temperature and concentration of liquid phase, even when a thermostable phase is used. Therefore the percentage liquid phase should not be greater than is practically necessary (12). Commonly the concentration of liquid phase used in GC-MS ranges from 0.5 to 5%.

A very useful standard column for the GC-MS instrument is a 1 to 2% methyl substituted silicon polymer such as SE-30 or OV-1 (13, 14). These liquid phases are both thermally stable for mass-spectrometry work up to 300°C with careful preconditioning. These liquid phases are classified as nonselective, and separations are based almost entirely on the molecular weight and shape of the molecule. The phase OV-17, a substituted silicon polymer containing an equal amount of methyl and phenyl groups, has properties similar to SE-30 and OV-1, but is classified as moderately selective. F-60, which is a siloxane polymer with p-chlorophenyl groups as well as methyl groups, has been found useful for high-temperature work. Highly efficient columns have been prepared and used to separate biologically important amines and for general clinical applications in the analysis of urine (15, 16). QF-1 and OV-210 are included among the liquid phases that show selective retention for alcohols, ketones, and esters but not for carbon–carbon unsaturation (17); QF-1 and OV-210 are thermostable up to about 250°C; QF-1 contains fluoroalkylsilicone groups and OV-210 equal amounts of methyl and trifluoropropylsilicone groups. A liquid phase for selective retention of carbon–carbon unsaturation, but without stereoselectivity for hydroxyl substitution is obtained when methyl phenyl siloxane polymers contain a relatively high percentage of phenyl substituents, such as, OV-25 (75% phenyl 25% methyl), which is thermostable to 300°C (18).

Linear polyester phases show selective retention for alcohols, ketones, esters, and carbon–carbon unsaturation. The most stable are those derived from highly hindered alcohols. Neopentyl glycol succinate (NGS) and ethylene glycol succinate (EGS) are two examples of this group; however, proper use of temperature is of the utmost importance. To obtain better thermal stability one can use the polymer EGSS, which is prepared from ethylene glycol, dimethyl siloxane, and succinic acid. This type of phase has been used for long-chain fatty acid analysis (19). Cyanoethyl methylsilicones (CNSi) and cyanopropylphenylmethyl silicones (OV-225), which are relatively thermostable substances, resemble polyester columns of low or moderate selectivity in their general behavior. In steroid research, ketones are easily separated from alcohols on a CNSi column (20). There are many procedures published on the deactivation and the subsequent coating of the support (21), but the procedure varies from one laboratory to another.

III. INTERFACE IN GC-MS COMBINATION

A molecule separator in the combination technique GC-MS serves two purposes. First, it reduces the outlet pressure from the GC column to a pressure suitable for the ion source region of the mass spectrometer, and secondly, it increases the sample to carrier-gas ratio when its input and output concentrations are compared. The separator ideally passes all the sample through without decomposition while all carrier gas is removed, with no delay time to degrade the separation of the GC. The efficiency of a separator can be measured by the determination of its yield Y and enrichment N and will, in this article, be defined as:

$$Y = \frac{Q_{MS}}{Q_{GC}} \times 100\% \qquad N = Y\frac{He_{GC}}{He_{MS}}$$

where Q_{MS} is the amount of eluate entering the MS, He_{MS} the flow of carrier entering the MS, Q_{GC} the amount of eluate leaving the GC, and He_{GC} the flow of carrier leaving the GC. Approximate operational data of the different types of separators are given in Table 1.

A. The Jet Separator

The first jet type separator used in gas chromatography—mass spectrometry was reported in 1964 (9, 22). The principle of the jet orifice was published by Becker (23). The carrier gas and the eluate from the column are

TABLE 1
Summary of Operational Characteristics of Separators

Type of separator	Yield, %	Enrichment	Working flow range, ml/min	Lag time, sec	Max working temperature, °C	Published, year	Ref. no.
Jet	50	40–86	1–60	none	350	1964	9,24
Porous glass	50	5–20	1–60	negligible	350	1964	10,33
Teflon capillary tubing	50	1–5	2–30	20–30	280	1966	38
Silicon rubber membrane	50	10^5	1–50	noticeable	250	1966	42
Porous Silver I		(2–100)	1–5	negligible	350	1967	39
Porous Silver II	50	4–24	4–18	negligible	250	1968	40
Variable conductance (slit)	30	10–60	1–100	negligible	350	1969	46
Silicon-coated silver membrane	40	10^5	~1–5	noticeable	250	1969	43
Porous stainless steel	50	(10)–108	7–35	negligible	250	1970	41
Combined membrane and porous silver	50	2–40	5–60	noticeable	250	1970	45

fed under pressure and viscous flow conditions to a nozzle from which the gas emerges in the jet periphery. A sharp-edged collector nozzle effects a separation, since the lighter component of the gas is enriched in the peripheral portion. Figure 2 shows a schematic drawing of the separator that was described in 1967 (24). The molecule separator is constructed of stainless steel and made in two stages in order to accommodate a higher flow rate of carrier gas. The geometric dimension of the orifices in the jets of the separator, as well as the distances between the jets, has been experimentally tested to obtain optimal performance. For practical reasons the diameter of the orifice of the first jet is about 0.1 mm. A smaller diameter of the orifice would make

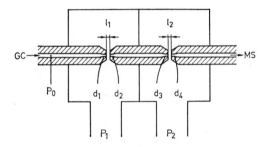

Figure 2. The jet molecular separator.

the construction more complicated and could easily become plugged with dust from the gas chromatography column. The dimensions (in mm) of the orifices and the distances used in one of the experimental molecule separators are as follows: $d_1 = 0.1$, $l_1 = 0.15$, $d_2 = 0.24$, $d_3 = 0.24$, $l_2 = 0.5$ and $d_4 = 0.3$ mm. Pressure P_0 varies between 300–700 mm Hg, P_1 between 10^{-4}–10^{-3} mm Hg for a He flow rate exceeding 10 ml/min. A carrier-gas flow rate from 5 to 60 ml/min can be accepted. It is possible to use a single stage separator for a helium flow rate below 10 ml/min (25). When using a two-stage separator at such a low flow rate, it is recommended that carrier gas be added after the column so that the first jet can operate at optimal pressure (26, 27). The efficiency of the molecule separator is dependent on the flow rate as well as the molecular weight. For methyl palmitate, experimental results indicate that more than 50% of the sample injected into the gas-chromatographic column passed the separator to the ion source. Some of the advantages in using a jet-type molecule separator in combination gas chromatography—mass spectrometry are as follows:

1. The small dimensions of the separator.
2. Mechanically stable construction of the separator.
3. Good separation of helium from sample.

4. Beam velocity greater than velocity of sound.
5. Efficiency almost independent of the temperature of the separator.
6. Low pyrolysis effect.
7. Low memory effect.

Practical results have shown that all organic compounds that can be studied by gas chromatography can also pass the jet molecule separator and be studied by the mass spectrometer. Typical applications of this separator as an interface in the GC-MS instrument can be found in the literature (28–32).

B. The Porous Glass Separator

The earliest published model of this device was made up of a fritted glass tube with a 5-μ pore size, a 0.1-mm constriction to the GLC column, and a

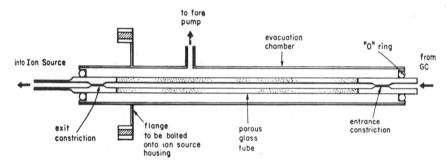

Figure 3. The porous glass separator.

0.12-mm constriction to the mass spectrometer (10). An improved version was published a year later (33). The construction with an 8-in.-long porous glass (8-mm OD and 4-mm ID) having an average pore diameter of 1 μ, an entrance constriction of 0.2 mm, and an exit constriction of 0.1 mm, was found to be the best on an experimental basis (33). Figure 3 shows a diagram of the improved version of the porous glass separator. As the effluent passes through the porous glass tube, the carrier gas is preferentially withdrawn through the wall and a part of the sample is passed into the ion source. The enrichment factor was found to be 50, by measuring the ion intensities of m/e 31 versus m/e 4 of a mixture of diethyl ether and helium introduced through the GC and the normal inlet to the mass spectrometer. The enrichment factor is related to the molecular weight, and roughly, at least 10% (but probably not more than 50%) of the sample injected into the GLC column reached the ion source. Several papers have been published in which the porous glass separator has been used (34–37).

C. The Teflon Capillary Separator

A schematic diagram of this device is shown in Figure 4. Certain polymer films exhibit varying degrees of permeability to gases. The effluent passes through a thin walled Teflon tubing that has a vacuum applied to the outer surface area, and the carrier gas is preferentially removed from the solute by virtue of its higher permeability through the Teflon tubing. This device was described in 1966 (38). The basic part of the interface system consists of a 7-ft-long thin walled (0.005 in.) Teflon capillary tubing (0.020-in. OD, 0.010-in. ID) with a surface area of 20 cm². A 4-ft coil of 0.010-in. ID × 0.020-in. OD stainless-steel tubing was then connected to each end for adequate reduction of pressure at the flow rate range 2–30 ml/min. The entire system was heated in a small aluminium block and the temperature

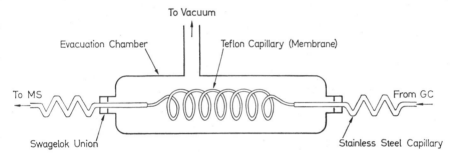

Figure 4. The Teflon capillary separator.

was closely controlled to 0.1°C with a proportional programmer. The enrichment factor as compared to a direct probe insertion of the sample varied from 40 to 70%. A temperature effect on the enrichment factor for this device was discovered, which could be used to optimize the system. At 200°C no sample was found to permeate through the tube, but at 300°C almost all the sample was pumped away before reaching the ion source.

D. The Porous Silver Membrane I

A silver membrane used to separate carrier gas from the eluants of a gas chromatograph was reported in 1967 (39). Figure 5 shows the schematic drawing of the device. The sample from the column enters a chamber of copper, which is enclosed within two porous membrane sheets. After preferential diffusion of the helium through the membrane, the remainder enters the ion source of the MS through a capillary needle. The pore size of the silver membrane is about 0.2 μ and the membrane consists of 80% free space. The

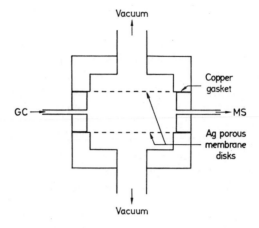

Figure 5. The porous silver membrane I.

enrichment factor, calculated with or without the separator dismantled from the ion source inlet, was reported to give a steady increase in efficiency from 2 or 3 at low mass, to a factor of more than 100 at mass 200. The separator is compact, bakeable, and has replaceable elements.

E. The Porous Silver Membrane II

In 1968 another silver-membrane-type separator was described (40). This separator had the advantage of minimizing the absorption effect reported for the diffusion-type separators. This device works in principle like the porous tube devices. It is constructed of standard swagelock fittings with a small silver membrane forced into one arm of a standard T-tube as shown in Figure 6. The silver membrane has a surface area of 0.1 cm² and a thickness

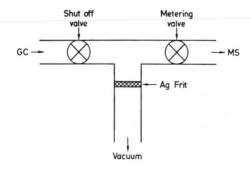

Figure 6. The porous silver membrane II.

of 0.002 in. with a maximum pore diameter of 3 μ. Satisfactory operation has been reported for saturated and olefinic hydrocarbons to C_{20}, and for saturated and unsaturated aldehydes, ketones and dicarbonyl compounds, and fatty acid esters to C_{22}. No preferential retention of olefinic compounds by the porous silver membrane has been noted. The separator temperature range is 100 to 250°C. A carrier gas flow rate of 10 ml/min or less is required.

F. The Porous Stainless-Steel Separator

This works in principle like the porous glass separator. Figure 7 shows the separator, which is constructed entirely of stainless steel (41). The porous tube is 4 in. long, $\frac{1}{2}$-in. O.D., and has a $\frac{1}{8}$-in. wall thickness and 0.1-μ mean pore diameter. The entrance constriction to the separator is made of

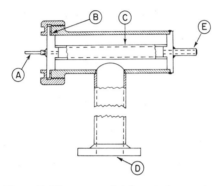

Figure 7. The porous stainless steel separator.

either $\frac{1}{16}$-in. stainless-steel capillary or an isolation valve. The exit of the separator is a $\frac{1}{4}$-in. stainless-steel tube and a variable metering valve. The latter is useful in providing a continuously variable and reproduceable exit constriction and for single adjustment of the ion source pressure to the desired value. The porous tube can be removed for cleaning by unscrewing the cap, and to change the effective length of the porous tube a shorter piece can be attached and the remaining length filled with a section of solid tubing. The separator efficiency, or yield, is reported to be 40–48% and the enrichment factor 108 with a He flow rate of 35 ml/min and 10 for 10 ml/min He flow. The simplicity and ruggedness of construction, high efficiency, and suitability for high-molecular weight compounds are the principle advantages reported in the use of silanized porous stainless steel as an interface material for a diffusion-type separator. The unit can be baked at high temperatures, rapidly heated and cooled, and easily disassembled for clean-ing. Since the separator can be fabricated and machined in a variety of dimensions, considerable flexibility and ingenuity can be exercised in the design and construction for any particular GC-MS.

G. Silicon Rubber Membrane

Figure 8 shows a schematic diagram of this separator, which was described in 1966 (42). This has a working membrane area of the order of 5 cm². It is fixed to a rigid support and sealed into the device by vacuum flanges and gaskets and is easily replaceable. The operating temperature range is from ambient to over 250°C. Helium and other carrier gases are

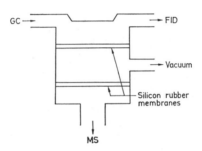

Figure 8. Silicon rubber membrane.

rejected by their low solutibility in the silicon membrane, while organic vapors diffuse through the membrane. With a one-stage device, a 90% transmission of the solute and a delay time shorter than 1 sec is reported. However, to be able to use a simple vacuum system, a two-stage device is recommended, which will give about 50% maximum transmission that is externally varied by a special valve used as a variable aperture between the two stages and the transmission to vary between 2–50% of the original amount of solute. Some of the reported advantages mentioned by the inventors are as follows:

1. GC outlet can be maintained at atmospheric pressure.
2. The performance is independent of the particular carrier gas (He, N_2, H_2, Ar) used.
3. Less than 0.5% of the carrier gas flow is pumped away by vacuum pumps.
4. A single stage will extract 90% of the organic material in the gas stream.

H. Silicon-Coated Silver Membrane

More recently an improved version of the Llewellyn-Littlejohn single-stage membrane separator was described. In this device the membrane consists of a 0.002-in. thick porous silver membrane with 0.2-μ holes. This was coated three times with a 2% solution of 60 mole% phenyl, 40 mole% methyl silxoy copolymer (43). The silicon film was found under microscopic inspection to be 0.0002-in. thick. To minimize bleeding from the membrane, it was conditioned at 250°C overnight, and very little bleed of silicon into the mass spectrometer was found at 225°C. One application with this separator has recently been published (44).

I. Combined Membrane and Porous Silver Separator

Recently, a description of a two stage molecule separator was published (45). Figure 9 shows a schematic view of the separator. The first stage

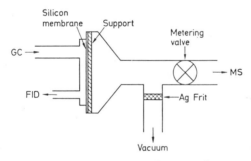

Figure 9. Combined membrane and porous silver separator.

contains a thin membrane of dimethyl silicon polymer, 0.001 in. in thickness, supported on a 20-cm² circular piece of sintered stainless steel welded to a stainless-steel ring. The second stage uses a silver frit installed in the leg of a tee between the first stage and the variable leak to the mass spectrometer. The advantages claimed in using this separator are (1) the outlet of the GC is at atmospheric pressure, (2) the silicon membrane can be designed to transmit more than 90% of most organic compounds by proper control of its surface area and operating temperature, and (3) the separator can be used with widely varying carrier-gas flow rates, permitting GC/MS operation with flow programming. The yield and enrichment of this separator have been carefully studied at different GC carrier-gas flow rates and at different temperatures of the separator. The yield of the two stages operating together varies from 40 to 60% and the enrichments between 2 and 40, depending on the flow rate of the gas chromatograph.

J. Variable Conductance Separator (Slit Separator)

This separator, reported in 1969, is made of stainless steel and has an outer diameter of 60 mm (46). The principle is shown in Figure 10. Here the diffusion occurs at two sharp circular edges about 20 mm in diameter. The carrier gas is pumped away outside and inside the two circular edges.

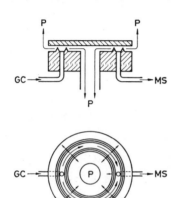

Figure 10. Variable conductance separator (slit-separator).

The distance between the edges and the cover plate is controlled by a spindle. When the slit is closed, gas flow reduction is less than 1 ml/min. With a medium slit opening of 30 μm it is about 60 ml/min. Therefore this separator is suitable for all types of columns. The practical sample yield through the separator is reported to agree very well with the theoretical calculation and is, for Ar/He, about 26% at a carrier gas flow of 13 ml/min and a flow into the MS of 1 ml/min. At 22 ml/min the yield is 13%, and at 37 ml/min it is about 8%. For higher molecular weights the yield is proportionally higher, and cholesterol ($M = 386$) with a helium flow rate of 30 ml/min gives a yield of 32%. Due to the very small interior surface of the separator, decomposition, absorption, and tailing effects are reported to be reduced to a minimum, and the observed differences between chromatograms recorded with and without the separator were negligible.

K. General Discussion

Only a few comparisons of different separators have been made and published during their six-year period of existence. Probably the reason for this is that in testing the different enrichment devices under their individual optimal working conditions, relatively great difficulties occur in constructing the test system for obtaining objective and accurate results. Several reviews of the various interfacial systems used in the coupling of gas chromatography with mass spectrometry have been published (47–50). Another report contains a comparison of the porous glass tube and the Teflon tube separators where the operating conditions and the advantages and disadvantages of the two systems are summarized in table form (51). One of these is shown as Table 2.

TABLE 2

Operating Characteristics of the Fritted-Glass-Tube and Porous-Teflon-Tube Molecular Separators

	Fritted glass	Teflon
Usable flow rates	1 to 100 ml/min	1 to 15 ml/min
Operating temperature	350°C[a]	280 to 330°C
Optimum pressure in ion source	$2–3 \times 10^{-6}$ torr	$1–2 \times 10^{-5}$ torr
Separation yield[b]	15–65%	50–90%
Enrichment[b,c]	5–20	1–5
Lag time[d]	Negligible	20–30 sec
Peak distortion	Negligible	Appreciable[e]

[a] Temperature should be as low as possible, but avoid condensation.
[b] Range for molecular weights 30–128.
[c] Flow rate 5 to 15 ml/min.
[d] Time for compound to pass through separator.
[e] Severe distortion with flow rates exceeding 15 ml/min.

A comparison between the porous glass separator and the jet separator has also been made (49). For the jet separator the overall sample yield was found to be 20% at 35 ml/min and also 20% for the porous glass, but in the latter it varied with the residence time in the separator. Improved separators have been compared with their previous version in some cases, such as the porous glass (35), the porous membrane (40), and the silicon membrane (43).

When columns with flow rates below 1 ml/min are used, a direct connection can usually be made to the mass spectrometer, but for flow rates of more than 1 ml/min, some type of sample enricher must be used. Otherwise, much of the sample will be lost in the necessary splitting procedure. However, the working condition for any molecule separator will vary in a more or less unfavorable way if flow rates other than the ones for which the separator is optimized are used. This is of particular importance when capillary columns are used, since each component of the eluate usually is present in small amounts.

Usually the separators, which are based on different permeabilities for carrier gas and heavier molecules, have a relatively large active surface area. This often leads to an increase in the dead volume and can cause loss of gas-chromatographic resolution by remixing of the eluted peaks. The absorptive loss in polar compounds is also increased with larger active surface area, but improvements can be made by silanizing the separator with bistrimethylsilylacetamide (52). For low-molecular-weight polar compounds, this was found to result in a decrease in the minimum amount of sample for a GC-MS combination with a porous glass separator (10^{-6}–10^{-9} gm). Another report states that recovery of microgram amounts of high boiling polar compounds was unsatisfactory and that methyl palmitate was almost completely retained by the separator (40). Here the frit glass separator performance was improved by treating it with trimethylchlorosilane.

High-resolution capillary columns were used together with the porous glass separator in 1966 (53, 54) and together with the jet separator in more recent reports (55, 56). With the porous glass separator as the interface between a high-resolution glass capillary column and mass spectrometer, the GC resolution changed from 236,000 to 150,000 theoretical plates—a loss in resolution of 32%. When the same type of column was connected to a jet separator as the interface to the mass spectrometer, the loss of GC resolution was reported to be only about 10 to 15%. The best result was obtained when only the second jet stage of the separator was used, but good results were also obtained when 5 ml/min extra helium was added to the effluent from the column and the complete two-stage jet separator was used. This coincides very well with data obtained in our laboratory as well as with data collected elsewhere (27). With the improved membrane separator, the loss of GC resolution is reported to vary between 40 and 60%, as compared with 50 to 80% from the earlier silicon rubber membrane separator (43).

When a modified membrane separator is used as a first stage and a silver frit as a second stage in a two-stage separator, the broadening of the chromatographic peaks is increased only 10 to 20% on passage through the dual-stage separator (45). However, for high-molecular-weight amines, the peak width may be increased as much as 50%. The silicon membrane stage in this separator was carefully tested and some of the most important points are as follows:

1. The residence time that the organic compound is in contact with the membrane is one of the primary factors determining the efficiency of the

membrane. For a given membrane area the residence time will decrease with increased flow rate.

2. A high carrier-gas flow rate requires a large membrane area.

3. A large membrane area means that a relatively large amount of carrier gas passes through the membrane.

4. The GC peak can be broadened by slow diffusion through the silicon membrane.

5. The membrane temperature should be maintained between 50 and 75°C below the boiling point of the material passing through.

6. The yields decreased linearly with temperature for all compounds studied except the methyl esters.

The membrane separator has not yet been widely used, but perhaps some of the new modifications to this separator can improve its usefulness.

When separators are constructed of metal, the chemical reactivity must be taken into account. In order to minimize this effect in the porous stainless-steel separator (41), the whole system, including the metal valve at the entrance and exit of the separator, was silanized prior to use. This coating was maintained through several 10-μl injections of Silyl-8 column conditioner. The frequency of injections is dependent on the temperature used. Good results were reported for cholesterol, and the background bleed from this type of silanization seemed to be negligible. Our experience with the jet separator is that chemical reactivity is minimized due to lack of dead volume and the high velocity of the gas through the jets. However, long metal capillary tubings connected between the gas-chromatographic column and the separator should be avoided, and the volume between the separator and the ion source should be kept as small as possible, although not so small that the pumping speed will be affected.

The tolerance range of the gas flow rate through the separator is of importance both for capillary and packed columns, and a max/min working flow ratio of five is desirable in order to obtain optimal GC separation as well as a minimum loss of sample in the separator. For all types of columns a high velocity of the carrier gas through the interface system minimizes the time the substance is in contact with its surface, and less absorption and decomposition can take place. The size of the dead volume of the interface is not a major problem if packed columns are used, but is serious for packed columns with lower flow rates of carrier gas. The loss of sample in the separator is of course more critical in the use of capillary column GC-MS, as the available amount of eluted substance from the column usually is in the nanogram range and can cause difficulty in obtaining high-quality mass spectra.

IV. MASS SPECTROMETERS USED IN THE GC-MS COMBINATION

To obtain optimal results from the combination gas chromatograph—mass spectrometer, care must be taken so that one of the instruments does not interfer with the other. However, here as well as in other analytical instruments, compromises must be made. The column, the carrier gas, and the flow rate must be chosen in such a way that the efficiency of the separator is high, but the pressure in the ion source and the vacuum system relatively low. If a flame ionization detector or other type of detector is used, a splitter must be connected to the end of the column. In this case the flow rate to the separator is independent of the total flow rate through the column, since it operates at atmospheric pressure. There must be an optimal flow for a given construction of the splitting system. At a higher flow rate, there is sample loss, and at a lower flow rate, air and hydrogen are pumped into the separator and the mass spectrometer, except when a rubber membrane or similar separator is used.

The total ion current of a mass spectrometer is often used to indicate the presence of sample in the GC effluent. Normally, the total ion current monitor is located after the source exit slit. Therefore the sensitivity of this measurement is dependent on slit width, that is, resolution. Also, the collector for the total ion current is only exposed to a portion of the ion beam. If the collector is placed prior to the source exit slit, the former problem is eliminated.

The mass spectrum of a GC component should be scanned as rapidly as possible. This is to ensure that the sample concentration remains relatively constant while the spectrum is being obtained. If the spectrum is obtained while the sample pressure is varying, the relative intensity of the resulting peaks on the mass spectrum will differ somewhat from the same sample run via direct insertion. When the scan speed is increased, the bandwidth of the recording system must also be increased, and thus a lower signal-to-noise ratio must be accepted. The problem of varying sample pressure can also be compensated for by physical, instrumental, or computer techniques if the total ion current variation is monitored during the scan. However, mass spectra obtained with varying sample pressure will not alter the effectiveness of identifying different compounds using computer library searching.

The time of flight mass spectrometer allows very rapid repetitive scanning in the microsecond range. Although this type of instrument has been used for many years, it has not been successful for a general combination with a gas chromatograph. This is probably due to the relatively low resolution of the mass spectrometer. In the case where the instrument has been used for this purpose, a capillary column is directly connected to the ion source of the mass spectrometer.

The quadrupole type of mass spectrometer has been used as a residual gas analyzer for several years, but during the last few years the instrument has been reconstructed for analysis of organic compounds. The advantages in this type of mass spectrometer are (a) the possibility of fast scans in the millisecond range, (b) the linear mass scale, and (c) the ease with which it can be used as a multiple ion detector. A disadvantage of the instrument is the intensity reduction at higher masses as compared to the magnetic-sector instrument. The quadrupole instrument has been used mostly in work with the direct probe. However, at several laboratories it is also used in combination with a GC system. In this mode it will probably, to a greater extent, be connected to capillary columns due to the possibility of very fast scan and relatively higher sensitivity at the low mass range. Very little is known about how much the resolution and intensity of the quadrupole will decrease after its prolonged use for organic analysis, but, it seems possible that its lifetime will be much shorter in view of its inability to maintain specified operating levels as compared to a magnetic mass spectrometer.

Low-resolution magnetic-sector mass spectrometers with fast scanning capabilities can be used with advantage to obtain complete mass spectra from GC effluents, both from capillary and packed columns either with direct connection or through a separator.

Owners of less modern mass spectrometers, who desired to convert the instrument into a combined GC-MS instrument, have often chosen the porous glass separator as the interfacial part, since this separator was one of the first commercially available. There have been more mass spectrometers equipped with the porous-type separator than any other type of separator, but the type of GC-MS instrument chosen is usually dependent on the application. Conversely, from published papers in different journals, it is apparent that the combined GC-MS instrument with the jet separator is used more often in biochemistry and medical research.

In the combination of gas chromatography and high-resolution mass spectrometry, only magnetic-sector instruments have been used. Such a combination was described briefly in 1964 (10) and more completely in 1965 (33). In these experiments a porous glass separator was used as an interface between a column and mass spectrograph with Mattauch-Herzog geometry. Using this method several mass spectra of different components could be exposed and identified. The exposure time for each spectrum was about 20 sec. The porous glass frit separator has also been used in GC-MS instruments where scanning of the high-resolution mass-spectral data was carried out. The data were recorded directly onto analog magnetic tape (57). A Nier-Johnson geometry instrument was used with a scan speed of 10 sec per decade. The analog tape was slowly replayed into a computer via an analog-to-digital converter with a resulting accuracy in mass determination of 10 ppm at a resolution of 10,000.

A GC-high resolution MS combination was described in 1968 (58). Again, a porous glass separator was used as an interface to a packed column and mass spectrometer with a resolving power set at 10,000. The scanning speed was limited to 40 sec/decade as an on-line computer was used. The combination of GC-high resolution MS has not yet been shown to be a practical working instrument for general applications, although several commercial companies today claim that existing instruments can do the job.

V. GC-MS APPLICATIONS

The number of applications of the combination gas chromatograph-mass spectrometer has increased greatly, and only a few examples will be mentioned here. Many further examples of this technique are discussed in Chapter 10, "The Application of Mass Spectrometry to Problems in Medicine and Biochemistry."

A complex mixture of tobacco smoke was analyzed using a glass capillary column (110 m × 0.42-mm ID) coated with polypropylene glycol and connected to a magnetic-sector mass spectrometer via a porous glass separator (53). The column was operated with a carrier-gas flow of hydrogen at 4 ml/min when connected to the combined GC-MS and 5.6 ml/min using a separate FID gas chromatograph. In this GC-MS combination a resolution of 150,000 theoretical plates was obtained versus 236,000 when a FID was used. The tobacco smoke was found to contain more than 168 components, which were identified from the combined data. Similar analysis of a heavier molecular fraction of the gas phase of tobacco smoke have been made using GC-MS and only GC with a glass capillary column made up of a 55 m × 0.35-mm ID coated with Emulphor O, and connected to a magnetic-sector mass spectrometer via a jet molecular separator (55). Figure 11 shows that in both cases the GC separation gave about 350 peaks of which the direct connection GC-MS identified 133 peaks. These are the major constituents of the extracted material as compared to the unidentified peaks. The GC-MS diagram shows less separation of smaller peaks when the neighboring peak is of high concentration. This and other effects such as the change in retention time for some of the peaks may be due to air passing through the column between the two runs (59).

The combined GC-MS instrument has been used in several investigations of naturally occurring compounds. In one investigation the microbial metabolites of 3β, 16α-dihydroxy-5α-pregnan-20-one were converted into their

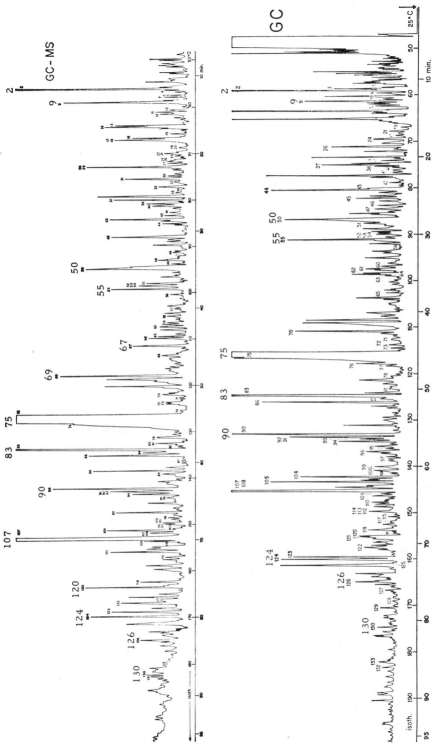

Figure 11. GC diagrams obtained by the injection of 4 μl tobacco mixture on a 55 m × 0.35 mm glass capillary column coated with Emulphor O. (a) Total ion current detection. (b) Flame ionization detection.

111

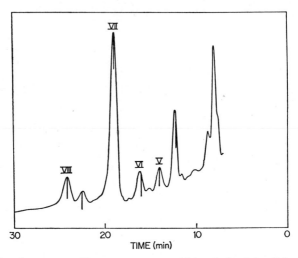

TIME (min)

Figure 12. Gas chromatographic—mass spectrometric analysis of the silyl ethers of microbial metabolites of 3β, 16α-dihydroxy-5α-pregnan-20-one.

trimethylsilylethers and injected onto a 3% QF-1 packed column at 190°C (60). The column was interfaced to a magnetic sector mass spectrometer through a jet separator. The gas chromatographic—mass spectrometric total ion current diagram is shown in Figure 12. The four compounds, which could be identified by their mass spectra and gas chromatographic retention times, are marked V–VIII in the figure. Figure 13 shows the mass spectra of (a) authentic 3α-hydroxy-5α,17α-pregnan-20-one trimethylsilyl ether and (b) the major compound (VII) of the total ion monitor record shown in

TABLE 3

Relative Intensities of Important Peaks in the Mass Spectra of Silyl Ethers of 3-hydroxy-17α- and -17β-pregnan-20-one Isomers

GC peak label	Configuration	Relative intensities of peaks at m/e, %							
		71	215	229	230	285	300	375	390
V	$3\alpha, 5\alpha, 17\alpha$	13	87	14	18	51	100	8	42
VI	$3\alpha, 5\alpha, 17\beta$	4	28	6	6	18	100	20	14
VII	$3\beta, 5\alpha, 17\alpha$	16	46	12	10	31	24	28	100
VIII	$3\beta, 5\alpha, 17\beta$	2	18	4	4	17	37	100	34
	$3\alpha, 5\beta, 17\beta$	2	26	9	19	27	100	26	3
	$3\alpha, 5\beta, 17\alpha$	7	46	13	100	12	74	3	12
	$3\beta, 5\beta, 17\beta$	2	15	5	7	25	100	7	5
	$3\beta, 5\beta, 17\alpha$	13	52	16	18	48	62	16	100

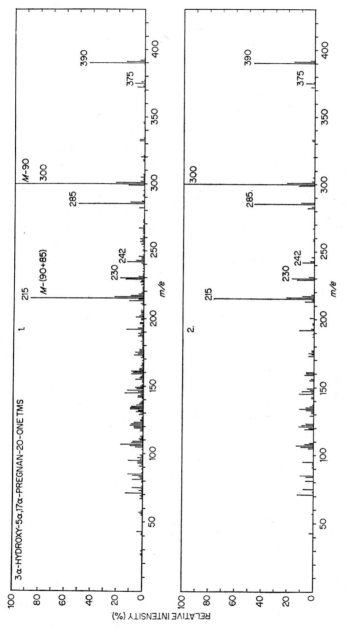

Figure 13. Mass spectra of the silyl ethers of authentic 3α-hydroxy-5α, 17α-pregnan-20-one (1) and compound 2 from feces.

113

Figure 12. Table 3 shows the differences in the relative intensities of characteristic peaks in the mass spectra of the possible isomers of the trimethylsilyl ether of 3-hydroxy 17α- and 17β-pregnan-20-one.

VI. SPECIFIC DETECTORS

In the event that it is desired to search for a particular known compound in a complex mixture, several procedures can be followed. The scan of the mass spectrometer can be cycled and an almost continuous recording of the effluents emerging from the GC column can be obtained. A sophisticated method using computer techniques has been developed to handle the large quantity of data produced by continuously cycled scanning (36). The computer was programmed to select a few ions in each MS scan and plot their intensities versus spectra number, thereby creating a specific detector for a given compound using a few very significant ions. Earlier, a less-expensive method had been developed in this laboratory, where three important ions of a molecule could be chosen by a three-way switching of the ion source accelerating voltages so that these ions would be focused at the detector slit when the magnetic field was held constant (61). By making this switching automatic and recording the detector-amplifier response on UV-paper, the same specific GC detection was obtained. The acceleration voltage switching method has one advantage over the complete spectra scan in that there are only one, two, or three ions that are in focus during the evaluation. The sensitivity is considerably increased, since the signal-to-noise ratio can be increased with higher gain and lower filtering frequency. This is due to the slower switching frequency as compared to the amplifier speed which is required for a fast scan. This very important gain in sensitivity has been used for extensive studies on drug metabolism (62). A disadvantage of this method is that when the accelerating voltage is changed, it affects the intensity relation between the ions in focus. Therefore an intensity-normalization factor must be calculated.

VII. COMPUTERS IN COMBINATION WITH GC-MS

The use of a computer system in combination with the GC-MS instrument is becoming increasingly important, and soon such systems will be included as a standard part of a GC-MS laboratory. An example of such a computer configuration is shown in Figure 14.

The magnetic disk memory shown demonstrates one possible solution to the fast storage of high speed data collected from the mass spectrometer (63). The size of the core memory can be varied but larger memories increase the possibility of making a more complete data evaluation before the data is stored on an external memory such as magnetic disk or a tape. Modern computers can continuously collect data and at the same time perform data reduction in order to reduce occupied core storage. The main advantages in computerization of the GC-MS data are as follows:

1. The establishment of the mass scale automatically with help of a reference compound or sensor for measuring the magnetic field.

2. Collection at a fast rate of the output data from the GC-MS instrument.

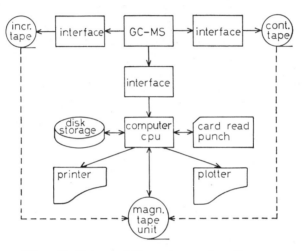

Figure 14. Data acquisition system configuration.

3. The background spectra can be subtracted from actual spectra, for example, column bleed from sample.

4. Presentation of the mass spectra in a normalized form and plotted as a bar graph a short time after the sample has been injected into the column.

Only a few papers have been published that discuss computers connected on-line to GC-MS instruments (63–65). However, several commercial companies have announced that complete on-line data-acquisition systems are available. Although it seems easier to connect a computer to a low-resolution rather than to a high-resolution mass spectrometer, there are several technical difficulties that nevertheless occur. To establish the mass scale in a low-resolution magnetic-sector MS, a reference compound cannot be scanned simultaneously with the studied compound, since peaks would interfere with normal fragments in the spectrum. One method of obtaining good reproduceability of the mass scale is to employ cyclic scanning of the magnetic field on a time basis (64). Here, the reference compounds must be introduced into the MS separately. A second method is to use a Hall-effect semiconductor as a sensor of the magnetic field and thereby establish the mass scale (63). At the same time one can have the computer check the accelerating voltage and calculate a normalization factor to the mass scale. This factor can also be obtained by the detection of known masses, such as m/e 28, 32.

Many high-resolution mass spectrometers have been connected on-line to computers, and the direct probe or heated inlet has been used for introduction of samples. However, there has been found only one publication concerning the gas-chromatographic column connected to a combined high-resolution mass spectrometer-computer (58). One reason is the difficulty in obtaining acceptable accuracy of exact mass numbers due to a poorly reproduceable scan function for rapid scanning of the magnetic field. For standard operation of a high-resolution mass spectrometer, a scanning time of 40 sec/mass decade is normally used for a resolution of 15,000.

To obtain accurate high-resolution data from a GC combination, the scanning time must be reduced to at least 6 to 8 sec/mass decade. In order to receive enough data information per mass peak for this scan speed, the sampling frequencies of the A/D converter should be in the order of 60 kHz. This in turn means that in a 7-sec scan, 420,000 data points will enter the computer. Therefore the computer must have a large core storage and/or be able to process the data at high speed.

VIII. SUMMARY

The combination of GC-MS has greatly improved the possibility of identifying the separated components from a GC column. The separator is an important part of this combination and should be carefully chosen. For a certain analysis one separator may be very favorable while for others the same device may not suffice. The most important factors to be considered in work with the separators are the dead volume, the pyrolysis effect, and the lag time. Different parameters of existing molecular separators are summarized in this article. Computer techniques for data retrieval are also briefly discussed. However, a more detailed picture of the use of computers may be found elsewhere in this book. (See the first four chapters of this volume.)

Further development of the GC-MS instruments will probably involve the development and use of highly efficient columns that possess restricted flow rates. These columns will yield a well separated and concentrated sample that may be passed either directly or through a separator into the mass spectrometer. Computers will also be used to create new approaches to the solutions of previously unsolved problems.

ACKNOWLEDGMENT

Support of the Swedish Medical Research Council (Grant No. B70-13X-171-06B) and the Swedish Board for Technical Development (Grant No. 69-410/U285) is gratefully acknowledged. We thank Mr. D. Gooch for his skilful assistance in reviewing this text and his many helpful suggestions. We also thank Dr. D. Brent for stimulating discussions in this work.

REFERENCES

1. J. C. Holmes and F. A. Morell, *Appl. Spectrosc.*, **11**, 86 (1957).
2. R. S. Gohlke, *Anal. Chem.*, **31**, 535 (1959).
3. R. S. Gohlke, Ibid., **34**, 1332 (1962).
4. L. P. Lindemann and J. L. Annis, ibid., **32**, 1742 (1960).
5. W. H. McFadden, R. Teranishi, D. R. Black, and J. C. Day, *J. Food Sci.*, **28**, 316 (1963).
6. J. A. Dorsey, R. H. Hunt, and M. J. O'Neal, *Anal. Chem.*, **35**, 511 (1963).
7. A. A. Ebert, ibid., **33**, 1865 (1961).
8. R. Ryhage, *Ark. Kemi*, **20**, 185 (1962).
9. R. Ryhage, *Anal. Chem.*, **36**, 759 (1964).
10. J. T. Watson and K. Biemann, ibid., **36**, 1135 (1964).
11. D. C. Locke and W. W. Brandt, in L. Fowler (Ed.), *Gas Chromatography*, Academic Press, New York, 1963, p. 55.
12. R. K. Levy, H. Gesser, T. S. Herman, and F. W. Hougen, *Anal. Chem.*, **41**, 1480 (1969).
13. C. J. W. Brooks, E. C. Horning, and J. S. Young, *Lipids*, **3**, 391 (1968).
14. D. C. DeJongh, T. Radford, J. D. Hribar, S. Hanessian, M. Bieber, G. Dawson, and C. C. Sweeley, *J. Amer. Chem. Soc.*, **91**, 1728 (1969).
15. C. E. Dalgliesh, E. C. Horning, M. G. Horning, K. L. Knox, and K. Yarger, *Biochem. J.*, **101**, 792 (1966).
16. P. Capella and E. C. Horning, *Anal. Chem.*, **38**, 316 (1966).
17. P. Eneroth, B. Gordon, R. Ryhage, and J. Sjövall, *J. Lipid Res.*, **7**, 511 (1966).
18. E. C. Horning, W. J. A. VandenHeuvel, and B. G. Creech, in D. Glick (Ed.), *Methods of Biochemical Analysis XI*, Interscience, New York, 1963.
19. E. C. Horning, K. M. Maddock, K. V. Anthony, and W. J. A. VandenHeuvel, *Anal. Chem.*, **35**, 526 (1963).
20. M. Makita and W. W. Wells, *Anal. Biochem.*, **4**, 204 (1962).
21. E. C. Horning, E. A. Moscatelli, and C. C. Sweeley, *Chem Ind. (London)*, 751 (1959).
22. E. Stenhagen, *Z. Anal. Chem.*, **205**, 109 (1964).

23. E. W. Becker, K. Bier, and H. Z. Burghoff, *Naturforsch.*, **10a**, 545 (1955).
24. R. Ryhage, *Ark. Kemi*, **26**, 26 (1967).
25. R. Ryhage, S. Wikström, and G. R. Waller, *Anal. Chem.*, **37**, 435 (1965).
26. A. Linnarsson, personal communication.
27. M. Novotny, *Chromatographia*, **2**, 350 (1969).
28. J-Å. Gustavsson and J. Sjövall, Exerpta Medica International Congress Series No. 184 Progress in Endocrinology, Proceedings of the Third International Congress of Endocrinology, Mexico, 1968.
29. M. G. Horning and E. A. Boucher, *J. Gas Chromatogr.* **5**, 297 (1967).
30. T. Luukkainen and H. Adlercreutz, *Biochim. Biophys. Acta*, **107**, 579–592 (1965).
31. B. Samuelsson and K. Samuelsson, *J. Lipid Res.*, **10**, 41 (1969).
32. E. Bayer and W. A. Koenig, *J. Chromatogr. Sci.*, **7**, 95 (1969).
33. J. T. Watson and K. Biemann, *Anal. Chem.*, **37**, 844 (1965).
34. H-Ch. Curtius, M. Müller, and J. A. Völlmin, *J. Chromatogr.*, **37**, 216–224 (1968).
35. S. P. Markey, *Anal. Chem.*, **42**, 306 (1970).
36. V. N. Reinhold and K. Biemann, *17th Annual Conference on Mass Spectrometry and Allied Topics, ASTM, Committee E-14, 1969*, Dallas, Texas, p. 115.
37. E. G. Perkins and P. V. Johnston, *Lipids*, **4**, 301 (1969).
38. S. R. Lipsky, C. G. Horvath, and W. J. McMurray, *Anal. Chem.*, **38**, 1585 (1966).
39. R. Cree, Pittsburg Conference on Analytical Chemistry and Applied Spectroscopy, March 1967.
40. M. Blumer, *Anal. Chem.*, **40**, 1590 (1968).
41. P. M. Krueger and J. A. McCloskey, *Anal. Chem.*, **41**, 1930 (1969).
42. P. Llewellyn and D. Littlejohn, *16th Annual Conference on Mass Spectrometry and Allied Topics, ASTM, Committee E-14, 1968*, Pittsburgh, Penna.
43. D. R. Black, R. A. Flath, and R. Teranishi, *J. Chromatogr. Sci.*, **7**, 284 (1969).
44. R. G. Buttery, R. M. Seifert, R. E. Lundin, D. G. Guadagni, and L. C. Ling, *Chem. Ind. (London)*, 490 (1969).
45. M. A. Grayson and C. J. Wolf, *Anal. Chem.*, **42**, 426 (1970).
46. C. Bruneé, H. J. Bültemann, and G. Kappus, see ref. 36, p. 121.
47. D. I. Rees, *Talanta*, **16**, 903 (1969).
48. W. S. Updegrove and P. Haug, *Amer. Lab.*, Febr. 1970, 8-30.
49. F. A. J. M. Leemans and J. A. McCloskey, *J. Amer. Oil Chem. Soc.*, **44**, 11 (1967).
50. J. T. Watson, in L. S. Ettre and W. H. McFadden (Eds.), *Ancillary Techniques of Gas Chromatography*, Interscience, New York, 1969.
51. M. A. Grayson and C. J. Wolf, *Anal. Chem.*, **39**, 1438 (1967).
52. W. D. McLeod, Jr. and B. Nagy, *Anal. Chem.*, **40**, 841 (1968).
53. J. A. Völlmin, I. Omura, J. Seibl, K. Grob, and W. Simon, *Helv. Chim. Acta*, **49**, 1768 (1966).
54. K. Grob and F. J. Burrus, Boncourt, Cic., *Beitr. Tobakforschung*, **13**, 403 (1966).
55. K. Grob and J. A. Völlmin, *J. Chromatogr. Sci.*, **8**, 218 (1970).
56. J. A. Völlmin, *Chromatographia*, **3**, 239 (1970).
57. W. J. McMurray, B. N. Greene, and S. R. Lipsky, *Anal. Chem.*, **38**, 1194 (1966).

58. J. R. Chapman, M. Barber, W. A. Wolstenholme, and E. Bailey, in C. L. A. Harbourn (Ed.), *Gas Chromatography 1968*, Institute of Petroleum, London, England, 1969.
59. K. Grob, personal communication.
60. H. Eriksson, J-Å. Gustafsson, and J. Sjövall, *Eur. J. Biochem.*, **6**, 219–226 (1968).
61. C. C. Sweeley, W. H. Elliott, I. Fries, and R. Ryhage, *Anal. Chem.*, **38**, 1549 (1966).
62. C-G. Hammar, B. Holmstedt, and R. Ryhage, *Anal. Biochem.*, **25**, 532 (1968).
63. B. Hedfjäll, P-Å. Jansson, Y. Mårde, R. Ryhage, and S. Wikström, *J. Sci. Inst.*, 1969 Series 2, Vol. 2, 1031.
64. R. A. Hites and K. Biemann, International Mass Spectrometry Conference, Berlin, Germany, Sept. 25, 1967.
65. W. E. Reynolds, J. C. Bridges, R. B. Tucker, and T. B. Coburn, *16th Annual Conference on Mass Spectrometry and Allied Topics, ASTM, Committee E-14, 1968*, Pittsburgh, Penna., p. 77.

An Application of Artificial Intelligence to the Interpretation of Mass Spectra*

B. G. BUCHANAN

Department of Computer Science, Stanford University, California

A. M. DUFFIELD

Department of Chemistry, Stanford University, California

and

A. V. ROBERTSON

Department of Organic Chemistry, The University of Sydney, N.S.W. Australia

I. Introduction 122
  A. Research on Artificial Intelligence 122
  B. Universal Logic of Structure Determination 124
  C. Overall Strategy of the DENDRAL Project 125
II. The DENDRAL Program 126
  A. The Need for a Linear Notation and Its Requirements 126
  B. The Topological Centroid as the Key to Map the Structure 128
  C. General Principles of DENDRAL Enumeration 131
  D. Further Enumeration of the Alkanes 133
  E. Extension for Unsaturation and Heteroatoms 137
  F. Complete Acyclic DENDRAL Canons 140

* This report is a summary of the current status of the Heuristic DENDRAL project conducted jointly by the Departments of Chemistry, Computer Science, and Genetics at Stanford University under the direction of Professors Carl Djerassi, Edward A. Feigenbaum and Joshua Lederberg. This research was financed by the Advanced Research Projects Agency (Contract SD-183), the National Aeronautics and Space Administration (Grant HGR-05-020-004), and the National Institutes of Health (Grant AM-04257). Much of the programming reported here was done by Mrs. Georgia Sutherland, Mr. A. B. Delfino, Dr. G. Schroll, and Dr. A. Buchs.

G. BADLIST and GOODLIST 142
H. Commentary 143
III. The Heuristic DENDRAL Program 144
A. The Relation of the Algorithm to the Computer Program 144
B. Planning 146
   1. The Preliminary Inference Maker 146
   2. Ketones 147
   3. Ethers 153
   4. Amines 155
   5. Compounds of the General Class R_1-X-R_2 162
C. Structure Generation 167
   1. The DENDRAL Algorithm Plus Heuristic Constraints 167
   2. Implementation of the Structure-Generator Algorithm 168
   3. Preventing Construction of Unlikely Molecules 169
   4. Specifying Required Substructures 169
   5. Rote Memory 170
D. Verification 170
   1. The PREDICTOR 170
   2. The EVALUATION Function 173
IV. Conclusion 174
References 176

I. INTRODUCTION

A. Research on Artificial Intelligence

Artificial intelligence research is aimed at exploring the boundaries of computer problem-solving capabilities. Many tasks that require some measure of intelligence when performed by humans can now be performed by computers; many more cannot. Computer scientists working in the area of artificial intelligence are attempting to develop techniques that will allow computers to solve more and more of the difficult intellectual problems.

The tasks that have proved most amenable to solution by computers over the last decade or two have been those characterized either by complex numerical operations or tiresome repetitiveness. Swift and accurate solutions for many problems once considered to be intellectually challenging can now be expected routinely from machines. There remain many problems, however, that are neither numerical nor repetitive but that require cognitive power of a different sort for their solution. Playing a game or determining an organic molecular structure are examples of nonnumerical complex problem solving.

How far can we go in making computers do intellectual work? Can they be programmed to prove difficult theorems, to make scientific discoveries, to perceive and respond to verbal patterns in order to carry on a sensible conversation? Can they do useful work that involves seeing the outside world and manipulating it—everyday jobs such as driving a car or constructing a house?

Goals like these have not been reached yet. Nevertheless, several computer groups are working at the present frontiers of such studies. Game-playing programs are of central significance in artificial intelligence, not because it is important that computers should play any game well but rather because it is an excellent means of evaluating our ideas about the processes of intelligence against real human performance. In chess, for instance, the path from the clearly defined starting point to a recognizable end pattern is beset with so many alternatives that only a small fraction can be fully explored. Humans deal with such situations by making value judgments about the options they think of, ignoring the rest. Finding the computer counterpart of human game-playing strategies proves to be an excellent opportunity for the logical analysis and understanding of our mental mechanisms. It leads to fundamental discoveries about the nature of human comprehension and thought processes, if only to frame more clearly or from new angles the questions that need to be posed before answers can be sought. The noun "heuristics" is used to embrace the tactics of the game, that is, the often ill-defined principles of good play that can guide a person or machine to a satisfactory solution but do not guarantee a solution.

Humans out-perform computers in formulating scientific hypotheses either to explain data or to make a prediction susceptible of experimental verification. Studies on the nature of scientific discovery require an understanding of our heuristics for generating hypotheses. The decision-making processes in this area of thought can be examined by heuristic programming methods. But to make progress on such a broad and general problem it is necessary to choose a particular scientific task involving inductive behavior and explore it as fully and deeply as possible. The ramifications include the mechanization of scientific thinking itself.

There are few sciences, however, whose structure is suitable and whose concepts have been sufficiently formalized to provide appropriate areas of investigation for heuristic programming methods. The hierarchy of concepts in genetics, for example, which range from purely chemical aspects of molecular biology through the single cell to the complete organism and then whole populations is simply too complex yet for the construction of general mathematical models of much value. Mathematization of a science implies its representation in some form of symbolic language, but there are many sciences in which ordinary language provides the only reasonable description at the moment.

Motivations like these led Lederberg to perceive that organic chemistry was a discipline ripe for formalization along such lines. In terms of manpower and publications it is the largest unified science. Huge parts of it deal with synthesis on the one hand and structure determination on the other. Both are the subjects of current artificial-intelligence studies, the pathfinder

for computized design of organic synthesis being Corey (1). Lederberg chose
to concentrate on structure determination. This well-defined task may be
phrased as follows. What molecule could have produced a given set of
experimental data? We make hypotheses as to which of many possibilities is
the correct structure and then design experiments to verify the hypothesis.
The hierarchy of concepts in this mental task is quite limited, but the range of
possible structures and variety of methods for obtaining structural data are
both extensive. Organic chemistry is already represented abstractly in
symbols, although translation into a form of notation suitable for manipulation by a computer was necessary. Spectroscopic data consisting of numbers
is a preferable format for computer use compared with other structural
information such as chemical reactivity. The expertise of Djerassi's group in
the Stanford Chemistry Department made structural determination from
mass-spectral data a most convenient and attractive area on which to focus.

Originally then, the computer programs described in Sections II and III
were designed as tools for studying scientific reasoning in general and as a
tool for the development of modes of communication between computer and
scientist. It was conceived at Stanford by Lederberg and Feigenbaum after
Churchman of the University of California, Berkeley, suggested studying
the design of inductive systems. Mass spectrometry was chosen as a suitable
task area to study in detail for a number of reasons: the problems are far
from trivial, unlike many problems approached in the early years of artificial
intelligence; it is an important area of chemical inquiry and therefore any aid
a computer could give would be valuable; mass-spectral analysis seems as
typically inductive as any other routine scientific task; and Lederberg's
algorithm in Section II solves a common computer programming problem of
finding a machine-readable representation for the primary concepts—in this
case for chemical structures.

B. Universal Logic of Structure Determination

The starting point of a structure determination is the molecular composition, readily provided these days by high-resolution mass spectrometry.
The next thing in logic would be to define the search-space in which the
answer must be found; in this context it is a systematic list of *all* the structures
having the composition. Every member of the set of isomers is a molecular
hypothesis that satisfies the data available so far. No other molecules do.
This central feature of the logic of structure determination has had to be
bypassed historically by chemists because they had no general procedure for
compiling such a list. Indeed, there have been only a dozen or so papers
attacking this ancient problem at the heart of organic chemistry. The first,
appearing in 1875, was by Cayley, professor of mathematics at Cambridge

University, who made an unsuccessful attempt to devise a formula yielding the number of alkanes (2).

The next step is to use whatever information is available in order to truncate the list of candidate structures as quickly as possible. Given enough data, all isomers but one will be excluded, and we regard it as the correct or proven structure. Such a problem is "fully determined": the identification rules are adequate enough to ensure a solution. The original data may leave several alternatives, whereupon we make predictions based on the differences between the remaining candidates and set out to provide experimental verification until the evidence leaves only one acceptable molecular hypothesis. A solution to the latter problem was incomplete until enough heuristics of a satisfactory kind had been accumulated to deal with it.

C. Overall Strategy of the DENDRAL Project

The overall strategy was therefore divided into two main objectives.

1. To base the computer program on an algorithm that generates an exhaustive, nonredundant list of all the structural isomers of a given chemical composition. Exhaustiveness is desirable in the generator, since the program should not be placed in the unenviable position of the working scientist who often fails to solve a problem because he fails to consider all possible answers. A nonredundant generator is especially important in a computer program that will run unmonitored for a considerable period of time. Not only is it a waste of time to generate previous structures, but also, the program might well fail to finish a problem if it devotes its time to repetition of past work. Implementation of the algorithm necessitated the invention of a new system of notation for representing organic structural formulas. Conceptually, the notation and the algorithm are inextricably linked. Not only is the notation the means by which humans and the computer communicate, but it also provides the machinery by which the algorithm executes the permutation of the various structural combinations of atoms. The program is called "DENDRAL" since its heart is a *dend*ritic *al*gorithm founded in the topological theory of tree graphs. Lederberg devised the algorithm and notation and had the DENDRAL program operational for acyclic structures in 1964 (3–5). Extension of the algorithm to cyclic compounds was not a simple matter. The symmetry elements of many ring systems pose extra problems in eliminating redundancy, and parts of the fundamental mathematics needed to classify their topology were lacking. Nevertheless, the foundations of cyclic DENDRAL are well set, and progress continues to be made towards perfection of the program (6–9). Its use for mass spectrometric applications is still in the embryonic phase, and so cyclic DENDRAL will receive little

attention in this chapter. Section II describes the acyclic DENDRAL algorithm and notation.

2. To devise a computer program that would perform an organic structure determination given a molecular formula and a mass spectrum (and other chemical data, if needed and available). This program is called "Heuristic DENDRAL" and it operates by using the known structure/spectrum correlations to constrain the DENDRAL isomer generator to produce (hopefully) a single isomer for that composition. Progress to date is described in Section III.

The Heuristic DENDRAL program has been written by computer scientists who, in collaboration with mass spectroscopists, attempt to formalize the rules of thumb constituting mass-spectral interpretation (the heuristics of the game, insofar as they can be discerned at the moment). Organic chemists have been unconcerned with and are notoriously inarticulate about their mental mechanisms during such tasks as structure determination. On the whole, they tend to be satisfied with getting the answer in as quick and practical way as possible, and then to move on to the next job. The professional ethic seems to be more concerned with the end result, and there has been little reflection (at least in published form) on the generality of the logical processes by which it was obtained. Systematic analysis and codification of these operations proves to be an exciting, amusing, and humbling experience having important and surprising ramifications.

These matters are emphasized because the successes and limitations of the DENDRAL project need evaluation in a context broader than whether the computer derived the correct structure of a compound from its mass spectrum. Some of the wider issues are touched on in Section IV, and for details the reader is referred elsewhere (7, 10).

II. THE DENDRAL PROGRAM

A. The Need for a Linear Notation and Its Requirements

Let us first recall the prime purpose of a structural formula in chemistry. It is a high-level abstraction that typically conveys only the atomic connectivity within the molecule—that is, which atoms are connected to which. It does not include explicit statements about other kinds of chemical information—for example, bond lengths, bond angles, the reactivity, the color, and the odor. Sacrifice of all such properties results in a compact, useful symbol of the connectivity that also acts as a label (distinct from the name) to identify the substance.

A prerequisite for the storage and manipulation of chemical structures within computers is a system of notation for representing structural formulas as a string of typewriter characters on one line, that is, a "linear notation." Various linear notations such as the Wiswesser, IUPAC, and Hayward systems have been devised, and their suitability for computer use was reviewed recently (11). None proved adequate for the DENDRAL project, and Lederberg devised a new system. His notation is, *inter alia*, as concise, as unambiguous, more general, more systematic, and easier to translate than the others.

The first aim of any system of linear notation is that the formulas representing a molecule should be unambiguous: there must be no confusion about which molecule is intended. This is a relatively simple problem, at least for acyclic structures. The string of symbols standing for the atoms and bonds is set out in the order of connectivity from some selected starting point. The starting point for simple structures is typically the end of a chain, producing, for example, the ordinary structure formula $CH_3.CH_2.CH_2.CH_3$ for *n*-butane. Subscripts are not permitted in a strictly linear notation and, they can be eliminated easily here by abstracting the symbols for hydrogen atoms, whereupon C.C.C.C becomes a linear formula for *n*-butane that no chemist would mistake for any other molecule.

Chain-branching is typically depicted in ordinary structural formulas with the connectivity in two dimensions, and it must be transformed into one-dimensional format to achieve linearity. Again, chemists are familiar with the device of using parentheses in simple cases for this purpose. There is no mistaking isobutane when it is portrayed as $CH_3.CH(.CH_3).CH_3$ or C.C(.C).C. Complex structures, however, require extensive nests of parentheses that become cumbersome and confusing. "Polish format" is a well-known alternative in computing that avoids nested parentheses, and had already been applied to the linearization of chemical formulas (12):

$CH_3.CH.\quad CH_3\quad CH_3$ C.C..   C   C   C.C..CC

isobutane, polish format   hydrogens abstracted   spaces abstracted

In the Polish system, the bonds issuing from a branched atom are placed immediately after the atom, and the attached groups follow in the same sequence. The arrows on the first formula for isobutane show the sequence relationship between each bond/atom pair following the branch point. The second formula again dispenses with hydrogen atoms, an abbreviation accommodated as readily by computers as humans. Spaces left between alkyl radicals in the first and second formulas for ease of human perception are irrelevant to the computer, and compression to the third formula is possible.

The above procedure amounts to making a linear map of the connectivity, and it must yield a linear formula that cannot represent another molecule. It will be unambiguous, but it will not be unique for the particular molecule. Alternative formulas are generated depending on which atom is chosen as the starting point and the order in which the others are mapped. For example, the following formulas for ethanol all convey the connectivity without ambiguity. The formulas are presented first using parentheses (where necessary) and then in compressed Polish format.

$CH_3.CH_2.OH$ $HO.CH_2.CH_3$ $CH_2(.OH).CH_3$ $CH_2(.CH_3).OH$

C.C.O   O.C.C   C..OC   C..CO

This illustrates the second aim for an acceptable linear notation, that there should be one, and only one, formula for each molecule. It is vastly harder to reach than the first aim. A solution was essential for the DENDRAL project to avoid the redundancy of alternative formulas for the same molecule in the list of isomers. The issue can be aptly phrased as follows. Is there a general mapping procedure that will select just one of the above Polish formulas for ethanol to be *a unique formula*? This is not a new problem. The same difficulty has plagued systems of organic nomenclature from the outset in selecting the correct names for compounds. One reason for our numerous, complex rules of nomenclature is the arbitrariness of the mapping procedure. For example, 3-methylpentane, 2-ethylbut-1-ene, 2-ethylbutan-1-ol, and 3-methylpentan-3-ol all share the same skeleton. This is not apparent from the names because the method of mapping the skeleton (for the alkane, locate the longest chain) changes with the introduction and position of functional groups (locate the longest chain containing the group). Linear notations previously developed (11) also lack a universal theoretical basis for their mapping strategy, and can only resolve the issue of uniqueness with *ad hoc* rules.

B. The Topological Centroid as the Key to Map the Structure

Lederberg found a mapping procedure that solves the problem. Exclusion of all chemical information but connectivity from structural formulas has an important consequence: molecular structures become susceptible to general analysis by the methods of topological graph theory. The "graph" in this branch of nonnumerical mathematics bears little relation to the curves and charts used to display data. Instead it is a formal diagram for analyzing connections between a number of entities (13). Graphs have two principal components: nodes (that can represent atoms of different types) and edges (that can represent bonds). Parameters such as length of edges, or shape, are irrelevant. Graphs that separate into two parts by cutting any edge are called "tree graphs" and correspond to acyclic structures. The cyclic graphs, like cyclic structures, need more cuts for separation. Readers may recollect encounters with the simple topology of cyclic graphs as mathematical puzzles: for example, finding a round trip through the graph that traverses each edge just once (Euler's problem of the Koenigsberg bridges) or else each node just once (the Hamiltonian circuit, which happens to yield a formal method of classifying cyclic graphs).

The search for the first step of a universal mapping procedure for structural formulas can hence be transformed into a topological question. Is there a unique place in a tree graph that can be used for the map origin, the fiducial point of the survey? The answer has been available for a century! In 1869 the mathematician Jordan showed (14) that any tree graph has a unique center called the "mass center" or "centroid." It is the point in the tree that provides the most evenly balanced allocation of nodes. For a graph having an odd number of nodes, the centroid is a node from which each branch carries less than half the total nodes (e.g., propane, the pentanes). For a graph of even node-count, the centroid may be a node or an edge, depending on the arrangement of nodes; if it is a node, the definition in the previous sentence applies (e.g., isobutane, 3-methylpentane); if it is an edge, then it joins halves of equal node-count (e.g. ethane, n-butane, 2-methylpentane). The definitions are illustrated in Table 1.

TABLE 1

Tree Graphs Corresponding to Some Alkane Skeletons[a]

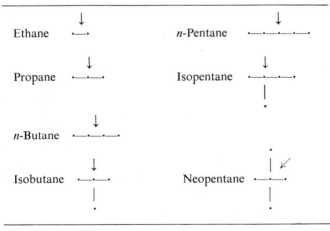

Arrows indicate centroids.

The kernel of Lederberg's breakthrough can now be stated in one sentence. *Use the unique centroid as the origin to map the system.* With the root of the tree uniquely planted, the ambiguity of alternative formulas based on the other starting points is eliminated. Technical terms for the two types of centroid are "leading bond" (abbreviated lb) if it is an edge, and "leading atom" (abbreviated la) if it is a node. Linear formulas describing the connectivity of the tree graphs above can now be developed as shown in Table 2.

TABLE 2
DENDRAL Notation for the Molecules in Table 1

| | | Linear DENDRAL notations | |
Molecule	Centroid	Extended	Compressed
Ethane	lb	. CH_3 CH_3	. CC
Propane	la, 2°C	CH_2 . . CH_3 CH_3	C . . CC
n-Butane	lb	. CH_2 . CH_3 CH_2 . CH_3	. C . CC . C
Isobutane	la, 3°C	CH . . . CH_3 CH_3 CH_3	C . . . CCC
n-Pentane	la, 2°C	CH_2 . . CH_2 . CH_3 CH_2 . CH_3	C . . C . CC . C
Isopentane	la, 3°C	CH . . . CH_3 CH_3 CH_2 . CH_3	C . . . CCC . C
Neopentane	la, 4°C	C . . . . CH_3 CH_3 CH_3 CH_3	C CCCC

Several points deserve comment. In the compressed formulas, it is useful to remember that adjacent letters (i.e., without dots in between) represent atoms not connected to each other.* Second, the idea of having a bond as a map origin (the lb) calls for slight mental reorientation, since it is a concept foreign to chemists and their systems of notation and nomenclature. Note an implicit difference between the symbols used for a leading bond and those for all other bonds. Although a simple dot is used as the symbol for all single bonds, that for a leading bond implies a connection between *two* following atoms or radicals (and these must be of equal node-count), whereas that for any other bond indicates just *one* attachment. Third, isopentane presents the first occasion in which the branches pendant from the centroid are not identical (two methyls and an ethyl group), and a rule is needed to order the sequence if a *unique* linear formula is to be preserved. The rule chosen (first canon of DENDRAL order) is that radicals of lower node-count (i.e., radicals containing fewer non-hydrogen atoms) precede those of higher node-count.

* The C_2 hydrocarbons represent trivial exceptions to this rule.

C. General Principles of DENDRAL Enumeration

Two implications of the extremely important centroid conception should now be apparent. It will first be noted from the alkanes above that the desired *unique* linear formulation has been achieved (given appropriate canons of precedence as well). Second, for any one composition, permutation through all possible centroids and pendant radicals in systematic order (implying more canons) necessarily generates the desired complete list of isomers. The list will be exhaustive, since consideration of all centroids must include the tree graph for every isomer. Redundancy is avoided by the canons of precedence that order the constituent branches of each tree uniquely.

TABLE 3
Centroids Possible for the Alkanes

Even node-count	Odd node-count
lb	la, 2°C
la, 3°C	la, 3°C
la, 4°C	la, 4°C

About 40 years ago, Henze and Blair discovered this permutation in their successful enumeration of the alkanes (15) and monovalent alkyl radicals (16). They were apparently unaware of Jordan's theorem, however, and Lederberg was the first to perceive the potential generality of their approach. They were not concerned with linear notations, their aim being simply to derive a recursive algebraic expansion to calculate the total number of compounds of a given carbon content.

In enumerating the alkanes, the natural chemical listing (a "dictionary ordering") would be to start with the straight-chain isomer and move systematically towards the most highly branched isomer, resulting in the following sequence for centroid permutation: lb; 2° la; 3° la; 4° la. Consideration of the centroid definitions above discloses that a primary carbon atom can never be a centroid, nor can a secondary carbon atom in molecules of even node-count. Table 3 summarizes the possible centroids.

The question "How many hexanes are there?" can now be answered by constructing all the tree graphs having six similar nodes and permuting through all centroids and radicals. The dendritic appearance of the algorithm is obvious from Scheme 1.

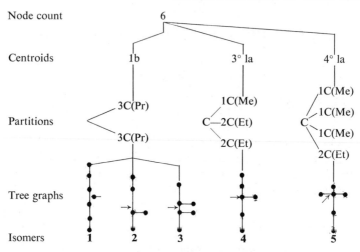

Scheme 1. DENDRAL enumeration of the five hexanes.

In DENDRAL notation, the dictionary list of the hexane isomers then has the following appearance (some spaces have been left in for easy comprehension).

1	.	C.C.C	C.C.C		
2	.	C.C.C	C..CC		
3	.	C..CC	C..CC		
4	C...	C	C.C	C.C	
5	C....	C	C	C	C.C

Given a 3° la centroid, the remaining five nodes must be allocated among three branches so that all branches have less than half the total nodes. There is only one such partition, yielding isomer **4**. Similarly, there is only one way to share five nodes among the four branches pendant from a 4° la, and isomer **5** results. The first three isomers, all stemming from the lb centroid, arise from the two arrangements possible for a three-node radical (*n*-propyl and *iso*propyl). Another canon of DENDRAL order is therefore required to establish the sequence of such radicals in the notation. For example, in isomer **2**, the rule puts *n*-propyl before *iso*propyl, thus excluding the alternative of .C..CC C.C.C. In more general form, when assessing different alkyl radicals of equal node-count, a node-by-node comparison is made moving outward from the centroid until the first point of inequality is reached, whereupon priority of rank in the radicals is decided by the rule that 2° precedes 3°C, and 3°C precedes 4°C. In other words, a straight chain always has first priority; and for two groups with the same number of branches, the one branching further away from the central root appears earlier. Readers may care to confirm that the ranking of butyl radicals, for example, is thus Bu^n, Bu^i, Bu^s, Bu^t.

A moment's thought reveals that the canon in the preceding paragraph is also the permutation sequence for isomers sharing the same centroid, that is, the means by which hexane isomers **1**, **2**, **3** were generated in that order. It also happens to be the sequence selected above for permuting the centroids. This illustrates an important feature of the DENDRAL program. The canons that dictate the linear formulation are just those needed for the most logical permutation through the DENDRAL tree of isomers. The permutation sequence is the same thing as allotting to each isomer a *unique* place in the dictionary list. When comparing the relative positions of two isomers on the list, a method for ranking separate molecules is needed, and it is reasonable that the method for imposing canonical order on radicals within one molecule should do this job too. *In short, one set of canons defines both a unique linear formula and a unique dictionary position.* Here is the reason for the earlier statement that the notation and algorithm are inextricably linked.

D. Further Enumeration of the Alkanes

The DENDRAL algorithm is illustrated further in Schemes 2 and 3, which depict, in slightly different ways, the enumeration of the heptanes and octanes.

For both 3° and 4° la centroids in the heptanes, there are two ways to partition the remaining pool of 6 atoms. The algorithm is completed by ramifying the tree at all 3-node branches using the two kinds of propyl group. This yields the expected 9-heptanes. Readers would find it instructive to set out the tree graphs and linear formulas in dictionary order.

In Scheme 3 the top row is based on a leading bond centroid, the middle row on a tertiary leading atom, and the last row on a quaternary leading atom. There are five partitions in which the residual nodes are distributed over two

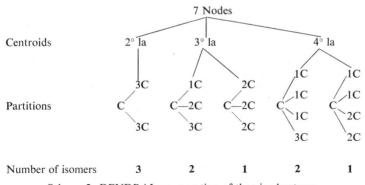

Scheme 2. DENDRAL enumeration of the nine heptanes.

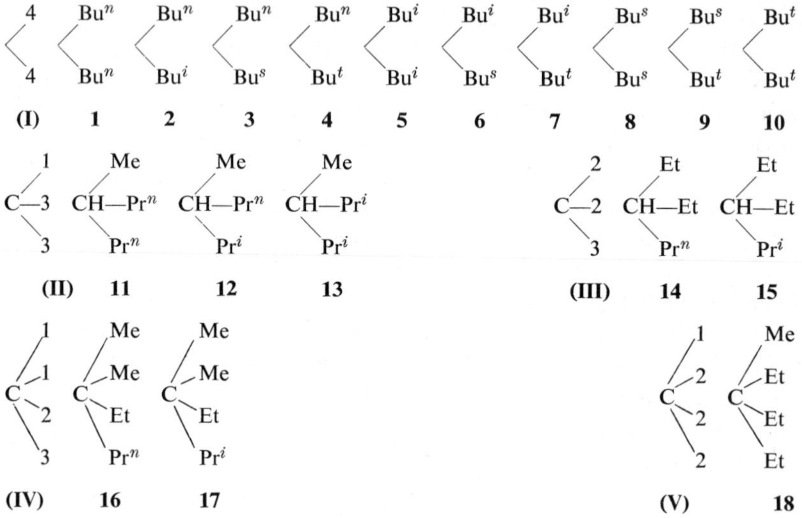

$$
\begin{array}{cccccccccc}
4 & Bu^n & Bu^n & Bu^n & Bu^n & Bu^i & Bu^i & Bu^i & Bu^s & Bu^s & Bu^t \\
4 & Bu^n & Bu^i & Bu^s & Bu^t & Bu^i & Bu^s & Bu^t & Bu^s & Bu^t & Bu^t \\
(I) & 1 & 2 & 3 & 4 & 5 & 6 & 7 & 8 & 9 & 10
\end{array}
$$

$$
\begin{array}{ccccc}
C\!\!<^{1}_{3} & CH\!-\!Pr^n(Me) & CH\!-\!Pr^n(Me) & CH\!-\!Pr^i(Me) & C\!\!<^{2}_{3}(Et) \quad CH\!-\!Et(Et) \quad CH\!-\!Et(Et) \\
 & Pr^n & Pr^i & Pr^i & Pr^n \qquad Pr^i \\
(II) & 11 & 12 & 13 & (III) \quad 14 \qquad 15
\end{array}
$$

(IV) 16   17   (V) 18

Scheme 3. DENDRAL enumeration of the 18 octanes.

branches (I), three branches (II and III), or four branches (IV and V). Permutation through all the possible subgraphs for 4-node branches (butyl radicals) and 3-node branches (propyl radicals) then subdivides the branches until the tree is fully grown. The 18 isomers are numbered in dictionary order. Inspection of the first 10 isomers, for example, shows how well DENDRAL order matches the organic chemist's intuitive ordering.

TABLE 4

Ways of Partitioning the Residual Nodes Among the Branches Emanating from the Possible Alkane Centroids

Centroids		Node count					
		6	7	8	9	10	11
1b		3	—	4	—	5	—
		3	—	4	—	5	—
2° 1a	$C\!\!<$	—	3	—	4	—	5
		—	3	—	4	—	5
3° 1a	$C\!\!-$	1	1 2	1 2	1 2 2	1 2 3	1 2 2 3
		2	2 2	3 2	3 2 3	4 3 3	4 3 4 3
		2	3 2	3 3	4 4 3	4 4 3	5 5 4 4
4° 1a	$C\!\!<$	1	1 1	1 1	1 1 1 2	1 1 1 2	1 1 1 1 2 2
		1	1 1	1 2	1 1 2 2	1 2 2 2	1 1 2 2 2 2
		1	1 2	2 2	2 3 2 2	3 2 3 2	3 4 2 3 2 3
		2	3 2	3 2	4 3 3 2	4 4 3 3	5 4 5 4 4 3

It can be seen there are really three levels of permutation in executing the DENDRAL algorithm for molecules. These apply to any class of compounds as well as the alkanes.

1. Permute through all centroids.

2. Permute through the various ways to allocate the remaining nodes among the possible branches. Table 4 summarizes the possible alkane partitions from the hexanes up to the undecanes.

3. Permute through all possible radicals. In the alkane case this means any branch containing more than two nodes.

For the examples given so far, step (3) is actually an enumeration of the alkyl radicals which, as Henze and Blair emphasized, is a prerequisite for alkane enumeration. The general concepts used for constructing a tree of molecules also apply for a tree of radicals, but there are noteworthy differences of detail. First, there is no problem about selecting a map origin. The

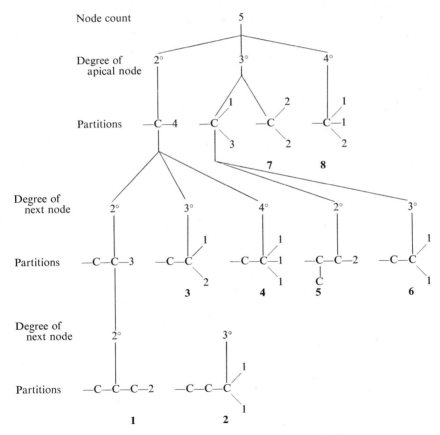

Scheme 4. Generation of the tree of pentyl radicals.

site of the free valence defines the starting point, and "apical node" is a useful technical term to designate this atom. The tree is expanded as follows.

1. Permute according to the possible degrees of the apical node.

2. Partition the residual nodes among the branches efferent from the apical nodes. The restriction that limits the number of nodes in any branch attached to a centroid does not apply here. All partitions are possible provided that each branch carries at least one node. If there is only one partition, then this branch terminates and yields one radical.

3. If there is more than one partition, iterate steps 1 and 2 through the next nodes until all branches are complete. The whole process is illustrated in Scheme 4, and the radicals are listed in dictionary order in Table 5.

TABLE 5
Formulas for the Eight Pentyl Isomers from Scheme 4

	Tree graph	First partition	Linear formula
1	—C—C—C—C—C	—CH_2—Bu^n	. C . C . C . C . C

2	—C—C—C—C
	         |
	         C
		—CH_2—Bu^i	. C . C . C . . CC

3	—C—C—C—C
	      |
	      C
	      |
	      C
		—CH_2—Bu^s	. C . C . . CC . C

4	—C—C—C
	      |
	      C
		—CH_2—Bu^t	. C . C . . . CCC

5	—C—C—C—C
	   |
	   C
		—CH—Pr^n
		   |
		   Me
			. C . . CC . C . C

6	—C—C—C
	   | |
	   C C
		—CH—Pr^i
		   |
		   Me
			. C . . CC . . CC

7	—C—C—C
	   |
	   C
	   |
	   C
		—CH—Et
		   |
		   Et
			. C . . C . CC . C

8	C
	|
	—C—C—C
	   |
	   C
		Me
		|
		—C—Et
		   |
		   Me
			. C . . . CCC . C

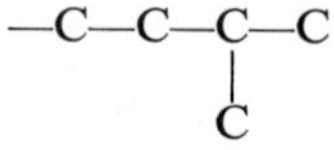

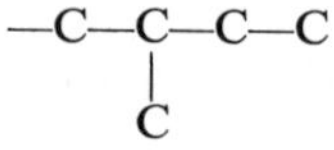

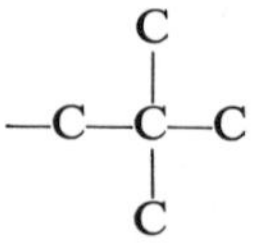

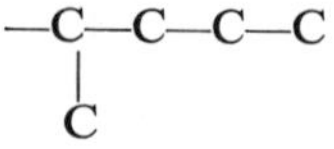

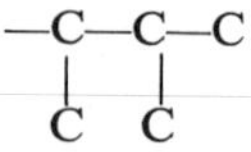

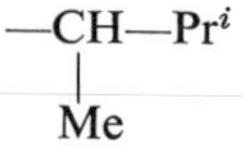

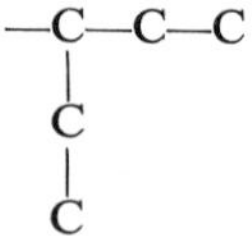

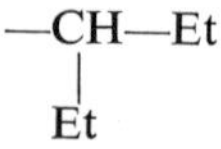

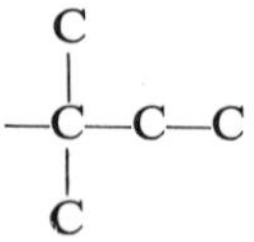

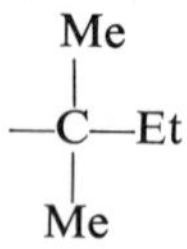

E. Extension for Unsaturation and Heteroatoms

Extension of DENDRAL to more complex organic structures is straight-forward, the only additions necessary being sufficient canons of precedence to maintain a unique formula and permutation sequence. Heteroatoms are simply other kinds of nodes that differ, *inter alia*, in connectivity from the saturated carbon nodes discussed so far. Their ordinary symbols can be used and the connectivity implied by the normal valence considerations. Acetylenic and olefinic atoms are simply carbon nodes attached to two or three other atoms respectively, and topologically could easily be given new symbols representing this type of "divalency" and "trivalency." This step has not been taken, however, and unsaturation is represented in the usual way. Hetero-atoms and unsaturation cause a great proliferation of both centroids and radicals.

The centroid definitions still define the first division of the tree (odd or even node-count, excluding hydrogen) and the second division by the number of branches efferent from the centroid. At this stage new rules are needed to take account of heteroatoms or unsaturation. The canons chosen divide the tree next by composition of the centroid (if it is a leading atom), and then by unsaturation. The sequence chosen for permutation through alternative compositions is C, N, O, P, S. By coincidence it is both alphabetical and increasing by atomic number. The sequence chosen for permuting unsatura-tion is single bond before double bond before triple bond, other things being equal. These canons are illustrated in Scheme 5. It is evident that a leading bond may be single, double, or triple, in that order. A leading atom may or may not carry a multiple bond. Leading atoms with only one efferent radical (e.g., C. or O:) do not appear because they are forbidden by the centroid definitions. Valence restrictions exert a powerful truncating force, eliminating entries such as O.:, O..., N..:, C...:. Initial compositions different from that in Scheme 5 will generate smaller or larger trees. For example, the alkane enumerations above are only those branches of Scheme 5 containing C and single bonds. An alkene tree would have just C and single and double bonds, and so on.

Completion of the full DENDRAL tree of isomers then requires partitioning of the residual atoms among the branches of each centroid, and generation of the tree of radicals for each. The residual composition avail-able for partitioning depends on the centroid. For centroid **1** in Scheme 5 it is C_nNO and one unsaturation; for **2** it is C_nNO and no unsaturation; for **7** it is saturated C_nO. Some centroids leave an identical residue (e.g., **4** and **5** have $C_{n-1}NO$ with no unsaturation). In simple cases such centroids may become equivalent owing to the limited number of partitions subsequently

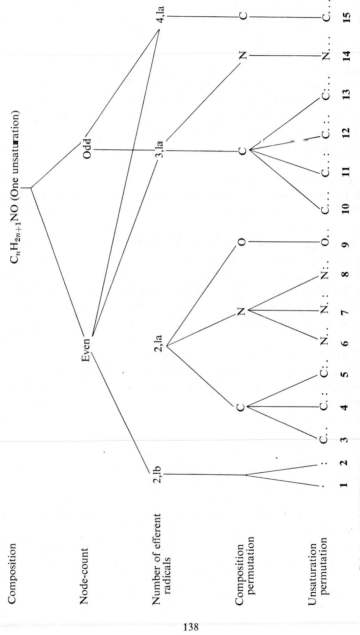

Scheme 5. DENDRAL permutations leading to the possible centroids for a composition heteroatoms and unsaturation.

possible, but not in complex cases (e.g., in isobutylene, C. . : C C C takes care of the only partition and C. :. C C C is then redundant and is eliminated).

Scheme 6 illustrates how the tree of radicals is developed for a composition $C_nH_{2n}NO$ connected by a single bond to the centroid (this bond is "efferent" from the centroid and "afferent" to the radical). Permutation proceeds first by degree of apical node,* then by its composition, and then

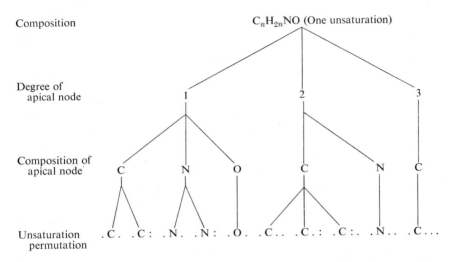

Composition $C_nH_{2n}NO$ (One unsaturation)

Degree of apical node   1   2   3

Composition of apical node   C   N   O   C   N   C

Unsaturation permutation   .C.   .C :   .N.   .N :   .O.   .C..   .C. :   .C :.   .N..   .C...

Scheme 6. DENDRAL permutations leading to the possible apical nodes for a radical containing heteroatoms and unsaturation, and attached to a centroid by a single bond.

by unsaturation. The result is the set of possible apical nodes. Further ramification (not shown) proceeds by partitioning the remaining elements of the graph among the branches and iterating the process through the next nodes (now the apical nodes of smaller trees) until, at the terminating twig, only one partition can be made.

Scheme 6 does not answer the question "How many radicals are there of composition $C_nH_{2n}NO$?" since the multiplicity of the afferent bond was already defined by the relevant centroid. The tree corresponding to this

---

* In dealing with the alkanes above, the "degree of the apical node" has been represented by the customary usage of 2°, 3°, and 4° for secondary, tertiary, and quaternary saturated carbon, respectively. These symbols are inappropriate for unsaturated carbon or heteroatoms, and henceforth *degree of apical node will be defined as the number of attached substructures* (efferent radicals).

question would include double bonds as afferent links in permuting the unsaturations, resulting in the following set of apical nodes for the radicals.

.C. , .C: , :C. , .N. , .N: , :N. , .O. , .C.. , .C.: , .C:. , :C.. , .N.. , .C...

(The initial dot in each serves as a reminder that radicals are being generated.)

F. Complete Acyclic DENDRAL Canons

DENDRAL order for complex acyclic structures has so far been described from the context of structure generation. The canons can be presented more compactly as a set of axioms to place two radicals in proper order within a linear formula. The following criteria are evaluated in turn. The first inequality that is encountered dictates the relative order of the two radicals.

1. *Node-count* (excluding H). A lower node-count (number of atoms) precedes a higher node-count.

2. *Composition* (excluding H). Atoms of different type within a radical are ordered in the sequence of C, N, O, P, S (this rule would need extension for other elements). When comparing two radicals of equal node count, the one with more carbon atoms will be first. If the carbon content is the same, the one with more nitrogens will be first, and so forth.

3. *Unsaturation.* For radicals having the same composition list, the more saturated one precedes the more unsaturated one. Any unsaturation in the afferent link is included.

4. *Apical node.* Two different radicals equal so far must be isomeric. The structures are analyzed starting at the apical nodes, the following criteria are evaluated in sequence, and again the first inequality decides the issue. Note that count, composition, and unsaturation are once more the basis for discrimination, although in connection with different things.

   a. *Degree.* The apical node with less branching (fewer efferent radicals) precedes the other.

   b. *Composition.* A carbon atom is the first apical node, then N, and then O, P, S.

   c. *Multiplicity of the afferent link.* A radical with a single bond leading to the apical node precedes one with a double bond, and so on.

5. If these criterial fail to determine the ranking, the radicals pendant from the two apical nodes are arranged in order and compared in pairs. The node-count of each pair is first inspected (this orders, for example, the two partitions in Scheme 2 based on a 4° la, the third pair being methyl versus ethyl); if need be, the radicals are then paired for composition, then unsaturation, whereupon the next node becomes a new apical node, and so on, recursively.

Stereoisomerism has not been mentioned yet. Further simple rules, and appropriate symbols in the linear formulas, readily incorporate the further bifurcations of the tree to take account of geometrical and optical isomers (3). Mass spectra are relatively insensitive to such structural subtleties and so there was no need to include them for Heuristic DENDRAL.

The general principles for comparing radicals encompass those needed for comparing molecules. Two molecules of different molecular formulas can be ranked by total node-count, then by composition (excluding H), then by degree of unsaturation (H needed here). Given the same molecular formula, the isomers are ranked initially by the centroids (e.g., Scheme 5) using once more the sequential tests of count (but now counting the number of efferent radicals), composition, and unsaturation. Given the same centroid, any difference is in the radicals.

All these concepts are best understood by working through some simple examples. Representative lists in dictionary order are given below. Readers may care to translate the DENDRAL notation into ordinary structural formulas, name each species, and construct the DENDRAL tree for each composition. It is all far easier to do than to describe.

One way of picturing structure generation is to construct the entire tree of isomers starting from the top, as in the schemes. The dictionary list is then

C_4H_8
. C.C C:C
: C.C C.C
C..: C C C

C_5H_{10}
C.. C.C C:C
C.: C.C C.C
C... C C C:C
C..: C C C.C
C.:. C C C.C

C_3H_8O
. C.C C.O
. C.C O.C
C...C C O

$C_4H_{10}O$
C.. C.C C.O
C.. C.C O.C
O.. C.C C.C
C... C C C.O
C... C C O.C
C... C O C.C
C.... C C C O

C_3H_6O
. C.C C:O
. C:C C.O
. C:C O.C
: C.C C.O
C..: C C O
C.:. C C O

$C_2H_6O_2$
. C.C O.O
. C.O C.O
. C.O O.C
. O.C O.C
C... C O O

$C_4H_{11}N$
C.. C.C C.N
C.. C.C N.C
N.. C.C C.C
C... C C C.N
C... C C N.C
C... C N C.C
N... C C C.C
C.... C C C N

the set of structures at the bottom of each branch, reading from left to right. The number of isomers for that composition is of course the number of entries in the list. A far less laborious procedure is to write down the first structure on the list in linear format. For a molecule this will be an unbranched chain with any heteroatoms or unsaturation as near to the ends as possible. This has the effect of putting the heteroatoms and unsaturation as far to the right as possible in the linear formula. The first structure is then transformed into the second by rearrangement of unsaturation and heteroatoms. This amounts to going *backwards* through the DENDRAL canons, and the second isomer will be the first new structure that results. It can then be transformed to the third, and so on. The "incrementing" process can be continued until the last possible permutation is reached. Inspection of the lists above illustrates the idea. Note how the centroids appear in sequence. A double bond starts on the right and is moved through to the left (C_5H_{10}); likewise for a heteroatom ($C_4H_{10}O$ or $C_4H_{11}N$). Given a double bond and a heteroatom, the former moves before the latter (C_3H_6O). This is the gist of the computer program, and other short cuts are noted in Section III.

The other main use for the canons is in converting an ordinary structural formula to linear notation. The steps involved here are as follows:

1. Strip hydrogen atoms from the skeleton.
2. Count the skeletal atoms.
3. Locate the centroid and code it in linear format.
4. Dissect each radical, node by node, according to the canons, by coding in linear format.
5. List the radicals in dictionary order after the centroid.

A final practical note to this section concerns DENDRAL input and output at the teletype. There is usually no symbol for a triple bond (: or ≡) on an ordinary teletype, and we have most commonly used "$" in working and in publication to represent a triple bond. For readability, a leading bond has been indicated by "<" to suggest the two branches, and the equal sign replaces the colon for indicating double bonds. Multiplicity of bonds has occasionally been indicated by the numerals "1, 2, and 3" instead of "., :, and $."

G. Badlist and Goodlist

In using the DENDRAL program, anything that can be done to truncate the potential list of isomers as early as possible results in much improved efficiency. So far, the structure generator described is simply a prolific topologist that achieves the original aims of compiling an isomer list, without exception or repetition. *Anything* permitted by valence will be there, including gem-diols, peroxides, hydroperoxides, hemiacetals (e.g., $C_2H_6O_2$ above), enols (C_3H_6O), and the like. What the program needs is some chemical common sense, a theory of chemical instability. A list known as BADLIST provides this facility. It is simply an input list of substructures to be excluded from that run. The program prunes the DENDRAL tree at every branch for which a BADLIST entry appears as the tree is being generated.

Conversely, if it is desired to restrict the output to a list of isomers all containing a given substructure(s), such a group(s) can be placed on another input list called GOODLIST. In this event the program ignores all other structures.

GOODLIST and BADLIST are not easy to implement by hand, but they are exceedingly effective in making the computer program smarter and more efficient. One disadvantage of the unrestricted DENDRAL output is that isomers having close structural relations are scattered in the list. For example, the alcohols for composition $C_4H_{10}O$ above are mixed up with the ethers. GOODLIST and BADLIST provide machinery to overcome this difficulty, and restrict and classify the output according to one's taste and interest at the moment.

H. Commentary

The main aim of Section II has been to describe the DENDRAL algorithm and notation in enough detail to execute it by hand in simple cases. Otherwise the STRUCTURE GENERATOR module of the Heuristic DENDRAL program can only be regarded as magic. The latter program is in fact described in Section III without using linear notation, in the hope that it will make for easier reading.

The DENDRAL program itself has of course many uses quite unrelated to mass spectrometry. For a start, it is of great interest in defining the boundaries of organic chemistry. Although with a given number of carbon atoms, the number of topologically possible alkanes may be enormous (15), the rate at which the numbers increase given some unsaturation and heteroatoms is staggering. BADLIST and GOODLIST can be used to explore the concept of the functional group as a device to truncate the list of possible isomers for a given composition. Such issues were discussed recently in a paper on the number of acyclic compounds containing C, H, O and N (17). The teaching of organic chemistry is likely to benefit, since isomerism can now be presented on sound topological principles. Early chapters in textbooks of the future may well be quite different.

III. THE HEURISTIC DENDRAL PROGRAM

A. The Relation of the Algorithm to the Computer Program

The heart of the whole computer program is Lederberg's DENDRAL algorithm described in Section II. Various additional features were added in order to (a) constrain structure generation to the most relevant molecules (with respect to a given problem) and (b) verify the molecules after generation. The overall design of the program is shown in Scheme 7. For convenience the program is segmented into three subprograms: planning, structure generation, and verification.

Section III will attempt to explain the function of each part of Scheme 7 without recounting details of the computer program itself. Heuristic DEN-DRAL is written in the LISP programming language and runs on the IBM 360/67 computer at Stanford University. Program details and strategy have been described elsewhere (7, 18–22).

In the planning phase the program examines a low-resolution mass spectrum together with the empirical formula of the molecular ion to determine the functional group present in the unknown molecule that produced that spectrum. All molecular structures subsequently generated must contain *all* the indicated substructures, and *no* substructures inconsistent with the mass spectrum. Low-resolution data were chosen to start with because of the ready availability of published low-resolution spectra. Nothing in the program inhibits the input of high-resolution data if available.

The STRUCTURE GENERATOR utilizes the functional group information to produce a list of all possible chemical structures that are consistent with the list of "good" and "bad" substructures.

In the verification phase the program uses the rules of organic mass spectrometry to predict significant features of the mass spectra of each candidate structure. Then the program compares the predicted and the original mass-spectral data to eliminate any candidate structure found to be incompatible with the given experimental data. This enables the candidate structures to be listed in order of their "plausibility" or estimated degree of confirmation.

Because the constraints that have been included in the program to limit the search space are heuristic, there is no guarantee that the correct structure will not be bypassed. Failure of a test run is of course the signal to modify the constraints within the program after a detailed analysis of the decision procedures of organic mass spectroscopists.

The program's ability to solve mass-spectrometry problems obviously depends on the amount of knowledge it has about this subject. Within the

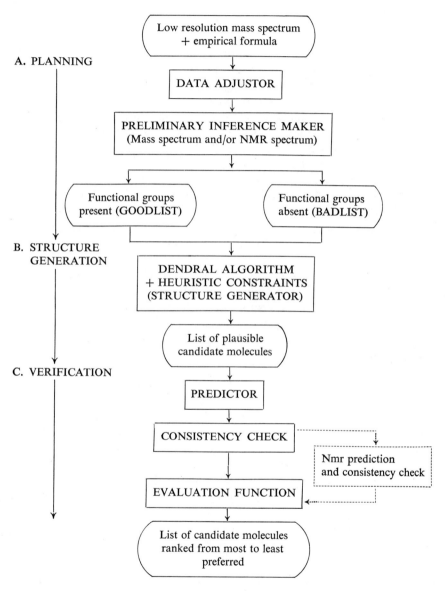

A. PLANNING

B. STRUCTURE
   GENERATION

C. VERIFICATION

Low resolution mass spectrum
+ empirical formula

DATA ADJUSTOR

PRELIMINARY INFERENCE MAKER
(Mass spectrum and/or NMR spectrum)

Functional groups
present (GOODLIST)

Functional groups
absent (BADLIST)

DENDRAL ALGORITHM
+ HEURISTIC CONSTRAINTS
(STRUCTURE GENERATOR)

List of plausible
candidate molecules

PREDICTOR

CONSISTENCY CHECK

Nmr prediction
and consistency check

EVALUATION FUNCTION

List of candidate molecules
ranked from most to least
preferred

Scheme 7. The original overall design of the heuristic DENDRAL program.

145

discussion of the program segments shown in Scheme 7 are details of the mass-spectrometry rules used in those segments. Since the most detailed rules of the program are rules for aliphatic ketones, ethers, thioethers, amines, alcohols, and thiols, these are discussed at length, especially in the first section on the planning segment. The STRUCTURE GENERATOR currently does not use any theory of mass spectrometry. In the verification stage the program has many general rules about mass-spectrometric processes as well as the specific rules about the classes above from the planning phase (but in a different form). By storing a large amount of knowledge about a few functional groups, we hope to show that the program can perform well for any groups, given similar knowledge. Examples of the program's ability are thus taken from these classes. One of the obstacles to extending the program is the amount of experimental work required to extend the mass-spectrometry rules to other functional groups.

The initial task for the Heuristic DENDRAL program was to interpret the mass spectra of aliphatic ketones (23). This choice was made in view of the vast amount already known about the modes of fragmentation of aliphatic ketones within a mass spectrometer (24). If the program had failed within this restricted and comparatively well understood class of organic compounds, the whole approach would not have seemed profitable for more complex organic systems. However, Heuristic DENDRAL has demonstrated its ability to solve problems of increasing degrees of complexity, as can be appreciated from its correct identification of unknown ethers, amines, thiols, alcohols, and thioethers from their mass and nmr spectra. As will be discussed, the latter physical parameter (nmr) has added an additional degree of problem-solving power to the program.

B. Planning

1. *The PRELIMINARY INFERENCE MAKER*

The original conception* of the preliminary inference maker was to ascertain correctly the nature of the functional group present in the unknown compound from the supplied low-resolution mass spectrum and the empirical formula† of the molecular ion. The rules it uses have been programmed.

---

* As will be explained later in this section, the *PRELIMINARY INFERENCE MAKER* was subsequently expanded to contain all the theoretical input. This meant that there was no theory left for the *PREDICTOR*, and this latter section of the program was then not used for the compounds discussed in Section B-5.

† As will be described later in this section, the program no longer requires either the molecular ion or the empirical formula be identified. Using heuristics, these parameters are recognized by the program.

They are, essentially, the same as those used by the organic mass spectroscopist for depicting the fragmentation modes initiated by various functional
groups. This information is then relayed to the STRUCTURE GENERATOR
by way of GOODLIST or BADLIST.

It is important for the PRELIMINARY INFERENCE MAKER to
inform the STRUCTURE GENERATOR of the functional group and as
much of its molecular environment as feasible because the number of possible
structures may be very large, as noted in Section II. For instance, the empirical composition consistent with an aliphatic ketone is also compatible with
monounsaturated alcohols and ethers. When one considers the incredible
number of possible structures (17) of composition $C_nH_{2n}O$ for, say, more
than five carbon atoms, it is apparent that as much truncation as possible
in the PRELIMINARY INFERENCE section of the program is desirable,
especially in view of the fact that each candidate structure must be scrutinized
later by the PREDICTOR. In point of fact, this last condition resulted in the
PRELIMINARY INFERENCE section of the program being so finely tuned
that no extra truncation could be achieved by the PREDICTOR.

2. *Ketones*

The primary rule for the identification of the keto function in aliphatic
molecules is based on alpha-cleavage around the keto group (23). The
program searches for two peaks in the original data (one of which should be
in excess of 10% relative abundance) such that their sum is equal to the
molecular weight of the unknown plus 28 amu. This rule originates from
expressing the following α-cleavage fragmentations (formation of *a* and *b*)
in the algebraic form:

$$M + 28 = a + b$$

$$R_2-C{\equiv}O^+ \xleftarrow{-R_1} R_1-CO-R_2 \xrightarrow{-R_2} R_1-C{\equiv}O^+$$

$$a \qquad\qquad M \qquad\qquad b$$

$$\Big\downarrow -CO \qquad\qquad\qquad\qquad \Big\downarrow -CO$$

$$R_2^+ \qquad\qquad\qquad\qquad\qquad R_1^+$$

$$c \qquad\qquad\qquad\qquad\qquad d$$

Furthermore, once the peaks represented by *a* and *b* are located in an unknown mass spectrum, Heuristic DENDRAL seeks confirmation of the
presence of the ketone group by identification of the two decarbonylation
product ions *c* and *d*.

In addition to the ions *a*, *b*, *c*, and *d* resulting from an initial α-cleavage
of the carbonyl group, Heuristic DENDRAL must be able to represent
fragments resulting from hydrogen rearrangement processes. In the case of

aliphatic ketones this incorporates the McLafferty rearrangement process, which can be mechanistically depicted as M $\rightarrow$ e (25). As is well known, this rearrangement is site specific, that is, a γ-hydrogen atom must be available for this fragmentation to occur. The product ions e and e' undergo further fragmentation to f provided they both contain a γ-hydrogen atom. Heuristic DENDRAL must be capable of correctly identifying whether a ketone has a γ-hydrogen atom, and if the answer is in the affirmative, then it must correctly formulate the rearrangement ions e, e', and f, bearing in mind that alkyl substitution may occur on the carbon atom adjacent to the carbonyl function.

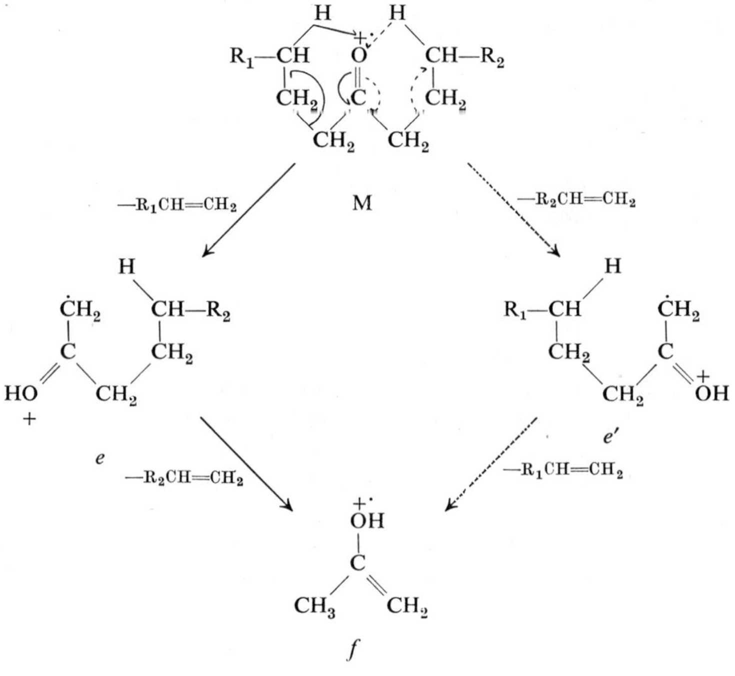

For more efficient operation of the PRELIMINARY INFERENCE MAKER on aliphatic ketones, it was found desirable to define the further substructures methyl ketone3 (**1**), ethyl ketone3 (**2**), *iso*propyl ketone3 (**3**), and *n*-propyl ketone3 (**4**). The name "ketone3" was used to remind us of the three carbons and the gamma hydrogen that are included in the substructures. These four substructures would then yield McLafferty rearrangement ions g, h, i, and j of masses 58, 72, 86, and 86 respectively. The species j emanating from *n*-propyl ketone3 would afford k (m/e 58) by the operation of a double McLafferty rearrangement process (26). Hence the ions g, h, i, and j serve to differentiate between the substructures **1**, **2**, **3**, and **4**, while the additional presence of k would distinguish between the partial structures **3** and **4**.

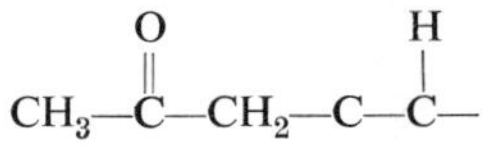

$$CH_3-\overset{\overset{\displaystyle O}{\|}}{C}-CH_2-\overset{\overset{\displaystyle H}{|}}{C}-C-$$

(1)

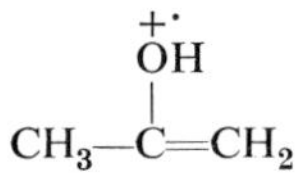

$$CH_3-\overset{+\cdot}{\overset{\displaystyle OH}{|}}{C}=CH_2$$

g, m/e 58

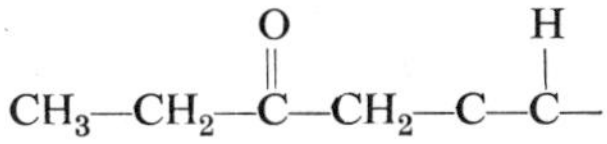

$$CH_3-CH_2-\overset{\overset{\displaystyle O}{\|}}{C}-CH_2-\overset{\overset{\displaystyle H}{|}}{C}-C-$$

(2)

$$CH_3-CH_2-\overset{+\cdot}{\overset{\displaystyle OH}{|}}{C}=CH_2$$

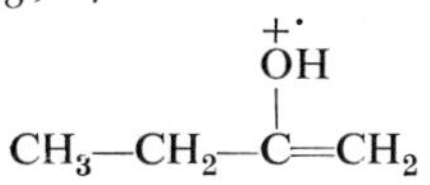

h, m/e 72

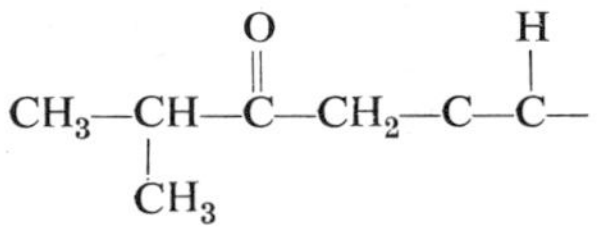

$$CH_3-\underset{\underset{\displaystyle CH_3}{|}}{CH}-\overset{\overset{\displaystyle O}{\|}}{C}-CH_2-\overset{\overset{\displaystyle H}{|}}{C}-C-$$

(3)

$$CH_3-\underset{\underset{\displaystyle CH_3}{|}}{CH}-\overset{+\cdot}{\overset{\displaystyle OH}{|}}{C}=CH_2$$

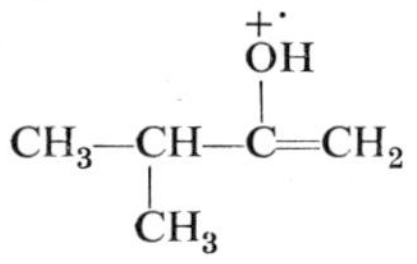

i, m/e 86

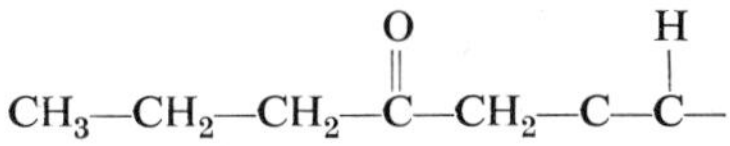

$$CH_3-CH_2-CH_2-\overset{\overset{\displaystyle O}{\|}}{C}-CH_2-\overset{\overset{\displaystyle H}{|}}{C}-C-$$

(4)

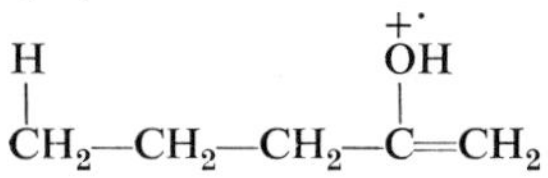

$$\underset{\underset{\displaystyle CH_2}{}}{\overset{\overset{\displaystyle H}{|}}{}}-CH_2-CH_2-\overset{+\cdot}{\overset{\displaystyle OH}{|}}{C}=CH_2$$

j, m/e 86

$$CH_3-\overset{+\cdot}{\overset{\displaystyle OH}{|}}{C}=CH_2$$

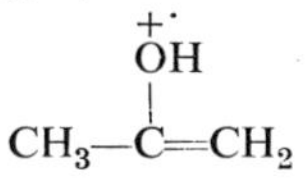

k, m/e 58

In addition to the theory of mass spectrometry stored in the PRE-LIMINARY INFERENCE MAKER, much of this material is duplicated in the PREDICTOR section of the program. Mass spectroscopists have observed the occurrence of ions corresponding to the McLafferty rearrangement product plus an additional hydrogen atom in aliphatic ketones possessing a linear chain of five or more carbon atoms (27). This process was introduced into the PREDICTOR section of Heuristic DENDRAL in the form that if a candidate structure from the STRUCTURE GENERATOR contained a linear chain of five or more carbon atoms, then it would predict an ion of mass corresponding to a McLafferty rearrangement plus one.

With this knowledge of the theory of mass-spectrometric fragmentation processes of saturated ketones, Heuristic DENDRAL interpreted the mass spectra of many unknown aliphatic ketones. The results for 12 ketones are summarized in Table 6. As shown from the ranking of the candidate structures, the correct answer scored well in all but two examples. In these two instances the heuristics failed. However, in the case of 4-nonanone, the program was subsequently amended in such a way that the correct answer appeared as the final output. The heuristic concerning the McLafferty rearrangement process was altered to read that if there were two potential McLafferty rearrangement ions, one originating from a carbon chain of five

TABLE 6

Results of the Computer Interpretation of Some Ketone Mass Spectra

Ketone spectrum	Number of acyclic structures				Ranking of candidates
	DENDRAL		Heuristic DENDRAL		
	Total isomers[a]	Total ketones[a]	After structure generation	After consistency check	
2-Butanone[b]	11	1	1	1	1st, 2-butanone
3-Pentanone[b]	33	3	1	1	1st, 3-pentanone
3-Hexanone[c]	91	6	1	1	1st, 3-hexanone
2-Methylhexan-3-one[c]	254	15	1	1	1st, 2-methylhexan-3-one
3-Heptanone[b]	254	15	2	2	Tie for 1st; 3-heptanone and 5-methylhexan-3-one
3-Octanone[b,c]	698	33	4	4	1st, 3-octanone
4-Octanone[c]	698	33	2	1	1st, 4-octanone
2,4-Dimethylhexan-3-one[c]	698	33	4	3	Tie for 1st; 2,4-dimethylhexan-3-one and 2,2-dimethylhexan-3-one
6-Methylheptan-3-one[b]	698	33	4	2	1st, 3-octanone; tied for 2nd, 6-methylheptan-3-one, 5-methyl-heptan-3-one, and 5,5-dimethyl-hexan-3-one
2-Nonanone[c]	1936	82	7	5	1st, 3-nonanone
3-Methyloctan-3-one[c]	1936	82	4	3	Consistency check eliminated correct structure because no McLafferty +1 peak was present in original mass spectrum
4-Nonanone[c]	1936	82	4	1	1st, 4-nonanone. The program was revised after results were first published (23) showing the correct structure eliminated

[a] Total isomers were computed with enol on BADLIST; the numbers therefore include ketones, aldehydes, unsaturated ethers (including enol-ethers), and all unsaturated alcohols except enols. Total ketones were computed by putting ketone (only) on GOODLIST.
[b] Literature mass spectrum: A. G. Sharkey, J. L. Shultz, and R. A. Friedel, *Anal. Chem.*, **28**, 934 (1956).
[c] Mass spectrum from Stanford University chemistry department.

or more methylene groups and the other from a primary hydrogen atom, the latter might not appear in the mass spectrum. The computer solution of a typical aliphatic ketone is presented below as Scheme 8. Section II of this article enables the reader to translate the DENDRAL notation into structural formulas.

(EXPLAIN (QUOTE C8H16O) S:Ø932Ø (QUOTE TEST5) (QUOTE JUL-27-68))
   *GOODLIST = (*N-PROPYL-KETONE3*)
*BADLIST = (*C-2-ALCOHOL* *PRIMARY-ALCOHOL* *ETHYL-ETHER2*
   METHYL-ETHER2 *ETHER2* *ALDEHYDE* *ALCOHOL* *ISO-PROPYL-
KETONE3* *ETHYL-KETONE 3* *METHYL-KETONE3*)

(JULY-4-1968 VERSION)
C *N-PROPYL-KETONE3*H6
MOLECULES   NO DOUBLE BOND EQUIVS
   1.   C=..   O   C3H7   CH2.C3H7 ,
   2.   C=..   O   C3H7   CH2.CH..CH3 CH3 ,

(SCORE (QUOTE TEST5) S:Ø932Ø)
JUL-27-68
1.) C211OC1C1CC1C1C1C$

((43. . 1ØØ.) (57. . 88.) (58. . 22.) (71. . 1ØØ.) (85. . 88.) (86. . 22.)
(128. . 14.))

2.) C211OC1C1CC1C11CC$

((43. . 87.) (57. . 1ØØ.) (58. . 8.) (71. . 87.) (85. . 1ØØ.) (86. . 4.) (1ØØ. . 4.
(128. . 16.))

*THIS CANDIDATE IS REJECTED BECAUSE OF (1ØØ.).

*LIST OF RANKED MOLECULES:

1. #1.
   S = 6.
   P = (57. 71. 43. 85. 86. 58.)
   U = NIL

* 1. #N MEANS THE FIRST RANKED MOLECULE IS THE NTH IN THE
      ORIGINAL NUMBERED LIST ABOVE.
   S = THE SCORE (HIGHEST = BEST) BASED ON THE NUMBER OF
      SIGNIFICANT PREDICTED PEAKS.
   P = THE LIST OF SIGNIFICANT PREDICTED PEAKS.
   U = THE LIST OF POSSIBLY SIGNIFICANT UNRECORDED PEAKS IN
      RESOLVING SCORING TIES (THE FEWER IN DOUBT THE BETTER).
DONE

Scheme 8. Heuristic DENDRAL'S interpretation of the mass spectrum of a typical aliphatic ketone (4-octanone).

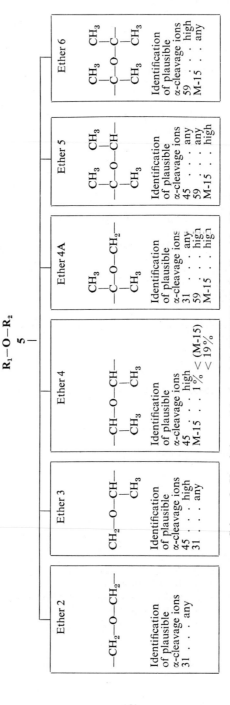

Scheme 9. Heuristics used by heuristic DENDRAL in ether identification.[a]

[a] The numbers in the names serve as reminders of the number of carbon atoms in the ether subgraphs.

152

3. *Ethers*

In the development of the Heuristic DENDRAL program, the PRE-LIMINARY INFERENCE MAKER, on being confronted with an unknown mass spectrum, searched for the possible presence of the aliphatic ether moiety by seeking affirmative answers to the following specific points. Peaks corresponding to the elimination of 17 and 18 amu from the molecular ion should be less than 2% relative abundance. This rule was necessary since the mass spectra of many aliphatic ethers (28) contain fragment ions equivalent to the expulsion from the molecular ion of a hydroxyl radical or of water. Since the empirical formula of a saturated ether is also consistent with that of a saturated alcohol, one can use the abundance of the M-17 and M-18 ions to differentiate between these two classes of compound. It is well known (29) that alcohols readily eliminate water from their molecular ions and that this process accounts for peaks in excess of 2% relative abundance. The program must also recognize two peaks in the mass spectrum of an aliphatic ether (5) corresponding to the mass of the alkyl fragments R_1 and R_2 as well as the alkoxy fragments OR_1 and OR_2.

$$R_1 - O - R_2$$
$$(5)$$

If the preceding conditions are met, Heuristic DENDRAL attempts to expand the ether subgraph C—O—C implied by 5 to any of the six subgraphs in Scheme 9. These additional constrictions arise from the well-known α-cleavage of ethers (30), which produce algebraic relationships of the following kind for, say, the ether 4 subgraph:

$$M + 72 = l + m$$

For the program to recognize the ether 4 subgraph, it must recognize two peaks in the unknown mass spectrum whose combined mass is equal to the molecular ion plus 72 mass units. For the substructures ether 2, ether 3, ether 4A, ether 5, and ether 6, the masses of the radicals duplicated in α-cleavage are 44, 58, 72, 86, and 100 amu respectively. Possible α-cleavage ions *l*, *m*, *o*, and *p* are shown in the fragmentation map for the ether 4 subgraph.

The α-fission ions *l*, *m*, *o*, and *p* formed by fragmentation of an ether 4 can then undergo rearrangement by hydrogen transfer to yield the ions *n*, *q*, and *r* respectively. The values depicted in Scheme 9 as "31 . . . high," "45 . . . high," and the like represent the masses of the rearrangement ion *n* and its analogs which result from decomposition of the initially produced α-cleavage ion.

Mass spectrometry alone cannot differentiate substitution of an alkyl chain in saturated ethers beyond the two α-carbon atoms. For this reason,

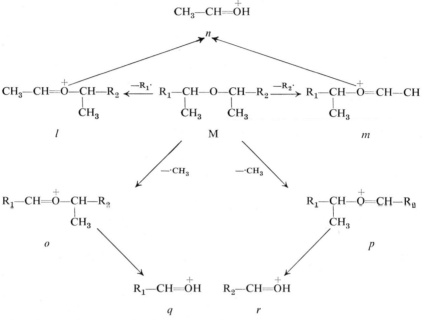

when Heuristic DENDRAL interpreted the mass spectrum of an unknown saturated ether, it tended to produce several (often 5 to 10) candidate structures from the STRUCTURE GENERATOR (31). Although in our experience the correct answer was always included, it became desirable to truncate this list in the PREDICTOR section of this program. The PRE-DICTOR, when relying only on mass-spectral theory, could not direct this truncation to any substantial degree. For this reason a PREDICTOR subroutine relying on nmr input data was developed and introduced into the program at the verification phase (Scheme 7). It should be stressed that the program can use nmr data if available but does not require this input to function.

The nmr program requires two conditions to operate: a list of candidate structures (from the CONSISTENCY CHECK function) and the nmr spectrum of the unknown compound. For each candidate structure an nmr spectrum is predicted, and this is compared to the experimental data. The chemical shift information must agree to within 0.3 ppm and the predicted resonance must display an identical multiplicity and integral value as the unknown. If the recorded nmr spectrum is a multiplet, then the predicted nmr spectrum must have signals within ±0.3 ppm of this chemical shift, and the value of the integrals must be identical. Should the signal require-ments not be satisfied between the predicted and the unknown nmr spectrum, then the disparity is marked and fed to the EVALUATION function. The best possible score is zero, which results when the two spectra are identical. Otherwise the score is the product of all the integral values of the unassigned signals multiplied by 0.75 for each multiplet. Hence, the lower the score for any candidate, the higher its plausibility.

As is evident from Table 7, the use of nmr spectroscopy, even in this elementary manner, results in unequivocal answers (31). Thus in problems of this degree of complexity (aliphatic ethers) it was clearly evident (see Table 7) that the use of mass spectrometry (to locate the class of compound and to produce a restricted list of viable structures) coupled with nmr spectroscopy (to truncate the list of candidate structures) was a powerful combination when programmed into Heuristic DENDRAL. The question remained, however, whether this combination in either this or other forms would prove successful for the interpretation of problems of greater complexity.

4. *Amines*

The next stage in the development of Heuristic DENDRAL was to challenge it with a task in which the total search space (i.e., the number of possible isomers) was enlarged. Aliphatic amines were chosen in view of the markedly increased number of isomers per nitrogen atom as compared to the number of possible ketone and ether isomers per oxygen atom. Second, aliphatic amines represent a class of molecules about which much mass-spectrometry theory has been accumulated (32, 33). In order for the program to succeed in this endeavor, in view of the expanded search space, more explicit heuristics than utilized so far in the PRELIMINARY INFERENCE MAKER became mandatory.

The results of Heuristic DENDRAL'S capability of adapting itself to the solution of amine mass spectra have been published (34) (see also Table 8). The basic change made to the program was to include all the details of mass spectrometric and nmr theory into the PRELIMINARY INFERENCE MAKER. Thus the strategy employed focused on obtaining as much trunca-tion of the search space as quickly as possible before presenting superatoms to the STRUCTURE GENERATOR for amplification to molecular structures. In practice, the heuristics employed were so selective that the PRELIMINARY INFERENCE MAKER, without assistance from the PREDICTOR, was able to accomplish dramatic pruning of the total number of isomers. A detailed flowchart of the decisions used by the PRELIMINARY INFERENCE MAKER is found in Scheme 10.

In order to approach the solution of low-resolution mass spectra of unknown saturated amines, it became necessary to define a complete set of possible amine subgraphs (i.e., superatoms*) that could be inferred from the

---

* A superatom is defined as a substructure with at least one free valence. In the present instance only linkages to other carbon atoms are allowed.

TABLE 7

Results of the Computer Interpretation of Some Saturated Ether Mass Spectra

Ether spectrum chosen[a]	Number of acyclic structures				Ranking of candidates
	DENDRAL		Heuristic DENDRAL		
	Total isomers[b]	Ethers[b]	After structure generation	After consistency check	
Methyl t-butyl ether	14	6	2	2	Correct structure ranked below ethyl n-propyl ether
Ethyl i-propyl ether	14	6	4	4	Correct structure ranked first
Ethyl i-butyl ether	32	15	2	2	Correct structure tied with ethyl n-butyl ether
Ethyl t-butyl ether	32	15	3	3	Correct structure ranked first[c]
Di-n-propyl ether	32	15	1	1	Correct structure ranked first[c]
Di-i-propyl ether	32	15	10	10	Correct structure ranked first[c]
n-Propyl n-butyl ether	72	33	2	2	Correct structure tied with n-propyl i-butyl ether
i-Propyl s-butyl ether	72	33	1	1	Correct structure ranked first
Di-n-butyl ether	171	82	3	3	Correct structure tied with n-butyl, i-butyl, and di-i-butyl ether
i-Butyl t-butyl ether	171	82	15	15	Di-t-butyl ether ranked first. Correct structure tied for second with i-propyl i-amyl ether
Ethyl n-heptyl ether	405	194	17	13	Correct structure tied with 12 other ethyl ethers
n-Butyl n-amyl ether	405	194	8	8	Correct structure tied with 7 other butyl amyl ethers
Di-n-amyl ether	989	482	10	10	Correct structure tied with 9 other di-amyl ethers
Di-i-amyl ether	989	482	10	10	Correct structure ranked first[c]

[a] Mass spectra used as "unknown" were taken from the literature (28).

[b] Total isomers comprise all alcohols and ethers. There are no unstable structures to put on BADLIST for a composition $C_n H_{2n+2}O$. Total ethers were computed with ether on GOODLIST.

[c] Nmr spectra correctly differentiated the correct structure from the other candidates.

156

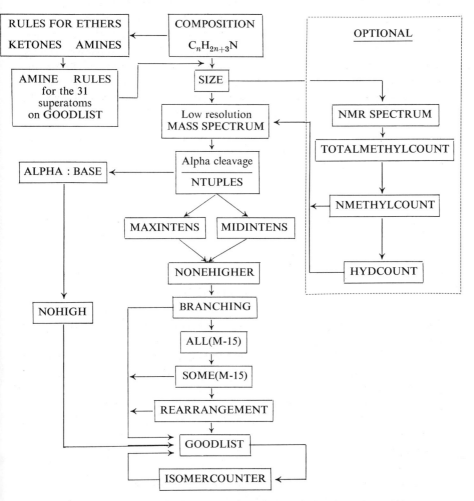

Scheme 10. Flowchart of the decisions made by heuristic DENDRAL in the solution of saturated amine mass spectra.

data. This is conveniently accomplished by using the letters M, S, P, and T to define the degree of substitution on the α-carbon atom of an aliphatic amine. In this convention P, S, and T represent primary (6), secondary (7), and tertiary (8) carbon atoms respectively, while M represents an unsubstituted α-carbon atom (i.e., methyl group). The canonical order of the symbols is M < P < S < T. With this imposed order, PMM is the triplet used to denote subgraph 9, although MPM and MMP are equivalent names. Using this nomenclature there are 31 superatoms that can be built by various combinations, within a saturated aliphatic framework, of M, S, P, and T.* Some additional examples of the representation of superatoms PPP and TSP (10 and 11) in this nomenclature are presented below.

$$H_2-N-CH_2- \qquad H_2-N-\overset{|}{C}H- \qquad H_2-N-\overset{|}{\underset{|}{C}}-$$

6: P   7: S   8: T

$$-CH_2-N(CH_3)_2 \qquad -CH_2-\overset{\overset{|}{C}H_2}{\underset{|}{N}}-CH_2- \qquad -\overset{\overset{|}{C}H_2}{\underset{|}{C}}-\overset{|}{N}-\overset{|}{C}H-$$

9: PMM   10: PPP   11: TSP

The strategy employed by Heuristic DENDRAL for the interpretation of saturated amine mass spectra (supplemented by nmr data if available) is diagrammatically represented in Scheme 10. All 31 superatoms are initially placed on GOODLIST and removed only when they fail to pass a heuristic decision. The first test each is subjected to is depicted in Scheme 10 as SIZE. Here the program decides whether the empirical formula (supplied with the low-resolution mass spectrum) contains enough carbon atoms to construct the smallest possible molecule from any of the superatoms by the addition of methyl radicals to all free valences. If a superatom requires more carbon atoms than are available to make this molecule, then that superatom is discarded.

The program introduces nmr spectroscopy into its decision-making processes at the beginning, that is, in the PRELIMINARY INFERENCE SECTION. Thus, immediately after it has normalized the amplitudes of the peaks in the mass spectrum and removed any improbable masses (e.g., M-4 through M-14), the program accepts nmr input data.† Basically, the program attempts to decide the total number of methyl groups attached

---

* In addition three more possibilities exist that are not superatoms but molecules, viz., M, MM, and MMM (methylamine, dimethylamine, and trimethylamine, respectively).

† The program does not require nmr input data and will proceed directly to a consideration of the mass spectrum if no nmr spectrum is available.

to carbon and to nitrogen and if possible the number of hydrogen atoms attached to the α-carbon atom of the unknown amine. This is accomplished by searching within certain limits of the chemical shift value for signals corresponding to these entities. If no integration curve is given with the nmr data, the program then attempts to use as much information as possible. For instance, the presence of a sharp singlet at δ 2.2 ppm would indicate the presence of one or more N-methyl groups, but the absence of an integral curve means that the program is unable to decide anything about the C-methyl region (below δ 1.2 ppm) of the spectrum.

When the program has completed its survey of the nmr data, its decisions are used to validate each of the superatoms on GOODLIST. When a superatom fails any test it is removed from GOODLIST. If the program was unable to define, for instance, the number of C-methyl groups from the nmr data due to the absence of an integration curve, then all superatoms automatically would pass the C-methyl test.

At the completion of the nmr-based decisions, the program confronts the mass spectrum. The first rule programmed into the mass-spectrometry section relates to the well-known propensity of aliphatic amines to undergo α-fission (32, 33) (see $\mathbf{12} \to s + t$).

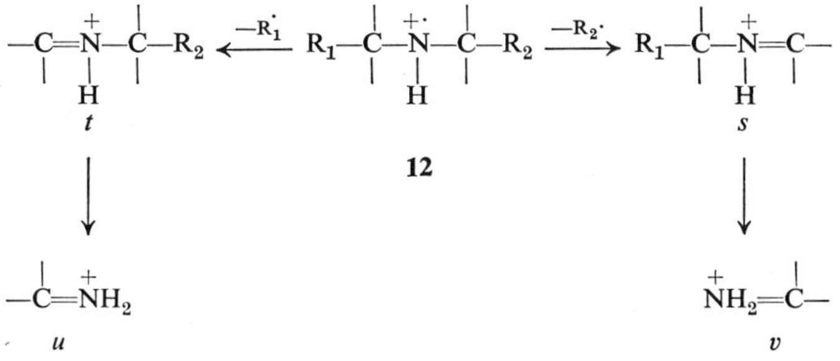

Decision ALPHA:BASE (Scheme 10) validates the presence of those superatoms with one free valence (for instance P, PM, and PMM), which should produce in their mass spectra a base peak at m/e 30, 44, and 58 respectively, resulting from α-cleavage. Furthermore, there should be no peaks in excess of 10% relative abundance (TEST NOHIGH) above half the molecular weight.

For superatoms having two or more α-cleavage ions, the program must identify a definite number of α-cleavage peaks in the mass spectrum. The lowest mass limit searched by the program for the identification of α-fission peaks is given by the expression molecular weight of the superatom plus $[(n - 1) \times 15]$, where n is the number of free valencies in that superatom. For example, PPP (10) has $n = 3$ and a mass of 56; hence in this case, the search begins at mass 86. To validate the presence of the PPP subgraph, the program must find sets of three peaks (the lowest possible mass being 86) whose sum is equal to $[(n - 1) \times \text{mol wt}) + 56]$. If none of these sets, called n-tuples, is identified, then the α-fission heuristic rejects that super-atom.*

The program rejects any n-tuple $(x_1, \ldots, x_n)$ that does not meet the following five conditions that are described in more detail in the original publication (34).

1. For superatoms with no possibility of rearrangement (e.g., PP and PPP), the sum of the intensities of the peaks $x_1, \ldots, x_n$ must exceed 70% relative abundance (test MAXINTENS in Scheme 10). For the classes of amines that exhibit intense ions from rearrangement processes (e.g., **12** → $s + t$ and $u + v$), the total intensity of any n-tuple must exceed only 30% relative abundance (operation of MIDINTENS in Scheme 10).

2. The highest mass number in the n-tuple, x_1, must be the most abundant ion in the spectrum between x_1 and M-1† (decision NONEHIGHER in Scheme 10).

3. The lowest mass in the n-tuple, x_1, corresponds to a more abundant ion than the next higher ion, x_j, in the n-tuple list, provided that they differ in size by at least three carbon atoms. If the difference in carbon number between x_i and x_j is greater than C_3, the program will let x_i be less abundant than x_j,‡ and this is embodied in the test BRANCHING in Scheme 10.

4. For any n-tuple containing only M-15 entries [Scheme 10, ALL (M-15)], the M-15 fragment in the unknown mass spectrum must be higher than a predetermined value.

---

* The term "n-tuple" is used to refer to a set of n mass numbers that correspond to n possible α-cleavage ions. For example, suppose that an unknown amine had a molecular weight of 171 and that the superatom TSP (**11**) was under scrutiny. The following sets of n-tuples would be constructed as valid parameters, one or more sets of which would have to be verified by the experimental data for TSP to pass this test.

> $n = 6$; molecular weight of TSP $= 53$; sum $= [(5 \times 171) + 53] = 908$
> lowest α-fission peak possible $= (5 \times 15) + 53 = 128$
> possible α-fission peaks (128 142 156)
> possible n-tuples: (156 156 156 156 156 128)
>   (156 156 156 156 142 142)

† The M-1 ion is not considered by the program to be an α-fission fragment. In addition, the M-15 ion, even if it is absent from the unknown mass spectrum, is allowed to be used in the construction of n-tuples provided that its mass together with the masses of the other peaks in the n-tuple list satisfied the equation used for their derivation.

‡ Should the difference be C_3 or greater, then the possibility exists that the higher mass ion in the n-tuple list could result from the expulsion of either a secondary or tertiary radical. It is well known in mass spectrometry that the expulsion of these types of radicals is preferred over the loss of a primary radical in α-cleavage process.

5. For any *n*-tuple containing one mass, x_i, resulting from the elimination of an alkyl radical, with all other entries being M-15, the peak x_i must exceed a predetermined value [decision SOME (M-15) in Scheme 10].

A superatom is rejected if all its possible α-cleavage ions (*n*-tuples) are eliminated. For the precise details of these five points, the reader is referred to the original literature (34).

At this stage of the program those candidates that remain on GOOD-LIST, with at least one surviving *n*-tuple, are then used to calculate the mass of the alkyl fragments lost in the α-cleavage process. The masses of the radicals lost are equal to the difference in mass between the *n*-tuple value and the molecular weight. These masses are referred to as a "partition" because they show how to allocate the remaining composition to the free valences of the superatom. A subroutine then calculates the number of isomers compatible with the structure of the superatom and its partitions. For instance, ethyl-1,3-dimethylbutylamine (13) belongs to the superatom class SP, and this was assigned the partition (15 15 57). This means that the superatom SP could be expanded into either 14 or 15 depending on how the two methyl and one butyl radical are assigned. The program chooses 15 rather than 14 on the basis of preferred rearrangements.

$$
\begin{array}{ccc}
\underset{\underset{\text{H}}{|}\ \underset{\text{C}}{|}\quad\ \underset{\text{C}}{|}}{\text{C—C—N—C—C—C—C}} &
\underset{\underset{\text{H}}{|}\ \underset{\text{C}}{|}}{\text{C}_4\text{H}_9\text{—C—N—C—C}} &
\underset{\underset{\text{H}}{|}\ \underset{\text{C}}{|}}{\text{C—C—N—C—C}_4\text{H}_9} \\
(13) & (14) & (15)
\end{array}
$$

The program has been successfully tested with a total of 93 saturated amines, and of this total 37 were supplemented by nmr spectra (34). A brief summary of its capability in handling these unknowns is recorded in Table 8. In all instances the correct solution to any problem was always found in the final output. Clearly, from Table 8 the program was able to function efficiently even with a molecule the size of tri-*n*-heptylamine (16) $C_{21}H_{45}N$, which contains 38,649,142 *a priori* aliphatic structures. The program,

$$(n\text{-}C_7H_{15})_3N$$
(16)

without nmr information, reduced this to 1938 candidates. Had an nmr spectrum (plus integral curve) been available, only the correct structure would have resulted. Other examples of the high degree of truncation of the total search space obtained with the program are evident from Table 8.

The strategy adopted for the solution of saturated amine mass spectra was to use as much theory as possible in the planning section of the program. When this is done no additional theory remains for the PREDICTOR, and since it could then add nothing, it was not used in the amine program.

Table 8
Results of the Computer Interpretation of Some Saturated Amine Mass Spectra

Amine spectrum chosen	Number of acyclic structures		
	Total isomers[a]	Using mass spectrum only	Using mass and nmr spectra
Methyl ethyl-*n*-propylamine	39	15	2
Dimethyl 2-butylamine	39	6	1
n-Propyl-*n*-butylamine	89	10	1
1,3-Dimethylpentylamine	89	16	4
1,5-Dimethylhexylamine	211	34	9
Methyl-*i*-propyl-*n*-butylamine	211	31	1
Ethyl di-*i*-propylamine	211	18	1
Methyl-*n*-propyl-*n*-butylamine	211	24	1
n-Propyl-*n*-hexylamine	507	42	1
3,3,5-Trimethylhexylamine	507	89	—[b]
Methyl-*n*-propyl-*n*-hexylamine	1238	46	1
Dimethyl 3-octylamine	1238	36	1
n-Butyl-*n*-hexylamine	1238	48	1
n-Amyl-*n*-hexylamine	3057	112	—
Tri-*n*-butylamine	7639	8	—
Di-*n*-heptylamine	48865	510	—
Tri-*i*-amylamine	124906	40	9
Methyl 8-hexadecylamine	321198	3471	—
n-Octyl-*n*-nonylamine	830219	6942	1
Dimethyl 8-hexadecylamine	2156010	14418	1
Tri-*n*-heptylamine	38649142	1938	—

[a] These numbers were computed by a subprogram developed by A. B. Delfino and A. Buchs.

[b] Blanks in the last column indicate that no nmr spectrum was available.

5. Compounds of the General Class R_1—X—R_2 (X = N, O; S, R_2 Possibly H)

Following Heuristic DENDRAL'S success in interpreting aliphatic amine mass spectra (34), it was logical to extend the program to the solution of the general heteroatomic class of compounds represented by $C_nH_{2n+v}X$ where v = the valence of X, which can be either N, O, or S (35). This expanded the program's capacity to the general solution of aliphatic ethers [the previous work (31) was not of a completely general nature], alcohols, thiols, and thioethers. In addition, for these compounds, it is now unnecessary to supply the empirical formula or the molecular ion to the program. Heuristic search enables the correct molecular ion to be deduced (even in test cases when it was deliberately removed from the mass spectrum), and the program decides what heteroatom is present on its evaluation of various ion series present. The empirical formula is then calculated from the inferred heteroatom and the molecular ion.

The first question to be answered by Heuristic DENDRAL is whether the unknown mass spectrum belongs to the "saturated aliphatic monofunctional" (or, for short, SAM) class of compound. This is achieved by generating a reduced mass spectrum, that is, by removing from the unknown mass spectrum those ion series known to be prevalent in these aliphatic compounds. For a spectrum to be accepted as a SAM candidate the reduced mass spectrum must contain no ion of greater abundance than 10% and the average ion abundance must be less than 3%.

At the next level of operation the program decides whether the heteroatom is more likely to be N, O, or S from its analysis of the ion series present in the unknown mass spectrum. The score achieved by each heteroatom is then ranked from highest to lowest, with the former representing the best degree of fit between that heteroatom and the unknown mass spectrum. For each heteroatom this score must exceed a predetermined level for that candidate to be accepted. The highest ranked heteroatom is then used to determine the mass of the most plausible molecular ion, and this in turn is used to calculate the empirical composition of the molecular ion.

Once the empirical formula of the unknown has been established, the program builds the complete set of superatoms (35) (see section d for the amine superatom representation) and retrieves the mass-spectrometric theory associated with these superatoms. Should one or more of the subgraphs be validated by the theory, the total inference process is complete and the program prints the result. However, if no subgraph is accepted, the program proceeds to calculate the molecular weight of the next higher homolog in that particular heteroatomic class. Heuristic DENDRAL is programmed to proceed through three validation runs for any heteroatom, and if no successes (superatom validation) are achieved, the next most preferred heteroatom is tested. If no superatoms are found to explain the data after all the heteroatoms have each been tested three times, the program assumes that the mass spectrum must result from a non-SAM compound.

The subgraphs built by the program must of course be a complete and nonredundant list, and this is achieved by using the symbols T, S, P, and M as described (34) for amines in Section D. Hence subgraphs with three symbols must represent tertiary amines; with two symbols, secondary amines, ethers, or thioethers; and with one symbol, primary amines, alcohols, or thiols.

The validation procedures (both nmr and mass spectrometric) for each

TABLE 9

Results of the Computer Interpretation of Some Saturated Alcohol and Ether Mass Spectra using the General Identification Program

Alcohol	Number of $C_nH_{2n+2}O$ isomers	Number of inferred isomers[a] A	B
n-Butyl	7	2	1
Sec-butyl	7	3	2
2-Methyl-2-butyl	14	1	1
n-Pentyl	14	4	1
3-Pentyl	14	1	1
2-Pentyl	14	2	1
3-Hexyl	32	2	1
3-Methyl-1-pentyl	32	8	4
n-Hexyl	32	8	1
3-Heptyl	72	4	1
3-Ethyl-3-pentyl	72	1	1
2,4-Dimethyl-3-pentyl	72	3	1
3-Methyl-1-hexyl	72	17	6
n-Octyl	171	39	1
3-Octyl	171	8	1
2,3,4-Trimethyl-3-pentyl	171	3	1
n-Nonyl	405	89	1
2-Nonyl	405	39	1
6-Ethyl-3-octyl	989	39	9
3,7-Dimethyl-1-octyl	989	211	41
2-Butyl-1-octyl	6045	1238	25
n-Dodecyl	6045	1238	1
3-Tetradecyl	38322	1238	1
n-Tetradecyl	38322	7639	1
n-Hexadecyl	151375	48865	1

Ether	Number of $C_nH_{2n+2}O$ isomers	Number of inferred isomers[a] A	B
Methyl-n-propyl	7	2	1
Methyl-iso-propyl	7	3	1
Methyl-n-butyl	14	2	1
Methyl-iso-butyl	14	2	1
Ethyl-iso-propyl	14	1	1
Ethyl-n-butyl	32	4	1
Ethyl-iso-butyl	32	4	2
Ethyl-tert-butyl	32	1	1
Di-n-propyl	72	2	1
n-Propyl-n-butyl	72	4	1
Ethyl-n-pentyl	72	3	2
Iso-propyl-sec-butyl	72	4	1
Iso-propyl-n-pentyl	171	3	1
Di-n-butyl	171	2	1
Iso-butyl-tert-butyl	171	34	1
Ethyl-n-heptyl	405	8	1
n-Butyl-n-pentyl	405	18	7
Di-iso-pentyl	989	10	1
Di-n-pentyl	989	125	2
Di-n-hexyl	6045	780	21
Bis-2-ethylhexyl	151375	780	1
Di-n-octyl	151375	780	1
Di-n-decyl	11428365	22366	1

[a] A, inferred isomers when only mass spectrometry is used; B, inferred isomers when the number of methyl radicals is known from nmr data.

164

TABLE 10

Thio and Thioether Mass Spectra using the General Identification Program

Thioether	Number of $C_nH_{2n+2}S$ isomers	Number of inferred isomers[a]	
		A	B
Methyl-ethyl	3	1	1
Methyl-n-propyl	7	1	1
Methyl-iso-propyl	7	2	1
Di-ethyl	7	1	1
Methyl-n-butyl	14	3	1
Methyl-iso-butyl	14	5	2
Methyl-tert-butyl	14	1	1
Ethyl-n-propyl	14	2	1
Ethyl-n-butyl	32	3	1
Ethyl-tert-butyl	32	1	1
Di-n-propyl	32	2	1
Methyl-n-pentyl	32	10	1
Di-iso-propyl	32	1	1
Ethyl-n-pentyl	72	4	1
n-Propyl-n-butyl	72	5	1
Iso-propyl-tert-butyl	72	1	1
n-Propyl-iso-butyl	72	3	2
n-Propyl-n-pentyl	171	4	1
Ethyl-n-hexyl	171	8	1
Di-n-butyl	171	5	1
Di-sec-butyl	171	3	1
Methyl-n-heptyl	171	21	1
Di-n-pentyl	989	12	1
Di-n-hexyl	6045	36	1
Di-n-heptyl	38322	153	1

Thiol	Number of $C_nH_{2n+2}S$ isomers	Number of inferred isomers[a]	
		A	B
n-Propyl	3	2	1
Iso-propyl	3	1	1
n-Butyl	7	3	1
Iso-butyl	7	3	1
Tert-butyl	7	1	1
2-Ethyl-2-butyl	14	1	1
3-Methyl-2-butyl	14	2	1
n-Pentyl	14	4	1
3-Pentyl	14	5	3
n-Hexyl	32	8	1
2-Hexyl	32	12	5
2-Methyl-1-pentyl	32	8	4
3-Methyl-3-pentyl	32	1	1
2-Methyl-2-hexyl	72	8	3
n-Heptyl	72	17	1
n-Octyl	171	39	1
2-Ethyl-1-hexyl	171	39	9
1-Nonyl	405	89	1
n-Decyl	989	211	1
n-Dodecyl	6045	1238	1

[a] A, inferred isomers when only mass spectrometry is used; B, inferred isomers when the number of methyl radicals is known from nmr data.

TABLE 11

Results of the Computer Interpretation of Amine Mass Spectra using the General Identification Program

Amine	Number of $C_nH_{2n+3}N$ isomers	Number of inferred isomers[a] A	Number of inferred isomers[a] B
n-Propyl	4	1	1
Iso-butyl	8	2	1
N-methyl-n-propyl	8	4	1
2-Pentyl	17	2	1
N-methyl-n-butyl	17	4	1
Di-n-propyl	39	8	1
Ethyl-n-butyl	39	6	1
Ethyl-n-pentyl	89	16	1
N-methyl-di-iso-propyl	89	15	3
1-Methylheptyl	211	34	1
Di-n-butyl	211	24	1
Di-ethyl-n-butyl	211	17	3
Tri-n-propyl	507	2	1
N,N-dimethyl-2-ethylhexyl	1238	156	9
n-Tetradecyl	48865	10115	1
Bis-2-ethylhexyl	321988	2340	24
Di-ethyl-n-dodecyl	321988	2476	1
N-methyl-bis-2-ethylhexyl	830219	2340	24
n-Octadecyl	2156010	48865	1
N-methyl-n-octyl-n-nonyl	2156010	15978	1

[a] A, inferred isomers when only mass spectrometry is used; B, inferred isomers when the number of methyl radicals is known from nmr data.

superatom of a given heteroatomic class are basically similar to those discussed in Section D for amines (34); for a comprehensive discussion the reader is referred to the detailed publication (35).

Typical results presented by Heuristic DENDRAL for the solution of unknown mass spectra of alcohols, ethers, thiols, thioethers, and amines are collated in Tables 9, 10, and 11 respectively. These tables demonstrate the ability of the program to identify the correct type of compound from a low-resolution mass spectrum and to truncate the *a priori* number of solutions (i.e., the total number of isomers) of any problem to a far smaller, and hence more manageable, number. The inherent power of mass spectrometry coupled with nmr spectroscopy for the solution of structural problems in organic chemistry is apparent from these tables.

C. Structure Generation

1. *The DENDRAL Algorithm Plus Heuristic Constraints*

The program that implements the DENDRAL concepts to compile lists of molecular structures is called the STRUCTURE GENERATOR. As noted before, the entire Heuristic DENDRAL system revolves around the STRUCTURE GENERATOR in the general case. Only when sufficiently detailed rules can be given to the PRELIMINARY INFERENCE MAKER (e.g., amines, ethers, alcohols, thiols, and thioethers) can structure generation be avoided. Yet the number of molecules it produces is too large to be useful for most problems, unless the generation process is severely constrained. Thus the algorithm is prevented from constructing molecules that are either undesired or incompatible with the given analytical data.

In short, the STRUCTURE GENERATOR is designed to solve the following problem:

GIVEN: 1. An empirical formula (each atom of which has already been assigned mass and valence).
2. A list of likely substructures.
3. A list of impossible substructures.

TASK: Generate all structures compatible with the given data. If there are no data-oriented lists of likely or impossible substructures, the program generates all structural variants of the given formula.

SUMMARY OF THE STRUCTURE-GENERATION PROCESS:

1. If some substructures are required (i.e., GOODLIST specifies one or more "superatoms"), change the given empirical formula to replace the atoms of the substructures with the superatom.
2. Start the algorithm working on the modified empirical formula.

3. During structure generation, prevent the algorithm from generating forbidden substructures (i.e., those items on BADLIST).

4. Print each molecular structure generated, after connecting radicals to superatoms, replacing superatom names by their structural representation and checking for forbidden substructures (for the last time).

These mechanisms are discussed briefly in the following sections from a programming point of view, the algorithm itself having been explained from a different perspective in Section II.

2. *Implementation of the Structure-Generator Algorithm*

The fundamental ideas behind the computer implementation of the DENDRAL algorithm have been to use the algorithm efficiently and at the same time maintain flexibility in the program. Since the actual encoding of the algorithm has been described in other publications (6, 7, 10), only the more general considerations will be taken up here. Needless to say, while the computer program follows the DENDRAL canons, there is also much "book-keeping" the program must do. Thus, this general explanation can hardly be taken as a literal description of how the program functions.

The program that embodies the algorithm can be thought of as a series of four steps, closely paralleling the algorithm itself:

1. List all possible centroids (in canonical order) together with the residual empirical formula for each.

2. For each centroid, break up the residual formula into the appropriate number of subformulas. A collection of these subformulas is referred to as a partition. List (in canonical order) all possible ways of making the appropriate number of subformulas.

3. Repeat the whole process for each subformula.

4. Stop when there is no residual formula.

One of the basic problems inherent in the STRUCTURE GENERATOR in the past has been the rigidity of the code embodying the canons of DEN-DRAL order. These canons, described in Section II, specified the canonical form of a structure, and thus the implicit generating sequence. Changing the generating order for structures was difficult because of the inflexibility of the program. For example, the suggestion to give preference to nonbranching structures was difficult to implement.

Reorganizing this part of the program made the generator into a table-driven program, by putting the DENDRAL canons (attributes and their values) on a list that is accessed to decide how to enlarge the structure that is being built. The features to be considered are the type of atom to choose as the centroid and the degree of branching on that node. For example, it might be desirable to see highly branched structures with heteroatoms near the center of the structure before any others in the output list. Or, if desired, unbranched structures could be eliminated from the output entirely merely by revising the one list.

3. *Preventing Construction of Unlikely Molecules*

The program's theory of chemical stability is such that unstable structures are never generated. Its "theory," however, is nothing more than a list of substructures, the presence of any of which makes a molecule undesirable for the particular problem at hand. This list is called "BADLIST" for obvious reasons.

BADLIST is also used by the program to pass information from the planning program to the STRUCTURE GENERATOR. When the planning program decides, for example, that no ether could have produced a given mass spectrum, the ether superatom (C—O—C) is put on BADLIST. Thus BADLIST constrains the algorithm from generating molecules that are poor solutions for reasons of incompatibility with mass spectral or nmr data.

The STRUCTURE GENERATOR avoids generating structures containing forbidden substructures by checking rigorously before attaching new atoms to a piece of structure. At every step in generating a radical, the program determines whether the atom and bond about to be attached to the partially built structure will include one of the forbidden substructures. Usually only part of the structure has been generated because unallocated atoms are added only to stable pieces. It is not necessary to wait until the whole molecule has been built to check for instability. Naturally if there are no structures on BADLIST, then there are no constraints on the output of implausible structures.

4. *Specifying Required Substructures*

The basic components used by the STRUCTURE GENERATOR are chemical atoms and their two properties, valence and atomic weight. These atoms are connected by bonds to form radicals that may in turn be connected with other atoms and radicals to form larger radicals and molecules.

The STRUCTURE GENERATOR can also treat complex structural fragments as atoms. Structures so treated have come to be known as "superatoms." The STRUCTURE GENERATOR replaces a group of atoms in the given empirical formula by the name of a corresponding superatom, and generates structures with the revised composition (including the superatom name). Only at output time (if then) do the constituent atoms of the superatom reappear. Three benefits arise from the use of superatoms.

1. The generation of isomers of a composition is faster because there are fewer atoms to be assigned in the composition.

2. Structural fragments essential to an explanation of a mass spectrum may be made into superatoms. All isomers will contain these substructures, and thus the output list is more relevant to the data.

3. Ring structures can now be generated by the previously acyclic STRUCTURE GENERATOR by specifying each different ring as a superatom. This is one way Lederberg's algorithm (4, 8, 9) for generating cyclic structures can be implemented, for the ring systems can be generated first, as specified by their own classification rules, and then bundled up as superatoms to which substituents are attached.

5. *Rote Memory*

The STRUCTURE GENERATOR can also build a dictionary that contains all the structural isomers of every empirical formula (and every subset of every formula) that has been encountered. This dictionary contains lists of radicals, saved under names that can be reconstructed from the compositions. Whenever structure generation is under way, the program first searches the dictionary to see whether the current composition has been encountered previously. If a dictionary entry exists for a composition, it is assumed to be an exhaustive list of all radicals that can be made from the composition. No further structure generation is performed; the dictionary list provides the output. This simple device can obviously save much time.

D. Verification

1. *The PREDICTOR*

In the course of designing the Heuristic DENDRAL program for formulating hypotheses to explain mass spectral data in the general case, it became apparent that the program needed a detailed theory of mass-spectral fragmentation processes. Although the STRUCTURE GENERATOR can suggest plausible candidate structures for explaining the data, it has no way of testing its candidates. A theory by which the computer could make verifiable predictions about each candidate would reduce the set of likely candidates. [Contrast this with the efficiency gained by putting all the detailed theoretical knowledge (Sections B–3 and B–4) in the PRELIMINARY INFERENCE MAKER]. Testing serves the dual purpose of eliminating hypotheses that are inconsistent with experimental data and ranking all the remaining hypotheses. So it is in the program: the PREDICTOR makes a testable prediction for each candidate molecule; the prediction is checked against experimental data by a CONSISTENCY CHECK; and the remaining structures are ranked by the EVALUATION FUNCTION.

The prediction made for each candidate molecule is the set of important peaks to be found in its mass spectrum. When this is insufficient to differentiate candidates [e.g., in aliphatic ethers (Section B–3)], a predicted nmr spectrum is also used, as mentioned earlier. In broadest outline, the mass-spectrum predictor calculates a spectrum (list of mass-intensity pairs) for a molecule in the following series of steps:

1. Calculate the mass of the molecular ion and an associated intensity, depending on the degree of unsaturation.

2. Determine the nature and extent of eliminations and rearrangements of the molecular ion.

3. Break a bond between a pair of adjacent atoms in the molecule, looking at each bond only once.

4. Calculate the masses of the two resulting fragments. Then calculate an intensity (on an absolute scale) for each fragment ion by estimating the probability that this bond will break and the probability associated with ionization of each fragment. (Double and triple bonds are considered to have zero probability of breaking.)

5. Determine the nature and extent of eliminations and rearrangements from each fragment ion and each daughter ion.

6. Add isotope peaks and peaks at Mass + 1, Mass + 2, to account for hydrogen addition to some fragment ions (optional).

7. Recycle through 3–6 until every bond in the molecule has been considered once.

8. Adjust the intensities to percent of the highest peak.

The following discussion elucidates the theory by explaining briefly what the program does in each of the above steps.

Step 1. The molecular ion will appear in the predicted mass spectra of all molecules except alcohols. The intensity associated with the molecular ion is a linear function of the degree of unsaturation of the molecule. This function is easily changed, but it is currently twice the product of all bond multiplicities (on an absolute scale). For example, in a molecule with all single bonds except for one double bond and one triple bond, the intensity of the molecular ion would be $1 \times (1 \times 1 \times \cdots \times 1 \times 2 \times 3) = 12$. Peak heights on this absolute scale are translated onto a 0–100 scale at the end.

Step 2. The PREDICTOR contains rules for rearranging ketones, ethers, amines, and the like as described previously. It also will carry out 1,2-elimination of water (thermal) and 1,3- and 1,4-elimination of water, HCl, and H_2S from the molecular ion whenever possible.

Steps 3, 4. After the program has considered the molecular ion it calculates the likelihood that the molecular ion will fragment at each of the bonds between atoms. Its theory says that only single bonds will break, thus skipping double and triple bonds in the molecule. Of the single bonds, it distinguishes carbon-carbon bonds from all others, usually assigning a higher probability to their cleavage. The probability that an ion will break at a given bond depends on the environment of the bond and the functional groups present in the molecule. The probability associated with the ionization of one or the other of the resulting fragments also depends on these features. Briefly, the program first checks for the presence of special cases mentioned above. If none of the special case conditions holds, it then determines both the probability that any particular bond will break and the probability of ionization for each resulting fragment by referencing the eight features for the general case.

The general rules of mass spectrometry in the program apply to any molecule whose functionality has not been described in detail. For instance, since the program has no specific knowledge of the cleavage patterns and rearrangements of amino acids it applies the general rules, while it applies specific rules to, say, aliphatic ketones, ethers, and amines. As mentioned earlier, it looks at each bond and decides whether either fragment (or both) is ionized and which secondary processes may occur. It attempts to assign a relative abundance to each ion, but its rules for this are weak. Although trivial from an expert's point of view, the features the program considers are described briefly below.

a. The multiplicity of the bond: no double or triple bonds will be broken. The types of atoms joined by the bond: C–C bonds are usually considered more likely to rupture than others. With C–X bonds, the frequency of C–X cleavage depends on the identity of X.

b. The multiplicities of the α-bonds: the program will not break vinyl bonds, but favors α-cleavage next to carbonyl groups.

d. The degree of substitution of the atoms joined by this bond: the higher the degree on the substitution on the α-carbon atom the greater the probability of C–X bond cleavage; for example, R_3C—X is favored over RCH_2—X.

d. The number and types of nonalkyl substituents on the atoms joined by the bond under consideration: bonds α to a heteroatom are more likely to break, with some heteroatoms having more influence (e.g., N) than others (e.g., O and S).

e. The multiplicities of the β-bonds: the program favors cleavage of allylic bonds.

f. The length of alkyl chains lost by fragmentation: losing a longer carbon chain increases the relative abundance of the ionized fragment.

h. The total number of carbon and noncarbon atoms in each fragment: the program assigns a higher probability of ionization to larger fragments, all things being equal.

Step 5. Since fragment ions, as well as the molecular ion, may eliminate neutral molecules or radicals to form more stable ions, the program must be able to predict the most significant occurrences. After the program calculates the mass of a fragment and the relative frequency of its ionization, it checks the fragment for possible elimination and rearrangement processes. Exactly the same procedure is used as for the molecular ion, but the list of characteristic subgraphs may be different depending on the functional groups present in the molecule. The processes that are currently known to the program in detail are those described above for the specific functional groups examined so far.

Steps 6–8. Thus the program examines each bond to calculate the probability of cleavage, and each fragment to calculate both the probability of ionization and the possibility of rearrangements. In addition, it has already calculated a molecular ion peak and has looked for the possibility of eliminations and rearrangements in the molecular ion. By the time it has finished, it should have calculated a list of mass-intensity pairs corresponding to the most significant peaks in the actual mass spectrum for the same molecule. At the very end, intensities are converted to percent of the highest peak.

For example, the program predicts the following spectrum for 4-octanone:

[(43 . 100) (57 . 88) (58 . 22) (71 . 100) (85 . 88) (86 . 22) (128 . 14)]

The molecular ion has mass 128. The two other even-numbered peaks of high intensity, 86 and 58, are the results of the single and double McLafferty rearrangements. The peaks at 85 and 71 result from α-cleavages, in each case with only the heteroatom-containing fragment retaining the charge. The peaks at 57 and 43 come from loss of carbon monoxide (mass 28) from each of the α-cleavage fragments.

2. The EVALUATION Function

After candidate structures have been generated by the STRUCTURE GENERATOR, the program needs some way to attach a degree of plausibility to each one. The PREDICTOR makes predictions for each one; the program must now decide the extent to which the predictions confirm or disconfirm each candidate hypothesis. A strictly numerical evaluation function can score predicted spectra on the basis of how much they "cover" the peaks in the original spectrum without adding spurious peaks—perhaps weighting various kinds of failures. But a numerical evaluation fails to account for the higher theoretical significance of some peaks over others. Thus a nonnumerical evaluation function has been used.

By the time the EVALUATION routines are invoked, the PREDICTOR has already marked various kinds of cleavages and rearrangements as being very significant from a theoretical point of view. For example, the results of α-cleavage in ketones, amines, and ethers are put on a list named SIGNIFI-CANT, together with the results of other theoretically significant peaks in the predicted spectrum. Evaluation is a subsequent two-step process: reject any candidate whose predictions are inconsistent with the original data, and rank the remaining candidates.

For each candidate molecule, the program looks to see whether all significant predicted mass points are represented in the original spectrum or not. If there is a peak at mass point x and its intensity level is higher than the expected intensity level from an isotope peak (1% of the intensity of the $x - 1$ peak multiplied by the estimated maximum number of carbon atoms in the $x - 1$ peak), then the next significant peak is considered.

If the significant peak is not present in the original spectrum, then this candidate is rejected entirely. For example, if the predicted spectrum shows rearrangement peaks at the wrong mass points, it should not warrant further consideration, since the theory is strongly violated by that candidate. When the evaluation routine decides that the original spectrum shows a significant peak only because this is an isotope peak, the candidate is also rejected.

The second step of the evaluation routine is to rank the remaining candidates, each of which accounts for some of the nonisotope peaks in the recorded spectrum, but not necessarily all. Each molecular structure is assigned a score equal to the sum of the significance values associated with peaks in its hypothetical mass spectrum. The best candidate is taken to be the one that accounts for the peaks with the highest combined degree of signifi-cance. The other remaining candidates can be ordered by their scores.

IV. CONCLUSION

The collaboration of chemists and computer scientists has produced a tool of some practical utility from the chemical viewpoint and a most interesting program from the viewpoint of artificial intelligence.

The DENDRAL project does succeed at emulating the inductive be-havior of the scientist in a sharply defined area of organic chemistry. The process of scientific induction, always mysterious because of its close connec-tions with human creativity, is notoriously difficult to implement on a com-puter. The DENDRAL program provides an excellent model to test the concept that induction is a process of efficient selection from the domain of all possible alternatives. On this theory, a more creative thinker is a person who somehow is able to perceive options and relationships that do not occur to others, and who has efficient heuristics for selecting the ones that will lead to a solution of the problem at hand. The search and evaluations are necessarily heuristic for complex problems. Once a hypothesis has been formed (discovered) by these poorly understood processes to explain a given set of data, the simpler business of testing the explanations can then proceed. The tests amount to logically deductive predictions from the hypothesis (and theory) that can be matched against further data. When no other hypothesis is concordant with the data, we have discovered a scientific explanation (solved the problem). It is rare to be able to compile a list of hypotheses that is rigorously exhaustive. More commonly we simply try to exclude those that happen to be evident. It will be noted that Heuristic DENDRAL embodies all these features, and some of the descriptions of the project emphasize this perspective (7, 10, 22).

Outside the specific context of mass spectrometry, the DENDRAL research group continues to address itself to interesting computer science and philosophical problems (36, 37). For example, how can computer programs be designed to solve complex scientific problems efficiently and still maintain enough flexibility to allow easy changes as the underlying scientific theory grows?

Within the context of chemistry, the general research strategy has been to demonstrate the ability of the program to deal with classes of compounds that were conceptually simple but present interesting and different problems. Thus after delineating the size of the space of chemical structures in which the chemist searches for solutions to analytical problems (17), the program has been adjusted to work with the mass spectra of aliphatic ketones (23), ethers (31), amines (34), and then to the general solution of aliphatic compounds [including alcohols, thiols, and thioethers of the general formula $C_nH_{2n+v}X$ ($v =$ valence of X)]. Whereas the ketones could be handled efficiently using only mass-spectral data, ethers required nmr data to remove some of the ambiguities remaining after the mass-spectrum analysis. Aliphatic amines presented a still greater challenge because of the larger number of isomers compatible with any amine mass spectrum. For this class, the program, therefore, had to have very precise mass-spectrometric identification rules, and had to use the nmr data before generating structures. In fact, the program did not really have to generate structures at all. Other classes of compounds to be studied will, we hope, present still further challenges.

Clearly the internal structure of the Heuristic DENDRAL program has evolved since the time it was working only with aliphatic ketones. The amine (34) and subsequent generalized work (35) emphasized the PRELIMINARY INFERENCE MAKER by including all theory in this part of the program.

It remains to be seen whether this approach will continue to be successful with problems of greater complexity.

As can be seen from the tables of results, the program performs within a small margin of error for the functional groups studied. The correct structure is rarely excluded from the end list of possible answers; moreover, the correct structure is often the only answer that the program finds plausible. In some rough comparisons with graduate students and post-doctoral fellows in mass spectrometry at Stanford University, the humans took more than ten times as long to work some of the same problems as the computer, with no better results (7). Clearly the development and inclusion of more detailed heuristics in the program has resulted in its being superior to humans in the admittedly limited domain of the interpretation of mass spectra of aliphatic compounds of general formula $C_nH_{2n+v}X$ (X = N, O, S).

Extension of heuristic DENDRAL to polyfunctional and cyclic compounds is under way. Preliminary work (38) on programming and mass spectrometry of monocyclic compounds indicated many of the problems to be met in extending the DENDRAL programs. One of the biggest problems is the present rudimentary state of mass-spectrometry theory. Even for the functional groups already examined, many arbitrary guesses had to be made before the computer could predict the significant peaks in the mass spectra of quite small molecules. The discipline forced on one by computer programming to distinguish clearly between things known versus things unknown results in important dividends. Not the least of these is the identification of areas in need of urgent attention by organic mass spectroscopists, such as aliphatic aldehydes (39).

There is no wish to restrict Heuristic DENDRAL to low-resolution mass spectra and nmr spectra. Information of any structural kind could also be used as input—for instance, infrared spectra. Similarly, specific mass-spectral subprograms such as metastable maps, molecular ion identification routines, high-resolution input, and the like would all be additions readily accepted by the program. The DENDRAL project is an evolving and continuing one, and its state at the present time simply reflects the relatively few man-years of time that have been devoted to it.

It is clear that the feasibility of using computers for this approach to mass-spectral analysis has been established, even though the effort of making the program run for every conceivable problem has not been attempted. The techniques of artificial intelligence do seem to provide useful tools for problem-solvers in science.

REFERENCES

1. E. J. Corey and W. T. Wipke, *Science*, **166**, 178 (1969).
2. A. Cayley, *Ber.*, **8**, 1056 (1875).

3. J. Lederberg, "DENDRAL-64, A System for Computer Construction, Enumeration and Notation of Organic Molecules as Tree Structures and Cyclic Graphs. Part I, Notational Algorithm for Tree Structures," NASA Report CR-57029, December 15, 1964. Copies are available from the author.

4. J. Lederberg, "Topological Mapping of Organic Molecules," *Proc. Nat. Acad. Sci. U.S.*, **53**, 134 (1965).

5. J. Lederberg, "Topology of Molecules," in the National Research Council's Committee on Support of Research in the Mathematical Sciences, *The Mathematical Sciences*, M.I.T. Press, Boston, Mass., 1969, p. 37.

6. G. L. Sutherland, "DENDRAL-A Computer Program for Generating and Filtering Chemical Structures," Stanford Artificial Intelligence Project Memo No. 49, February 15, 1967.

7. B. G. Buchanan, G. L. Sutherland, and E. A. Feigenbaum, "Toward an Understanding of Information Processes of Scientific Inference in the Context of Organic Chemistry," in B. Meltzer and D. Michie (Eds.), *Machine Intelligence 5*, Edinburgh University Press, 1970 (also available as Stanford Artificial Intelligence Project Memo No. 99, September, 1969).

8. J. Lederberg, "DENDRAL, A System for Computer Construction, Enumeration and Notation of Organic Molecules as Tree Structures and Cyclic Graphs. Part II, Topology of Cyclic Graphs" (NASA Report CR-68898, December 15, 1965); *Part III, Complete Chemical Graphs: Embedding Rings in Trees"* (March 13, 1968). Copies are available from the author.

9. J. Lederberg, "Systematics of Organic Molecules, Graph Topology and Hamilton Circuits. A General Outline of the DENDRAL System," NASA Report CR-68899, January 12, 1966. Copies are available from the author.

10. J. Lederberg and E. A. Feigenbaum, "Mechanization of Inductive Inference in Organic Chemistry," in B. Kleinmetz (Ed.), *Formal Representation of Human Judgment*, Wiley, New York, 1968 (also available as Stanford Artificial Intelligence Project Memo No. 54, August 2, 1967).

11. M. F. Lynch, *Endeavour*, **27**, 68 (1968).

12. H. Hiz, *J. Chem. Doc.*, **4**, 173 (1964); S. H. Eisman, *J. Chem. Doc.*, **4**, 186 (1964).

13. O. Ore, *Graphs and Their Uses*, New Mathematics Library, Random House, New York, 1963.

14. C. Jordan, *Journal für die reine and angewandte Mathematik*, **70**, 185 (1869).

15. H. R. Henze and C. M. Blair, *J. Amer. Chem. Soc.*, **53**, 3077 (1931). See also D. Perry, ibid., **54**, 2918 (1932).

16. H. R. Henze and C. M. Blair, *J. Amer. Chem. Soc.*, **53**, 3042 (1931).

17. J. Lederberg, G. L. Sutherland, B. G. Buchanan, E. A. Feigenbaum, A. V. Robertson, A. M. Duffield, and C. Djerassi, *J. Amer. Chem. Soc.*, **91**, 2973 (1969).

18. J. McCarthy, P. W. Abrahams, D. J. Edwards, T. P. Hart, and M. I. Levin, *LISP 1.5 Programmer's Manual*, M.I.T. Press, Boston, Mass., 1966.

19. J. Allen, "PDP-6 LISP (LISP 1.6)," Operating Note No. 28, Stanford Artificial Intelligence Project, August 28, 1967.

20. B. Buchanan, G. L. Sutherland, and E. A. Feigenbaum, "Heuristic DENDRAL: A Program for Generating Exploratory Hypotheses in Organic Chemistry," in B. Meltzer and D. Michie (Eds.), *"Machine Intelligence 4,"* Edinburgh University Press, Edinburgh, 1969 (also available as Stanford Artificial Intelligence Project Memo No. 62, July 26, 1968).

21. E. A. Feigenbaum, B. G. Buchanan, and J. Lederberg, "On Generality and Problem Solving," in B. Meltzer and D. Michie (Eds.), *Machine Intelligence 6*, Edinburgh University Press, 1971.

22. J. Lederberg, G. L. Sutherland, B. G. Buchanan, and E. A. Feigenbaum, "A Heuristic Program for Solving a Scientific Inference Problem: Summary of Motivation and Implementation," Stanford Artificial Intelligence Memo No. 104, November 1969.

23. A. M. Duffield, A. V. Robertson, C. Djerassi, B. G. Buchanan, G. L. Sutherland, E. A. Feigenbaum, and J. Lederberg, *J. Amer. Chem. Soc.*, **91**, 2973 (1969).

24. H. Budzikiewicz, C. Djerassi, and D. H. Williams, *Mass Spectrometry of Organic Compounds*, Holden-Day Inc., San Francisco, Calif., 1967, Chapter 3.

25. F. W. McLafferty, *Anal. Chem.*, **31**, 82 (1959).

26. H. Budzikiewicz, C. Fenselau, and C. Djerassi, *Tetrahedron*, **22**, 1391 (1966).

27. W. Carpenter, A. M. Duffield, and C. Djerassi, *J. Amer. Chem. Soc.*, **90**, 160 (1968).

28. F. W. McLafferty, *Anal. Chem.*, **29**, 1782 (1957).

29. Reference 24, Chapter 2.

30. Reference 24, Chapter 6.

31. G. Schroll, A. M. Duffield, C. Djerassi, B. G. Buchanan, G. L. Sutherland, E. A. Feigenbaum, and J. Lederberg, *J. Amer. Chem. Soc.*, **91**, 7440 (1969).

32. R. S. Gohlke and F. W. McLafferty, *Anal. Chem.*, **34**, 1281 (1962).

33. Reference 24, pp. 297–303.

34. A. Buchs, G. Schroll, A. M. Duffield, C. Djerassi, A. B. Delfino, B. G. Buchanan, G. L. Sutherland, E. A. Feigenbaum, and J. Lederberg, *J. Amer. Chem. Soc.*, **92**, 6831 (1970).

35. A. Buchs, A. B. Delfino, A. M. Duffield, C. Djerassi, B. G. Buchanan, E. A. Feigenbaum, and J. Lederberg, *Helv. Chim. Acta*, **53**, 1394 (1970).

36. C. W. Churchman and B. G. Buchanan, *Brit. J. Phil. Sci.*, **20**, 311 (1969).

37. E. A. Feigenbaum, in *Artificial Intelligence: Themes in the Second Decade* (Proceedings of the 1968 IFIP Congress), Edinburgh University Press, Edinburgh, 1968.

38. Y. M. Sheikh, A. Buchs, A. B. Delfino, G. Schroll, A. M. Duffield, C. Djerassi, B. G. Buchanan, G. L. Sutherland, E. A. Feigenbaum, and J. Lederberg, *Org. Mass. Spectrom.*, **4**, 493 (1970).

39. R. J. Liedtke and C. Djerassi, *J. Amer. Chem. Soc.*, **91**, 6814 (1969).

Newer Ionization Techniques

HENRY M. FALES

Laboratory of Chemistry, National Heart and Lung Institute, National Institutes of Health, Bethesda, Maryland

I. Introduction 179
II. Field Ionization 180
III. Ion-Molecule Reactions (Chemical Ionization) 190
IV. Radiation Sources 204
V. Lasers 206
VI. Metastables 207
VII. Photoionization 209
VIII. Electrojet Methods 210
IX. Ion-Bombardment Methods 212
References 213

I. INTRODUCTION

In the vast majority of mass spectrometers* in use today as organic analytical instruments, ions are produced by passing a stream of electrons through the vaporized sample. The method has many very real advantages. Thus it is a simple matter to produce an intense beam of electrons by heating a filament, and the energy of the electrons may be easily varied by altering the electric field through which they are accelerated. The considerable ($\sim$1–2 eV) energy range that the beam encompasses due to the heated filament may cause difficulties in accurate measurement of appearance potentials, but it is unlikely that this factor is a disadvantage from an analytical viewpoint. The heat produced by the filament is a minor problem that can be largely overcome by proper source design. A more important disadvantage is that the processes by which energy is transferred during collision and the events leading to fragmentation are subject to theoretical

---

* Mass spectrometers, since they invariably are used to disperse ions, might be more properly referred to as *ion* mass spectrometers. This qualification may well become necessary in the future if practical methods are found to study dispersion of molecules according to mass alone.

interpretation in only the most trivial cases. Complex rearrangements of whole carbon skeletons occur, with hydrogen nuclei migrating over several bonds; neutral molecules are expelled from the centers of complex ions and, most annoyingly, several structures of the same empirical formula may be present at a given m/e value.

It is only through the extensive efforts of a wide group of chemists that the details of these processes are beginning to be described, much less explained on a theoretical basis. Fortunately, in many cases it appears that some localization of reactive site (radical or charge) occurs, so that within a given structural class enough guidelines may be derived to enable the chemist to predict fragmentation. The validity of these correlations is heavily dependent on the number of reference samples from which it has been derived, but unusual effects are noted almost daily in the literature, and accurate interpretation is still something of an art. For these reasons, the mass spectrometrist will be receptive to other approaches to ion production whenever they show promise of disclosing additional information through the control of new parameters.

In the following chapter some of the more recent methods of ion production have been investigated, particularly field and chemical ionization, with an eye towards their potential in the identification and structural elucidation of complex organic compounds.

II. FIELD IONIZATION

For several years, those of us interested in the uses of mass spectrometry for the elucidation of structures of complex organic compounds have been aware of the apparent potential of field-ionization mass spectrometry. The occasional appearance in the literature of spectra obtained in this mode, where intense molecular ions or $M + 1$ ions are visible, continues to entice the organic mass spectrometrist. For structural purposes, the importance of identifying the molecular ion distinctly can scarcely be overestimated.

It is clear however, that operation of a spectrometer in the field-ionization (FI) mode is not entirely routine, at least with commercially available apparatus. This author is aware of several commercial systems that have proved unsatisfactory as received. Furthermore, the technique is still in an early stage of development, as evidenced for example, by the enormous enhancement of intensity, heretofore a serious drawback of the FI mode, recently obtained by Beckey, who activates the emitter (1) with benzonitrile under carefully controlled conditions.

Nonetheless, the results that can be obtained with careful attention to detail are so striking as to encourage many to acquire equipment even in the present state of development, and the method merits serious consideration as an alternative to electron bombardment.

Field ionization is the outgrowth of early experiments in the area of field-emission microscopy, particularly by Müller (2) and Gomer (3) and their co-workers. In field *emission* (not field *ionization*) an electron is removed from a metal surface by the application of a high electric field. The reason for this may be understood by referring to Figure 1 (4), where the zero level of energy arbitrarily is set to correspond to an electron infinitely removed from the surface of a metal. Such abstraction requires application of a certain amount of energy corresponding to the "work function" ϕ ($\sim$3–5 eV for

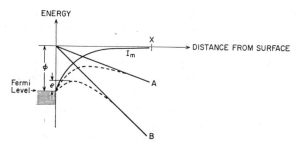

Figure 1. Potentials near a metallic surface.

most metals). To supply this energy, in the absence of an electric field, the metal must be heated to $\sim$1500°K. The work function is analogous to the ionization potential of a gas, and it may be further familiar to mass spectroscopists in connection with ion-source or vacuum-tube filaments, which often have their effective work functions lowered by the addition of small amounts of rare earths (Th, etc.), so that the required electron emission can be achieved at lower filament temperatures.

Although the physical boundary of the metal is abrupt, the potential at small distances from the surface does not rise as sharply because of two facts (5). In the first place, quantum mechanical principles suggest that there is a finite probability of finding the electron just outside the surface of the metal as well as within it, even without the application of extra energy. Second, the removal of an electron from an atom in the metal surface will leave behind a positive charge whose interaction with the electron will drop off with the square of the distance. These factors combine to lower the potential of an electron when it is very near the surface. The resulting "image potential" is given schematically by line *Im* in Figure 1.

If in addition an externally applied field is applied between the metal and some point X as represented by the linear potential gradient A (Figure 1), the resultant effective potential that an electron encounters near the surface of the metal is the combination of these several potentials as shown by the dotted line. Application of a field that is still stronger (B) may further lower the potential hill so that finally even ambient thermal energy (e) may be sufficient to allow an electron to be lost from the surface.

Due to the very small distance (10^{-8} cm) between an electron and the nucleus in an atom, the field required for such a field-induced dissociation event to succeed is strong indeed: $5 \text{ V}/10^{-8}$ cm or 5×10^8 V/cm, and in practice application of such intense fields is very difficult to achieve. One approach is to form the metal into a tip or edge by growing very fine "whiskers" of metal from a surface by the application of a high field in the presence of certain organic compounds. The field in the vicinity of the tip of such a whisker may be very large (although it falls off rapidly with distance), since it is equal to the applied voltage divided by the radius of curvature of the tip.

Fortunately, it it not necessary to apply the full field theoretically required for electrons to pass over the potential hill in Figure 1 in order to obtain electron emission from such a metal. Heisenberg's uncertainty principle dictates that an electron of given momentum inherently possesses finite probability of being found on either side of the resultant potential hill in Figure 1. Such emission occurs whenever the equation 2 $(2m/h^2)^{1/2}(\phi^{3/2}/Fe) \simeq 1$ holds (6), where m = momentum of the electron, h = Planck's constant$/2\pi$, ϕ = work function, F = applied field, e = electrostatic charge. The actual energy required for the so-called penetration of an electron "through" the energy barrier (tunneling) depends on the width of the hill that is equal to ϕ/Fe (7), so that lowering the work function of a metal also improves electron emission by tunneling.

If one applies a field of the required strength to an emitter tip under conditions such that the evolved electrons are registered on a surrounding phosphor screen, a highly enlarged image of the tip will be observed. Since the emission of electrons varies with the work function (ϕ) and local field intensity, which in turn varies along different edges of the crystal, a contrast outline of the crystal surface is observed on the fluorescent screen. This is the principle of the field-emission microscope.

When the polarity of the applied field is reversed, metal from the tip may be torn off as positive ions if the potential is high enough, even to the extent that tungsten will evaporate at liquid-hydrogen temperatures. It was during investigations such as these that Müller (8) noticed unusually sharp figures developing on the screen when hydrogen gas at 10^{-3} torr was admitted to the apparatus. By modifying the system so that the species responsible for the figures could be mass analyzed, Inghram and Gomer (9) showed that H_2^+ and H^+ were present. When helium was substituted for hydrogen, it became apparent that the figures being observed varied with the history and type of metal in the tip. It is considered that the helium atoms are field ionized in the vicinity of the positively charged tip and then accelerated along lines of force that reflect the local anisotropies of the tip itself. A magnified image of the tip is therefore produced on the fluorescent screen in a manner similar to field emission, only utilizing helium ions rather than electrons. Individual surface atoms of tungsten and platinum are easily resolved, and the ultimate resolution is reportedly better than 4 Å. No applications of the field ionization microscope to organic substances have been reported to date, but it would seem to be a fruitful area for research.

Formation of field ions in this manner occurs by a process wholly analogous to field emission, except that in the former an electron is withdrawn

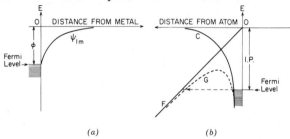

(a)   *(b)*

Figure 2. Potentials near an atom at a distance from a metallic surface.

from a molecule of gas (via a similar tunnelling process) by the high positive potential of a nearby metal tip and this is the same process utilized in the field ion-mass spectrometer sources. Figure 2a describes the *pseudo* image potential (Ψ_{im}) experienced by an electron in an atom as it approaches close to a metal surface (10). This effect is due to the nature of the charge distribution within the atom itself. The positive charge of the nucleus of an atom will induce a negative charge in the metal, and similarly the electrons of the atom will develop a positive charge within the metal. As long as the nucleus is some distance from the metal, the two charge effects will cancel. However, at the moment of actual transfer of an electron from an atom to the metal, the time interval involved will be too short for a nucleus to change its position. Under these conditions the interaction of the migrating electron with the nearby positive charge in the metal will reduce the energy required for electron transfer by a finite amount.

On the other hand, if an atom is far removed from the metal surface, the valence electron experiences the well-known Coulomb potential, part of which is represented by C in Figure 2b, where IP is the ionization potential of the atom. In the vicinity of an electric field, represented by the linear voltage gradient F, the Coloumb potential will be altered as shown in G. If the field is high enough, sufficient thermal energy may be available even at room temperature to allow a valence electron to tunnel through the potential hill, resulting in spontaneous ionization or autoionization of the atom or molecule (11). At lower fields the atom or molecule may approach close enough to the surface of the metal for the Ψ-image potential (Figure 2a) to add its effect (Figure 3), and an electron may now be found in the metal without having acquired energy sufficient to pass over the potential hill (tunneling), provided only that the ground electronic state of the atom is at least as high as the Fermi level (ground state) of the metal. Levels lower than this are already occupied and cannot accommodate the new electron. The critical distance (Xc) (12) at which this ionization may occur satisfies the relation $Xc = (IP - \phi)/F$, and for typical substances using moderate values of field, this distance is of the order of 5 Å.

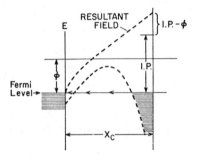

Figure 3. Potentials when an atom nears a metallic surface.

From this equation it can be seen that compounds with lower ionization potentials may approach closer to the tip surface before being ionized. Likewise, tips constructed of metals with higher work functions allow closer approach prior to ionization from the gas phase. Application of higher fields should also allow closer approach to the metal tip of a field ion source, but experimentally it is found that the opposite is true, since under these conditions autoionization of the molecule occurs at distances of as much as 100 Å from the tip.

If in addition the vapor is polar in nature, condensation near the tip may occur at the critical distance (Xc) due to polarization forces. It has been calculated (13) that under these conditions of electric field, an increase in density by a factor of 10^7 will occur for water at 23°C, so that even if the pressure in the ion chamber were only $\sim 10^{-3}$ torr, condensation would occur. Ionization of a second substance in the milieu of such a film will be particularly facile and should take place at much lower applied fields since (a) the ionization potential is lowered by $1/k^2$ where k is the dielectric constant of the film and (b) ionization may take place via ion-molecule reactions with ions already present within the film itself. The latter process is apparently the source of the $M + H^+$ and $M + H^+ \cdot (H_2O)_n$ ions so often encountered in FI spectra. Under conditions of film formation, lower applied fields often suffice to cause field ionization, and it has been observed that even mechanically weaker metals such as silver may be utilized as satisfactory FI emitters (14) without being destroyed.

In general, then, two types of processes may be expected under field-ionization conditions. The first originates from molecules undergoing ionization and subsequent field-induced dissociation at some distance from the tip, under the application of high fields. Such processes are obviously unimolecular, as is the case in the usual electron-impact sources. On the other hand, at lower field, especially with more polar compounds, these processes will be less prevalent than those due to reactions taking place at the surface of the thin film surrounding the field-ionization tip or blade. Here one will observe the results of intermolecular processes and the spectra may resemble those observed in chemical ionization (ion-molecule) mass spectra.

Fortunately, several techniques are available for distinguishing among these phenomena besides the obvious increase in mass provided by attachment of species from the film. In the first place, most FI spectrometers are constructed so that the final accelerating voltage is achieved by the addition of fixed retarding potentials placed in the flight path of the ion as it is repelled by the high positive voltage on the emitter tip. Since the field drops sharply in the vicinity of the tip due to its shape, any process that occurs at some distance from the tip ($\sim$100 Å or more) (whether it be due to autoionization and dissociation, or fragmentation that has been delayed for a time after the molecule was ionized near the tip surface) will result in the particle experiencing less total accelerating voltage. Ions from such events will therefore be displaced toward lower mass on the mass scale. At very high fields, it may be expected that *all* molecules will autoionize before reaching the tip, and a series of broad peaks tailing strongly toward lower masses will be observed unless some effort is made to collimate the ions according to energy.

Apparently, no general rules have evolved correlating peak shapes (tailing) with chemical structure under FI conditions; thus methanol shows tailing from the molecular ion but not from the fragment CH_2OH^+ (15), suggesting that the latter ion was formed by fragmentation occurring very near the tip. On the other hand, in the case of ethanol (16), the molecular ion and $M + H^+$ ions are sharp, and tailing *is* observed in the CH_2OH^+ fragment ion, indicating that decomposition has been delayed for some time after ionization. Beckey has termed ions leading to such tailing "fast metastables," since it has been calculated that they should result from decompositions occurring within $\sim$5 $\times$ 10^{-14} sec (roughly the period of a C–C vibration) after ionization (17). They are thus distinguished from ordinary metastables decomposing $> 10^{-10}$ sec after ionization, which are observed as usual at $m^* = m_d^2/m_p$ where m_d is mass of the daughter ion and m_p is mass of the parent ion. Recently Beckey and his co-workers (18) have demonstrated how such "fast metastables" can supply very graphic evidence confirming the statistical theory of mass spectrometry, since lifetimes derived from the peak tail show a continuous variation, and discrete states are not observed. Conversely, when temperature studies of such peaks are made, a qualitative picture of the distribution of internal energies transferred to the ions by the FI process can be obtained using the statistical theory of mass spectra to supply rate constants. Catalytic processes occurring at the tip can be identified by varying the temperature of the tip itself (19): an increase in temperature should reduce $M + H^+$ ions and other ions that result from fragmentation processes occurring in adsorbed films near or on the tip. Changing the tip to a less catalytically active material may also reduce such effects. Finally, application of higher fields may remove the initial ionization site to a position more remote from the tip. Under these conditions, peaks due to reactions of the molecular ion with an adsorbed film [e.g., formation of N_2H^+ from N_2^+ and an H_2O film (20)] may disappear.

While there is a great temptation to study catalytic processes using the field-ionization technique, Beckey has pointed out (21) that caution must be used in extrapolating results to field-free conditions, since application of high electric fields can profoundly alter chemical reactions. On the other hand, he proposes that "high-field chemistry" be further investigated with this tool.

Beckey, in extensive applications of field ionization to mass spectrometry, noted that $C_2H_5^+$ is the most intense ion in the FI mass spectra of hydrocarbons. In fact, it was found to be slightly more intense than the molecular ion for C_6–C_8 n-alkanes when platinum and gold tips were used (22, 23). He assigned this peak to field dissociation in the gas phase, rather than a process involving ionization and decomposition at the tip followed by desorption of the fragment ion under the influence of the field. His theory is supported by the fact that gold and platinum give identical results. The latter might be expected to be more active if a catalytic adsorption process were involved.

From an analysis of the peak shape of the $C_2H_5^+$ ion as well as other strong FI ions, Beckey concludes that ions that are intense in FI spectra tend to be formed by dissociation very shortly *after* field ionization near the tip. The driving force for this dissociation is thus the action of the electric field on the ion rather than being due to catalytic reactions on the tip followed by desorption of the product ions under the influence of the applied field. Beckey further calculates the charge distribution (24) that would occur with n-alkane ions as they align themselves along the lines of force at the tip of an electrode in the presence of a high electric field. His calculations imply that dissociation occurs most easily at the place where the positive charge is most readily localized. In the case of n-alkanes, this is between the second and third carbon atoms most remote from the emitter tip (Figure 4) and is in accord with the observed spectra.

As mentioned previously, in order for a peak to be observed in FI, dissociation must occur rapidly ($\sim 10^{-14}$/sec) after ionization. This is different from conditions existing in electron impact (EI) sources where dissociation is initiated by electrons of 70 eV. In such sources, residence times of up to 10^{-6} sec may elapse in the ion chamber before acceleration. In this case, the spectra obtained reflect the fact that sufficient time has elapsed to produce more nearly statistical energy fluctuations, so the locus of the initial positive charge is less easily found.

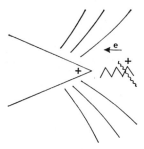

Figure 4. Alignment of a hydrocarbon near a field ion tip.

Beckey gives several rules (25) concerning the nature of the orientation of molecules in the vicinity of the tip and the type of ions to be expected. Comparison of molecules such as pentanol and di-n-propyl ketone show that factors such as field strength and temperature must be included in any analysis. Dissociation of a bond under the influence of the field (field dissociation) is brought about by polarization of the bonding electrons and occurs only if the positive charge may be located at a bond remote from the tip; otherwise the field acts to stabilize the bond. For example, in the case of di-n-propyl ketone, orientation of the carbonyl group in the vicinity of the tip due to the C–O dipole moment localizes the charge at the carbonyl carbon rather than at the C_2 carbon atom, so dissociation of the C_2–C_3 bond does not occur. In the case of n-pentanol, at low temperature the largest peak is that due to $C_2H_5^+$, because the OH group orients toward the tip, allowing charge to develop at the penultimate carbon atom remote from the tip as in the case of n-alkanes. As the temperature is increased, the pentanol molecule will more often be aligned in a position contrary to the field and parallel to the tip surface. The positive charge may not develop at the C_2–C_3 bond, and fission at other locations may occur. In fact, the abundance of the resonance stabilized $CH_2\!\!=\!\!OH^+$ ion is greatly increased.

In the case of carbohydrates (26), Krone and Beckey call attention to the peaks at $(M + 1)^+$, which are even more intense than those at M^+. In this case the bulk of the ions are formed in the adsorbed state where proton transfer can easily occur. The spectra are similar to those obtained by chemical ionization methods, although much less energy seems to be involved, since less fragmentation is observed. Good spectra were obtained from the disaccharide octamethyl-cellobiose. In this case the second most intense peak was due to the molecular ion.

Beckey (27) has recently shown that by actually dipping the highly activated emitter tip in a glucose solution, field desorption of ions, presumably formed on the tip itself, is possible. The ion at $(M + 1)^+$ was the most intense ion in the spectrum, with the dehydration peaks normally present under FI conditions nearly absent. This technique suggested also by Beynon (28), has enormous potential and further developments are impatiently awaited.

Wanless and Glock (29) have compared FI spectra with EI for a series of aliphatic hydrocarbon isomers as well as olefins and thiols, calling attention to the use of metastables for identification purposes. The "gain" in intensity of the molecular ions was also very clear, as expected from the grossly different energies imparted to the molecules by field ionization ($\sim$1–2 eV) and electron impact ($>$10 eV). In such comparisons care must be taken to specify the temperature of the ion source, since merely by operating at low temperatures in the EI mode, Spiteller (30) has observed that $n\text{-}C_{30}H_{62}$ shows the molecular ion as the second most intense peak in the spectrum!

Beckey has proposed (31) and modified (32) a rule concerning interpretation of differences between FI and EI spectra based on the aforementioned differences in time elapsing between ionization and acceleration. Thus little or no activation energy is required for dissociation of a C–C bond at 10^8 V/cm, and simple cleavage can occur within the residence time available near the tip (10^{-14}–10^{-12} sec). On the other hand, rearrangement processes, because of their larger entropy requirements, will be observed only as broad metastable ions in FI, since sufficient time will be available only during the relatively long flight time after acceleration. In comparing EI and FI spectra, a peak absent or of very low relative abundance ($\sim$0.03%) in FI, but relatively intense ($\sim$20%) in EI, may indicate the existence of a rearrangement process, while a peak of greater intensity in FI ($\sim$5%) will usually be due to direct bond rupture. Thus, ethyl benzene loses a methyl group in both FI and EI, but the product is probably the rearranged tropylium ion only in EI.

For similar reasons it is considered that the rupture of two bonds (e.g., the McLafferty rearrangement) is very unlikely to result in an intense ion in FI, although a metastable may be seen for the event. In actual fact, many such ions are present in the FI spectra of polar materials, but in these cases it is considered that the first bond is broken while the compound is in an absorbed film and the second is broken during field desorption. It may well be that ions from such film-controlled processes will also be found to be relatively intense in the chemical-ionization (CI) spectra of such compounds. To mimic the FI process even more closely in chemical ionization, it would be desirable to utilize the compound under investigation as its own reactant gas, since this is often the condition occurring at the tip in FI.

Application of FI to nucleosides appears very promising, since intense M^+ and $(M-H)^+$ as well as $(base + H^+)$ and $(base + 2H)^+$ fragments are observed (33). Steroid acetates show large $(M-HAc)^+$ peaks (34), although this fragmentation is less pronounced than under EI conditions. In fact, these spectra are very similar to those we have observed for similar compounds run in the chemical ionization mode using CH_4. Chlorinated insecticides, on the other hand, do not produce intense $(M + H^+)$ ions in chemical ionization with methane, but Damico et al. (35) have found them to provide quite satisfactory molecular ions in an FI source. These authors also call attention to the absence of ions representing loss of CO in FI, perhaps because insufficient time is allowed for rearrangement reactions.

Simple aliphatic amines also show large $(M + 1)^+$ ions as well as the expected $CH_2NH_2^+$ ions which, as in the case of fragmentation of n-pentanol to give CH_2OH^+, has been related to orientation of the molecule on the emitter surface (36).

In a study of alcohols, Beckey and Schulze (37) again observed $(M + 1)^+$ and $(2M + 1)^+$ ions due to polymerization processes. They further distinguish between "quick" and "slow" losses of water, the latter being evidenced by intense metastable ions presumably arising from a rearrangement process involving greater steric requirements. It would be interesting indeed to pursue these experiments with specifically deuterium-labeled materials to see if the two mechanisms could be disentangled.

Methyl esters (38) apparently show the molecular ion almost exclusively in FI, while hydroxy esters also show $(M - 18)^+$ ions as well as various intense metastable peaks.

Gomer has pointed out (39) that field-ionization sources may allow the study of excited states by adjusting the tip voltage so that field ionization will occur only from an excited state produced through irradiation of the sample. Knowing the velocity of the particles from the applied voltage, the lifetime of an excited state might be calculated.

Recently Beckey (40a) has compared the FI spectra of a series of isomeric decanes with their chemical ionization spectra. Invariably, there was more molecular ion intensity in FI than quasimolecular ion $[(M-1)^+]$ in CI, but it was apparent that no series of rules governing the observed peak intensities could be formulated for FI as they have been for CI (40b). The decanes are a somewhat limited class of compounds however, and it would be interesting to compare a wider series of compounds to delineate the advantages and disadvantages of both techniques.

In summary, it appears that FI spectra obtained in different laboratories, depending as they do on field strength, emitter style and material, emitter history (poisoning and conditioning), as well as source and emitter temperature, may be somewhat difficult to compare with one another and especially with EI spectra. This should not diminish their utility, particularly in ascertaining the molecular weight of an unknown. The determination of correlations between fragmentation pathways and molecular structure at present seems to be even less straightforward than electron impact, since catalytic processes and ion-molecule reactions are involved.

III. ION-MOLECULE REACTIONS (CHEMICAL IONIZATION)

In principle at least, converting a molecule to an ion by collision with an externally prepared ion would seem to offer many advantages. Thus, the heat of formation of many ions is known with accuracy, and their reactivities are either known or can be predicted with some confidence. Cross-sections for their ionization of other molecules can be high, and they may be allowed to react in the ion chamber for relatively long periods of time compared to EI, affording maximum conversion.

On the other hand, ensuring that a given ion will react with an unknown substance presents certain difficulties. Commonly an electron beam is used as a primary source of ionization to prepare the desired reactant ion, and it is usually necessary to use a relatively high pressure of gas in order to produce sufficient primary ions. For this reason some ions may not be easily accessible in high concentration. For example, the primary ions of methane, CH_4^+ and CH_3^+, are rapidly consumed by reaction with methane at 1 mm to produce CH_5^+ and $C_2H_5^+$ (41). Furthermore, when more than one ion is produced in the plasma, it is difficult to determine which is responsible for ionizing the substance under analysis. The most obvious way out of this situation is also the most technically complex. It consists of producing primary ions in one mass spectrometer and then allowing the beam at a selected m/e to enter the ionization chamber of a second analytical spectrometer where it encounters the vapor of a substance under analysis. Besides all of the problems associated with running two spectrometers at once, it is apparent that it will be difficult to produce intense ion currents by this method. Such tandem spectrometers, using a retarding voltage to slow down the ions as they exit from the first spectrometer, have been designed in all states of complexity (42a).

More recently, ion cyclotron resonance (ICR) spectroscopy has added a new dimension to such studies. In this technique (42b), the ions are produced in the usual way by electron impact, but instead of being forced out of the source region and dispersed according to mass, they are selectively made to drift into a region where they orbit at the cyclotron frequency $\omega = eH/mc$ where e = charge, H = magnetic field, m = mass, and c = velocity of light. Under these conditions the particles can be studied according to mass by noting the adsorption of energy from an electric field oscillating at frequency ω.

Ion-molecule reactions are studied in this apparatus by noting any variation in the spectrum as the pressure is changed in the cyclotron region. Furthermore, by applying a second radiofrequency field fixed at the resonance frequency of a given ion, this ion may be "heated." Variations in its reactions with the surrounding gas may then be easily studied as the first radiofrequency field is scanned. Very complete pictures of the overall ion-molecule reactions may be obtained, and the main drawback of the method at present is that the mass range seems limited to ~ 200. A second very disturbing feature of the method is that abundances of isotopic ions are incorrect, that is, $^{35}Cl/^{37}Cl = 4.8/1$ instead of $3/1$ (42a). At higher pressures this is apparently due to charge-transfer reactions between the parent ions, but at 10^{-7} it is ascribed to an unknown experimental effect. From the standpoint of structural analysis, the method differs from chemical ionization in that the ions under analysis are formed at very low pressures as in EI spectrometry. The effect of high pressure in CI is achieved in ICR through the very long effective path length (10–100 m) travelled by the ions as they circle in the cyclotron region, where collisions become numerous even when the gas is at 10^{-5} torr. Analytically, one might study either the properties of an unknown ion with known gases in the cyclotron region or vice versa. Except for the above-mentioned mass limitations, the method seems to combine the advantages of both tandem and CI methods.

As an analytical tool, the tandem spectrometer (but not ICR) is prohibitively complex, and in recent years several investigators, notably Field and Munson (43) and their co-workers, have advocated the use of a relatively simple modification of a standard mass spectrometer for the study of ion-molecular reactions. In this design, the ion source is enclosed so that the reactant gas, mixed with the compound under investigation, can escape only through two small orifices: an electron entrance hole (0.05×3 mm) and an ion exit slit (0.05×5 mm). Since considerable gas escapes even with orifices of this size, the ion-source housing must be provided with sufficiently powerful roughing and diffusion pumps to prevent arcing of the accelerating voltage.

Furthermore, the filament must be protected from reactive gases, and the pressure must be low enough to prevent loss of ions through their collision with escaping gas in the region in front of the ion chamber and in the analyzer. Differential pumping between the source and analyzer is imperative. A source of this design has been built for our AEI-MS902 spectrometer by Marvin Vestal of the Scientific Research Instrument Company of Baltimore, Maryland.

In a source of this type, a mixture of ions is formed, depending on the nature of the reactant gas. The reactions of these ions with an additive may be studied by noting variations of their intensity with pressure or other variables. Such a system has been termed "chemical ionization" by Munson and Field (44) to accentuate the essentially chemical nature of the process, that is, electron, proton, and alkyl transfer between molecules and ions. It is not "chemi-ionization," a process that involves metastable atomic states (see below); the emphasis in chemical ionization (CI) is on the interaction of low-velocity ions and molecules. Although it is necessary to understand any variation in the types of ions that are produced by the reactant gas at different pressures, attention is focused on the mass spectra of an additive, usually present only in trace amounts, as it is ionized and fragmented by the action of the reactant ions. For this reason, the mass scale is usually reproduced only from the point at which intense reagent gas ions cease to be observed. In fact, much of the ion current is due to ions of the reagent gas itself, and for this reason most ion beam monitors are quite useless; in the case of the MS-902 this meter is pinned in CI experiments. Vaporization of a sample is therefore best monitored through short scans on the U.V. recorder or oscilloscope.

In the pressure range commonly encountered in such a source (0.1–2 mm), gases are good conductors and it is therefore necessary that the reagent gas in the source at 8 kV be insulated from ground. We have accomplished this simply by admitting the gas to the source through a capillary leak in a glass tube, the higher pressure side at ground potential being maintained at $\frac{1}{2}$–5 atm. At this pressure an arc is not sustained and insulation is accomplished.

Most of the early work in chemical ionization spectrometry involved studies of the spectrum of methane at higher pressures (45a) as well as H_2 (45b, 46) and (rare gas + H_2) (47) systems. This work was logically extended first to studies of higher hydrocarbons and then simple aliphatic alcohols, esters, aromatic hydrocarbons, and cycloalkanes. Work in this field has been reviewed in several places (43, 44, 48–50).

In general, three types of process may be observed, depending on the reagent gas: charge exchange, proton transfer, and alkyl transfer. Rare gases such as He, Ne, Ar, Kr, and Xe have rather high ionization potentials,

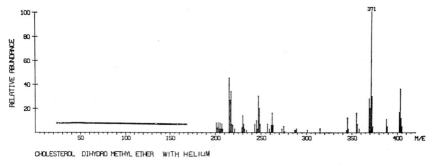

CHOLESTEROL DIHYDRO METHYL ETHER WITH HELIUM

Figure 5. Dihydrocholesterol methyl ether, using helium as a reagent gas.

and about the only process one expects is that of charge exchange followed by rather extensive fragmentation of the molecular ion. For molecules added at very high dilution to rare gases such as helium, this is apparently true (51). But with larger samples of molecules such as dihydrocholesterol methyl ether (Figure 5), the situation is not nearly as straightforward. Thus the largest ion in the molecular weight region is m/e 403 or $(M + 1)^+$, just as it is in the case of methane (Figure 6). The peak at m/e 402 (M^+) due to charge exchange is more intense than with methane, as expected, but the source of $(M + H)^+$ ions is puzzling. The presence of water has been minimized, and a peak at m/e 18 is barely visible under EI conditions. The only source of protons then, is the compound itself, and it must be concluded that, although the steroid is initially ionized by charge exchange with helium, secondary collisions with other steroid molecules occur as the substance progresses through the ion chamber causing proton transfer reactions to occur. The result is that the spectrum strongly resembles that obtained using methane.

When a strong base such as cassine is used, proton transfer becomes even more marked, and the ion at m/e 298 $(M + 1)^+$ is the strongest peak

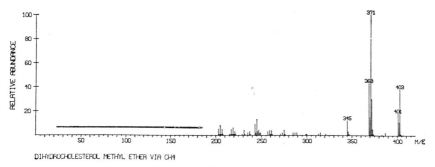

DIHYDROCHOLESTEROL METHYL ETHER VIA CH4

Figure 6. Dihydrocholesterol methyl ether, using methane as a reagent gas.

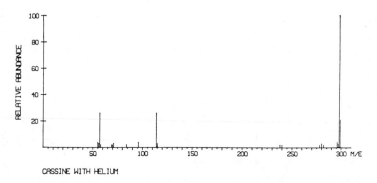

CASSINE WITH HELIUM

Figure 7. Cassine, using helium as a reagent gas.

in the spectrum (Figure 7) just as when methane is employed (Figure 8). Other differences between these two spectra are as yet unexplained. These spectra serve to emphasize the fact that an ion formed in such a source must undergo many collisions before escaping, not only with the reagent gas, but also with other molecules of the same compound. For this reason it is anticipated that isotope-labeling studies will be very difficult to perform. A

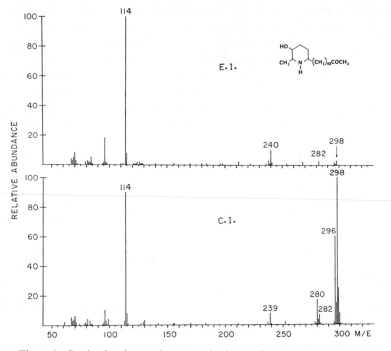

Figure 8. Cassine by electron impact and using methane as a reagent gas.

corollary of these observations is that it will be very difficult to obtain pure charge-exchange spectra from such a source.

Proton-rich gases, then, tend to form ions that transfer protons to other molecules. This is perhaps to be expected in the case of the very strong Brønsted acid CH_5^+, but the same reaction also occurs in the case of the t-butyl ion, forming isobutene as a reaction product. Nonetheless, the t-butyl ion, generated from isobutane or neopentane by electron impact, is considerably less acidic than CH_5^+, as has been demonstrated by Field (52, 53) using benzyl acetate. At $\sim100°$ the protonated benzyl acetate is the most prominent ion using i-butane, whereas the benzyl ion is the most intense if methane is used. At $\sim200°$ the spectra were similar with both gases and nearly the same as with methane at low temperatures. This is presumably a consequence of the fact that in a higher pressure source, sufficient collisions occur after ionization to allow the various ionic species to be at thermal equilibrium with the surrounding gas. The same cannot be said for electron-impact modes where the ion is rapidly expelled (10^{-6} sec) after formation. This facet of CI spectrometry allows its use in the determination of thermodynamic parameters through the use of Hammett plots and the like, and this appears to be an especially fruitful area for study. On the other hand, it is true that more complex, higher-molecular-weight compounds require high temperatures merely for their volatilization. Any differences effected by comparative use of methane and isobutane will be minimized under such conditions.

The extensive studies by Field and his co-workers on the action of methane-produced ions has resulted in the ability to predict quite accurately the nature of spectra obtained with this gas. In general, only in the case of normal and branched chain hydrocarbons are peaks observed at $(M - 1)^+$, and it is possible that they result from initially formed $(M + 1)^+$ ions, which rapidly lose H_2. Even in the case of cycloparaffins, ions are observed at $(M + 1)^+$, and in molecules of any complexity, $(M + 1)^+$ ions nearly always predominate. Subsequent losses, usually of neutral molecules, then occur to maintain the even-electron character of the initial ions. Intense odd-electron ions are rare, far more so than in electron impact.

Fewer peaks are present in CI spectra, and one usually observes clusters differing by two mass units, rather than one, for the above reason. In many stable systems containing nitrogen, keto groups, or aromatic rings, such as the tranquilizer diazepam (Figure 9), the quasimolecular ion, $(M + C_2H_5)^+$ and $(M + C_3H_5)^+$, will be the only ions observed above 1% intensity. In the case of quinine (54) (Figure 10, satellite peaks removed), the only other ion of importance is $(M + H - H_2O)^+$. This spectrum is typical of most basic drugs containing an alcoholic hydroxyl group.

Usually the methyl ion is totally consumed by the large excess of methane at ~1 mm, and as a rule no ion is observed at $(M + 15)^+$ *at this pressure.*

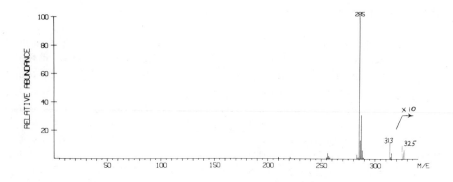

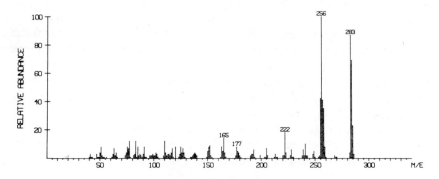

Figure 9. Diazepam by electron impact and using methane as a reagent gas.

The presence of such ions in a spectrum strongly suggests a homologous impurity.

In EI spectrometry, one of ie main reasons for the failure of a molecule to exhibit a molecular ion is associated with cleavage alpha to heteroatoms, one of the fragments being stabilized by electrons from the adjacent heteroatom:

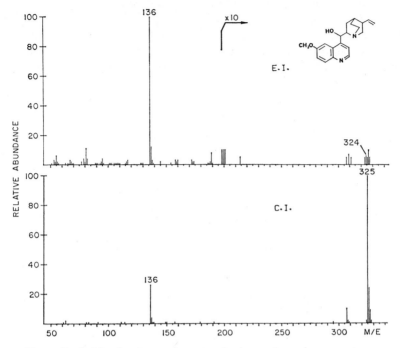

Figure 10. Quinine by electron impact and using methane as a reagent gas.

Any source of electrons will suffice, and a similar process is responsible for fission adjacent to benzyl and allyl groups.

$$\left[R_1-\underset{\underset{R_3}{|}}{\overset{\overset{R_2}{|}}{C}}-\overset{\overset{R_4}{|}}{C}=CR_5R_6 \right]^{+\cdot} \longrightarrow R_1{}^\cdot + \left[\overset{\overset{R_2}{|}}{C}\equiv\overset{\overset{R_4}{|}}{\underset{\underset{R_3}{|}}{C}}-CR_5R_4 \right]^{+}$$

In the above cases it is assumed that an electron has been lost from the most electronegative center by bombardment and that charge is localized mainly on the heteroatom-containing moiety, which is therefore observed in the spectrum. However, factors affecting stabilization of the radical $R_1^\cdot$ will also considerably facilitate cleavage. An extreme example is seen in the case

ephedrine

of ephedrine (Figure 11), where adjacent carbons *each* contain a heteroatom and, in addition one carbon is attached to an aromatic ring that helps to stabilize the radical $[C_6H_5CHOH]^{\cdot}$. As a result an intense ion is observed at m/e 58 (Figure 11), dominating the spectrum. In these cases the transition is from an odd-electron molecular ion to the combination of a radical and an

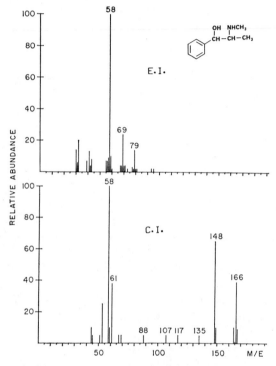

Figure 11. Ephedrine by electron impact and using methane as a reagent gas.

even-electron ion, that is, division of radical and charge on two separate fragments.

In CI with methane a different situation exists. The odd-electron character of the initial ion is lacking, and decomposition will take place in a manner depending on the energy transferred by the reagent ion and *the site at which it becomes attached*. It appears to be useful to consider that several sites are in fact protonated by the CH_5^+ ion in each case, although this is by no means proven. Thus, in the case of ephedrine, protonation of the nitrogen atom would be followed by loss of CH_3NH_2 to give the ion at m/e 135, but its intensity is relatively weak, since the only stabilization possible is from the adjacent oxygen or the aromatic ring. On the other hand, protonation of the hydroxyl leads to loss of water, the resulting ion being stabilized by both the ring and the adjacent nitrogen atom. The intensity of this ion, at m/e 148, is therefore high.

$$
\begin{array}{c}
\overset{\displaystyle H_2O^+}{|} \quad \overset{\displaystyle NHCH_3}{|} \\
Ar-CH-CH-CH_3
\end{array}
\longrightarrow
\begin{array}{c}
\qquad \overset{\displaystyle NHCH_3}{|} \\
Ar-\overset{+}{C}H-CH-CH_3
\end{array}
\longleftrightarrow
+\left\langle \bigcirc \right\rangle =C-\overset{\overset{\displaystyle NHCH_3}{|}}{}CH-CH_3
$$

$$
\begin{array}{c}
H \diagdown \quad \nearrow CH_3 \\
\underset{\overset{\displaystyle +}{N}}{} \\
Ar-CH\text{—}CH-CH_3
\end{array}
$$

The most intense ion, at m/e 58, is probably formed as a result of protonation of the ring:

$$
H^+\left\langle \bigcirc \right\rangle -\overset{\overset{\displaystyle OH}{|}}{C}H-\overset{\overset{\displaystyle CH_3}{|}}{C}H-NHCH_3 \longrightarrow \overset{H}{\underset{H}{}}\left\langle \bigcirc \right\rangle =CH-OH + \overset{\overset{\displaystyle CH_3}{|}}{C}H=\overset{+}{N}H-CH_3
$$

Thus the most prominent process takes place via protonation at the least basic site, emphasizing that the stability of the resulting ion, that is,

$$
\overset{\overset{\displaystyle CH_3}{|}}{C}H=\overset{+}{N}H-CH_3
$$

is the most important factor; in other words, protonation of the ring is more often followed by decomposition than is protonation at other sites. Nearly all of the primary events in CI mass spectrometry may be visualized by this approach. Yet, secondary processes also occur, that is, loss of a mole of water *after* loss of another fragment. In the case of ephedrine, loss of water from the ion at m/e 135 (loss of CH_3NH_2) leads to the ion at m/e 117. These processes require formation of new bonds in both charged and uncharged fragments, and the entropy requirement of these processes should be higher. Since the rates of such processes should be slower, metastable ions can be expected. In actual fact, many metastables are observed in CI spectra, even for primary processes.

Results on a wide variety of amines and other compounds (55) show that far less α-cleavage is to be expected in CI than in EI, and this is probably one of the most important features of the method.

Loss of other small molecules is common—for example, CH_3OH, C_2H_5OH, NH_3, H_2S, and $HCOOH$—but to date we have not observed loss of CO_2 or NO_2. In the case of certain dimers, loss of a monomer (56) can be observed, and the degree to which it occurs can often be associated with steric factors in the original compound.

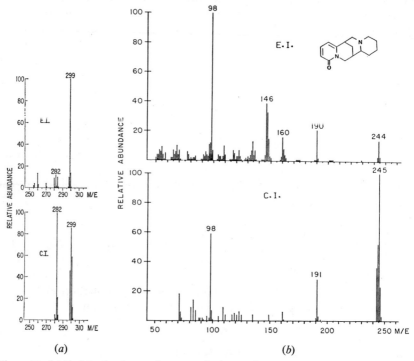

Figure 12. (a) Codeine by electron impact and using methane as a reagent gas. (b) Anagyrine by electron impact and using methane as a reagent gas.

Neither have we observed an ion that could be ascribed to $(M + 1 -$ methyl$)^+$, $(M + 1 - $ alkyl$)^+$, $(M + 1 - $ hydroxyl$)^+$, or the like, and the presence of such a species seems to be good evidence for a contaminant. Perhaps the one exception to this is the occasional appearance of an ion at M^+ that may be mistaken for an EI molecular ion. Considering the pressure in the source, it seems unlikely that such ions are in fact due to electron impact on the compound itself, but are rather due to charge-exchange. Two noteworthy examples are codeine (53) and anagyrine (54) (Figures 12a and 12b, respectively).

Chemical ionization methods with methane seem to be ideally suited to studies of drugs and drug metabolism because these materials typically involve many heterocyclic systems containing nitrogen. Such drugs are usually relatively stable since they must withstand storage, and yet they are reasonably small in molecular size since they must penetrate vital membranes. Figures 13 and 14 compare CI and EI spectra of a common barbiturate, butalbital (57) on one hand and the tranquilizer, promethazine, on the other. The advantages and disadvantages of both methods are immediately apparent.

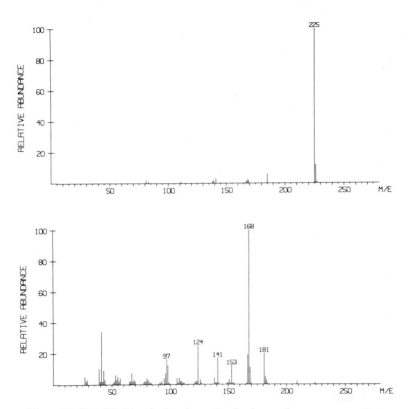

Figure 13. Butalbital by electron impact and using methane as a reagent gas.

It cannot be assumed that such clear-cut definition of the molecular weight will always be observed with methane as a reactant gas, however. Figure 15 shows cholesterol acetate, and the loss of acetic acid is seen to be particularly facile. The advantages over EI are slight indeed. In fact, whenever the only sites accessible to the proton are especially susceptible to fragmentation (ROH, ROAc, etc.), this result can be expected. One can only hope that the use of other gases, or negative ion chemical ionization, will be of some value in such cases. Other steroids containing keto groups or aromatic rings are quite satisfactory (58), exhibiting intense M + 1 ions. The extent of loss of water from the keto function has been correlated with structure.

Peptide analysis may be facilitated by CI methods. Cleavage occurs at the peptide bond, giving acyl ions, which can be reconstructed to the original sequence (Figure 16). Ions are usually present for the corresponding protonated amine from each fragment, providing a double check on the sequence (59).

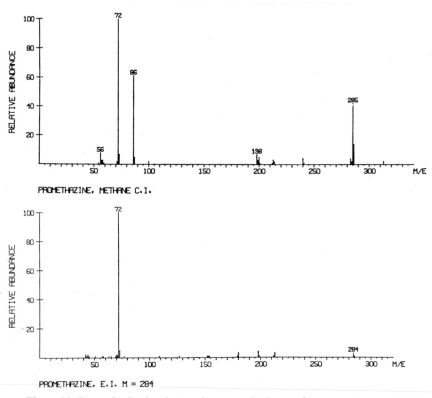

Figure 14. Promethazine by electron impact and using methane as a reagent gas.

Amino acids represent one of the outstanding successes of the CI mode, since they show little or no molecular ions in EI, even when they are protected as esters (60). In CI, ions are invariably present at $(M + 1)^+$, and fragmentation processes such as loss of $COOH_2$, NH_3, H_2O, H_2S, and the like are all distinctly observed (e.g., Figure 17) and easily correlated with structure (61a).

In summary, it can be seen that chemical ionization with protonic gases can often supplement EI spectra in a very useful fashion. Moreover, it is

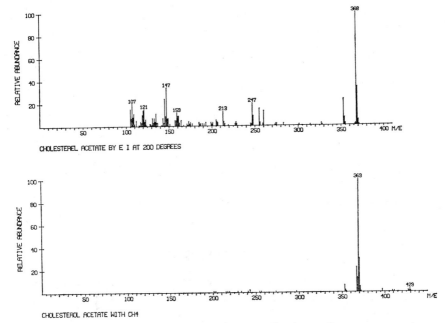

CHOLESTEREL ACETATE BY E I AT 200 DEGREES

CHOLESTEROL ACETATE WITH CH4

Figure 15. Cholesterol acetate by electron impact and using methane as a reagent gas.

never necessary to choose between the two systems, since EI spectra may easily be obtained in a CI source by turning off the reagent gas. Some sacrifice in sensitivity of the source for EI may be necessary, but in our experience the difference is minimal.

Increasingly we have found it very useful to analyze an unknown substance by both methods; it is surprising how often the data from one supplements the other. Finally, we can vouch for the fact that operation of a spectrometer in the CI mode is no more troublesome than under EI conditions (44).

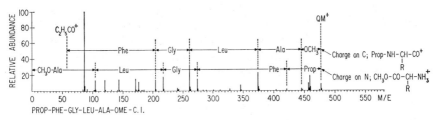

PROP-PHE-GLY-LEU-ALA-OME - C. I.

Figure 16. Prop-Phe-Gly-Leu-Ala-OMe using methane as a reagent gas.

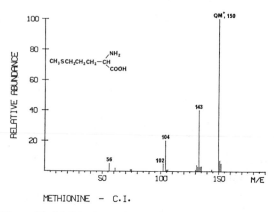

METHIONINE — C.I.

Figure 17. Methionine, using methane as a reagent gas.

IV. RADIATION SOURCES

It is somewhat surprising to observe that relatively little use is made of the differences in electron impact spectra caused by lowering the electron energy. Often this simple method is used to identify molecular ions, but the extensive treatises on correlations between structure and spectra contain little information on this important variable. One example of its utility concerns the observation that in simple dialkyl ethers, β-cleavage becomes important only at lower voltage (61b).

Presumably the lack of such data is partly due to the tedium involved in replotting data from several mass spectra, but as more laboratories become equipped with data-handling systems, such studies will probably become more common.

On the other hand, one may speculate on the use of even higher energy particles such as electrons accelerated to the meV range. Kebarle and Godbole (62) have shown that identical spectra (except for the presence of more multiply charged species) are obtained with 30 and 10,000 eV electrons in the cases of acetone, n-butane, benzene, and ethylene. These workers also confirmed, at higher energies, Mohler's observation (63) that the total ionization is approximately proportional to the number of valence electrons, within a homologous series. As the authors point out, Bethes' equation (64) for the ionization cross-section involving removal of a valence electron by a high-speed projectile suggests that the only pertinent factor relating to the properties of the projectile is the square of its charge. For this reason α-particles (He^{++}) should be $\sim$ four times as efficient as electrons of the same velocity.

Melton and Rudolph (65) used such an α-emitter source consisting of a film of 2×10^6 d.p.s. of Po^{208} on their repeller and were able to obtain 4×10^{-16} amp at 2×10^{-4} mm of acetylene. Contrary to expectations, these authors noted much more molecular ion $(C_2H_2)^+$ and less cleavage using Po^{208} than when 75 V electrons were used in the same source, but Kebarle et al. (62) suggest that this difference may have been due to δ rays accelerated by stray fields.

The real advantage in using radioactive (or high-energy) sources would seem to be in eliminating the hot filament and its associated electronics. On the other hand, most modern ion sources are so constructed as to shield the sample fairly effectively from the filament. Films of organic matter accumulating on such a radioactive plate would rapidly lower its efficiency. Radioactive contamination from decay products and the metal itself can be an annoyance if not a health hazard. Those of us who have used argon ionization detectors in gas chromatography can attest to this problem, and this is perhaps the prime reason that *flame* ionization detectors have found such wide acceptance. In one experiment in our laboratories, an attempt to utilize a ^{63}Ni source at higher pressures was abandoned due to its tendency to contaminate the ion chamber when heat was applied.

An additional advantage of high-energy radiation is in connection with higher-pressure $(0.01 - 1$ atm) sources, where an α-particle, for example, may still be effective after passing several mm deep into the gas. Aquilanti *et al.* (66) have used a 12 Curie tritium-laden (0.1 meV electrons) tantalum source for studying reactions of H_3^+, D_3^+, and A^+ with traces of small hydrocarbons. They obtained ion currents of 3.5×10^4 ion/sec at 0.1 torr and observed intensity-pressure variations that were reasonable based on the work of Field *et al.* (67) using 300 V electrons. In our experience, tritium-laden foils appear to lose much of their activity on heating to $\sim 200°$, and gas-chromatograph ovens using them are required to have an automatic cut-off at this temperature. In any case, the use of 12 curies of tritium would be prohibited in many laboratories.

Wexler *et al.* (68) used a Van de Graaff generator to supply 0.03 μA of 2 meV protons that enter and leave a source chamber through thin nickel windows. In contrast to other radioactive sources for like particles, such a system provides a thin collimated beam whose dimensions and location can be simply calculated. Also, the complications caused by secondary electrons, which are often emitted from ion-chamber surfaces when they undergo bombardment by radiation from radioactive sources, are avoided. The authors calculate that at 1-torr pressure, a proton of 2 meV energy would encounter 90 molecules within the 1-cm path length to the ion exit slit. In a study of methane at this pressure, their results were also essentially the same as those obtained by Field and Munson (69) regarding the abundancies of CH_5^+, $C_2H_5^+$, $C_3H_5^+$, and it seems that there is little advantage to this method of ion production for analytical purposes.

V. LASERS

The use of lasers to vaporize and/or ionize molecules in mass spectrometry is intriguing. Progress to date has been slow, at least as applied to organic substances. The effective temperature obtained at the focus point is difficult to control, and the ions that are produced last for less than 1 msec, so that only pulsed systems such as the (Bendix) Time-of-Flight spectrometer may be employed to analyze the ions.

Lasers have been successfully used to determine the nature of adsorbed films on metals, since water, CO_2, and small alkyl fragments may be readily identified even under pyrolytic conditions (70).

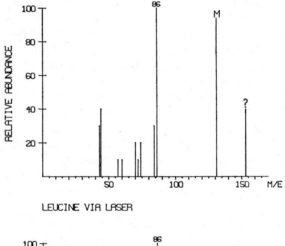

LEUCINE VIA LASER

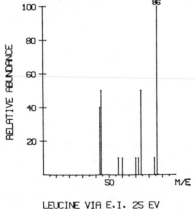

LEUCINE VIA E.I. 25 EV

Figure 18. Leucine analyzed by laser compared to electron impact.

Vastola and Pirone (71) have even successfully obtained molecular ions from a series of aromatic hydrocarbons and other substances pressed in pellets covered with graphite flakes to improve their absorptive properties. While their success with this series may not be surprising, since these materials have great thermal stability, they also successfully obtained a parent ion from leucine (Figure 18) along with an unexplained ion at even higher mass. Assurance that this ion was indeed due to leucine was provided by the presence of the M-45$^+$ ion customarily observed in EI spectra of amino acids. The EI spectrum of leucine (Figure 18) shows no parent ion, and this is evidence that the process of ionization by laser action is quite different from that of electron impact. Even as a tool to effect volatilization the method clearly shows considerable promise; for example, if a way can be found to selectively excite to the point of rupture the hydrogen bonds (3300 cm^{-1}) holding a peptide crystal lattice together, one might expect volatilization to occur.

VI. METASTABLES

The use of metastable molecules such as A* or He* to produce ions is intuitively attractive because such metastables are relatively easy to produce, either by the action of radioactive particles or electron bombardment, and the cross-section for ionization by such particles is very large (72). This is the primary process that occurs in argon-ionization gas-chromatograph detectors, although the primary ions so formed are usually accelerated by a high voltage to produce intense secondary ionization in order to maximize the detector current.

Penning ionization, $A* + M \rightarrow A + M^+ + e$, is distinguished from chemi-ionization (*not* chemical ionization), $A* + M \rightarrow AM^+ + e$. As Cermak (73, 74) points out, this type of ionization is fundamentally different from ion-molecule or electron-molecule reactions, since it involves inter-actions between *uncharged* particles. Thus the force fields in the vicinity of the interacting species are weak, mainly being due to Van der Waal's forces. Collision complexes may be long-lived and the Franck-Condon principle may no longer apply.

Furthermore, the state of excitation of the newly formed ion may not be directly correlated with the excess energy of the metastable molecule, since the electron can carry off a significant fraction of the energy. In fact, a whole field of study surrounds the measurement of the energy of the emitted electron in order to determine the electronic state of the ion, a process analogous to the Auger effect. The amount of energy transferred to a molecule from a meta-stable equals the excitation energy of the metastable less the sum of the ionization potential of the molecule ionized and the kinetic energy of the released electron.

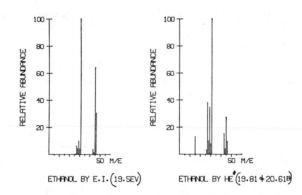

Figure 19. Ethanol by electron impact and using helium metastables.

In one apparatus (73, 74), a gas to be converted to metastables (usually an inert gas such as He, Ne, Ar, Kr, or Xe) is admitted at $\sim 10^{-4}$–10^{-3} mm to an electron beam whose potential is held 1–2 eV below the ionization potential but above the metastable excitation potential. Any ions fortuitously produced in this region are removed by a charged grid through which the metastable molecules drift to reach the collision chamber where they are allowed to react with the sample gas, admitted separately. Thus Figures 19 and 20 (74) compare the spectra of ethanol ionized by He* (19.81 and 20.61 eV) and Ar* (11.54 and 11.72 eV) with electron impact ionization of similar energy (19.5 and 11.8 eV, respectively). Although some quantitative differences exist, particularly at lower energies, the spectra are very similar, and it must be concluded that the overall ionization process is more like electron impact than might have been predicted.

Much more work has appeared on chemi-ionization processes such as

$$He^* + He \rightarrow He_2^+ + e$$
$$Xe^* + Hg \rightarrow XeHg^+ + e$$

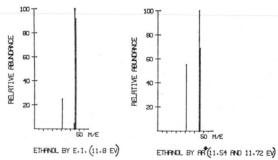

Figure 20. Ethanol by electron impact and using argon metastables.

probably because of the interest inherent in the formation of "compounds" of the inert and noble gases, but it is difficult to see how this process can be made to yield structural information on more complex molecules.

VII. PHOTOIONIZATION

Energy can be transferred to a molecule via collision with a photon as well as by electron impact. The range of energies required to ionize most molecules is in the vacuum ultraviolet. It is difficult to procure windows transparent in this region, although LiF is sometimes utilized, since it transmits radiation of 11.8 eV (1050 Å). To obtain higher energies, an electrical discharge is maintained in a gas (He or H_2) in a small tube adjacent to the source, with no restriction between the tube and the ion chamber. Differential pumping is essential with this method. For accurate work a monochromator must also be employed to select the desired wavelength. By this means, very accurate ionization and appearance potentials can be obtained, and in fact it seems difficult to justify the use of any other source for such measurements, considering the vast amount of debate and uncertainty revolving around the values obtained using conventional electron-impact sources.

However, the ion currents obtained with photoionization (PI) sources are of low intensity, and the difficulties surrounding the construction of such sources are not trivial (few if any are commercially available), so that one may reasonably ask what advantages it may provide over an electron impact source in regard to providing structural information. To this author's knowledge there have been no accounts showing major differences in relative ion abundances comparing photoionization to EI at the same nominal energy. A recent study of complex molecules using PI has confirmed this fact. (Personal communication, Richard Timmons, Catholic University, Washington, D.C.) Certainly a photoionization source may be run at a lower temperature because of the absence of a filament. This may be advantageous in the case of very labile compounds, even though EI sources are also available today that may be cooled to some degree. The only difference to be expected should arise from inhomogeneity in the energy of the electron beam, and this would certainly have a relatively minor effect on the spectrum.

Nonetheless, Brion and Hall (75) have shown how dramatic the temperature effect can be in affecting the spectrum of OsO_4 as well as a series of t-butyl-cyclohexanols and their mesylates. They ascribe the differences to pyrolytic effects from the hot filament itself, since the source ion-chamber was at room temperature in both cases. It would seem that improved baffling of the filament might reduce such effects.

In view of the lower sensitivity of PI sources, it seems doubtful that they will find great use in the analysis of more complex molecules. The advantage of lower temperature of operation is of less use in such cases since it is usually necessary to heat the sample merely to effect volatilization.

VIII. ELECTROJET METHODS

In recent years considerable effort has been spent in attempting to extend the upper mass limit, both by increasing the mass range of the mass spectrometer, and by finding methods by which materials of increasingly high molecular weight may be volatilized and ionized. In theory, as a molecule becomes larger in size and increases in complexity, ions may be more stable insofar as their internal energy may become distributed among more sites. A more important problem is our inability to volatilize high-molecular-weight polar compounds. Recourse has been made to volatile derivatives such as silyl ethers. Unfortunately such adducts often increase the complexity of the problem because of the unusual (and interesting) fragmentation pathways they incur (76). In theory, at least, there is no need to volatilize a sample in order that it be ionized by an electron beam. For example, Dole has shown (77) that if a material is dissolved and then sprayed from a negatively charged jet, one or more electrons may be accumulated on the individual droplet. If the concentration of solute is adjusted so that each droplet contains one molecule, the solvated ion may contain one or more negative charges, and these ions will be suitable for dispersion by any mass-spectrometric analyzer.

In order to gain information about the solute ion itself, means must be provided to remove the varying numbers of solvent molecules. If the material is merely allowed to expand into a vacuum, excessive cooling will result and ionic clusters of rather large size will result. Milne (78) *et al.* have recently proposed the use of such clusters, ionized by the action of an electron beam, as a possible mass marker for mass-spectrometer systems. However, Dole and his co-workers in recent papers (77) have shown that when the clusters are accelerated through an electric field in the presence of an inert gas such as nitrogen, heat transfer can take place and the major portion of the solvent is removed. The remaining solvent can be removed by the use of a skimmer system, with the result that the beam of desolvated ions, now traveling at supersonic velocities, can be mass analyzed by allowing them to pass through an evacuated space toward a plate in front of which a repeller grid has been adjusted so as to repel solvent molecules of low mass. As the repeller potential is increased, ions of successively higher energy (whether due to higher mass or lower charge) will be refused admittance to the collector plate. Sharp breaks should also occur in the case of a single species according to the number of charges that it has accumulated, and these breaks should occur at evenly spaced values of repeller voltage. If means can be found to permit the accumulation of no more than unit charge on the molecular species, the breaks would correspond to variations in the molecular weight of components of the solution. Dole has applied his method to solutions of polystyrene of number average molecular weight of 51,000 in a mixed benzene-acetone solvent. Curves were indeed observed with inflections corresponding to varying numbers of negative charges. The breaks were unusually sharp, considering that polystyrene of 51,000-number average molecular weight is undoubtedly a mixture of various species. The authors consider that this may be due to a so-called "mach focusing" effect within the jet, resulting in some monochromatization of the beam. Furthermore, the authors find that the polystyrene of 51,000 molecular weight forms a dimer as the most abundant species, while in the case of the polystyrene of 411,000 molecular weight, the tendency is to form triply charged monomers. As the authors point out, this experiment provides only the mass-to-charge ratio in a given case. If some method could be found to transfer these macroions to a mass-spectrometer system capable of better resolution at this extraordinarily high molecular weight, the method would undoubtedly find wide application. Of course, the system even as it stands is also applicable to substances of ordinary molecular weight. In this connection, mass spectroscopists have always seemed to be annoyed that it should be necessary to convert a charged substance such as a quaternary ammonium salt to a neutral molecule in order to render it volatile, only to have it reconverted to positive ions within the ion chamber. It seems logical that some extension of this type of device would be found suitable for abstracting counter-ions, e.g., a chloride ion from a quaternary ammonium chloride, leaving the quaternary ammonium ion in its free state. It seems clear that this is an area for extremely profitable research in the future. Recently there have been reports of the extension of the method to the biopolymer, lysozyme (79). Even in its present state, the method does make possible crude mass determinations on substances of such high molecular weight as to be inaccessible by other means. For example, in the cases alluded to previously, Dole calculates the vapor pressure of the polyethylene polymer of 14,000 amu to be only 10^{-480} at. The highest mass range to which any mass spectrometer currently available extends is approximately 8000 amu, but in this mode of operation its sensitivity is necessarily extremely low. In fact, in recent years we have successfully obtained a spectrum on a compound of mass 3628 amu (80). In this case the material happened to be extremely volatile because of the presence of large numbers of fluorine atoms, and we were able to produce ions in the ordinary way, viz electron bombardment. Extension to masses of 50,000 amu or greater would seem to require fields inaccessible at present except perhaps by superconducting magnet techniques that so far have not been applied to the field of mass spectrometry. Again, assuming that a method could be found for the production of ions of this extraordinarily high mass, the question of resolution enters. One can achieve resolutions of the order to 70,000–100,000 using extremely small slits, but the sensitivity is correspondingly low. At present it seems that some wholly new principle will have to be discovered before such methods become practical.

IX. ION-BOMBARDMENT METHODS

Ions can be used in place of electrons to form new ions, and the reaction products may be investigated by carrying out such a reaction in the ion chamber of a chemical ionization mass spectrometer as described previously. Another approach is to investigate the fragmentation products of the primary ions when it collides with a neutral molecule. Jennings (81) has used this method recently in the case of benzene and other substances by utilizing metastable defocusing techniques available in double-focusing mass spectrometers. In this system, ions of benzene are produced in a conventional electron impact ionization source at voltages near the ionization potential so that these ions constitute the major charge carrier in the system. The ions then pass through the flight tube into the electrostatic analyzer region where the pressure has been arbitrarily raised to about 10^{-5} torr, thus insuring that collisions will occur. Some of these collisions result in decomposition of the benzene molecular ion into fragment ions. Because of simple energy considerations, the newly formed ions of lower mass will continue in the same direction as the original benzene molecular ions. However, because of their lower mass and consequent lower energy, they will be more sharply deflected by the analyzer plates and will thus not be able to pass through an intermediate slit where they ordinarily enter the magnetic-sector region. However, if the electrostatic sector is adjusted to a lower voltage, the fragment ions may now pass into the magnetic sector, where they are analyzed like any other ions. Knowing the original mass of the benzene molecular ion and the voltage necessary to allow the daughter ions to pass through the intermediate slit, one can readily calculate the mass and approximate the intensity of the fragment peaks. It is indeed interesting to note that spectra from ions formed in this manner are not wholly unlike those produced in the conventional electron-impact method. As Jennings points out, this serves to emphasize the fact that collision processes have certain features in common, whether the colliding species be an electron or a neutral molecule. This procedure might best be conducted in a tandem mass spectrometer where one could study the relative stabilities toward collision of the various ions formed in the ordinary electron-impact method. For that matter, the whole process might even be reversed by passing molecules from a supersonic jet into a plasma of charged ions maintained in a stationary condition by an oscillating electric field (cf. discussion on ion cyclotron resonance, above). Another advantage of the method would seem to be that the ion beam so formed has a fixed original velocity and can be considered highly monochromatic in contrast to electrons prepared in the conventional beam. One might expect to observe quite different fragmentation spectra depending on the accelerating voltage employed. This may even conceivably be a new way to determine appearance potentials of fragment ions with a higher degree of accuracy.

REFERENCES

1. H. D. Beckey, E. Hilt, A. Maas, M. D. Migahed, and E. Ochterbeck, *J. Mass Spec. Ion Phys.*, **3**, 161 (1969).
2. R. H. Good and E. W. Müller, "Field Emission," in S. Flügge (Ed.), *Encyclopedia of Physics*, Vol. XXI, Springer-Verlag, Berlin, 1956.
3. R. Gomer, *Field Emission and Field Ionization*, Harvard University Press, Cambridge, Mass., 1961.
4. Figures 1, 2, and 3 are based on illustrations from Refs. 2 and 3 and are intended only to be qualitative representations.
5. Reference 3, p. 10.
6. Reference 3, p. 8.
7. Reference 3, p. 7.
8. E. W. Müller, *Z. Phys.*, **131**, 136 (1951); E. W. Müller, *Ergeb. Exact. Naturwiss.*, **27**, 290 (1953).
9. Reference 3, p. 83.
10. Reference 3, p. 69.
11. E. W. Müller, *Z. Phys.*, **131**, 136 (1951).
12. Reference 3, p. 70.
13. H. D. Beckey in R. M. Elliott (Ed.), *Advances in Mass Spectrometry*, Vol. 2, Macmillan, New York, 1963, p. 11.
14. P. Schissel, Ref. 3, p. 82.
15. M. G. Inghram and R. Gomer, *Z. Naturforsch*, **10a**, 863 (1955).
16. H. D. Beckey, *Naturwiss.*, **45**, 259 (1958).
17. Reference 13, p. 22.
18. H. D. Beckey, H. Hey, K. Levsen, and G. Tenschert, *J. Mass Spec. Ion Phys.*, **2**, 101 (1969).
19. P. O. Schissel, communication at the Field Emission Symposium, McMinnville, Oreg.
20. H. D. Beckey, *Z. Naturforsch*, **14a**, 712 (1959).
21. Reference 13, p. 16.

22. H. D. Beckey, *Z. Analyt. Chemie*, **170**, 359 (1959).

23. H. D. Beckey in W. L. Mead (Ed.), *Advances in Mass Spectrometry*, Vol. 3, The Institute for Petroleum, London, 1964, p. 48.

24. Reference 23, p. 52.

25. Reference 23, p. 53.

26. H. Krone and H. D. Beckey, *Org. Mass Spec.*, **2**, 427 (1969).

27. H. D. Beckey, *J. Mass Spec. Ion Phys.*, **2**, 500 (1969).

28. J. H. Beynon, *Mass Spectrometry and Its Application to Organic Chemistry*, Elsevier, 1960, p. 23.

29. G. G. Wanless and G. A. Glock, Jr., *Anal. Chem.*, **39**, 2 (1967).

30. M. Spiteller-Friedman, S. Eggers, and G. Spiteller, *Monatsh.*, **95**, 1740 (1964); see also G. Spiteller and M. Spiteller-Friedman, in R. Bonnett and J. G. Davis (Eds.), *Some Newer Physical Methods in Structural Chemistry*, United Trade Press, London, 1967, p. 121.

31. H. D. Beckey, *J. Mass Spec. Ion Phys.*, **1**, 93 (1968).

32. H. D. Beckey, H. Hey, K. Levsen, and G. Tenschert, *J. Mass Spec. Ion Phys.*, **2**, 101 (1969).

33. P. Brown, G. R. Pettit, and R. K. Robbins, *Org. Mass Spec.*, **2**, 521 (1969).

34. H. D. Beckey and H. Knöppelm, *Naturforsch*, **21a**, 1020 (1968)

35. J. N. Damico, R. P. Barron, and J. A. Sphon, *J. Mass Spec. Ion Phys.*, **2**, 161 (1969).

36. H. D. Beckey and G. Wagner, *Z. Naturforsch*, **20a**, 169 (1965).

37. H. D. Beckey and P. Schulze, *Z. Naturforsch*, **21a**, 214 (1966).

38. W. K. Rohwedder, *15th Annual Conference on Mass Spectrometry and Allied Topics, ASTM, Committee E-14*, 1967, Denver, Colorado, p. 131.

39. Reference 3, p. 101.

40a. H. D. Beckey, *J. Amer. Chem. Soc.*, **88**, 5333 (1966).

40b. F. H. Field, M. S. B. Munson, and D. A. Becker, *Advances in Chemistry*, No. 58, 167 (1966), American Chemical Society, Washington, D.C. 20004.

41. V. A. Tal'roze and A. K. Lyubimova, *Dokl. Akad. Nauk. SSSR*, **86**, 909 (1952)

42a. J. H. Futrell and T. O. Tiernan in P. Ausloos (Ed.), *Fundamental Processes in Radiation Chemistry*, Interscience, New York, 1968, Chap. 4, p. 199.

42b. J. L. Beauchamp, L. R. Anders, and J. D. Baldeschwieler, *J. Amer. Chem. Soc.*, **89**, 4569 (1967).

43. F. H. Field, *Accounts Chem. Res.*, **1**, 42 (1968).

44. M. S. B. Munson and F. H. Field, *J. Amer. Chem. Soc.*, **88**, 2621 (1966).

45. (a) *cf.* Ref. 41; (b) D. P. Stevenson, and D. O. Schissler, *J. Chem. Phys.*, **23**, 1353 (1955); (c) F. H. Field, J. L. Franklin, and F. W. Lampe, *J. Amer. Chem. Soc.*, **79**, 2419 (1957).

46. D. P. Stevenson in F. W. McLafferty (Ed.), *Mass Spectrometry of Organic Ions*, Academic Press, New York, 1963, p. 589.

47. F. F. Moran and L. Friedman, *J. Chem. Phys.*, **39**, 2491 (1963).

48. F. H. Field in E. Kendrick (Ed.), *Advances in Mass Spectrometry*, Vol. 4, Institute of Petroleum, London, 1968, p. 645.

49. J. H. Futrell and T. O. Tiernan in P. Ausloos (Ed.), *Fundamental Processes in Radiation Chemistry*, Interscience, New York, 1968, Chap. 4, p. 171.

50. G. S. King, dissertation, Chemistry Dept., Imperial College of Science and Technology, London S.W.7, England, May 1970.
51. F. H. Field, personal communication.
52. F. H. Field, *J. Amer. Chem. Soc.*, **91**, 2827 (1969).
53. See Ref. 55, p. 3683.
54. H. M. Fales, H. A. Lloyd, and G. W. A. Milne, *J. Amer. Chem. Soc.*, **92**, 1590 (1970).
55. H. M. Fales, G. W. A. Milne, and M. L. Vestal, *J. Amer. Chem. Soc.*, **91**, 3682 (1969).
56. H. Ziffer, H. M. Fales, G. W. A. Milne, and F. H. Field, *J. Amer. Chem. Soc.*, **92**, 1597 (1970).
57. H. M. Fales, G. W. A. Milne, and T. Axenrod, *Anal. Chem.*, **42**, 1342 (1970).
58. J. D. Baty, H. M. Fales, and G. W. A. Milne, in preparation.
59. A. A. Kiryushkin, H. M. Fales, T. Axenrod, E. J. Gilbert, and G. W. A. Milne, *Org. Mass. Spec.*, **5**, 19 (1971).
60. K. Biemann, J. Seibl, and F. Gapp, *J. Amer. Chem. Soc.*, **83**, 3795 (1961).
61a. G. W. A. Milne, T. Axenrod, and H. M. Fales, *J. Amer. Chem. Soc.*, **92**, 5170 (1970)
61b. C. Djerassi and C. Fenselau, *J. Amer. Chem. Soc.*, **87**, 5747 (1965).
62. P. Kebarle and E. W. Godbole, *J. Chem. Phys.*, **36**, 302 (1962).
63. F. L. Mohler, L. Williamson, and H. M. Dean, *J. Res. Nat. Bur. Standards*, **42**, 235 (1950).
64. H. Bethe, *Ann. Physik.*, **5**, 325 (1930).
65. C. E. Melton and P. S. Rudolph, *J. Chem. Phys.*, **30**, 847 (1959).
66. V. Aquilanti, A. Galli, A. Giardini-Duidoni, and G. G. Volpi, *J. Chem. Phys.*, **43**, 1969 (1965); V. Aquilanti and G. G. Volpi, ibid., **44**, 2307 (1966).
67. F. H. Field, J. L. Franklin, and M. S. B. Munson, *J. Amer. Chem. Soc.*, **85**, 3575 (1963).
68. S. Wexler, A. Lifshitz, and A. Quattrochi in P. J. Ausloos (Symposium Chairman), *Ion Molecule Reactions in the Gas Phase*, Advances in Chemistry Series No. 58, Amer. Chem. Soc., Washington, D.C., 1966, p. 193.
69. F. H. Field and M. S. B. Munson, *J. Amer. Chem. Soc.*, **87**, 3289 (1965).
70. L. P. Levine, J. F. Ready, and G. E. Bernal, *J. Appl. Phys.*, **38**, 331 (1967).
71. F. J. Vastola and A. J. Pirone, *Adv. Mass Spectrometry*, **4**, 107 (1968).
72. P. F. Little in S. Flugge (Ed.), *Encyclopedia of Physics*, Vol. XXI, Electron Emission Gas Discharges 1, Springer-Verlag, Berlin, 1956, p. 661.
73. V. Cermák, *J. Chem. Phys.* **44**, 3781 (1966).
74. V. Cermák and Z. Herman, *Coll. Czech. Chem. Commun.*, **30**, 169 (1965).
75. G. E. Brion and L. D. Hall, *14th Annual Conference on Mass Spectrometry and Allied Topics, ASTM, Committee E-14*, 1966, Dallas, Texas, p. 338.
76. G. H. Draffan, R. N. Stillwell, and J. A. McCloskey, *Org. Mass Spec.*, **1**, 669 (1968).
77. M. Dole, L. L. Mack, R. L. Hines, R. C. Mobley, C. D. Ferguson, and M. B. Alice, *J. Chem. Phys.*, **49**, 2240 (1968).
78. T. A. Milne, *J. Mass Spec. Ion Phys.*, **3**, 153 (1969).
79. E. Neher, M.Sc. thesis, University of Wisconsin (1967).
80. H. M. Fales, *Anal. Chem.*, **38**, 1058 (1966).
81. K. R. Jennings, *J. Mass Spec. Ion Phys.*, **1**, 227 (1968).

Mass–Spectral Studies Employing Stable Isotopes in Chemistry and Biochemistry

MARVIN F. GROSTIC

The Upjohn Company, Kalamazoo, Michigan

AND

KENNETH L. RINEHART, JR.

*Department of Chemistry, University of Illinois,
Urbana, Illinois*

I. Introduction		218
II. Biosynthesis		221
	A. Oxygen-18	221
	B. Nitrogen-15 and Oxygen-18	236
	C. Nitrogen-15 and Carbon-13	237
	D. Deuterium	240
III. Metabolism		249
IV. Studies of Mass-Spectral Fragmentations		255
	A. Oxygen-18	256
	B. Nitrogen-15	259
	C. Carbon-13	261
V. Studies of Organic Reaction Mechanism		268
	A. Oxygen-18	269
	B. Nitrogen-15	273
	C. Nitrogen-15 and Deuterium	274
	D. Carbon-13	275
	E. Oxygen-18 and Deuterium	275
	F. Deuterium	275
VI. Structure Studies		279
	A. Oxygen-18	279
	B. Nitrogen-15	280
	C. Deuterium	281
References		283

I. INTRODUCTION

The arrangement of material in a review is, of necessity, arbitrary, and we have organized the present chapter to reflect our own particular interests. Thus, applications of stable isotopes in mass spectrometry are described first for biochemistry, then for organic chemistry, and we have chosen to exclude nearly all applications to inorganic chemistry. Within the sections, ^{18}O studies are presented first, followed by ^{15}N studies, ^{13}C studies, and ^2H studies, an order reflecting the degree of use of the first three isotopes. While deuterium has been the most widely used stable isotope, we have placed it last in our discussion order, since considerations of space forced us to omit it from the section dealing with studies of mass spectral fragmentations. This subject is considered in more detail in Chapter 2.

Emphasis has been placed on the more recent literature, with little discussion of work prior to 1966 except where such discussion may have been required for a more complete picture. General treatments of the use of stable isotopes are found in the books by the Djerassi group (1–3), the recent book by Beynon et al. (4), the two reviews by Viallard (5, 6), and the ^{13}C review by Meyerson and Fields (7). Some biochemical applications have been reviewed by Van Lear and McLafferty (8) and by Rinehart and Kinstle (9). Readers are referred to these books and reviews, since all the studies noted in them are not included in the present review. Some applications of stable isotopes to biochemical problems are discussed in Chapter 10.

We have chosen to omit studies involving radioactive nuclei, as well as those involving only natural abundance of isotopes, even though such isotopes as ^{35}Cl and ^{37}Cl, ^{79}Br and ^{81}Br, and ^{32}S and ^{34}S are of characteristic abundance and provide excellent and distinct labels for mass spectrometry. For this reason geological dating studies are excluded. We have also usually excluded studies in which only the *amount* of incorporation of one heteroatom was determined, in order to stress the more sophisticated applications of mass spectrometry, where the position of incorporation has been determined or two or more heteroatoms are involved. Studies involving combustion or degradation techniques, in which the mass spectrometer was used only to analyze H_2O/D_2O, $H_2^{18}O/H_2O$, $^{15}NN/N_2$, $C^{18}O/CO$, or $CO^{18}O/CO_2$ ratios, have been slighted in favor of mass spectra determined on the entire compound of interest, a derivative of it, or a degradation product. Discussions of mass spectral fragmentations are restricted to those that are used to locate the label.

Throughout the discussion we have employed the convention of referring to the fragment weight of the unlabeled ion (rather than the labeled ion) unless otherwise stated. For example, the tropylium ion ($C_7H_7^+$) would be listed as the m/e 91 ion, while its ^{13}C-labeled counterpart would be listed as m/e 91 (^{13}C, 92).

Unless otherwise stated, all spectra discussed are low-resolution mass spectra. The interpretations of these spectra are frequently not straightforward, because the compound being studied is always less than 100% labeled and the contribution of the unlabeled material to the spectrum being studied must be subtracted. The problem can be amplified due to weak molecular ions; however, field-ion (10, 11) or low-voltage studies may help to alleviate some of these difficulties. A minimum enrichment of 5% of the stable isotope is usually required for low-resolution studies. Part of the difficulty in interpreting low-resolution spectra is the frequent occurrence of isobaric ions (ions of the same nominal mass). In subtracting unlabeled low-resolution spectra, one must assume either ionic homogeneity or a high degree of reproducibility for labeled and unlabeled compounds' spectra.

High-resolution techniques can often separate isobaric ions and allow their compositions to be determined. This technique, then, does not require isobaric purity, since the labeled and unlabeled ions can be separated and their ratios calculated. Table 1 shows the practical mass limits of this technique at a

TABLE 1

Doublets Encountered in High-Resolution Stable-Isotope Studies

Doublet	$\Delta M \times 10^3$	m/e limit (resolution = 30,000)
H_2—D	1.548	47
CH—^{13}C	4.467	134
CD—^{13}CH	2.922	88
^{13}CH—N	8.109	243
NH—^{15}N	10.789	324
$^{13}CH_2$—^{15}N	18.896	567
^{15}NH—O	13.021	391
ND—O	22.262	668
CH_2D—O	34.838	1045
$^{13}CH_3$—ND	9.657	290
$^{15}NH_3$—^{18}O	24.424	733
NH_2D—^{18}O	33.665	1010
^{13}CN—$C^{15}N$	6.320	190

resolution of 30,000 involving the stable isotopes of interest. The H_2—D doublet can be separated only up to mass 47, whereas, the CH_2D—O doublet can be resolved up to mass 1045. Readers are referred to the Beynon (12, 13) and McLafferty (14) books for information on other doublets of interest as well as for the resolution required to separate them.

While we have excluded radioisotopes in this review, it should be noted that a completely different, novel method has recently been described for locating a carbon label in fragment ions and should be mentioned (15). In this method, employing ^{14}C, the ions are collected on a graphite plate, which is used to expose a photographic film, which is in turn compared to the standard photoplate employed with an instrument of Mattauch-Herzog geometry. Only ions containing ^{14}C are observed on the second plate (auto-radiograph). Exposures are long (16 hours) to measure ^{14}C and, of course, accurate quantitation is impossible.

The chemical introduction of stable isotopes into organic compounds has been discussed elsewhere. General synthetic methods of introducing deuterium have been reviewed by Djerassi et al., as well as specific reference to ^{18}O labeling (2). The apparent ease of exchange reactions have made them especially attractive synthetic approaches. In particular, the base-catalyzed deuterium exchange adjacent to carbonyl groups offers insight about its environment, and ^{18}O exchange in the carbonyl itself offers structural information. The exchange of ^{2}H and ^{18}O on a gas chromatographic column has been described recently (16). All is not completely trouble-free, however, because the investigator must study the reverse exchange under the experimental conditions, in the atmosphere, and in the mass spectrometer itself. Bursey and McLafferty have reviewed some of the techniques of stable isotope labeling (17), but in practice a great deal of ingenuity is required, since one is usually interested not only in the introduction of a specific label but also in its introduction at a specific position.

In the final analysis, the attractiveness of using stable isotopes in mass spectrometry depends on how well the method competes with other methods and, in particular, with use of radioactive isotopes. For example, the widespread use and applications of ^{14}C labeling has resulted in the availability of a much larger number of commercially available compounds so labeled. The 1967 Isotope Index lists 42 ^{18}O, 91 ^{15}N, 63 ^{13}C, and about 300 ^{2}H labeled compounds. In sharp contrast, the same index lists almost 1400 ^{14}C-labeled compounds (18). (The total number is listed in each case with no consideration as to the percent enrichment or specific activity.) While in general the ^{13}C-labeled compounds cost more, no real cost comparison is meaningful without considering such parameters as the sensitivity of the method, the enrichment (or specific activity) required, and the effect of any needed chemical modifications. The initial expense may be only part of the cost and, here again, the ingenuity of the researcher plays a major role in determining what labeled starting material to use and what reaction(s) to employ to give the best yields of the desired labeled product.

II. BIOSYNTHESIS

Biosynthesis and metabolism are, of course, interrelated, in that the biosynthesis of any compound can be considered the metabolism of the substrate. However, the purpose of a study usually differentiates the two, and in this section we include only those studies that involve compounds found along normal animal or plant biosynthetic pathways. Under the heading metabolism we shall discuss studies dealing with compounds, especially drugs, that are foreign to the organism employed in the study.

A. Oxygen-18

Studies of steroid biosynthesis have been greatly aided by the use of ^{18}O and by mass spectrometry. Nakano and his co-workers reported studies on the mechanism of side-chain cleavage during steroid-hormone biosynthesis by rat testicular and adrenal enzymes (19, 20), on the location of the hydroxyl groups introduced into steroid molecules by these enzymes (21), and on the fate of the molecular oxygen required for the side-chain cleavage of cholesterol by endocrine enzymes (22).

The steroids involved in the side-chain cleavage studies were incubated in an $^{18}O_2$-enriched atmosphere, and 17α-^{18}O-hydroxylated progesterone was incubated under an O_2 atmosphere. The molecular ion region was analyzed for the ^{18}O incorporation and/or exchange. The 17α-hydroxyprogesterone (2) isolated from the incubation contained one atom of ^{18}O per molecule, while progesterone (1) showed neither incorporation nor exchange of ^{18}O atoms during the course of the incubation. This result also suggested that the ^{18}O incorporated into the 17α-hydroxyprogesterone was present in the 17α-hydroxyl group. Furthermore, the results suggested that the oxygen atom in the 17-keto group of androstenedione (3) and the 17β-hydroxyl group of testosterone (4) (both products derived from 17α-hydroxyprogesterone by side-chain cleavage) originates from the oxygen atom of the 17α-hydroxyl group in 17α-hydroxyprogesterone. This fact was indeed established when the mass spectra of androstenedione (3) and testosterone (4) obtained from an analogous incubation of 17α-hydroxyprogesterone under an ^{18}O-enriched atmosphere did not show any incorporation of ^{18}O atoms into the molecules.

Similar incubations and mass-spectral analysis were applied to a study of the side-chain cleavage of cholesterol (5) (20, 22). Cholesterol (5), 20α-hydroxycholesterol (6), 22-hydroxycholesterol (7), and $20\alpha,22R$-dihydroxycholesterol (8) were incubated, and the immediate product pregnenolone (9), as well as its further metabolite progesterone (1), were examined by mass spectrometry. The progesterone fractions derived from cholesterol (5) and

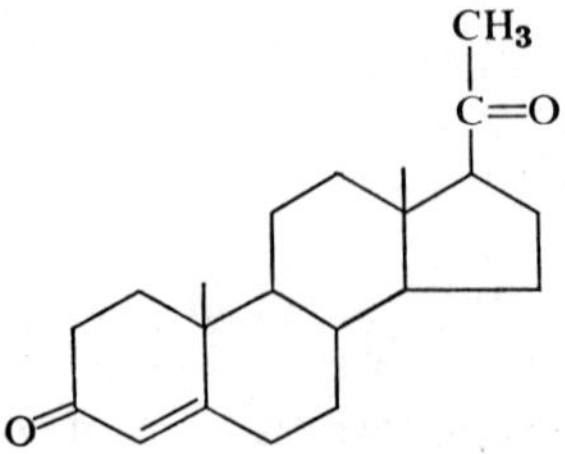

(1)

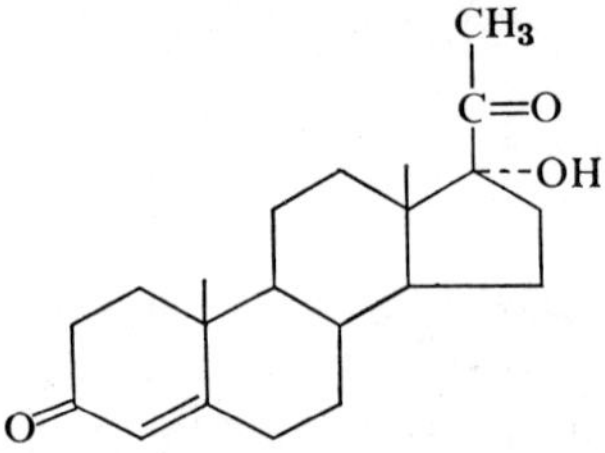

(2)

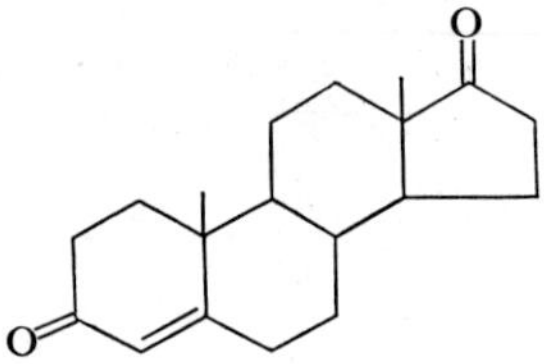

(3)

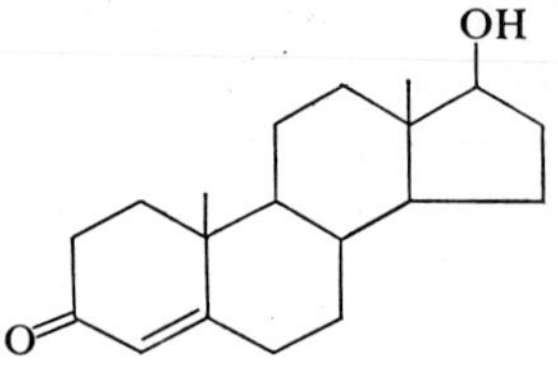

(4)

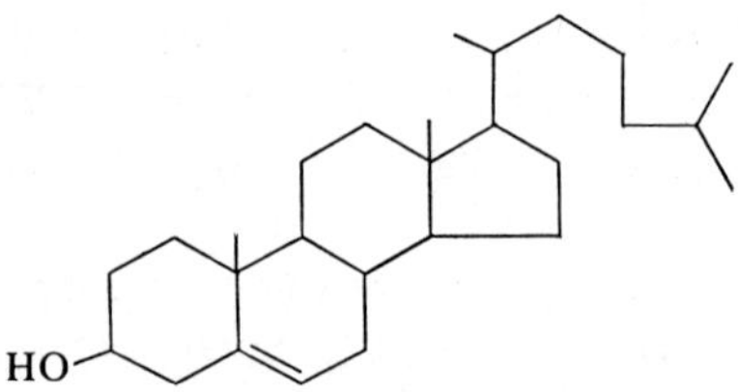

(5)

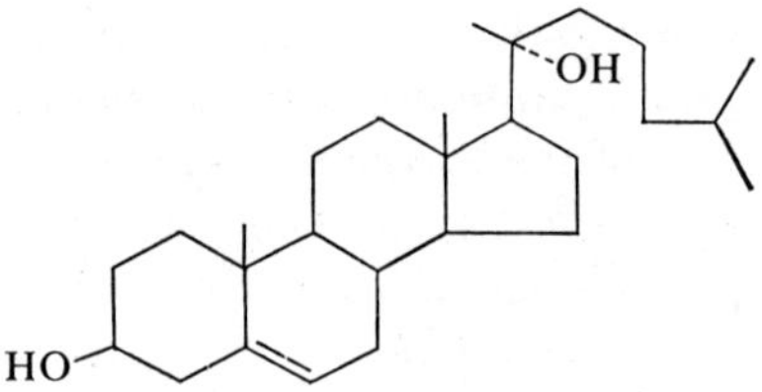

(6)

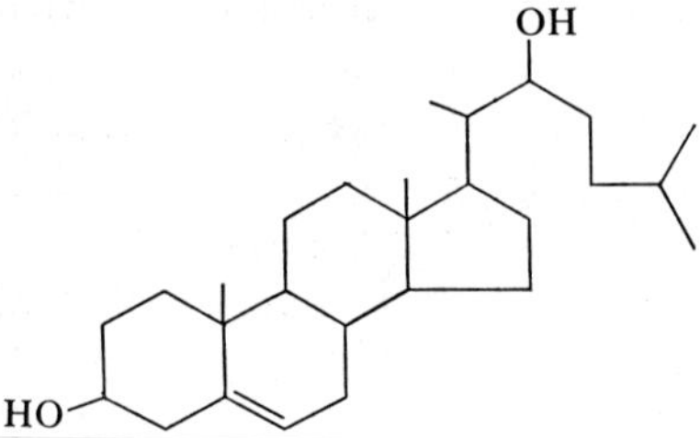

(7)

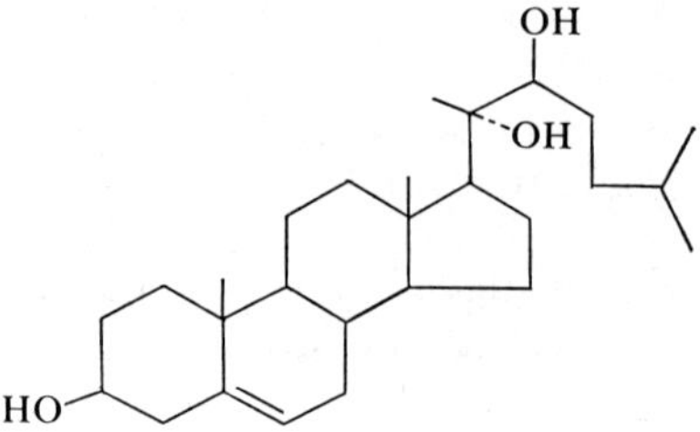

(8)

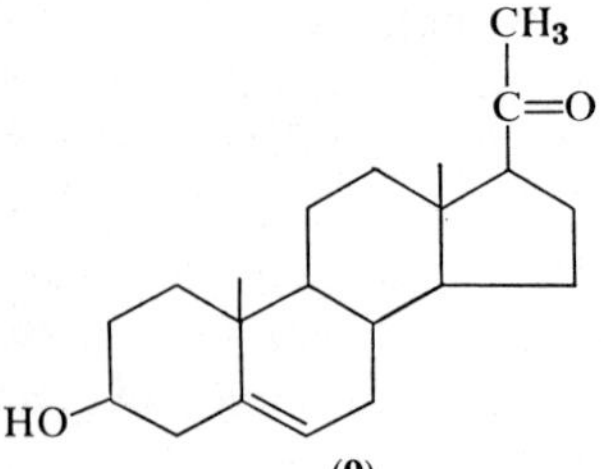

(9)

222

22-hydroxycholesterol (7) were found to contain an atom of ^{18}O per steroid molecule, while that derived from 20α-hydroxycholesterol (6) and 20α,22R-dihydroxycholesterol (8) did not show any ^{18}O in the molecule. In addition, the pregnenolone fraction derived from 20α,22R-dihydroxycholesterol (8) did not show any ^{18}O in the molecule.

A closer examination of the mass spectra of the molecules containing incorporated ^{18}O indicated the location of the label (21, 22). The mass-spectral fragment ions studied were those at $M - H_2O$ for the hydroxy compounds, those at $M - CH_2CO$ (from loss of ketene from the A-ring) for the Δ^4-3-ketosteroids, and those at $M - HOAc$ for the acetate derivatives.

The 17α-hydroxyprogesterone (2) produced under an $^{18}O_2$ atmosphere from progesterone (1) by means of microsomal enzymes of testis was found to contain the ^{18}O atom in the 17α-hydroxy group, since the dehydrated fragment ion did not show ^{18}O.

The $M - C_2H_2O$ fragment ion in the mass spectrum of labeled testosterone (4) produced from progesterone (1) under the same conditions still retained the ^{18}O, showing that the label was in the 17β-hydroxy group. Likewise, the same fragment ion in the mass spectrum of labeled progesterone derived from cholesterol (5) also retained the label and showed that the ^{18}O was in the 20-oxo group.

Finally, the $M - HOAc$ fragment ion in the spectrum of the acetate derivative of the labeled pregnenolone (9) produced from 20α,22R-dihydroxy-cholesterol (8) under $^{18}O_2$ by adrenal enzymes did not contain the label. These results allowed the following conclusions:

1. 17α-Hydroxypregnene-C-17,C-20 lyase in testicular tissue required molecular oxygen.

2. Progesterone-17α-hydroxylase required molecular oxygen, which was incorporated into the 17α-hydroxyl group of 17α-hydroxyprogesterone.

3. The introduction of an oxygen atom from molecular oxygen at the 20α-position of cholesterol was essential for side-chain cleavage.

4. The oxygen atom of the 20-oxo group originated from the oxygen of the hydroxyl group introduced in the 20α-position of cholesterol prior to cleavage.

In another study, Nakano, Sato, and Tamaoki studied the incorporation of molecular oxygen into the acetate of testosterone (4) produced aerobically from progesterone (1) by *Cladosporium resinae* (23). The molecular-ion region of the testosterone acetate so produced showed the incorporation of one atom of ^{18}O. The $M -$ ketene fragment ion also showed the presence of the ^{18}O atom. This acetate was chemically deacetylated to testosterone and its mass spectrum showed the same features, proving that the ^{18}O atom had been incorporated into the 17β-position of the steroid.

Van Tamelen *et al.* reexamined the conversion of squalene (**10**) to lanosterol (**11**) by rat liver preparations *in vitro* (24, 25). Squalene-2,3-oxide-^{18}O (**12**) was synthesized and incubated with the homogenates of rat liver. The incorporation of the ^{18}O atom into the lanosterol formed enzy-

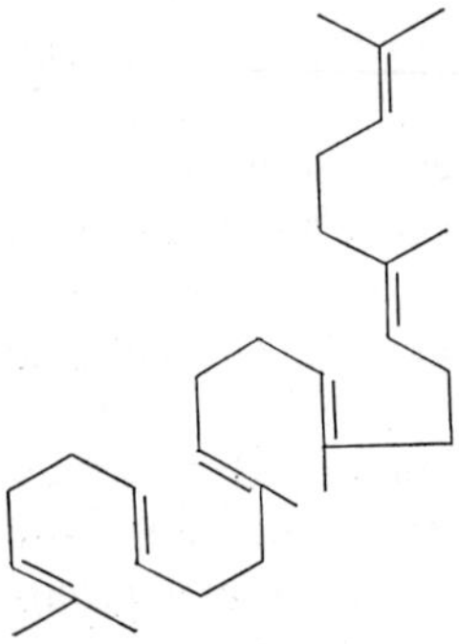

(10)

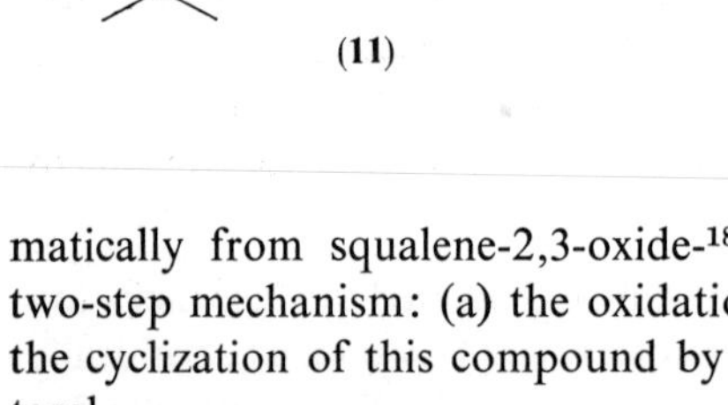

(11)

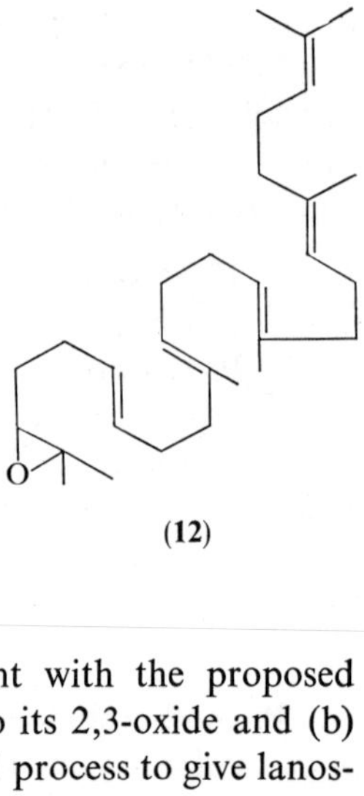

(12)

matically from squalene-2,3-oxide-^{18}O was consistent with the proposed two-step mechanism: (a) the oxidation of squalene to its 2,3-oxide and (b) the cyclization of this compound by an acid-initiated process to give lanosterol.

Ryhage and Samuelsson studied the incorporation of oxygen from ^{18}O-labeled oxygen gas during the formation of prostaglandin E_1 (**13**) from 8,11,14-eicosatrienoic acid (**14**) by a homogenate of vesicular glands (26). The important fragmentations of di-O-methyl-prostaglandin E_1 ethyl ester (**14a**) biosynthesized in an $^{18}O_2$ atmosphere are shown below.

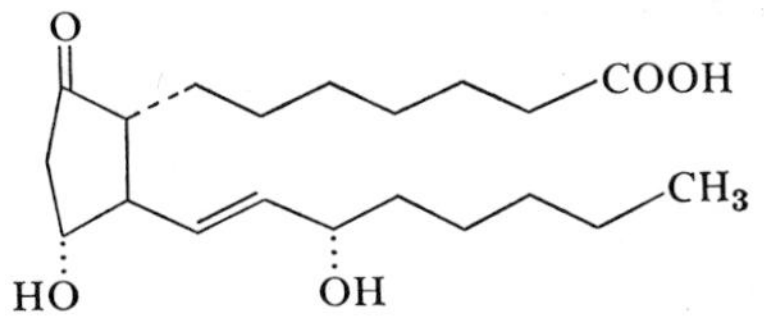

(13)

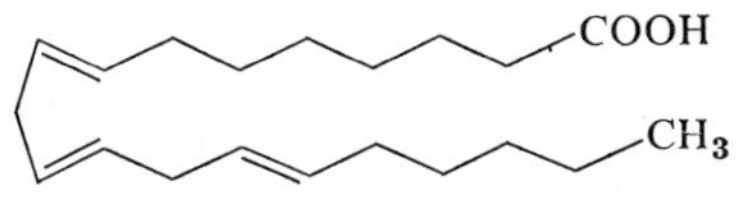

(14)

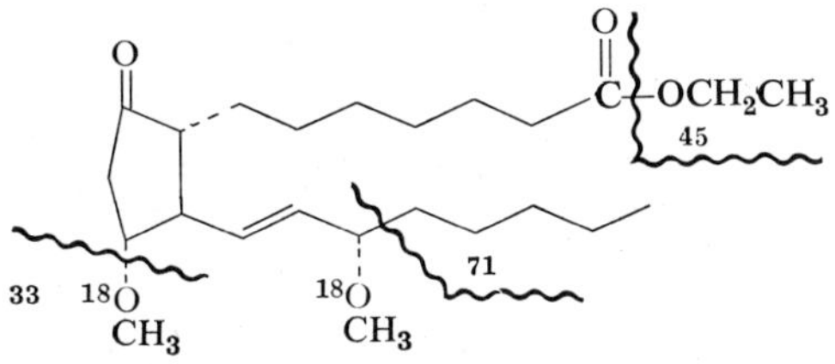

(14a)

The ions retaining the two methoxy groups and the keto group (M and M − 71) were four units higher in the compound biosynthesized in $^{18}O_2$ as compared with the reference. Furthermore, elimination of both methoxy groups (as CH_3OH) gave ions that had the same m/e values for both compounds.

Itada has investigated the mode of molecular oxygen attack in the cleavage of catechol (15) to *cis,cis*-muconic acid (16) catalyzed by pyrocatechase (27). This cleavage reaction imposed two questions: (a) whether

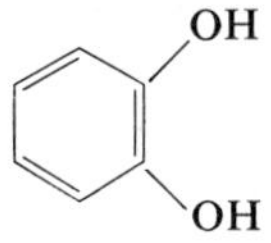

(15) 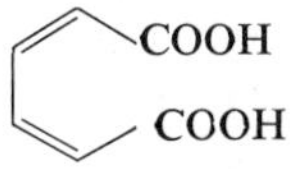   (16)

the two oxygen atoms incorporated were located in one carboxyl or two, and (b) whether the oxygenation involved equilibration among the three molecular species $^{16}O_2$, $^{16}O^{18}O$, and $^{18}O_2$ by cleavage of the O–O bond. Four mechanisms could result in the incorporation of two oxygen atoms from nonequilibrated $^{18}O_2$ into the product:

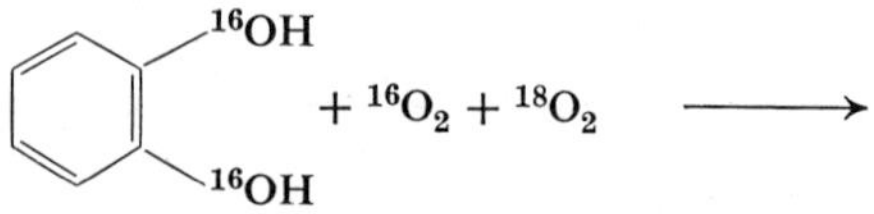

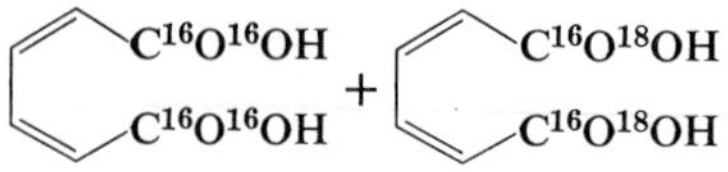

$$
\begin{array}{c}
^{16}OH \\
+ \; ^{16}O_2 + {}^{18}O_2 \longrightarrow \\
^{16}OH
\end{array}
$$

$C^{16}O^{16}OH$ + $C^{16}O^{18}OH$ (without equilibration)   Scheme [1] 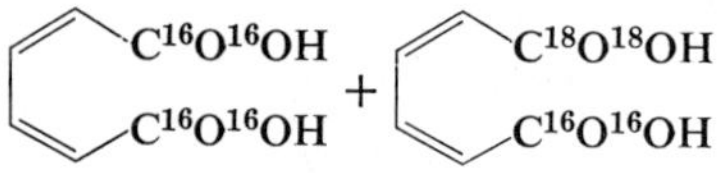
$C^{16}O^{16}OH$ $C^{16}O^{18}OH$

$C^{16}O^{16}OH$ + $C^{18}O^{18}OH$ (without equilibration)   [2] 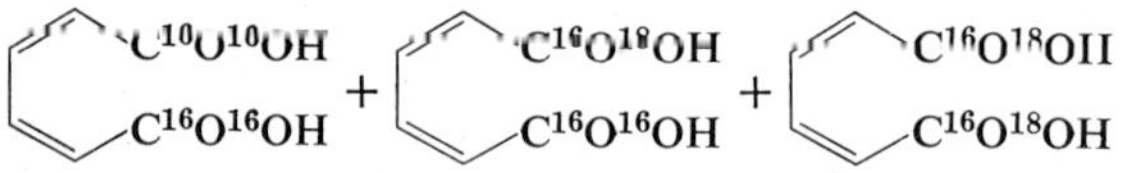
$C^{16}O^{16}OH$ $C^{16}O^{16}OH$

$C^{16}O^{16}OH$ + $C^{16}O^{18}OH$ + $C^{16}O^{18}OH$ (with equilibration)   [3] 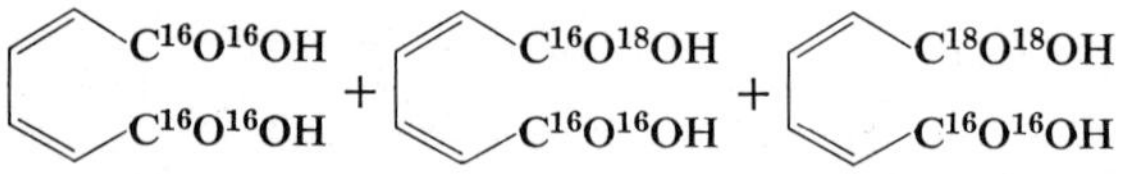
$C^{16}O^{16}OH$ $C^{16}O^{16}OH$ $C^{16}O^{18}OH$

$C^{16}O^{16}OH$ + $C^{16}O^{18}OH$ + $C^{18}O^{18}OH$ (with equilibration)   [4]
$C^{16}O^{16}OH$ $C^{16}O^{16}OH$ $C^{16}O^{16}OH$

The conclusions of this study were based on the intensities of the parent (M), M − OCH_3 and M − $COOCH_3$ ions in the mass spectra of the dimethyl esters of *cis,cis*-muconic acid, as summarized in Table 2. The data unequivocally support scheme (1), that is, one molecule of oxygen contributes one oxygen atom in forming each of the two carboxyl groups of the product.

TABLE 2
^{18}O-Enrichment of *cis, cis*-Muconic Acid

me	Un-labeled		^{18}O-labeled	Relative enrich-ment	Expected for (1)	(2)	(3)	(4)
Parent	170	100	100	100	100	100	100	100
	172	2.15	4.56	2.41	2.86	2.86	56.1	56.1
	174	0.01	26.8	26.7	26.9	26.9	8.09	8.09
M − OCH_3	139	100	100	100	100	100	100	100
	141	1.95	17.4	15.4	15.5	15.5	40.3	40.3
	143	0.17	13.7	13.5	13.3	13.3	3.54	3.54
M − $COOCH_3$	111	100	100	100	100	100	100	100
	113	0.86	27.4	26.5	27.8	1.19	28.1	21.1
	115	0.13	1.25	1.12	0	11.7	0	3.05

Studies on the origin of the oxygen atoms in extracellular tetraacetyl-phytosphingosine (17) produced aerobically by *Hansenula ciferrii* have been reported by Thorpe and Sweeley (28). The studies were carried out with ^{18}O-labeled water and molecular oxygen. The exact masses and empirical

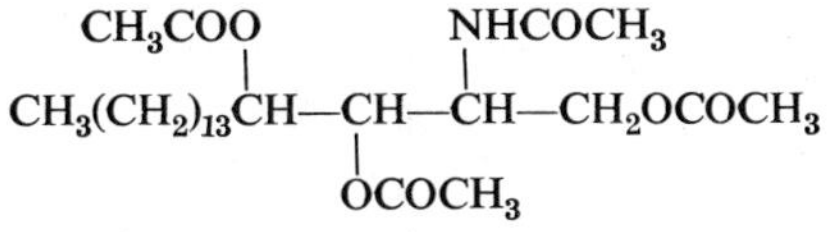

$$
\begin{array}{c}
\text{CH}_3\text{COO} \qquad\quad \text{NHCOCH}_3 \\
| \qquad\qquad\qquad | \\
\text{CH}_3(\text{CH}_2)_{13}\text{CH—CH—CH—CH}_2\text{OCOCH}_3 \\
| \\
\text{OCOCH}_3
\end{array}
$$

(17)

formulas of some of the fragment ions present in the mass spectrum of the product are listed in Table 3. The ions at m/e 84, 144, and 145 were used for interpretation of the position of the label introduced during the growth of the yeast culture in an $^{18}O_2$ enriched water.

The mass spectrum of 17 did not contain any ions which would distinguish labeling of the oxygens on C-3 and C-4. The mass spectrum of N-acetyl-1,3,4-tri-O-trimethylsilylphytosphingosine (18, M = 575) did contain fragment ions that could be used for this purpose, however. A prominent peak at M − 174 results from simple cleavage between C-2 and C-3 with positive charge retention on the long-chain fragment. With the phytosphingosine derivative, this fragment ion $[\text{CH}_3(\text{CH}_2)_{13}\text{CH(OSiMe}_3)\text{CHOSiMe}_3]^+$ is observed at m/e 401. Cleavage between C-3 and C-4 gives a fragment ion $[\text{CH}_3(\text{CH}_2)_{13}\text{CHOSiMe}_3]^+$, which occurs at m/e 299 (M − 276). This latter cleavage is unique with phytosphingosine, since there is no analogous M − 276 peak in the mass spectra of sphingosine or dihydrosphingosine derivatives. Comparisons of the relative ^{18}O abundance in the two fragment ions at m/e 299 and 401 were therefore considered to be of diagnostic value in distinguishing ^{18}O incorporation into the hydroxyl groups on C-3 and C-4 of phytosphingosine. This study showed that none of the oxygen atoms of tetraacetylphytosphingosine was derived from molecular oxygen, while all but one of the oxygen atoms were derived from, or exchanged with, oxygen of the water molecules in the medium.

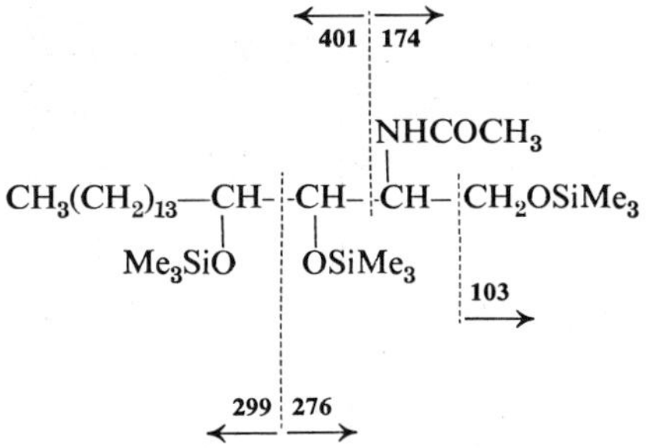

$$
\begin{array}{c}
\overset{\longleftarrow \;\; \vdots \;\; \longrightarrow}{} \\
401 \;\vdots\; 174 \\[4pt]
\text{NHCOCH}_3 \\
| \\
\text{CH}_3(\text{CH}_2)_{13}\text{—CH- CH- CH- CH}_2\text{OSiMe}_3 \\
| \qquad\quad\;\; | \\
\text{Me}_3\text{SiO} \qquad \text{OSiMe}_3 \\
\text{103} \longrightarrow \\[4pt]
299 \;\vdots\; 276 \\
\overset{\longleftarrow \;\; \longrightarrow}{}
\end{array}
$$

(18)

TABLE 3

Exact Masses and Empirical Formulas of Selected Fragment Ions from 17

Nominal m/e ratio	Empirical formula	Exact mass		$\Delta \times 10^3$	Possible structure
		Calculated	Found		
84	C_4H_6ON	84.0449	84.0439	−1.0	$[CH_3CONHCH{=}CH]^+$
144	$C_6H_{10}O_3N$	144.0661	144.0653	−0.8	$[CH(NHCOCH_3)CH_2OCOCH_3]^+$
145 (doublet)	$C_6H_9O_4$	145.0501	145.0494	−0.7	$[CH(OCOCH_3)CHOCOCH_3 + H]^+$
	$C_5{}^{13}CH_{10}O_3N$	145.0697	145.0688	−0.9	Mono-^{13}C of m/e 144
	$C_6H_{11}O_3N$	Not found in spectrum			
292	$C_{19}H_{34}ON$	292.2640	292.2649	+0.9	$CH_3(CH_2)_{12}CH{=}CHCH{=}CHCH{=}C^+$ $\quad\quad\quad\quad\quad\quad\quad\quad\quad$ | $\quad\quad\quad\quad\quad\quad\quad\quad CH_3CONH$
305	$C_{20}H_{35}ON$	305.2719	305.2739	+2.0	$M - 3 \times$ acetic acid
352	$C_{21}H_{38}O_3N$	352.2851	352.2846	−0.5	$CH_3(CH_2)_{13}CH{=}CC^+HNHCOCH_3$ $\quad\quad\quad\quad\quad\quad\quad$ | $\quad\quad\quad\quad\quad\quad\quad OCOCH_3$

228

In a study of the mechanism of the action of adrenal tyrosine hydroxylase, which might effect hydroxyl migration, Daly and his co-workers have
determined the position of ^{18}O in the dopa (19) formed from tyrosine (20) (29).

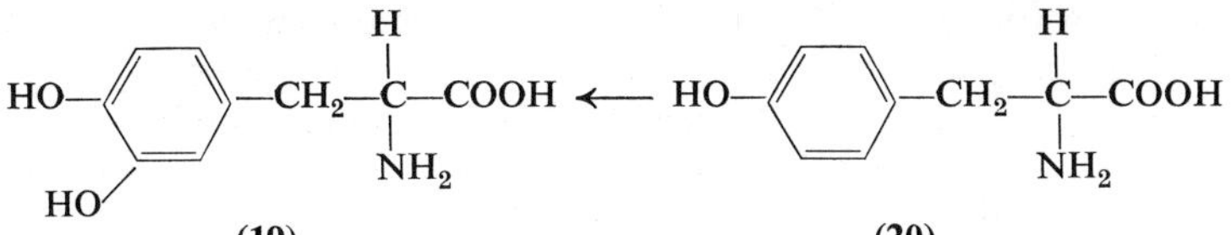

(19)   (20)

The reaction shown below illustrates this possible hydroxylation-induced
migration.

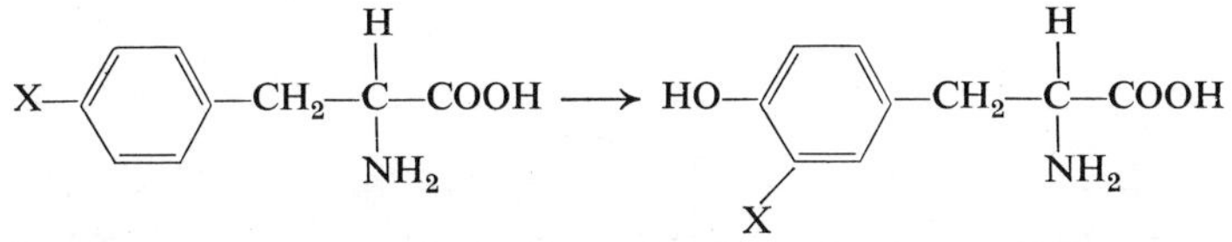

The study included tyrosine 4-^{18}O, hydroxylated in an $^{16}O_2$ atmosphere
as well as unlabeled tyrosine hydroxylated in an $^{18}O_2$-enriched atmosphere.

The ^{18}O content of the phenolic group of tyrosine was obtained from the
ratio of the peak at m/e 109 to the peak at m/e 107 (21). The ^{18}O content of
dopa was similarly calculated from the ratio of the peak at m/e 125 to the
peak at m/e 123 (22). The position of the ^{18}O was determined by enzymatically
converting dopa to 3-methoxytyramine and periodate oxidation of this
material to the o-quinone and methanol. Data on the incorporation of ^{18}O

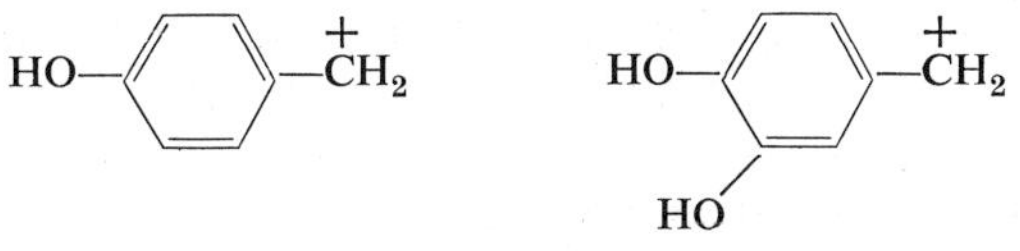

(21)   (22)

precursors are found in Table 4 and data from the ^{18}O assay on the 3-methoxy
group in Table 5. It is clear from these results that atmospheric oxygen was
the source of the 3-hydroxyl oxygen of dopa and that virtually none of the
3-hydroxyl was derived from the 4-hydroxyl of tyrosine.

The origin of the hydroxyl oxygen atom introduced during the hydroxylation of long chain compounds by a species of Torulopsis was studied by Heinz,
Tullock, and Spencer (30). The study involved the incubation of methyl
oleate (23) with whole cells in the presence of $^{18}O_2$ or $H_2^{18}O$. The resulting
17-hydroxy compound was converted to the methyl ether (24) for mass
spectral analysis of the characteristic peak at m/e 59 due to ion 25.

TABLE 4

Incorporation of ^{18}O into Dopa from Tyrosine-4-^{18}O and from $^{18}O_2$
with Adrenal Tyrosine Hydroxylase

Atom-percent excess ^{18}O in the phenolic group of tyrosine	Percentage $^{18}O_2$	Atom-percent excess ^{18}O in the phenolic group of dopa[a]
59	0	33
59	0	48
59	0	48
59	0	41
59	0	43
0	~80	58
0	96	82
0	90	77

[a] Dopa was prepared using tyrosine hydroxylase. The ^{18}O contents were determined by mass spectrometry.

The results of the study showed that (a) the oxygen atom introduced on hydroxylation is derived from molecular oxygen and not from water and (b) this oxygen atom is not lost on glycoside formation. Additional incubations involving esters of [18-$^2H_3$], [16, 18-$^2H_5$], [17-$^2H_2$], [17D-^2H], and [17L-^2H] octadecanoates revealed that no deuterium atoms at C-16 and C-18 were lost,

TABLE 5

Position of ^{18}O in Dopa Formed from Tyrosine-4-^{18}O or from $^{18}O_2$
with Adrenal Tyrosine Hydroxylase

Source of ^{18}O in dopa	Atom-percent excess ^{18}O in a phenolic group of dopa	Atom-percent excess ^{18}O in methanol	Percentage ^{18}O in 3-position[a]
$^{18}O_2$	58	48	83
$^{18}O_2$	81	73	90
Tyrosine-4-^{18}O	33	3	9
Tyrosine-4-^{18}O	45	3	7

[a] The amount of ^{18}O in the 3-position of dopa was determined by enzymatic conversion to 3-methoxytyramine and subsequent periodate oxidation to methanol and dopachrome. The ^{18}O content of the methanol was then determined by mass spectrometry.

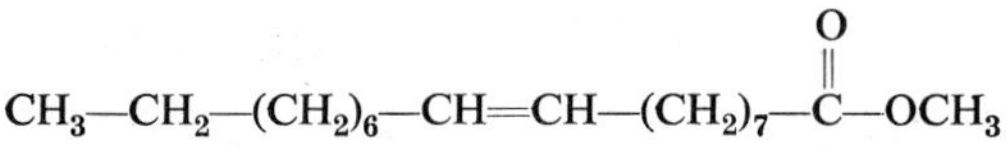

$$CH_3-CH_2-(CH_2)_6-CH=CH-(CH_2)_7-\overset{\overset{\displaystyle O}{||}}{C}-OCH_3$$

(23)

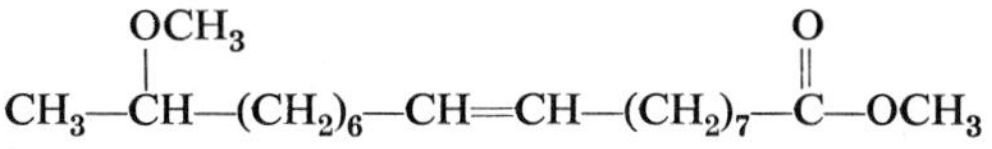

$$CH_3-\overset{\overset{\displaystyle OCH_3}{|}}{CH}-(CH_2)_6-CH=CH-(CH_2)_7-\overset{\overset{\displaystyle O}{||}}{C}-OCH_3$$

(24)

$$CH_3-\overset{\overset{\displaystyle +OCH_3}{||}}{CH}$$

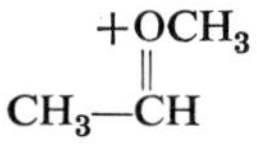

(25)

but the 17L-deuterium atom was. The authors concluded that most probably no unsaturated intermediates were involved and that the 17L-hydroxy acid is produced by displacement of an L-hydrogen atom with retention of configuration.

Caprioli and Rittenberg studied pentose synthesis in *Escherichia coli* using [1-¹⁸O]-, [2-¹⁸O]-, and [6-¹⁸O] glucose and [2-¹⁸O] fructose as the sole carbon sources (31). Growth was terminated in midexponential phase and the nucleic acids were isolated and degraded to the nucleoside level. The distribution of ¹⁸O in the nucleosides was determined by mass spectrometry.

Figure 1 shows the partial mass spectral fragmentation of adenosine (26)

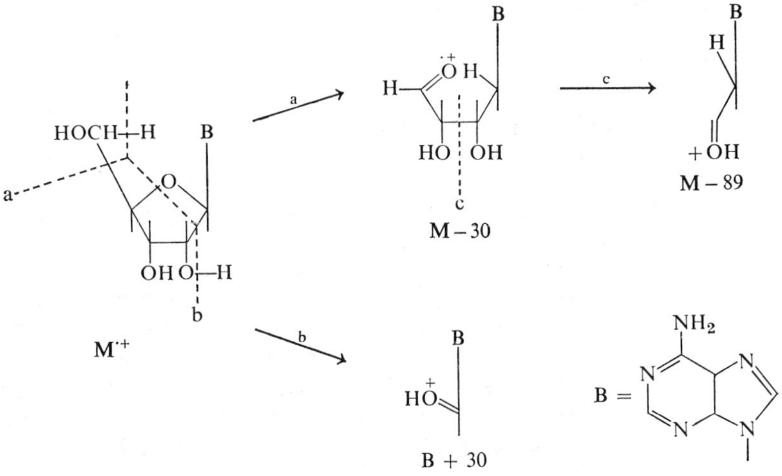

Figure 1. Partial fragmentation of adenosine (26).

and illustrates the typical peaks used in the study of the free nucleosides. Figure 2 shows the partial mass spectral fragmentation of pentakis(trimethylsilyl)guanosine (28) and illustrates the typical peaks used in the study

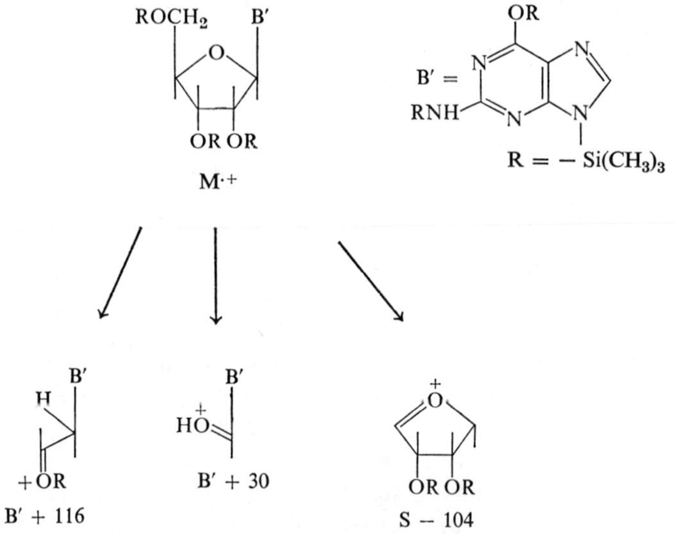

Figure 2. Partial fragmentation of pentakis(trimethylsilyl)guanosine (28).

of the trimethylsilyl derivatives of some of the nucleosides. The exact masses of some of the ^{18}O isotope peaks are listed in Table 6.

TABLE 6.
Exact Masses of Some ^{18}O Isotope Peaks in Nucleoside Spectra

Compound	Peak	Elemental composition	Theoretical mass[a]	Measured mass
Adenosine (26)	$(M + 2)^+$	$C_{10}H_{13}N_5O_3{}^{18}O$	269.1010	269.1021
	$([M - 30] + 2)^+$	$C_9H_{11}N_5O_2{}^{18}O$	239.0904	239.0906
	$([M - 89] + 2)^+$	$C_7H_8N_5{}^{18}O$	180.0771	180.0775
	$([B + 30] + 2)^+$	$C_6H_6N_5O_2{}^{18}O$	166.0614	166.0612
Deoxyadenosine (27)	$(M + 2)^+$	$C_{10}H_{13}N_5O_2{}^{18}O$	253.1061	253.1080
	$([M - 30] + 2)^+$	$C_9H_{11}N_5O^{18}O$	223.0953	223.0964
	$([B + 30] + 2)^+$	$C_6H_6N_5{}^{18}O$	166.0614	166.0607

[a] Calculated using ^{12}C = 12.0000 amu.

The ^{18}O abundance of the ions in the mass spectra of the nucleosides and derivatives are listed in Table 7, and the ^{18}O distributions in the

TABLE 7
^{18}O Abundance of Ions in Mass Spectra of Nucleosides

Nucleoside	Peak X	Oxygen atom(s) present in ion	^{18}O Concentration in ions (atoms % excess)			
			[1-^{18}O]-Glucose expt	[6-^{18}O]-Glucose expt	[2-^{18}O]-Glucose expt	[2-^{18}O]-Fructose expt
Adenosine (26)	M$^+$	2' + 3' + 4' + 5'	20.2	31.8	11.3	13.7
	(M − 30)$^+$	2' + 3' + 4'	0.3	0.6	11.8	13.6
	(M − 89)$^+$	2'	0.4	0.7	4.8	4.5
	(B + 30)$^+$	4'	0.0	0.3	6.6	9.4
Deoxyadenosine (27)	M$^+$	3' + 4' + 5'	20.0	31.8	7.1	9.3
	(M − 30)$^+$	3' + 4'	0.0	0.3	6.7	9.9
	(B + 30)$^+$	4'	0.4	0.6	7.2	9.8
Guanosine (Me$_3$Si)$_5$ (28)	M$^+$	2' + 3' + 4' + 5'	20.7	32.4	11.7	14.6
	(B' + 116)$^+$	2'	0.0	2.8	—	4.4
	(B' + 30)$^+$	4'	0.0	1.6	8.5	8.0
	(S − 104)$^+$	2' + 3' + 4'	0.0	0.7	11.9	15.4
Uridine (Me$_3$Si)$_3$ (29)	(M − 15)$^+$	2' + 3' + 4' + 5'	19.3	32.1	11.1	13.6
	(S − 104)$^+$	2' + 3' + 4'	0.0	2.5	12.2	14.8
Cytidine (Me$_3$Si)$_3$ (30)	(M − 15)$^+$	2' + 3' + 4' + 5'	19.2	29.6	10.0	14.2
	(S − 104)$^+$	2' + 3' + 4'	0.0	2.1	10.5	15.2
	(B″ + 30)$^+$	4'	0.0	0.7		10.6
Deoxycytidine (31)	M$^+$	3' + 4' + 5'	20.2			
	(M − 30)$^+$	3' + 4'	0.6			

TABLE 8

Summary of ^{18}O Distributions in the Nucleosides

Position of oxygen atom	^{18}O Concentration (atom percent excess)[a]			
	[1-^{18}O]-Glucose[b] expt	[6-^{18}O]-Glucose[c] expt	[2-^{18}O]-Glucose[d] expt	[2-^{18}O]-Fructose[e] expt
2'	0	1	5	5
3'	0	0	0	0
4'	0	1	7	9
5'	20	31	0	0

[a] Average value.
[b] 57.3 atom-percent excess ^{18}O.
[c] 48.6 atom-percent excess ^{18}O.
[d] 37.6 atom-percent excess ^{18}O.
[e] 36.5 atom-percent excess ^{18}O.

nucleosides are summarized in Table 8. Based on these data, the following conclusions were reported by the authors:

1. When [1-^{18}O]- and [6-^{18}O]-glucose were used as the carbon sources, 35 and 64%, respectively, of the original label of the hexose appeared in the 5'-oxygen atom of the nucleosides. No other oxygen atoms were labeled.

2. When [2-^{18}O]-glucose and [2-^{18}O]-fructose were used as the substrates, the ribosides were similarly labeled, with approximately 14% of the original label of the hexose in the 2' position and 22% in the 4' position (the deoxyribosides contained ^{18}O only in the 4'-oxygen atom).

3. Both the oxidative and nonoxidative pathways operate simultaneously to produce pentose phosphate.

4. The major portion (about 70%) of the pentose in the nucleic acids was synthesized via the nonoxidative pathway and the remainder via the oxidative pathway.

5. The evidence suggested that the enzyme aldolase in E. coli, in contrast to that of mammalian muscle, cleaves fructose 1,6-diphosphate without the obligatory loss of the C-2 oxygen atom.

Adenosine triphosphates labeled with ^{18}O were used as substrates by Follmann and Hogenkamp for deoxyadenosylcobalamin-dependent ribonucleotide reductase of Lactobacillus leichmannii in the presence of dihydrolipoate as reductant (32). The ^{18}O distribution in the deoxyadenosine phosphates formed was analyzed by mass spectrometry after their conversion to deoxyadenosine (27).

Table 9 shows the isotope-containing fragments obtained from 2'-^{18}O- and 3'-^{18}O-adenosine and from both 2'-deoxyadenosine preparations derived from 2'-^{18}O- and 3'-^{18}O-ATP. In Table 9, large isotope peaks are evident two

TABLE 9

^{18}O Distribution and Relative Peak Heights of Fragments in the Mass Spectra of Adenine Nucleosides (Intensities as % of the Molecular Ion M$^+$)

Nucleoside	Ion	Mass	Containing 0 atoms	Nucleoside labeled in 2'	Nucleoside labeled in 3'
Adenosine[a]	M$^+$	267	2', 3', 4, 5'	100	100
	(M + 2)$^+$	269		37	52
	(M − 30)$^+$	237	2', 3', 4',	150	135
	[(M − 30) + 2]$^+$	239		58	70
	(M − 89)$^+$	178	2'	163	120
	[(M − 89) + 2]$^+$	180		60	(0.62)[b]
Deoxyadenosine	M$^+$	251	3', 4', 5'	100	100
	(M + 2)$^+$	253		(1.39)[b]	24
	(M − 30)$^+$	221	3', 4'	100	150
	[(M − 30) + 2]$^+$	223		(1.05)[b]	37

[a] The ATP-^{18}O synthesized from these two nucleosides was diluted with unlabeled ATP prior to enzymatic reduction.

[b] Calculated values for the naturally occurring M + 2 isotope peak.

mass units higher than the normal ion if ^{18}O is present in the fragment. The data in Table 9 clearly show that the adenosine preparations used for the synthesis of 2'- and 3'-^{18}O-ATP were labeled solely in the 2' and 3' position, and (a) reduction of 2'-^{18}O-ATP to dATP results in complete loss of the isotope label and (b) reduction of 3'-^{18}O-ATP results in the 3'-^{18}O-dATP product with the same isotope content.

Table 10 shows the percent excess ^{18}O in the substrates and products of the ribonucleotide reductase reactions. From these results it is clear that

TABLE 10

Atom-Percent Excess ^{18}O in Nucleotides Before and After Reduction with Ribonucleotide Reductase from *L. leichmannii*[a]

Substrate	Percentage	Product	Percent
$2'$-^{18}O-ATP	16.1	*d*ATP	0.0
$3'$-^{18}O-ATP	19.1	$3'$-^{18}O-*d*ATP	18.6
$2'$-^{18}O-ATP	18.0	$2'$-^{18}O-ATP	18.8

[a] The values were calculated from the mass spectra of the corresponding nucleosides (Table 7) and are an average from 6–10 different spectra. Average error, ±0.5%.

with $2'$-^{18}O-ATP as substrate, all the label was lost from the $2'$ position. In another experiment it was shown that if dihydrolipoate is omitted from a reaction mixture, ribonucleotide reductase does not catalyze exchange of ^{18}O between $2'$-^{18}O-ATP and water.

From these studies the authors concluded that displacement of the hydroxyl function at the $2'$ carbon of a ribonucleotide by hydrogen proceeds in a concerted S_N1 type reaction in which the carbon-oxygen bond is not broken without concomitant reduction of the substrate.

B. Nitrogen-15 and Oxygen-18

Shaw and McCloskey studied the utilization of various labeled compounds by *Penicillium atrovenetum* in the biosynthesis of β-nitropropionic acid (32) (33). The four significant peaks used for the mass-spectrometry study are shown for the methyl ester (33, M = 133). The ion at m/e 102 was greatest interest because it was the only major ion still retaining the nitro group, and measurements of the intensities at m/e 103 and 104 relative to m/e 102 were used to calculate the ^{15}N and ^{18}O content. The authors reported the following conclusions:

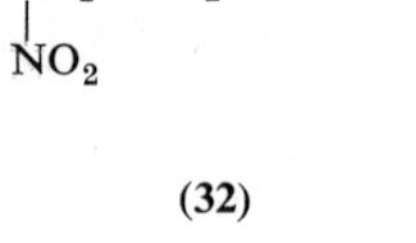

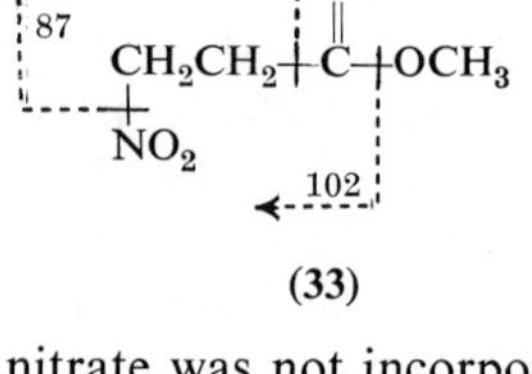

1. The label from ^{18}O-potassium nitrate was not incorporated into the nitro group.

2. The amino group of aspartic acid was utilized in preference to ammonium ion for the synthesis of the nitro group.

3. Label from tartaric acid, which promotes β-nitropropionic acid synthesis, was not incorporated into the nitro group.

C. Nitrogen-15 and Carbon-13

The biosynthesis of slaframine (34) was studied by mass-spectral techniques by Rinehart *et al.*, employing ^{15}N and ^{13}C (34). Figure 3 shows

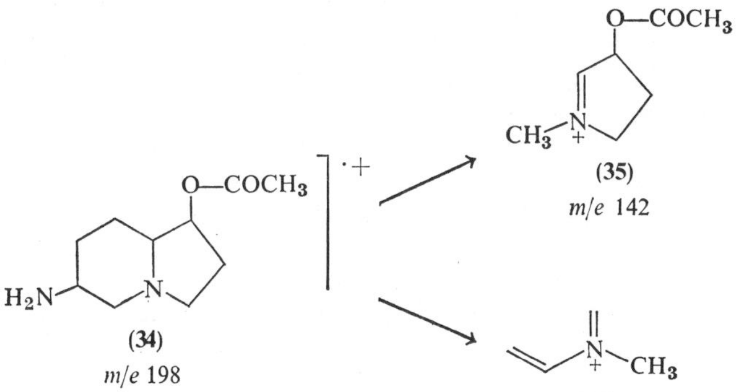

Figure 3. Mass-spectral fragmentations of slaframine.

the mass-spectral fragmentation of slaframine important for this study. The molecular ion allowed the calculation of total isotope incorporation, and ions 35 and 36 allowed calculation of the ^{15}N incorporation at the bridgehead nitrogen as well as in the carbon atoms shown. Both lysine-α- and -ε-^{15}N as well as lysine -ε-^{13}C were used as the substrates in these studies, with the mass-spectral results showing that both the α-amino and ε-amino nitrogens were incorporated specifically into the bridgehead nitrogen, but the ε-amino nitrogen was incorporated only half as well. The ε-^{13}C label appeared in the fragment at m/e 70, localizing its origin.

The biosynthesis of gliotoxin (37) was studied by Bose *et al.*, using ^{15}N- and ^{13}C-labeled compounds and the organism *Trichoderma viride* grown under aerobic conditions (35). Since gliotoxin was found to decompose thermally in the heated inlet system, the aromatic dethio compounds 38 and 39 were used for the mass-spectral studies.

TABLE 11
Labeling of Gliotoxin by Substrates

Substrate	Expt	Isotope enrichment[a], %
^{15}N-Glycine	1	8.9
	2	2.8
	3	8.3
	4	8.3
^{15}N-Phenylalanine	1	4.4[b,c]
	2	1.3
	3	0.6
	4	3.9
	5	4.0
2-^{13}C-Glycine	1	1.4
	2	2.5
1-^{13}C-Glycine	1	2.4
^{13}C-Formate	1	1.0
	2	2.5

[a] Based on compound **38** unless noted otherwise.
[b] Based on compound **39**.
[c] Due to the small quantity of material available, the error is probably quite large. In other experiments the error is about ±0.2.

TABLE 12
High-Resolution Mass Spectrum of Compound 38; Selected Peaks

m/e	Compound	Fragment
226	$C_{13}H_{10}N_2O_2$	$[M]^+$
199	$C_{12}H_9NO_2$	$[M - HCN]^+$
198	$C_{12}H_{10}N_2O$	$[M - CO]^+$
197	$C_{12}H_7NO_2$	$[M - NCH_3]^+$
	$C_{12}H_9N_2O$	$[M - CHO]^+$
170	$C_{11}H_{10}N_2$	$[M - 2CO]^+$
144	C_9H_6NO	
143	C_9H_5NO	
	$C_{10}H_9N$	
130	C_9H_8N	
129	C_9H_7N	
116	C_8H_6N	
115	C_8H_5N	

Table 11 shows the total percent isotope enrichment of gliotoxin as measured from the spectra of compounds **38** and **39**, while the compositions of some of the various fragments as determined by high-resolution mass spectrometry are listed in Table 12. It was assumed that the structure of the m/e 115 fragment corresponded to **40**. The isotope incorporation data for

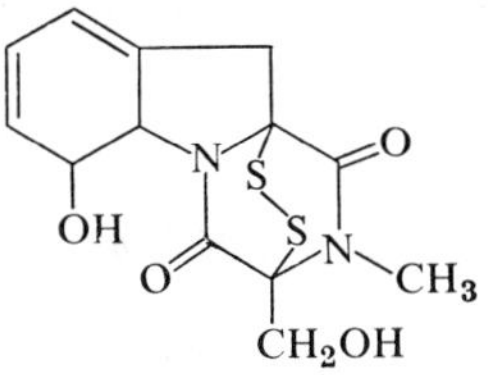

(37)

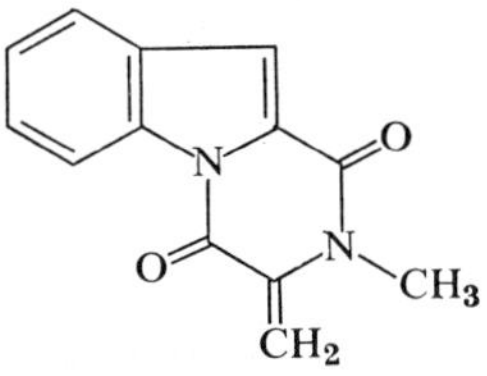

(38)

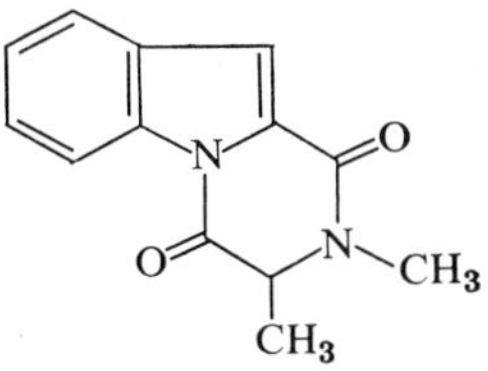

(39)

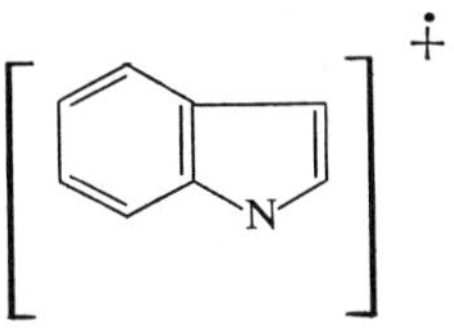

(40)

various fragments are listed in Tables 13 and 14. These data indicate that when ^{15}N-glycine is added to the medium, both nitrogen atoms in gliotoxin

TABLE 13
Incorporation Data on Compound **39**

Fragment, m/e	Fragment	Compound **39** from ^{15}N-glycine $\Delta(\pm 0.2)$, %
228	M^+	8.9
213	$[M - CH_3^+$	8.9
143		2.3
115		2.3

are labeled, while the carbon atoms of glycine do not appear to be incorporated into the indole portion.

TABLE 14

Incorporation Data on Compound **38**

Fragment, m/e	^{15}N-Phenyl alanine	^{15}N-Glycine	$2\text{-}^{13}C\text{-}$ Glycine	$1\text{-}^{13}C\text{-}$ Glycine
		Compound **38** from substrate shown		
226	1.3	2.8, 10.4	1.4	2.4
143	1.0	0.8, 3.5	0.3	0.0
115	0.8	0.6, 3.2	0.0	0.7

Waller, Ryhage, and Meyerson reported on the mass spectrometry of biosynthetically labeled ricinine (**41**) (36). While the biosynthetic pathway

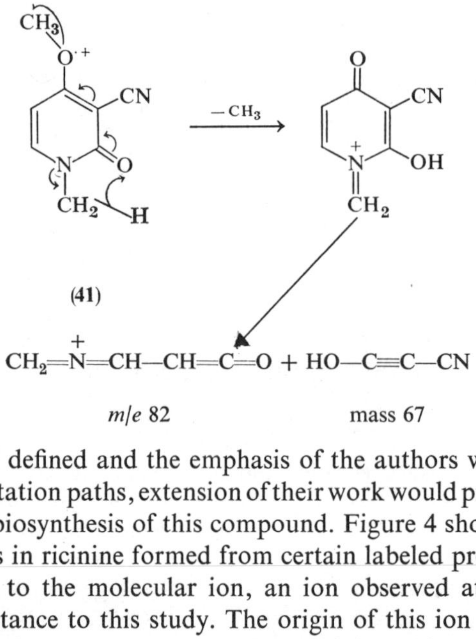

(41)

$$CH_2\!\!=\!\!\overset{+}{N}\!\!-\!\!CH\!\!-\!\!CH\!\!=\!\!C\!\!=\!\!O \ + \ HO\!\!-\!\!C\!\!\equiv\!\!C\!\!-\!\!CN$$

m/e 82　　　　　　　　　　mass 67

was already well defined and the emphasis of the authors was on the mass-spectral fragmentation paths, extension of their work would provide additional insight into the biosynthesis of this compound. Figure 4 shows the positions of isotopic labels in ricinine formed from certain labeled precursors.

In addition to the molecular ion, an ion observed at m/e 82 was of particular importance to this study. The origin of this ion is shown below, illustrating how the mass spectrum might aid in a biosynthesis study. In addition to the carbon incorporations shown in Figure 4, it was found that the nitrogen from ^{15}N-aspartate becomes the ring nitrogen of ricinine.

D. Deuterium

Studies on the biosynthesis of cholesterol involved deuterium as the stable isotope. Popjak *et al.* studied the mechanism of squalene biosynthesis

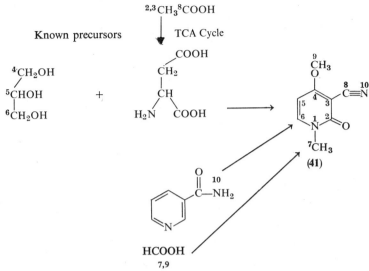

Figure 4. Labeling pattern of ricinine formed by *Ricinus communis* L. plants from isotopically labeled precursors.

from farnesyl pyrophosphate and from mevalonate (37). Squalene was biosynthesized from DL-mevalonate-5,5-D_2-2-^{14}C (42) using the S_{10} preparations of rat-liver homogenates. The farnesyl pyrophosphate was prepared

(42a)   X = D, Y = ^{14}C
(42b)   X = D, Y = ^{12}C

biosynthetically from 42. The bulk of the "squalene pool" was degraded by ozonolysis in order to obtain the four central carbon atoms as succinic acid. The resulting succinic anhydride and dimethyl succinate were analyzed by mass spectrometry. The fragmentation patterns of succinic anhydride are illustrated in Figure 5.

The calculations from the spectra led to the conclusion that the squalene succinic anhydride contained unlabeled, D_2 and D_3 molecules in the approximate proportions of 5:12:54; there were no molecules containing four deuterium atoms and probably insignificant numbers containing one deuterium atom. The complete mass-spectrometric analysis showed that the

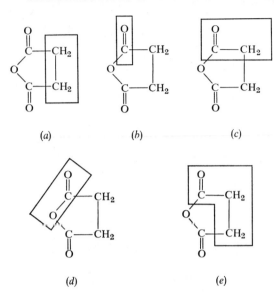

(a) (b) (c)

(d) (e)

Figure 5. Fragmentation of succinic anhydride in the mass spectrometer. (a) m/e 28;
(b) m/e 28; (c) m/e 42; (d) m/e 44; (e) m/e 56.

deuterio squalene sample gave mostly trideuterio succinic acid, requiring the labeling in the center of the squalene to be asymmetrical, $-CHD-CD_2-$.

The chemical synthesis of $1-{}^3H_2-2-{}^{14}C$ and $1-D_2-2-{}^{14}C$-*trans-trans*-farnesyl pyrophosphate and their utilization in squalene biosynthesis was also reported by Popjak *et al.* (38). Mass spectrometry was used to check the assumption from earlier results that the farnesyl pyrophosphate formed from mevalonate-5,5-D_2, as an intermediate in squalene biosynthesis, should contain six atoms of deuterium, two of which were attached to C-1 of farnesol (43). The farnesol obtained from the deuterium-labeled farnesyl-${}^{14}C$

$$CH_3-C=CH-CH_2-CH_2-C=CH-CH_2-CH_2-C=CH-CH_2OH$$
$$\quad\;\; CH_3 \qquad\qquad\quad CH_3 \qquad\qquad\quad CH_3$$

(43)

pyrophosphate gave a molecular ion at m/e 228, indicating the presence of six atoms of deuterium in the molecule. A large peak at m/e 195 represented $M - 33$ ion from loss of a $-CD_2OH$ group. Therefore, the earlier assumption was correct.

Scott and his co-workers studied the biosynthesis of vindoline (44, $R^1 = R^2 = H$) from mevalonic-5-D_2 lactone (42b) to confirm radiochemical experiments on the nature of the "C_{10} precursor" (39). The principal fragments of vindoline are shown in Figure 6. The mass spectrum of the

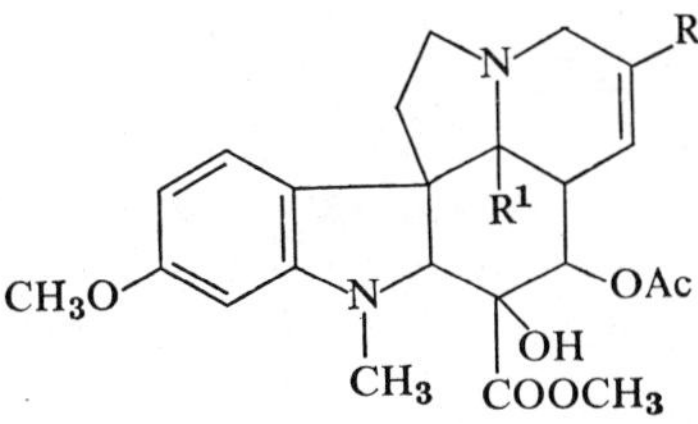

(44)

biosynthesized vindoline (**44**, $R^1 = R^2 = D$) showed both deuterium atoms in the molecular ion peak (m/e 458) and in fragments a, d, and e (Figure 6).

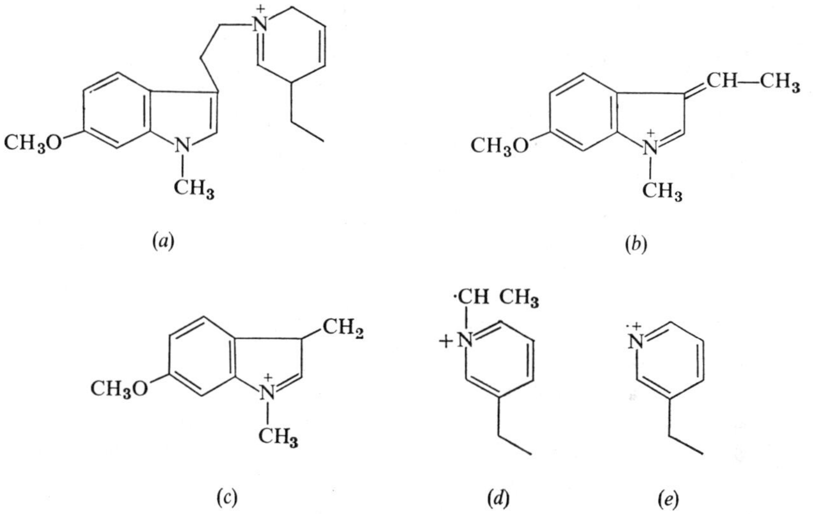

(a)

(b)

(c)

(d)

(e)

Figure 6. Principal fragments of vindoline. (a) m/e 296; (b) m/e 188; (c) m/e 174; (d) m/e 135; (e) 107.

None of the deuterium label was found in the tryptophan-derived segments (ions b and c, Figure 6). These results are compatible with the mevalonoid dimerization to the logical C_{10} precursor, geraniol (**45a**). Incorporation studies using deuterio geraniol (**45b**) supported this conclusion.

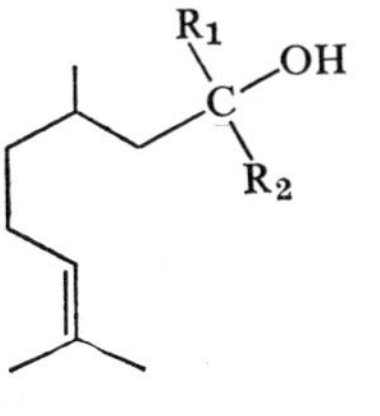

(45a) $R_1 = R_2 = H$
(45b) $R_1 = R_2 = D$

Mass spectrometry was used by Schroepfer to study the microbiological conversion of oleic acid to 10D-hydroxystearic acid in a medium enriched in deuterium oxide (40). The information concerning the location of the deuterium in the hydroxystearate was obtained by analysis of the mass spectra of the deuterated 10-hydroxystearate. A major peak in the spectrum of methyl 10-hydroxystearate (46) was observed at m/e 201, representing the $\overset{+}{HO}$=CH(CH$_2$)$_8$COOCH$_3$ fragment. This peak was observed at m/e 202 in the spectra of the methyl esters of the deuterated product (47) and the known

$$CH_3(CH_2)_7\underset{\underset{OH}{|}}{CH}(CH_2)_8COOCH_3 \qquad CH_3(CH_2)_7\underset{\underset{OH}{|}}{CHCDH}(CH_2)_7COOCH_3$$

(46)   (47)

DL-10-hydroxystearic acid-10-^{2}H (48). Likewise, the base peak in the

$$CH_3(CH_2)_7\underset{\underset{OH}{|}}{CD}(CH_2)_8COOCH_3$$

(48)

spectrum of 46 was observed at m/e 169 (m/e 201 minus 32) and at m/e 170 in the spectra of 47 and 48. Another prominent peak in the spectrum of 46 that was used for this study was that at m/e 172, assigned to the [(CH$_2$)$_8$-COOCH$_3$ + H]$^{.+}$ fragment. This peak was observed at the same position in the spectrum of 48 but at m/e 173 in the spectrum of 47. This provided proof that the D-label was on C-9 of the molecule.

Chemical removal of the hydroxyl and stereospecific microbiological conversion of the resulting stearic acid to oleic acid with retention of the deuterium atom showed that 47 was the 9L-deuterio compound.

Schroepfer and his co-workers studied the conversion of Δ^8-cholesten-3β-ol (49) to Δ^7-cholesten-3β-ol (50) by rat-liver microsomes (41). When the

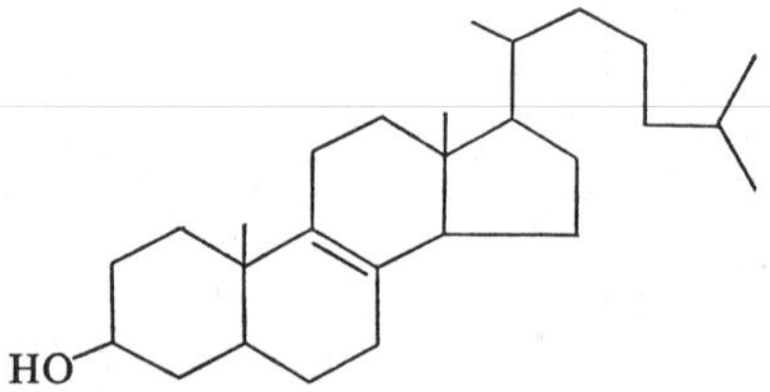

(49) 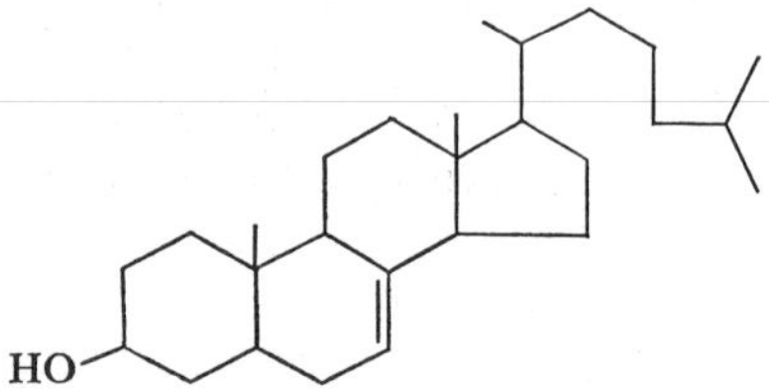   (50)

bioconversions were carried out in a deuterated medium, GLC-MS analysis of the trimethylsilyl ether derivatives of the Δ^7 product showed that this process involved the uptake of one atom of hydrogen from the solvent.

Deuterium isotope effects in the fermentation of hexoses to ethanol (Embden-Meyerhoff-Parnas glycolytic pathway) have been the subject of a very complex study by Katz *et al.* (42a). Unlabeled *D*-glucose and *D*-mannose were separately fermented by *Saccharomyces cerevisiae* in deuterium oxide to give mainly CD_3CD_2OD (33% from glucose, 37% from mannose), CHD_2CD_2OD (29, 31%), and CH_2DCD_2OD (23, 17%), and fully deuterated mannose and glucose-6-d_2 were separately fermented in water to give mainly CH_3CH_2OH (17, 50%), CH_2DCH_2OH (34, 9%), and CHD_2CH_2OH (36, 41%). The products were analyzed by proton magnetic resonance and mass spectra of the ethyl acetates (42b). The mass-spectral analyses were utilized only in the molecular-ion region, to give the distribution of deuterated species, while the label was located in methyl or methylene groups from the nmr spectra.

The isotopic compositions of the ethanol samples suggested the conclusions that (a) exchange with the medium of the methylene group of the product ethanol occurred at the triose phosphate level during the sequence triose phosphate isomerase-aldolase-glyceraldehyde diphosphate dehydrogenase, and (b) the incorporation of H or D from the medium into the methyl group of ethanol requires the presence of both phosphoglucose and phosphomannose isomerase during fermentation of either mannose or glucose, plus further exchange by pyruvate kinase.

Several studies have involved the use of methionine-CD_3 as a tracer in biosynthesis. Lenfant, Zissman, and Lederer studied the biosynthesis of the ethyl side chain of stigmasterol derivatives by the slime mold *Dictyostelium discoideum* (43). Mass spectrometry of the isolated stigmast-22-en-3β-ol (**51a**) proved without a doubt that five deuterium atoms had been incorporated into the molecule. The reference mass spectrum of the nondeuterated acetate derivative (**51b**) contained the molecular ion at *m/e* 456 and an intense peak

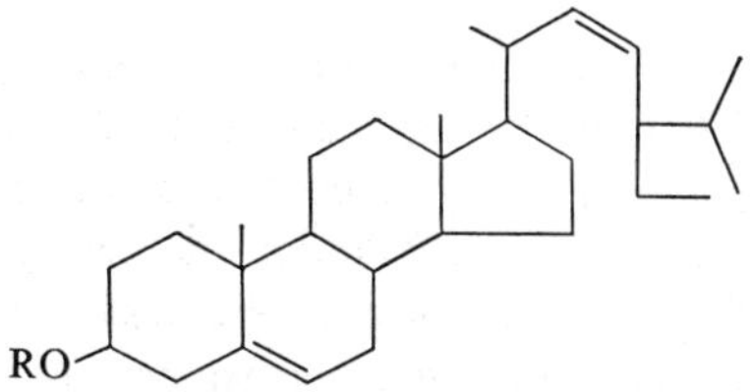

(**51a**) R = —H

(**51b**) R = —C(=O)—CH₃

at *m/e* 344, which was assigned structure **52**. In addition, the spectrum showed a small peak at *m/e* 427 (M − 29) apparently due to the loss of the side chain ethyl group and a peak at *m/e* 413 (M − 43) due to the loss of the isopropyl

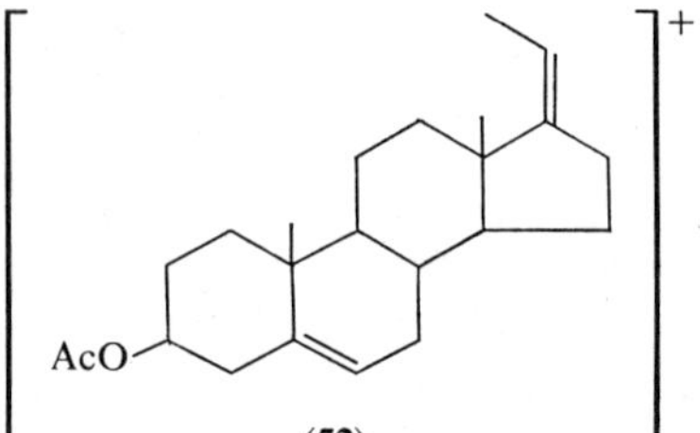

(52)

group. The acetate derivative of the deuterated sample from the methione-CD_3 feeding experiments showed the molecular ion at m/e 461 (456 + 5) corresponding to an incorporation of five deuterium atoms in the molecule. The peak at m/e 344 remained unshifted while the m/e 413 (M − isopropyl group) shifted to m/e 418 and the m/e 427 (M − ethyl group) remained unshifted. Therefore, all five of the incorporated deuterium atoms are on the side-chain ethyl group, and two successive C-methylation steps transfer the methyl group of methionine as a unit.

Jackman et al. used methionine-CD_3 to prove that the methyl group of methionine is the source of the ring-methyl group of vitamin K_2 (53) and of the C-methyl and O-methyl groups of the ring of ubiquinone (54) in

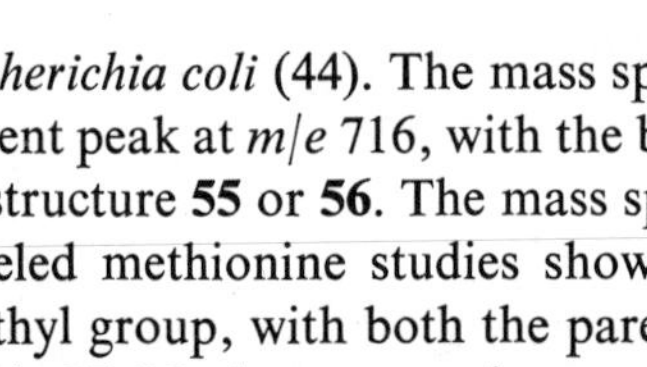

(53) 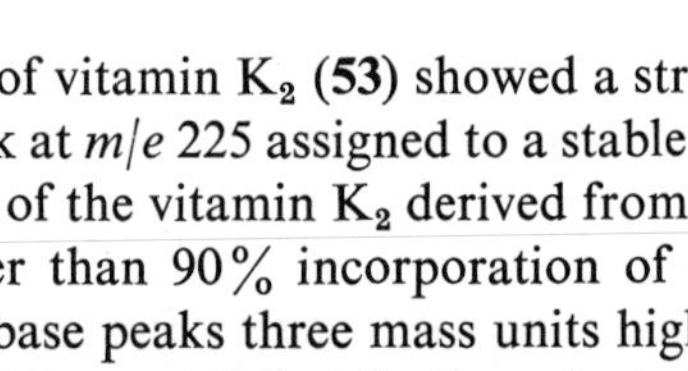   (54)

Escherichia coli (44). The mass spectrum of vitamin K_2 (53) showed a strong parent peak at m/e 716, with the base peak at m/e 225 assigned to a stable ion of structure 55 or 56. The mass spectrum of the vitamin K_2 derived from the labeled methionine studies showed better than 90% incorporation of one methyl group, with both the parent and base peaks three mass units higher, Table 15. Nuclear magnetic resonance spectra proved that the three deuterium atoms were on the ring C-methyl group.

The mass-spectral results for the ubiquinone study are shown in Table 16. These results indicated the incorporation of nine deuterium atoms (three methyl groups) to the extent of 99% or greater. The base peak (m/e 235 for the unlabeled compound; m/e 244 for the labeled) was attributed to an ion analogous to 55 or 56 generated from vitamin K_2. Again, nuclear magnetic resonance spectra proved that the nine deuterium atoms were

TABLE 15

Mass-Spectral Data for Vitamin K_2 [MK-8]
Isolated from Wild-Type Cells and Cells
Grown with [Me-$^2H_3$]Methionine

	Abundance	
m/e	Unlabeled	Labeled
719	—	27.3
716	27	<3
228	—	100[a]
225	100[a]	14
222	5	—

[a] Base peak of spectrum.

TABLE 16

Mass-Spectral Data for Ubiquinone [Q-8]
Isolated from Wild-Type Cells and Cells
Grown with [Me-$^2H_3$]Methionine

	Abundance	
m/e	Unlabeled	Labeled
735	—	32
726	22	<0.3
244	—	100[a]
235	100[a]	<1

[a] Base peak of spectrum.

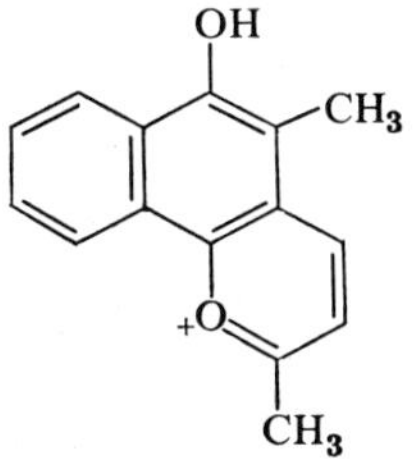

(55)

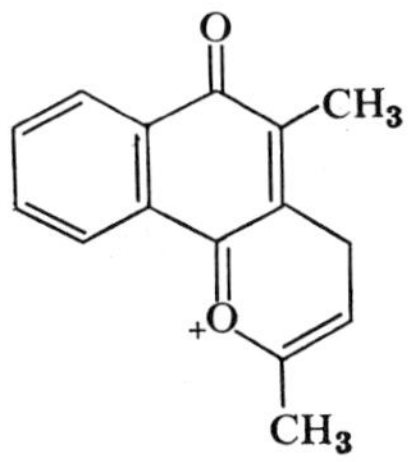

(56)

247

incorporated into the *C*-methyl and the two *O*-methyl groups of the quinone nucleus.

Argoudelis, Eble, Fox, and Mason showed that the S—CH_3 and N—CH_3 groups of the antibiotic lincomycin (**57**), as well as the C—CH_3

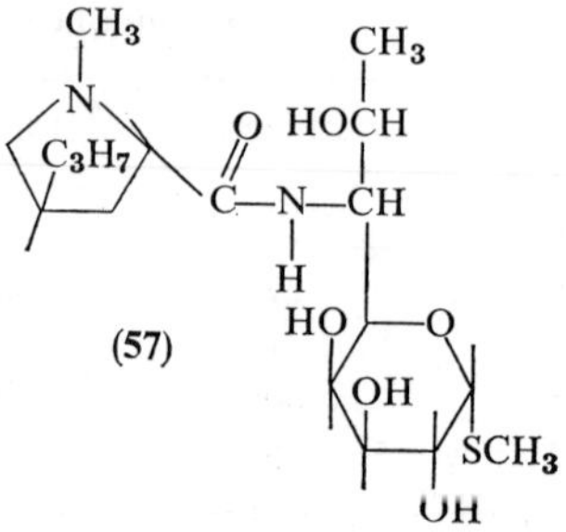

(**57**)

group present in the amino acid moiety, are derived from C_1 donors, using CD_3-methionine in the fermentation process (45). The mass-spectral data are summarized in Table 17. The parent and M—SCH_3 ions, along with the ions

TABLE 17

Mass Spectrum of Lincomycin Obtained by Feeding L-[CD_3]Methionine

	m/e (Relative abundance)			
M at	406 (48)	409 (18)	412 (20)	415 (13)
		$-SCH_3$ ↓ $-SCD_3$		
M − SCH_3 at	359 (58)	362 (25)	365 (20)	
		↓		
58 at	257 (58)	260 (22)	263 (21)	
		↓		
59 at	126 (57)	129 (24)	132 (20)	

58 and **59**, were pertinent to the study. The molecular ions observed at

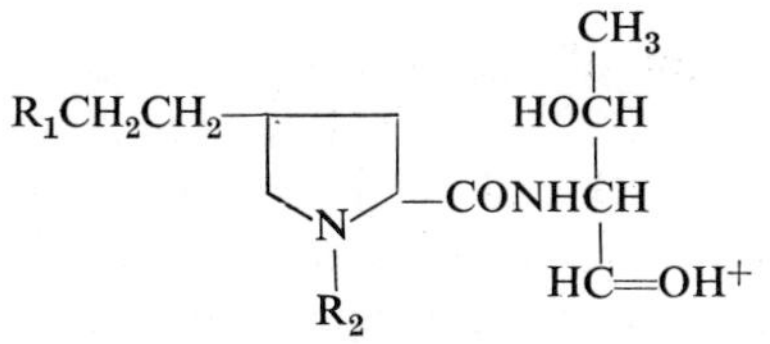

$$\text{R}_1\text{CH}_2\text{CH}_2\!\!\underset{\substack{|\\ \text{R}_2}}{\boxed{}}\!\!\!\!\underset{\text{N}}{}\!\!-\text{CONH}\overset{\begin{array}{c}\text{CH}_3\\ |\\ \text{HOCH}\\ |\end{array}}{\text{C}}\text{H}$$

$$\text{HC}\!=\!\text{OH}^+$$

$$\text{R}_1\text{CH}_2\text{CH}_2\!\!\underset{\substack{|\\ \text{R}_2}}{\boxed{}}\!\!\!\!\underset{\overset{+}{\text{N}}}{}$$

(58) 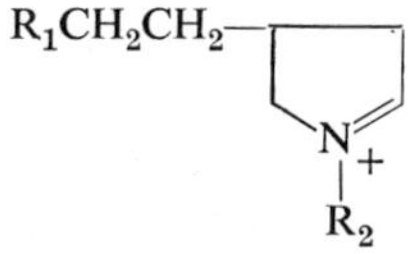   (59)

58a (m/e 257); **59a** (m/e 126): $\text{R}_1 = \text{CH}_3$; $\text{R}_2 = \text{CH}_3$
58b (m/e 260); **59b** (m/e 129): $\text{R}_1 = \text{CH}_3$; $\text{R}_2 = \text{CD}_3$; or $\text{R}_1 = \text{CD}_3$; $\text{R}_2 = \text{CH}_3$
58c (m/e 263); **59c** (m/e 132): $\text{R}_1 = \text{CD}_3$; $\text{R}_2 = \text{CD}_3$

m/e 490, 412, and 415 and the M $-$ $\text{SCH}_3(\text{CD}_3)$ ions observed at m/e 359, 362, and 365 indicated that a maximum of three CD_3 groups were incorporated, one an S—CD_3 group. Ions **58** and **59** showed that both the N—CH_3 and C—CH_3 groups of the propylhygric acid moiety of the antibiotic molecule had also incorporated CD_3 groups.

III. METABOLISM

As we noted earlier, studies reviewed in this section will generally involve compounds that are foreign to the normal animal or plant bio-synthetic pathway. Such compounds are often metabolized (altered) by the plant or animal in order to rid itself of the foreign material, in which case the metabolic reactions are referred to as detoxification mechanisms. Mass spectrometry has proven to be a powerful tool in such studies because of its high sensitivity for detecting small amounts of metabolites and because many metabolic changes are only slight modifications of the original material.

Goldman and Milne have studied the mechanism of the defluorination of fluoroacetate (46). The reaction

$$\text{XCH}_2\text{COO}^- + \text{OH}^- \rightarrow \text{X}^- + \text{HOCH}_2\text{COO}^-$$

(where X = F, Cl, and I) is catalyzed by the enzyme haloacetate halido-hydrolase.

The mass spectrum of the glycolic acid product (**60**) shows four ions

$$\text{HOCH}_2\text{COOH}$$
(**60**)

important in the metabolism study. The base peak of the spectrum is observed at m/e 31 and is due to the stabilized ion $\text{CH}_2 \overset{+}{=} \text{OH}$. The other three ions are the molecular ion at m/e 76, the ion at m/e 44 ($\text{COO}^{\cdot+}$), and the ion at m/e 45

$(O = C = \overset{+}{O}H)$. The mass spectrum of the glycolic acid derived from fluoroacetate in a medium containing 30% $H_2^{18}O$ showed additional peaks at m/e 33 ($CH_2^{18}OH^+$) and m/e 78 (^{18}O molecular ion). Therefore, it was concluded that the hydroxyl group of glycolate is derived from water. No evidence was found for the reversibility of the dehalogenation or exchanges between product and water.

Two halidohydrolases were found by Goldman, Milne, and Keister in extracts of a pseudomonad isolated from soil (47). Both enzymes catalyze a variety of dehalogenations, with one induced by chloroacetate, the other by dichloroacetate. Mass spectrometry was used to identify the products and to determine the position of ^{18}O in the product(s) as described previously. The ^{18}O atom incorporated from $H_2^{18}O$ was found in the 2-hydroxyl group of the 2-hydroxybutyrate formed in the enzymatic dehalogenation of 2-chloro-butyrate.

Holtzman, Gillette, and Milne found that one hydroxyl oxygen was derived from air in the enzymatic formation of 1,2-dihydronaphthalene-1,2-diol (61) from naphthalene (62) (48). The mass spectrum of a sample obtained

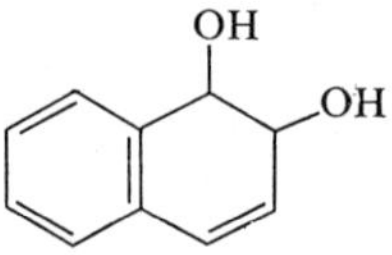

(61) 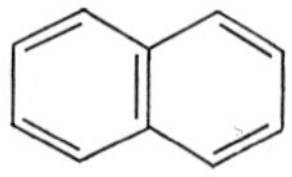   (62)

from mixtures in air showed the molecular ion of the diol at m/e 162, while the spectrum obtained from the diol formed in ^{18}O-^{18}O-enriched atmospheres showed a marked increase in the abundance of the ion at m/e 164, but no ion at m/e 166, which suggests the presence of only one atom of ^{18}O in the 1,2-dihydronaphthalene-1,2-diol. These conclusions were confirmed by accurate mass measurement (Table 18). The results are consistent with the hypothesis that liver preparations convert naphthalene to 1,2-dihydro-1,2-epoxy

TABLE 18

Accurate Mass-to-Charge Ratios in the Mass Spectrum of ^{18}O-1,2-Dihydronaphthalene-1,2-diol

m/e Found	Formula	m/e Calculated
162.0685	$C_{10}H_{10}^{16}O_2$	162.0687
164.0726	$C_{10}H_{10}^{16}O^{18}O$	164.0723

naphthalene, which in turn reacts with either hydroxyl ion to form 1,2-dihydronaphthalene-1,2-diol or glutathione to form S-(1,2-dihydro-2-hydroxyl-1-naphthyl) glutathione. The authors were able to show by nuclear magnetic spectroscopy that **61** was the *trans* diequatorial diol (49).

An inducible oxygenase that opens the pyridine ring during the metabolic degradation of Vitamin B_6 by a soil bacterium has been described by Snell *et al.* (50). The enzyme was found to catalyze the two reactions shown. Compound **64** was isolated as its dimethyl ester; the high mass region of the mass spectra of unlabeled **64** and of compound **64** formed in the presence

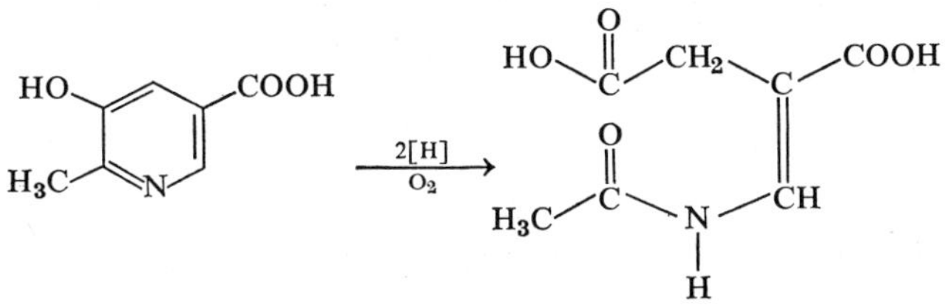

(63)   (64)

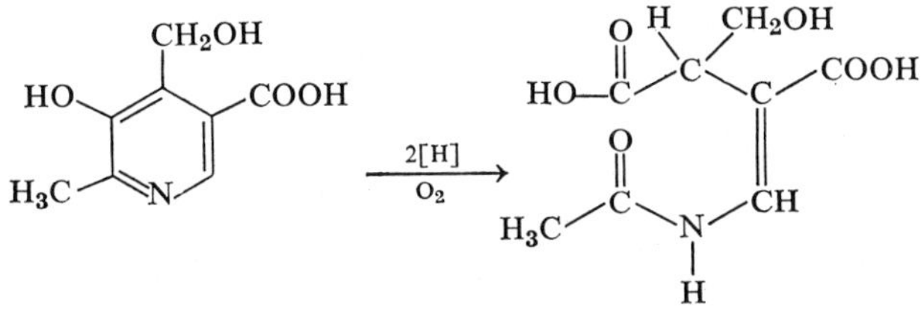

(65)   (66)

of $^{18}O_2$ were both measured. The two prominent fragment peaks at m/e 173 and 156 were interpreted in terms of the processes shown in Figure 7. Thus the measurement of the relative peak heights at m/e values of 219, 217, and 215 gave the total incorporation into the molecule; the peak height ratios of m/e 175:173 and 158:156 gave the position of any isotopic oxygen introduced into the molecule.

The mass spectrum of the dimethyl ester produced in the $^{18}O_2$ experiment shows not only a substantial peak at m/e 217 (M + 2), but also a definite peak at m/e 219 (M + 4). Therefore, some of compound **64** produced under these conditions contained two atoms of ^{18}O. [Experiment 4 (Table 19) showed that no exchange of oxygen between water and compound **64** occurred after its formation.] The summarized results of the tracer experiments (Table 19) were used to calculate the incorporations shown in Table 20. These results show that the ^{18}O isotope is incorporated predominantly into the acetyl group but also into the newly formed carboxyl group of the

TABLE 19

Extent of ^{18}O Incorporation into Compound **64**

Experiment	Reaction medium	m/e 219:217:215	175:173	158:156
1	$^{16}O_2$, $H_2^{16}O$	1.9:100	1.5:100	1.2:100
2	82.8%[a] $^{18}O_2$, $H_2^{16}O$	54:404:100	14.3:100	376:100
3	$^{16}O_2$, 4.1%[b] $H_2^{18}O$	3.3:100	3.1:100	1.5:100
4	Control[c]	2.0:100	Not measured	
Theoretical		1.5:100	1.1:100	0.9:100

[a] Obtained by mass spectrometric analysis of the residual oxygen.

[b] By dilution of a commercial sample.

[c] For the control experiment, the enzymatic reaction was allowed to go to completion in $H_2^{16}O$; water was then removed by lyophilization and the resulting solid was dissolved in 4% ^{18}O-enriched water. The mixture was then stirred for four hours at room temperature and prepared for analysis in the same manner as other samples.

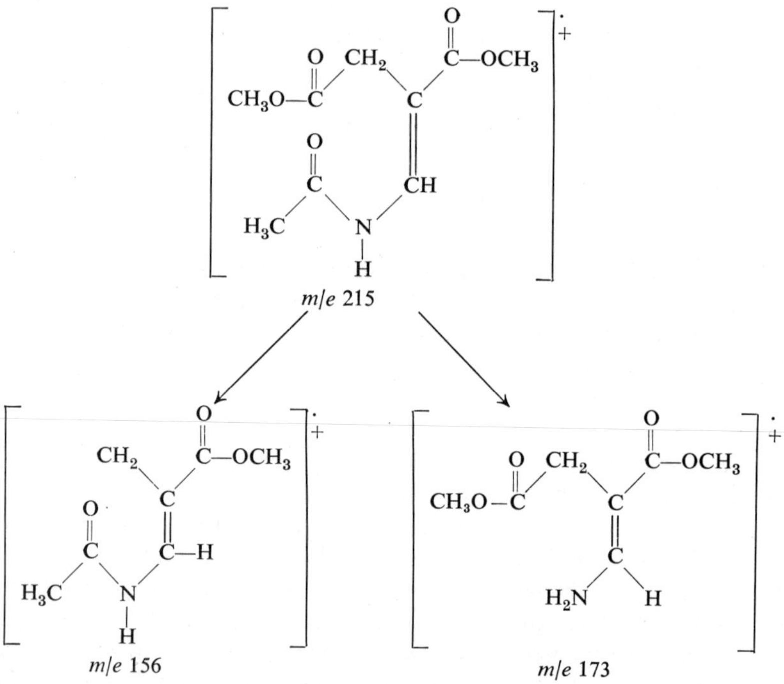

Figure 7. Fragmentation pattern of the dimethyl ester of **64**, which accounts for major peaks at m/e 173 and 156.

252

TABLE 20
Incorporation of ^{18}O into Compound **64** and Its Fragments

Incorporation from	^{18}O Incorporated into		
	Compound **44**	CH_3CO-	$HOOC-CH_2-$
	atoms		
$^{18}O_2$	1.10	0.95	0.14
$H_2{}^{18}O$	0.33	0.07	0.38

product. The postulated mechanism for the action of this enzyme is shown in Figure 8.

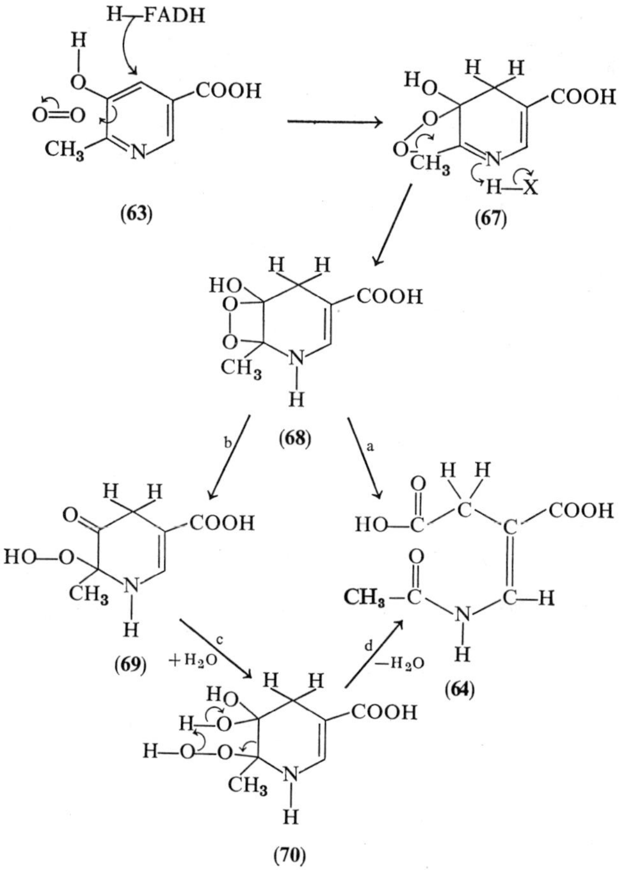

Figure 8. Postulated mechanism for the action of oxygenase.

In the course of a study of germ-free rats, the 21-dehydroxylation of steroids by rat intestinal microorganisms was studied by Eriksson, Gustafsson, and Sjövall (51). Pregnane-3,20-diols (**76** and **79**) were major products of the metabolism of $3\beta,21$-dihydroxy-5α-pregnan-20-one (**71**) in incubations with caecal microorganisms. However, this conversion did not proceed by dehydroxylation of 5α-pregnane-$3\beta,20\beta$ (or 20α) 21-triol (**72** or **73**), since no deuterium-labeled diol was formed from the incubation of a mixture of 20α (and 20β)-^{2}H labeled pregnanetriols. Thus C-21 dehydroxylation was presumed to take place from **71**, to give **74**, followed by reduction at C-20. Figure 9 summarizes these conversions of **71**.

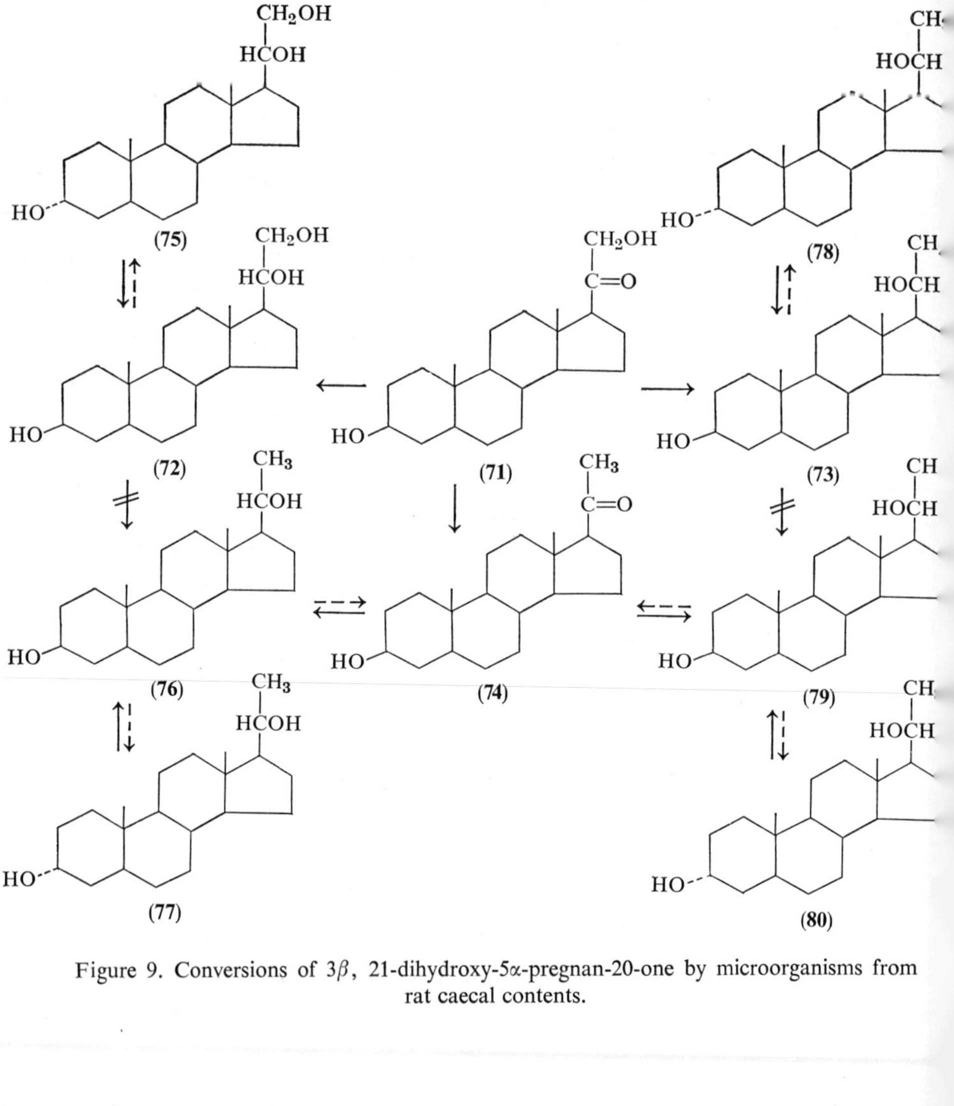

Figure 9. Conversions of 3β, 21-dihydroxy-5α-pregnan-20-one by microorganisms from rat caecal contents.

IV. STUDIES OF MASS-SPECTRAL FRAGMENTATIONS

Much of our present knowledge of the course of mass-spectral fragmentations rests on evidence derived from labeled compounds. The great bulk of
these studies have involved the use of deuterium, because it is cheap and
relatively easy to introduce. Deuterium can be extremely useful as a label,
since it alone can indicate which hydrogen atom or atoms are transferred in a
rearrangement. On the other hand, its rearrangement can be misleading in
assigning cleavage sites. In any event, its uses have been reviewed extensively
in the books of the Djerassi group (1–3), whose own exhaustive studies have,
of course, provided much of our present knowledge of fragmentation
mechanism. In the limited space available in the present review we have,
therefore, chosen to avoid mass-spectral fragmentations of deuterium-
labeled compounds and to concentrate instead on other isotopes. Readers
are referred to Chapter 11, to the books noted above, and others (4, 52) for
studies involving deuterium. We have also chosen to omit ion-molecule
reaction studies, feeling that these constitute a field unto themselves. They
have, in any event, been reviewed extensively elsewhere (53).

For a different reason we have neglected much of the earlier work
employing ^{15}N and [18]O in studies of mass-spectral fragmentations, the reason
being that these isotopes were employed to mark fragments where a high
resolution spectrum would have defined the fragments at least as well. For an
isotope to have utility in today's studies of mass-spectral fragmentations,
there must be at least two like atoms requiring differentiation by isotopic
labeling. A good example of the use of theoretically unnecessary labeling to
establish a fragmentation is found in an early report of loss of CO_2 from
phthalimide (**81**) (54), in which ^{15}N was used to distinguish those fragments
containing nitrogen.*

* The loss of CO_2 from phthalimides was reported again some three years later
(55, 56) by authors who used *N*-methyl and *N*-aryl markers instead of [15]N.

Other examples of labeling studies where high resolution would have served as well (had a high-resolution spectrometer been available) are a study of the fragmentation of N-methyl-4-pyridone-^{15}N (57), and a study of the fragmentation of 2-n-hexyl-2-decenal-1-^{18}O (58).

A. Oxygen-18

Labeling with heteroatoms has been rather little employed in mass-spectral fragmentation studies for the reason noted earlier: in high-resolution studies the heteroatom itself provides a label. However, some compounds contain more than one of the same heteroatom, and there a labeled hetero-atom is useful. Although much less used than ^{13}C (see below), ^{18}O has been used somewhat more as a label than ^{15}N, perhaps due to the greater number of compounds containing two or more oxygen atoms. A particularly elegant use of ^{18}O is provided by the Brown and Djerassi study (59) of aryl carbonate fragmentation with loss of carbon dioxide. Labeling of both ether oxygen atoms individually showed that at 12 eV it is the methyl group that migrates.

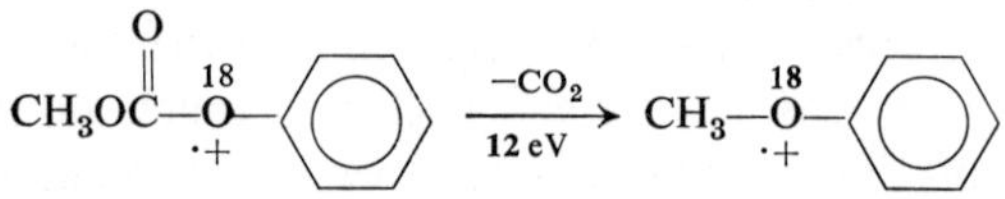

(83)   (84)

On the other hand, in diaryl carbonates (85), both aryl groups can migrate, and here ^{18}O was employed as a quantitative analytical tool in establishing the charge requirements for migration. For example, in extreme cases, exclusive migration of phenyl (versus p-nitrophenyl) and of 2,6-dimethyl-phenyl (versus phenyl) was observed and a reasonable fit was obtained of a Hammett plot versus σ^+.

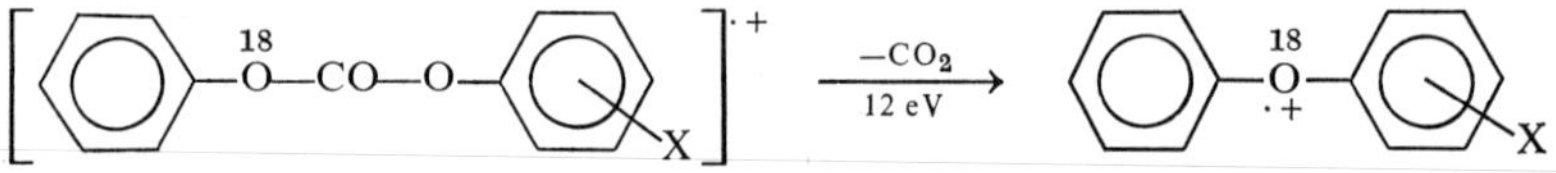

(85)

and

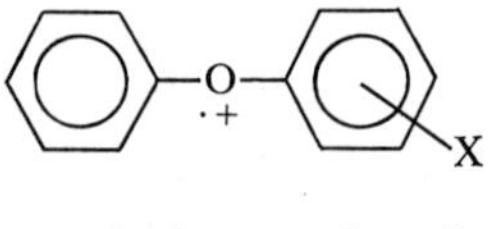

A low-resolution study of mass-spectral carbon monoxide loss from dibenzofurans and methoxybiphenyls has been reported by Pring and Stjernström (60). Results for the methoxybenzofurans (86) were quite clear, involving initial loss of the "terminal" oxygen and subsequent loss of the furan oxygen, as shown. However, interpretation of spectra for the hydroxy-benzofurans (87) indicated the $C_{11}H_7O^+$ ion had lost 20% of its ^{18}O label,

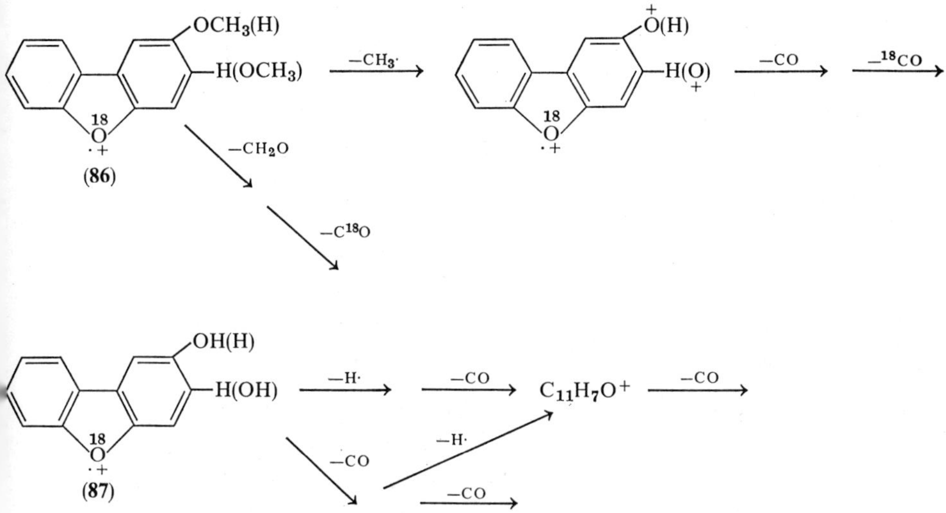

OCH$_3$(H)
-H(OCH$_3$)
^{18}O
$\cdot+$
(86)

$\xrightarrow{-CH_3\cdot}$

$\overset{+}{O}$(H)
-H(O)
^{18}O
$+$

$\xrightarrow{-CO}$ $\xrightarrow{-^{18}CO}$

$-CH_2O$

$-C^{18}O$

OH(H)
-H(OH)
^{18}O
$\cdot+$
(87)

$\xrightarrow{-H\cdot}$ $\xrightarrow{-CO}$ $C_{11}H_7O^+$ $\xrightarrow{-CO}$

$-CO$

$-H\cdot$

$-CO$

requiring at least two fragmentation mechanisms. Formation of a methoxy-dibenzofuran intermediate (89) was also indicated as a fragmentation route from the trimethoxybiphenyl 88. In this, *ca.* 50% of the ^{18}O was lost. This probably reflects a nonspecific loss of methyl in the first step, a conclusion in agreement with the authors' statement that "partial" loss of label had occurred in formation of the $C_{12}H_9O_2^+$ ion.

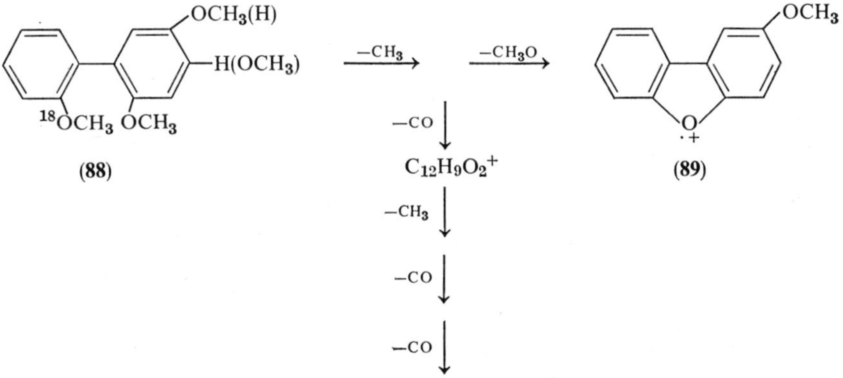

OCH$_3$(H)
-H(OCH$_3$)
^{18}OCH$_3$ OCH$_3$
(88)

$\xrightarrow{-CH_3}$ $\xrightarrow{-CH_3O}$

OCH$_3$
O
$\cdot+$
(89)

$-CO$

$C_{12}H_9O_2^+$

$-CH_3$

$-CO$

$-CO$

A very interesting study by Kinstle and Chapman employed a number of ^{18}O- as well as ^{13}C-labeled derivatives of the 2-phenoxybenzotropone 90 (61).

The spectrum of the 1-^{13}C derivative showed that only 20% of the label was lost in the first elision of carbon monoxide and 80% in the second, indicating an equilibrium between two benzotropones, **90** and **91**. This was confirmed

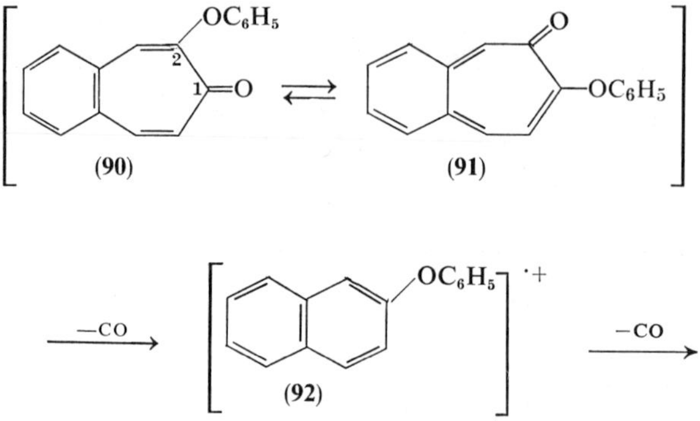

by ^{18}O labeling of the oxygen atoms (20% 1-^{18}O lost with the first CO versus 80% 2-^{18}O). In addition, about half the ^{18}O was lost from both ^{18}O-labeled compounds in the loss of OH. The fragmentation shown was suggested and the proposed intermediate, 2-phenoxynaphthalene-2-^{13}C (**92**), was shown to lose ^{13}CO.

Oxygen-18 has also been used as a tracer in following the fragmentation of bis(trimethylsilyloxy) fatty acid esters, derivatives that have proved useful in locating double bonds in unsaturated acids (127). Among the fragmentations characterized are those shown for the derivative **93** prepared from oleic

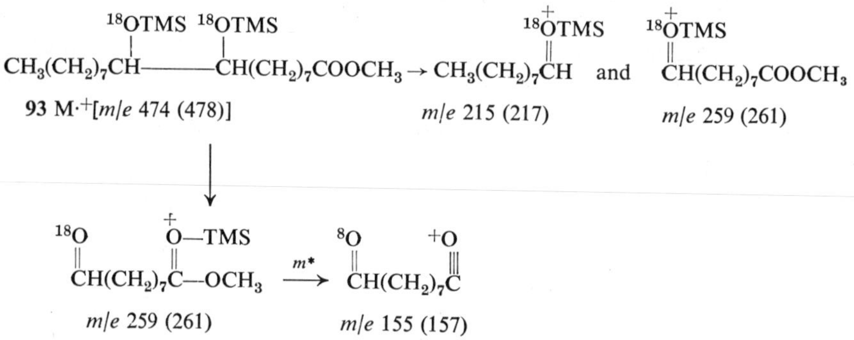

acid. Oxygen-18 was introduced by the oxidation of methyl oleate with ^{18}O-labeled osmium tetroxide. The fragmentations were confirmed in some cases by labeling with nonadeuteriotrimethylsilyl groups.

The ^{18}O- and 4,4-^{2}H$_2$-labeled derivatives of **94** were prepared, and their mass spectra were compared to those of unlabeled **94** in order to assign the

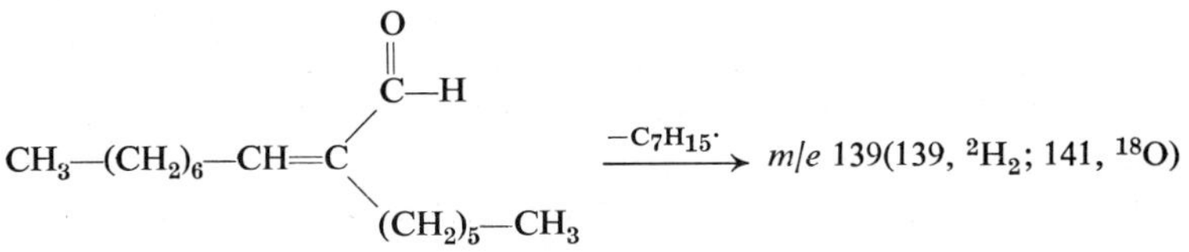

$$CH_3-(CH_2)_6-CH=C\Big\langle\begin{matrix}C-H \;(O)\\(CH_2)_5-CH_3\end{matrix} \qquad \xrightarrow{-C_7H_{15}\cdot} \quad m/e\ 139(139,\ ^2H_2;\ 141,\ ^{18}O)$$

(94) m/e 238 (240, $^2H_2$; 240, ^{18}O)

origin of fragmentations. These results were then used as aids in the structure proof of some α,β-unsaturated aldehydes from Valencia orange peel oil (58). Five of the seven compounds reported $M\cdot^+(m/e\ 238)$ were assigned structures as derivative of acrolein as follows: α-hexyl-β-heptyl **(94)** α-hexyl-β-octyl, α-heptyl-β-heptyl, α-octyl-β-heptyl, and α-hexyl-β-nonyl. Structures of the other two were inferred to be α-octyl-β-octyl- and α-heptyl-β-nonyl-acrolein.

B. Nitrogen-15

An example of the use of ^{15}N in mass-spectral studies may be found in the work of Schimbor (62, 63), who followed the very complex fragmentation of the N,N'-diacetyl derivative of the diaminosugar neosamine C, employing ^{15}N label at C-6 as well as deuterioacetyl, O-methyl and O-ethyl labeling. Of the numerous fragmentations observed, those shown in Figure 10 serve

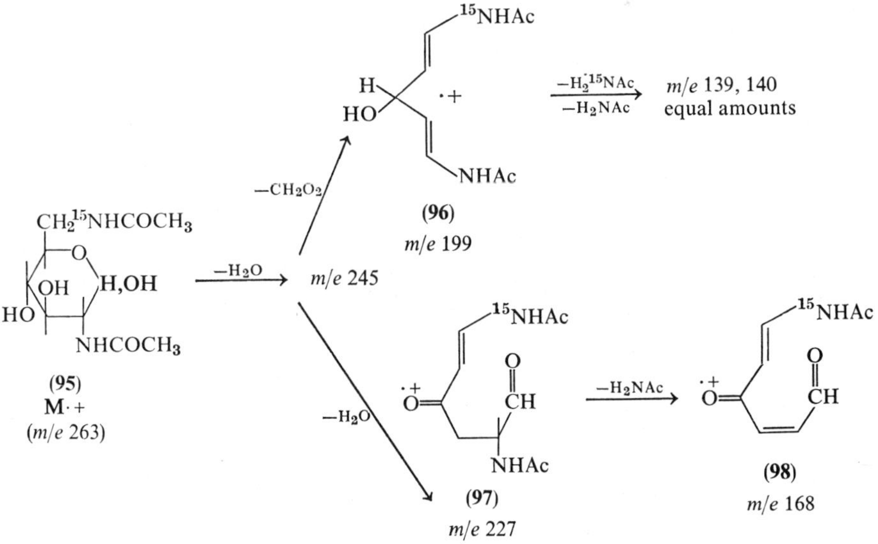

Figure 10. Fragmentation of N,N'-Diacetylneosamine C-6-^{15}N (**95**).

to illustrate the use of ^{15}N. Readers are referred to the original reference for a more complete discussion. Of particular interest in the present discussion are the specific loss of the C-2 acetamide from the m/e 227 ion (97) and the equal loss of C-2 and C-6 acetamide from the m/e 199 ion (96), suggesting a symmetrical structure for (96).

Another study employing ^{15}N to distinguish between two nitrogen atoms was reported by Bose et $al.$ (64). Among other things, this study provided a clear-cut argument for phenyl migration on the one hand (to give 101 and

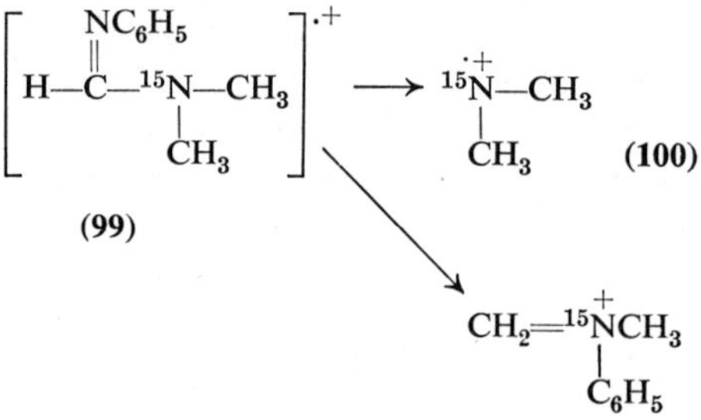

$$\left[\begin{matrix} & NC_6H_5 \\ & \| \\ H\!\!-\!\!C\!\!-\!\!^{15}N\!\!-\!\!CH_3 \\ & | \\ & CH_3 \end{matrix}\right]^{\cdot+} \longrightarrow \begin{matrix} ^{15}\overset{\cdot+}{N}\!\!-\!\!CH_3 \\ | \\ CH_3 \end{matrix} \quad (100)$$

(99)

$$\begin{matrix} CH_2\!\!=\!\!^{15}\overset{+}{N}CH_3 \\ | \\ C_6H_5 \end{matrix}$$

(101)

103b) and methyl migration on the other (to give 103a).

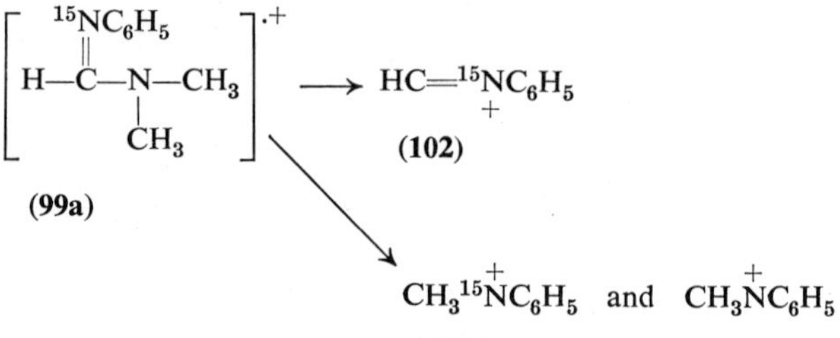

$$\left[\begin{matrix} & ^{15}NC_6H_5 \\ & \| \\ H\!\!-\!\!C\!\!-\!\!N\!\!-\!\!CH_3 \\ & | \\ & CH_3 \end{matrix}\right]^{\cdot+} \longrightarrow \begin{matrix} HC\!\!=\!\!^{15}NC_6H_5 \\ + \end{matrix}$$

(99a)   (102)

$$CH_3{}^{15}\overset{+}{N}C_6H_5 \quad \text{and} \quad CH_3\overset{+}{N}C_6H_5$$

(103a)   (103b)

A study of β-lactams, more particularly of penicillins (e.g., 104), employed ^{15}N to label the amide nitrogen (65). Of the fragments reported only that at m/e, 159 (105) contained the labeled nitrogen.

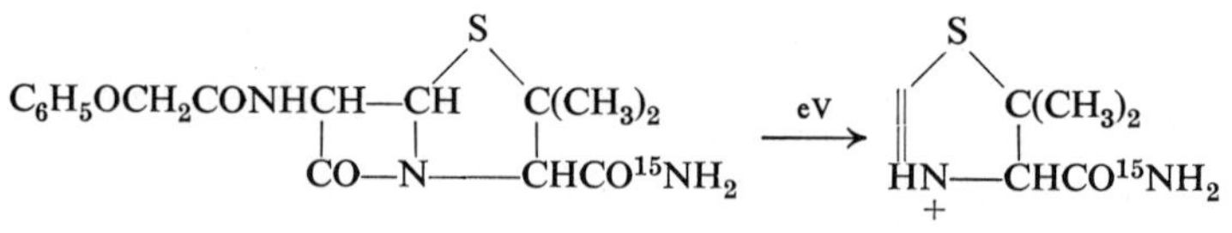

$$\begin{matrix} & & S \\ & & /\backslash \\ C_6H_5OCH_2CONHCH\!\!-\!\!CH & & C(CH_3)_2 \\ & | & | & | \\ & CO\!\!-\!\!N\!\!-\!\!\!-\!\!CHCO^{15}NH_2 \end{matrix} \quad \xrightarrow{eV} \quad \begin{matrix} & S \\ & /\backslash \\ & C(CH_3)_2 \\ \| & | \\ HN\!\!-\!\!CHCO^{15}NH_2 \\ + \end{matrix}$$

m/e 349 (350) m/e 159 (160)

M·+   (105) (^{15}N)

(104) (^{15}N)

C. Carbon-13

Except for deuterium, ^{13}C has been by far the most widely employed of the stable isotopes. A problem that has of its very nature required the use of ^{13}C labeling is the "toluene problem." Some time ago it was recognized that the $C_7H_7^+$ ion formed from toluene (and other C_7H_8 hydrocarbons) could not be adequately represented as a benzyl cation (108), and it was proposed that this was likely to be a tropylium ion (107) (66, 67), as shown in the following equation. This conclusion was derived in part from the low-

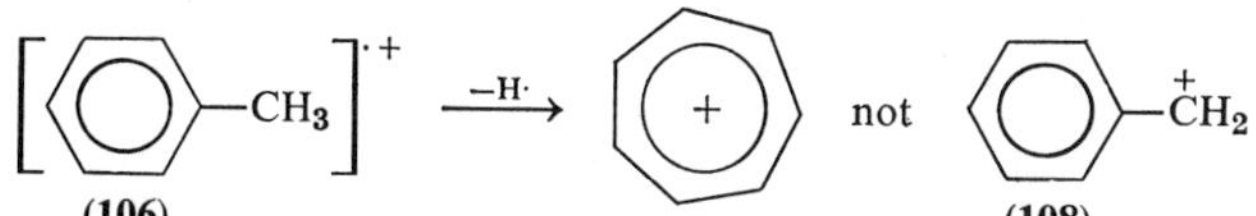

resolution spectrum of toluene-α-^{13}C (68) as well as the spectra of various deuteriated toluenes, since roughly a statistical amount of ^{13}C is lost in the fragmentation

$$C_7H_7^+ \xrightarrow{\ -C_2H_2\ } C_5H_5^+$$

the principal cleavage of the $C_7H_7^+$ ion, a result in agreement with a tropylium ion structure for $C_7H_7^+$ but not with a benzylium ion.

A more interesting question remained, however: whether the α-^{13}C atom retained some identity as a result of having been inserted specifically between C-1 and C-2 of the toluene ring, or whether it had lost all identity as a result of complete mixing of the carbon atom of the tropylium ion. To solve the problem it was necessary to synthesize toluene-α,1-$^{13}C_2$ (69, 70). If randomization were complete, the mass spectrum of this compound should contain $C_5H_5^+$, $^{13}CC_4H_5^+$, and $^{13}C_2C_3H_5^+$ ions in the ratio 1:10:10, while if C-1 and C-α remained bonded to one another, the ratio would be 1:2:4. The complications of interpreting spectra of this labeled compound required that the high-resolution technique be employed. Moreover, both toluene-α-^{13}C and toluene-1-^{13}C (impurities, together with unlabeled toluene, in the dilabeled compound) had to be prepared so that their spectra could be subtracted. However, in the end, the results (Table 21) agreed much better with a randomly dilabeled tropylium ion than with a vicinally dilabeled ion (70).

More recently, these techniques of dilabeling and high resolution have been applied to $C_7H_7^+$ ions from other sources—benzyl-α,1-$^{13}C_2$ bromide, bibenzyl-α,α',1,1'-$^{13}C_4$, and benzyl-α,1-$^{13}C_2$ alcohol (71). In spite of some experimental difficulties (mostly thermal decomposition), the spectra of these compounds also are in accord with a randomly dilabeled tropylium ion. The

TABLE 21

Relative Intensities of Ions from Toluene-α, 1-$^{13}C_2$

		Relative intensity	
		Calculated for tropylium ion	
Ion	Found	Random	1,2-
C_5H_5	0.060	0.048	0.143
$^{13}CC_4H_5$	0.457	0.476	0.286
$^{13}C_2C_3H_5$	0.483	0.476	0.576

spectrum of benzyl alcohol (109) is dominated, however, by the following less complicated fragmentation, demonstrated by appropriate labeling.

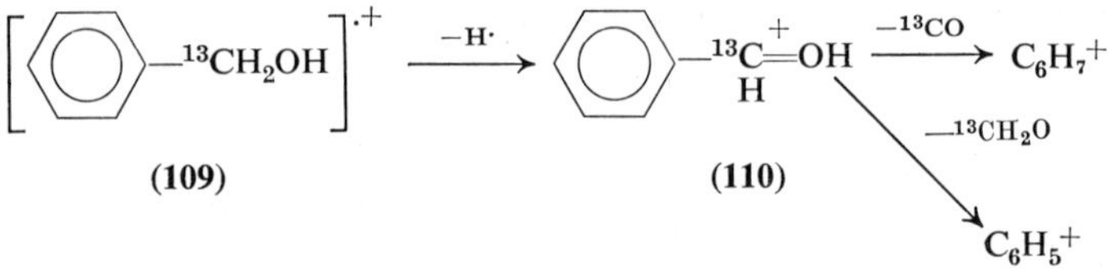

The results with dilabeled toluene have spurred study of a number of related systems. Aniline, phenol, and thiophenol have been of special interest due to their relationship to toluene. In these compounds only one label is required at high resolution, since the heteroatom itself provides a second label. Aniline has been most extensively investigated—by two groups, both at low resolution (69, 72) and high resolution (73, 74).

The low-resolution results with aniline-1-^{13}C (111) clearly demonstrated that the major fragmentation—loss of HCN from the parent ion—occurred without mixing of carbon atoms (69, 72). Similar results were obtained for the $C_6H_6N^{.+}$ ion from acetanilide-1-^{13}C (72). However, sulfanilamide-1-^{13}C behaved differently (72); metastable ions and rough quantitative data indicated that the $C_6H_6N^+$ ion lost both H^{13}CN and HCN, suggesting an azatropylium structure for that ion. Careful study of the aniline-1-^{13}C spectrum at high resolution (73, 74) allowed interesting observations to be made regarding the smaller peaks in the m/e 60–70 region. Data for those ions derived from the odd-electron parent ion indicated that it had not undergone rearrangement, but similar study of ions derived from the even electron M − 1 ion indicated it has rearranged, presumably to an azatropylium ion (112).

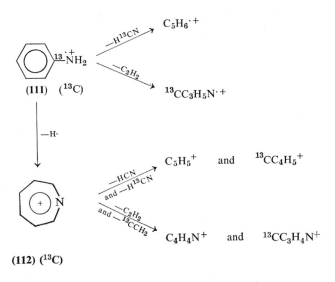

Acetanilide also showed this dual nature—even-electron ions rearranged, odd-electron ions unrearranged (74). The low-resolution results for sulfanilamide's even-electron $C_6H_6N^+$ ion were confirmed at high resolution, and extended to the $C_6H_6N^+$ ion from *p*-nitroaniline (74). The generalization implied for this series that even-electron ions rearrange but odd-electron ions do not is contradicted, however, by recent studies (75) on phenyl azide-1-^{13}C, in which the $C_6H_5N^{\cdot+}$ ion (formally phenyl nitrene) was assigned a rearranged structure since it lost, in part, labeled acetylene and unlabeled hydrogen cyanide. Unfortunately, only low-resolution data were available.

In an extensive low-resolution study (76) of nitrogen heterocycles that preceded the work on aniline and its derivatives, 1-methylisoquinoline (**113**), 2- and *N*-methylindole (**115** and **117**), and *N*-methylpyrrole (**120**), labeled with ^{13}C in individual compounds at the carbon atoms indicated by arrows, were all demonstrated to undergo considerable mixing of label in the fragmentation $M^{\cdot+} \xrightarrow{-H\cdot}, \xrightarrow{-HCN},$ in accord with rearranged structures for the even-electron $M - H$ ions, perhaps the ring-expanded forms shown (**114, 116, 118, 119, 121**).

The high-resolution studies carried out with aniline-1-^{13}C have recently been extended to phenol-1-^{13}C and thiophenol-1-^{13}C, with the results shown in Table 22 (71). Phenol does not undergo extensive loss of hydrogen, and almost no mixing of label is observed for loss of carbon monoxide or acetylene, fragmentations of odd-electron ions. On the other hand, fragmentation of thiophenol reveals considerable randomization of label regardless of the nature of the ion involved. That sulfur-containing compounds give

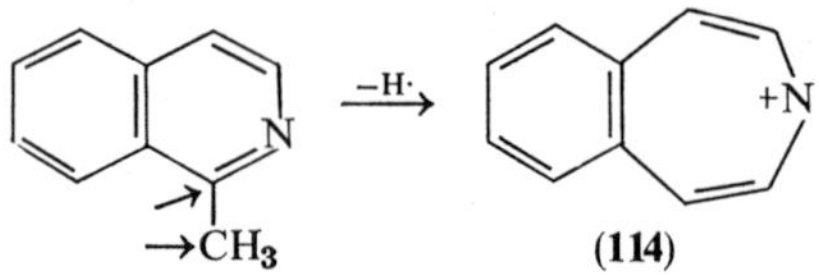

(113) → (114)

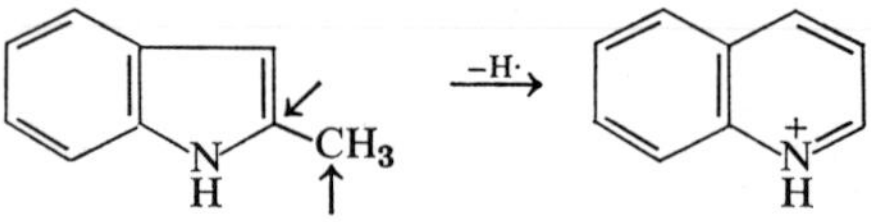

(115) → (116)

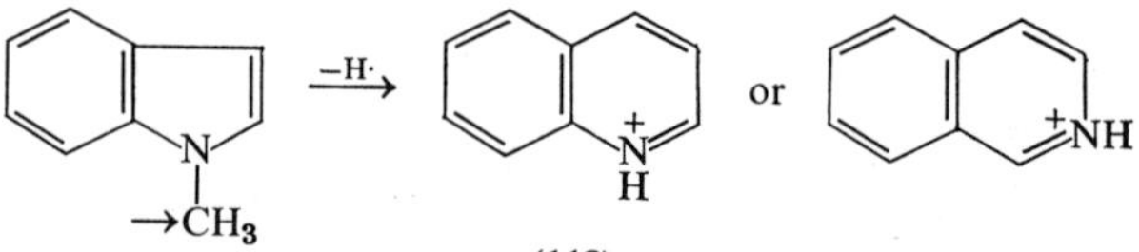

(117) → (118) or (119)

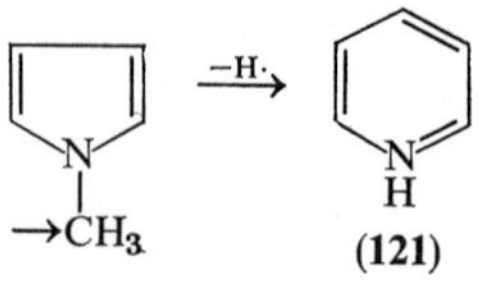

(120) → (121)

especially complex fragmentation patterns was demonstrated also for diphenyl-1-^{13}C sulfide (77). The results at low resolution suggested considerable randomization in loss of CS and CHS. Evidence for at least partial ring expansion was provided in a low-resolution study of 2-n-hexylthiophene-α-^{13}C, in which the C_3H_3, C_4H_5, CHS, and C_2H_2S ions were all somewhat labeled (78).

A second vexing problem that has been extensively studied with ^{13}C labeling is the loss of a methyl group from hydrocarbon ions. A particularly striking case involves loss of methyl from trityl ions; in the fragmentation shown, 96% of ^{13}C was retained in the $C_{18}H_{12}$ ion (79). Although not pointed out by the authors, this is very close to the result calculated for complete randomization of all 19 carbon atoms in the trityl cation. A similar result was obtained for loss of methyl from the odd-electron parent ion.

TABLE 22

Retention of Label in Ions from Phenol-1-^{13}C and Thoiphenol-1-[13]C

| Ion[a] | No ring expansion | Oxa (or thia) tropylium ion | | Found[b] | |
		1,2-	Random	O	S
C_5H_6 ·+	1.000	0.500	0.167	0.973	0.299
[13]CC_4H_6	0.000	0.500	0.833	0.029	0.702
C_4H_4X	0.000	0.200	0.333	0.203	0.346
[13]CC_3H_4X	1.000	0.800	0.667	0.798	0.655
C_4H_2O	0.000	0.200	0.333	0.000	—
[13]CC_3H_2O	1.000	0.800	0.667	1.000	—
C_3H_2S	0.000	0.250	0.500	—	0.121
[13]CC_2H_2S	1.000	0.750	0.500	—	0.879
C_5H_5 +	1.000	0.500	0.167	1.000	0.445
[13]CC_4H_5	0.000	0.500	0.833	0.000	0.555
C_3H_3O	0.000	0.250	0.500	0.055	—
[13]CC_2H_3O	1.000	0.750	0.500	0.944	—
C_3HS	0.000	0.250	0.500	—	0.025
[13]CC_2HS	1.000	0.750	0.500	—	0.975
CHS	0.000	0.500	0.833	—	0.780
[13]CHS	1.000	0.500	0.167	—	0.220

[a] ·+ = odd-electron ions; + = even-electron ions.
[b] O = phenol-1-^{13}C; S = thiophenol-1-[13]C.

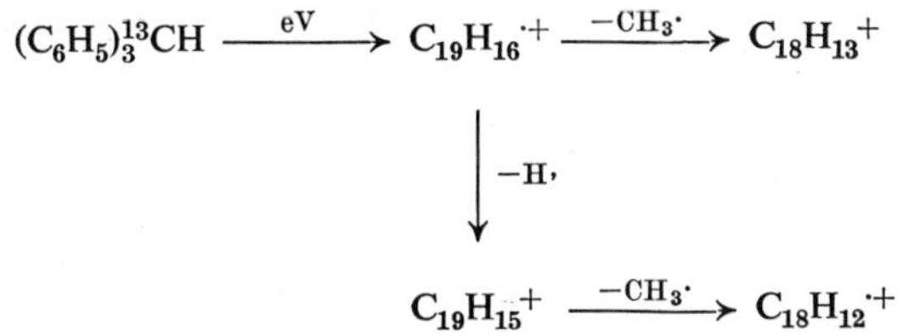

$$(C_6H_5)_3^{13}CH \xrightarrow{\text{eV}} C_{19}H_{16}^{·+} \xrightarrow{-CH_3^·} C_{18}H_{13}^+$$

$$\downarrow -H^·$$

$$C_{19}H_{15}^+ \xrightarrow{-CH_3^·} C_{18}H_{12}^{·+}$$

Loss of methyl in stilbenes and related compounds has been extensively studied at low resolution by Johnstone. In the fragmentations shown for 1,2-diphenyl-1-chloroethane (whose spectrum is essentially that of stilbene), about 50% of the ^{13}C was lost as methyl from the labeled compound (122) (80). However, for the 3- and 4-pyridyl analogs of stilbene (123 and 124),

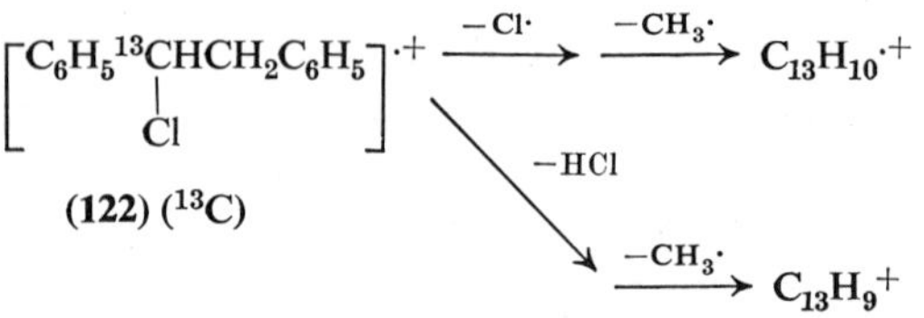

$$\left[C_6H_5{}^{13}CHCH_2C_6H_5\atop \quad\quad Cl\right]^{\cdot+} \xrightarrow{-Cl\cdot} \xrightarrow{-CH_3\cdot} C_{13}H_{10}{}^{\cdot+}$$

(122) (^{13}C)

$$\xrightarrow{-HCl}$$

$$\xrightarrow{-CH_3\cdot} C_{13}H_9{}^+$$

much less ^{13}C was lost with CH_3 (20% and none, respectively) and in the loss of methyl from the molecular ion of 1,4-diphenyl-1,3-butadiene-1-^{13}C, no ^{13}C was lost (81).

$$[R—CH={}^{13}CH—C_6H_5]^{\cdot+} \xrightarrow{-CH_3\cdot} C_{12}H_8N^+$$

(123) R = 3-C_5H_4N
(124) R = 4-C_5H_4N

The carbon atom lost as methyl was presumed to be the central carbon of 1,3-diphenylpropene since the 3-^{13}C-labeled compound **(125)**, lost no

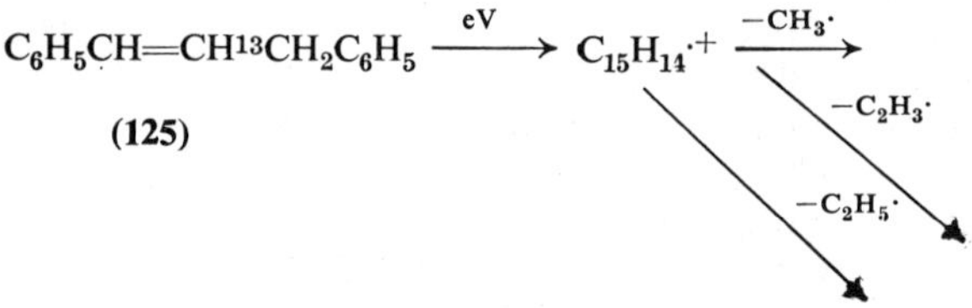

$$C_6H_5CH=CH^{13}CH_2C_6H_5 \xrightarrow{eV} C_{15}H_{14}{}^{\cdot+} \xrightarrow{-CH_3\cdot}$$

(125)

$$\xrightarrow{-C_2H_3\cdot}$$

$$\xrightarrow{-C_2H_5\cdot}$$

label (82). On the other hand, 60.4% of the ^{13}C was lost in the elision of C_2H_3 (vinyl radical), and 38.5% of the ethyl radical lost contained ^{13}C.

Loss of methyl from butene occurs with considerable loss of identity of individual carbons. This was demonstrated (83) in the low-resolution spectrum of 1-butene-4-^{13}C, which retained 75% label in the $C_4H_8{}^{\cdot+} \xrightarrow{-CH_3\cdot}$ $C_3H_5{}^+$ fragmentation. This was interpreted as involving a methylcyclopropyl radical ion as intermediate. Similar skeletal isomerization (25% isomerization) had been demonstrated earlier (84) for the loss of methyl from propene-1-^{13}C.

The studies described already in this section have addressed themselves to identification of ring expansion and loss of methyl. In addition, a number of studies of carbon monoxide and carbon dioxide losses have employed ^{13}C as label. Johnstone and Millard have demonstrated that the 2'- and 2-carbonyl groups are successively lost as CO in the fragmentation of the pyranocoumarone **126**, employing both the pyranocoumarone-2-^{13}C and -2',4'-$^{13}C_2$ derivatives (85). A similar study on the 2-^{13}C derivative of the furanocoumarone **127** indicated it to lose carbon monoxide from C-2, as expected (86).

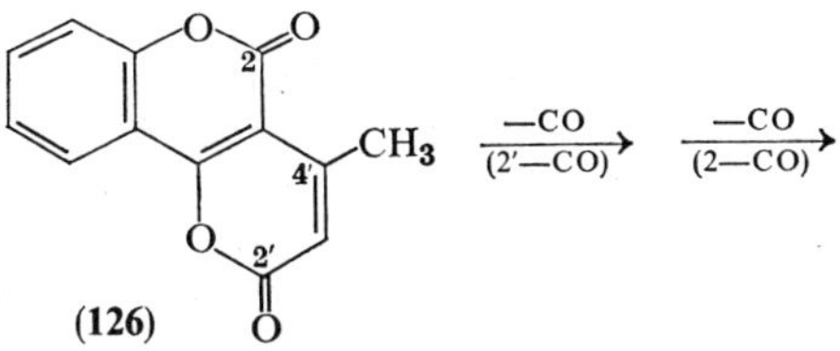

(126) $\xrightarrow[(2'-CO)]{-CO}$ $\xrightarrow[(2-CO)]{-CO}$

(126)

(127) $\xrightarrow[(2-CO)]{-CO}$

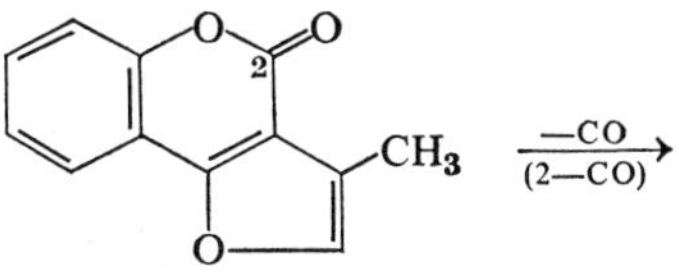

(127)

A study by Kinstle employed ^{13}C to follow the rather complex fragmentations of the nitrone **128** (87). Loss of unlabeled carbon monoxide was proposed to come from the cyclized intermediate **129**, but a different path, via **130**, was necessary to explain the more important fragmentations to the $C_6H_5N^{\cdot+}$ and benzoylium ions.

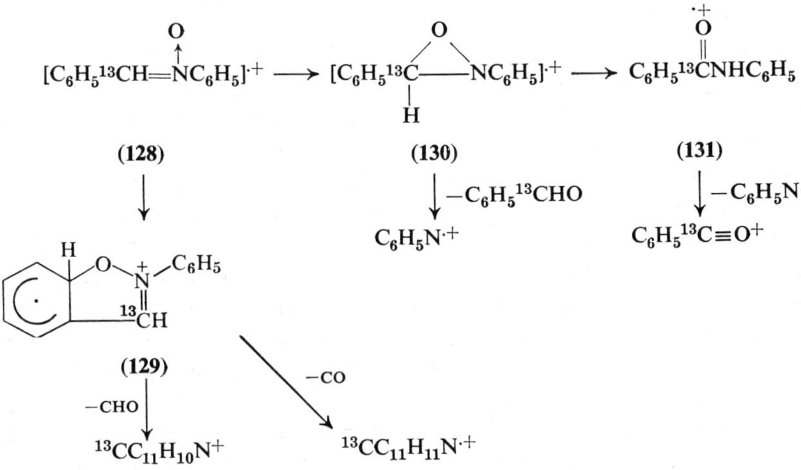

$$[C_6H_5{}^{13}CH{=}NC_6H_5]^{\cdot+} \longrightarrow [C_6H_5{}^{13}C\overset{O}{\underset{H}{\diagdown\diagup}}NC_6H_5]^{\cdot+} \longrightarrow C_6H_5{}^{13}\overset{\cdot+}{\overset{O}{\overset{\|}{C}}}NHC_6H_5$$

(128)   **(130)**   **(131)**

(128) $\downarrow$

(130) $\downarrow -C_6H_5{}^{13}CHO$

$C_6H_5N^{\cdot+}$

(131) $\downarrow -C_6H_5N^{\cdot}$

$C_6H_5{}^{13}C{\equiv}O^{+}$

(129)

$-CHO \downarrow$

$^{13}CC_{11}H_{10}N^{+}$

$\xrightarrow{-CO}$ $^{13}CC_{11}H_{11}N^{\cdot+}$

A study by Bose employing the 2-^{13}C derivative of the β-lactam **132** demonstrated that the amide carbon was lost as CO_2 (88), while a number of labeled derivatives (^{13}C, ^{18}O) were employed by Manhas (89) to locate the source of the carbon dioxide lost from N-trifluoracetylglycine methyl ester **(133)** as being the carboxyl group of the ester.

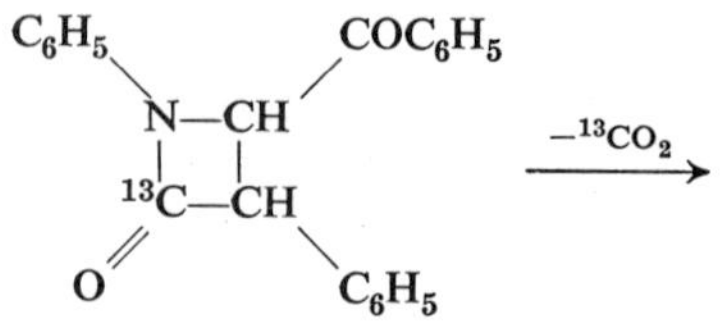

$$
\begin{array}{c}
C_6H_5 \qquad\qquad COC_6H_5 \\
\diagdown\qquad\qquad\diagup \\
N—CH \\
|\qquad| \qquad\qquad \xrightarrow{\ -^{13}CO_2\ } \\
^{13}C—CH \\
\diagup\qquad\diagdown \\
O \qquad\qquad C_6H_5
\end{array}
$$

(132)

$$
\begin{array}{c}
O \\
\parallel \\
CH_3OCCH_2NHCOCF_3 \xrightarrow{\ -CO_2\ }
\end{array}
$$

(133)

V. STUDIES OF ORGANIC REACTION MECHANISM

In the preceding section we discussed the uses of isotopic labeling in studying fragmentations of organic compounds in the mass spectrometer. These fragmentations are, of course, organic reactions, albeit under extreme conditions of vacuum and energy input. However, we have excluded them (and ion-molecule reactions as well, for the reasons detailed above) from the present section in order to focus attention here on the many applications of mass spectrometry in helping to establish the mechanisms of ordinary organic reactions. This has been a classically fruitful area for mass spectrometry, as is attested by the numerous references to application of the technique in textbooks dealing with organic reaction mechanisms (90–93). Examples abound: ^{18}O in esterification, ester hydrolysis, ozonolysis, and peracid oxidations, ^2H in the Cannizzaro reaction, in the aldol reaction and in hydrogenation, ^{15}N in the Fischer indole synthesis. Many of the early studies involved simple conversion of the compounds to CO_2, N_2, H_2, or H_2O for analysis in an isotope ratio spectrometer. Most recent examples content themselves with analyzing the molecular ion only, for retention or loss of label, though a few follow the isotope through the reaction product's mass-spectral fragmentations. As in other sections of the present review, we shall stress recent examples.

In addition to those investigations in which the labeling isotope is either retained or lost, a rather different type of study can also provide useful information. This involves the relative rates of reactions of labeled and unlabeled compounds, taking advantage of the kinetic-isotope effect. Classical examples would include studies of oxidation rates with chromium (VI) and manganese (VII), of rates of elimination reactions, and of the rates of aromatic substitution reactions (90–93). Since mass-spectral aspects of such studies usually involve no more than a determination of the isotope ratio of the molecular ion, and since kinetic-isotope-effect studies are nearly completely restricted to deuterium-hydrogen comparisons, we have chosen to omit studies dealing with that use of isotopes. However, we note here for the record the considerable interest in deuterium-isotope effects in mass-spectral fragmentations (95–98), in related photochemical reactions (99), and in other reactions (100, 101).

A. Oxygen-18

This isotope has been the most used in studies of organic reaction mechanisms, with the exception of deuterium. Hydrolysis remains a reaction of considerable interest. For example, exchange of the oxygen of nitrobenzene does not occur in $H_2^{18}O$ under either basic or acidic conditions (102). In a second example, basic hydrolysis of an ethylenediamine chelate in water-^{18}O indicated cobalt-oxygen bond cleavage (displacement), rather than carbon-oxygen cleavage (hydroxide addition) (103).

$$[Co(H_2NCH_2CH_2NH_2)_2(OCOAr)_2]^+ \xrightarrow[H_2^{18}O]{OH^-} \xrightarrow{H^+} ArCOOH$$

(134) unlabeled

Studies of phthalates indicate a more complex pattern. The ethanolysis of phthalic acid (135) proceeded with addition of ethanol-^{18}O and subsequent acyl-oxygen bond fission. On the other hand, reaction of diethyl phthalate (136) with *tert*-butyl alcohol gave alkyl-oxygen cleavage under sulfuric acid catalysis (104). The di-*t*-butyl phthalate (137) was analyzed at m/e 148 (150) according to the fragmentation shown.

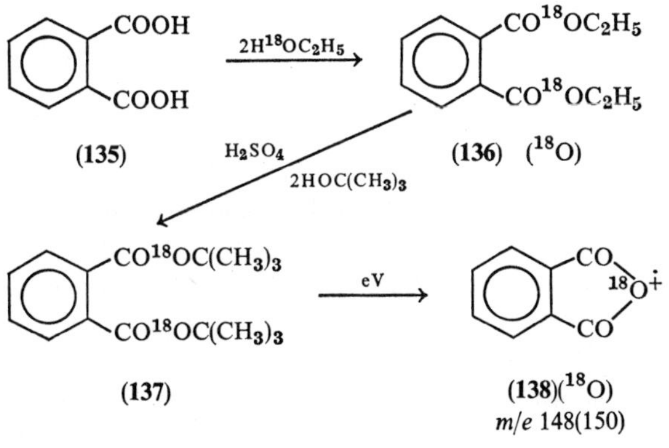

Oxygen-18 was employed in an elegant study of the geometry of the methylcyclopropylcarbinyl cation. Since the rate of racemization (k_{rac}) of the optically active alcohol equalled the rate of incorporation (k_{exc}) of ^{18}O

from water-^{18}O, the carbonium ion was judged to have a plane of symmetry like that in ion **140** (105). Deuterium oxide treatment was employed to demonstrate lack of an olefinic intermediate.

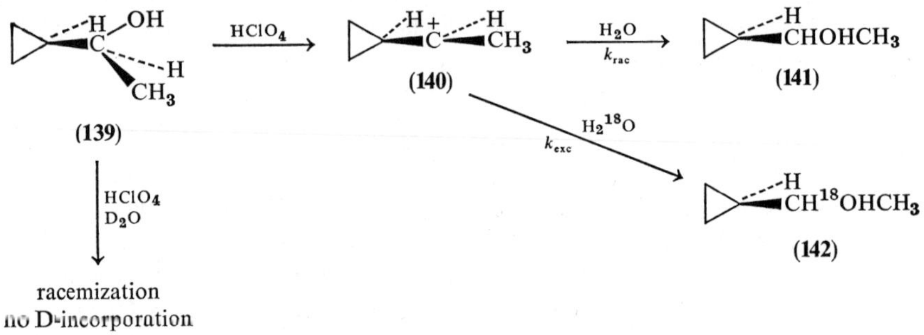

racemization
no D-incorporation

Hydrolysis of β-methylstyrene oxide (**143**) in water-^{18}O was employed to

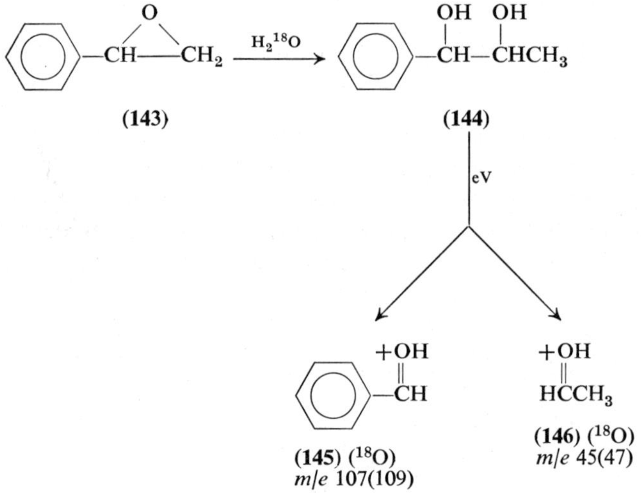

demonstrate the direction of oxide ring opening (106). In acid only the benzyl-oxygen bond was cleaved, but in base the direction of opening depended on the oxide's stereochemistry—30% benzyl-oxygen cleavage for the *cis* epoxide, 90% for the *trans*. Analysis of the position of incorporation was effected from the fragmentation peaks at m/e 107 (109) and 45 (47).

Another subtle use of label was involved in explaining the observation that pyrolysis of the phthalate **149** gave some 1,2-dimethylcyclohexene, which could not have arisen from the usual concerted *cis* ester pyrolysis mechanism; an ion pair intermediate (**153**) was proposed (107). This would require partial exchange of two ester oxygens in recovered phthalate at an intermediate time period, and this was indeed confirmed by reducing the ester to an alcohol (155) enriched in ^{18}O.

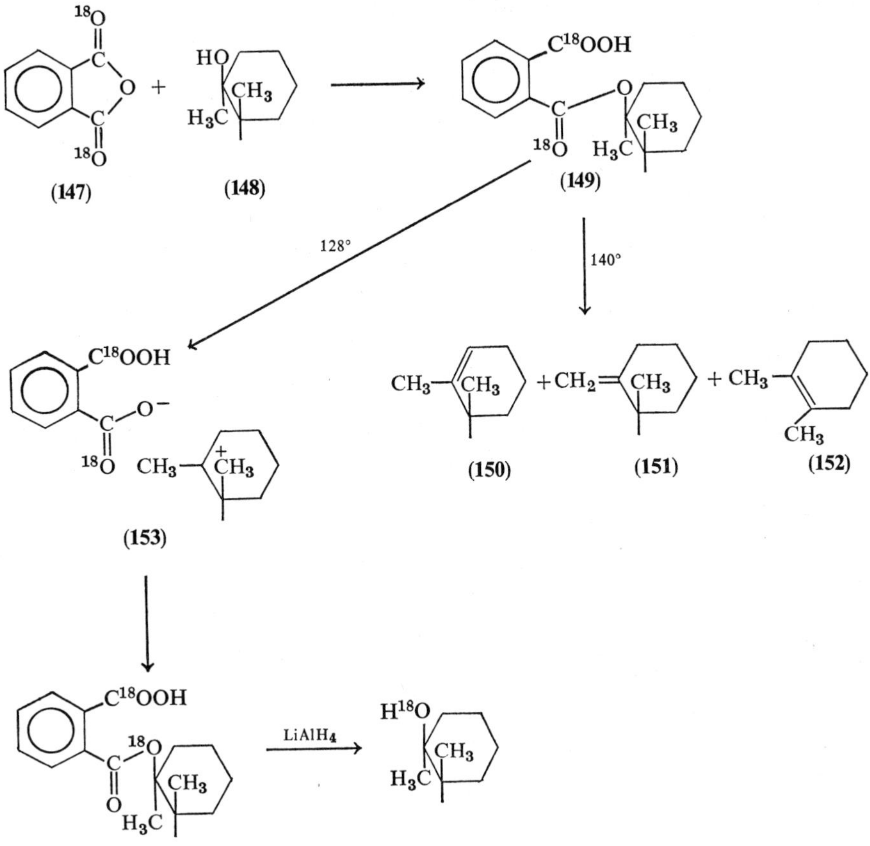

(147) (148) (149)

(153)

(150) (151) (152)

(154) (155)

Another general type of reaction that has been much studied by labeling techniques is oxidation. A specific reaction investigated is ozonolysis. In one study, stilbene (both *cis* and *trans* isomers gave similar results) was ozonized in the presence of benzaldehyde-^{18}O (108). The ozonide (156) was isolated and analyzed mass spectrometrically according to the fragmentation indicated. The sum of the ^{18}O excess in the ions at m/e 122 and m/e 106 must be the total excess in 156, while the ^{18}O in the ion at m/e 196 is the excess in the ether (C–O–C) oxygen of the ozonide. Since the excess at m/e 196 was equal to that at m/e 106 plus that at m/e 122, all the excess was assigned to the ether oxygen. This would be in accord with the mechanism and intermediates shown in brackets, but not with others postulated earlier.

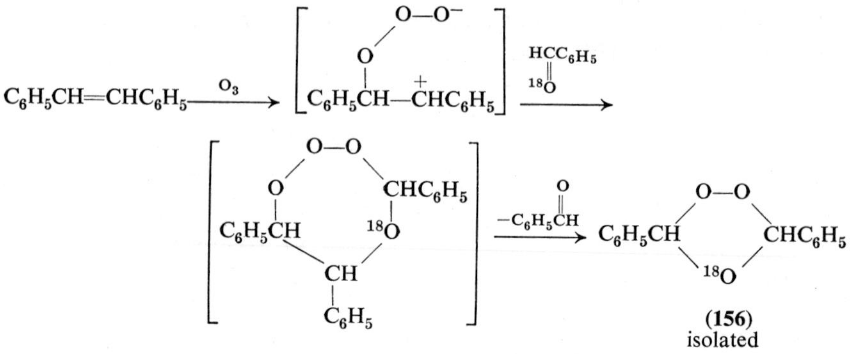

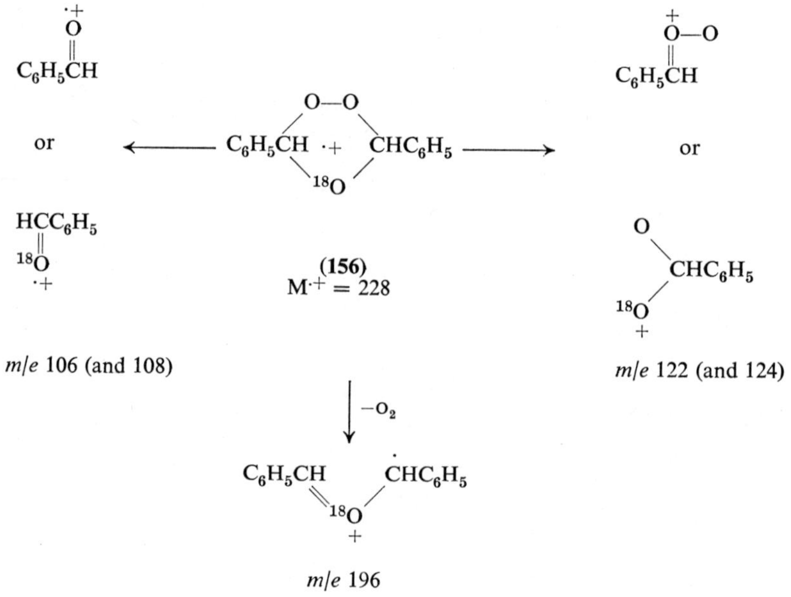

$$C_6H_5CH{=}CHC_6H_5 \xrightarrow{O_3} \left[C_6H_5\overset{\overset{\displaystyle O{-}O^-}{|}}{CH}{-}\overset{+}{C}HC_6H_5 \right] \xrightarrow[{}^{18}O]{HCC_6H_5} \longrightarrow$$

$$\left[\begin{array}{c} C_6H_5\overset{\overset{\displaystyle O{-}O}{\diagup}}{C}H \quad {}^{18}O \diagdown CHC_6H_5 \\ \underset{\overset{|}{C_6H_5}}{CH} \end{array} \right] \xrightarrow{-C_6H_5\overset{\overset{\displaystyle O}{\|}}{CH}} C_6H_5CH \overset{O{-}O}{\diagup}\ {}^{18}O\ \diagdown CHC_6H_5$$

(156)
isolated

$$C_6H_5\overset{\overset{\displaystyle \cdot{+}}{O}}{\overset{\|}{C}}H$$

or ⟵ $C_6H_5CH \overset{O{-}O}{\underset{{}^{18}O}{\cdot{+}}} CHC_6H_5$ ⟶ or

$$\overset{HCC_6H_5}{\underset{\cdot{+}}{{}^{18}O}}$$

(156)
$M^{\cdot+} = 228$

$$C_6H_5\overset{\overset{\displaystyle \overset{+}{O}{-}O}{\|}}{C}H$$

$$\overset{O}{\underset{{}^{18}O}{\diagdown}}CHC_6H_5 \quad {}^{+}$$

m/e 106 (and 108)   *m/e* 122 (and 124)

$$\Big\downarrow -O_2$$

$$C_6H_5CH \overset{\diagdown}{\underset{{}^{18}O}{\diagup}} CHC_6H_5 \quad {}^{+}$$

m/e 196

Oxidation of methane by ozone under irradiation in the presence of labeled oxygen demonstrated that the oxygen in methanol and formaldehyde comes from ozone under these conditions, rather than from oxygen (109).

$$CH_4 + O_3 + {}^{18}O_2 \xrightarrow[87\,°K]{2537\,Å} CH_3OH,\ CH_2O,\ H_2$$
unlabeled

Oxidation of polyformaldehyde by labeled oxygen indicated a partial introduction of ^{18}O into the products, however (110).

$$(OCH_2)_nO \xrightarrow{{}^{18}O_2} H_2{}^{18}O + HC^{18}O_2H$$
partially labeled

Oxidation of the sugar thiocarbonate (157) by chlorine with a small

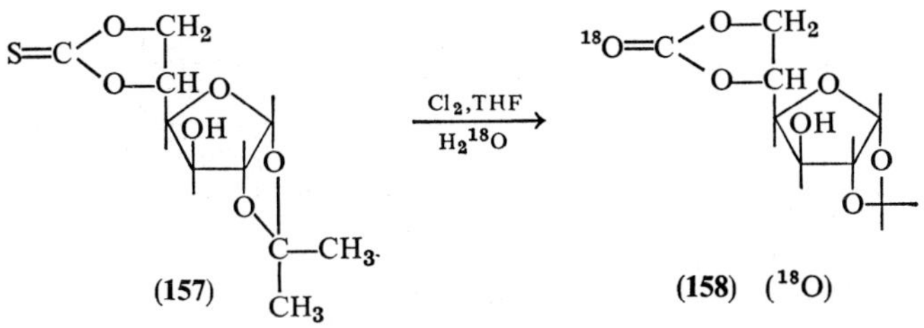

$$S=C\diagdown \begin{matrix} O-CH_2 \\ O-CH \end{matrix} \xrightarrow[H_2{}^{18}O]{Cl_2,THF} {}^{18}O=C\diagdown \begin{matrix} O-CH_2 \\ O-CH \end{matrix}$$

(157)   (158) (^{18}O)

amount of added water-^{18}O showed the new oxygen in 158 to be derived from the water (111), though it is difficult to see where else it could come from.

B. Nitrogen-15

A good illustration of the use of this element in attacking mechanistic problems is found in the work of Leonard and Zwanenburg (112), who established the direction of addition of acetonitrile (ring opening) to the aziridine 159. Hydrolysis of the product gave acetamidoisobutyrophenone

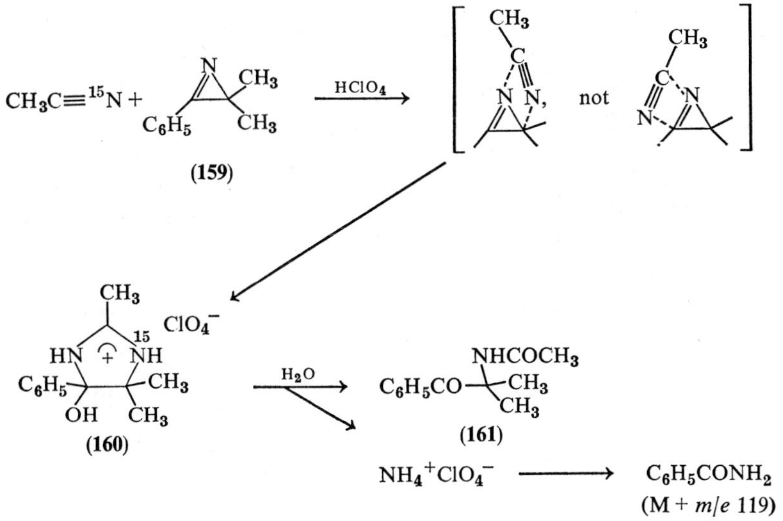

$$CH_3C\equiv{}^{15}N + \text{(159)} \xrightarrow{HClO_4} \left[\ \right] \quad \text{not} \quad \left[\ \right]$$

(159)

(160)

$$\xrightarrow{H_2O} C_6H_5CO-C\diagdown \begin{matrix} NHCOCH_3 \\ CH_3 \\ CH_3 \end{matrix}$$

(161)

$$NH_4{}^+ClO_4{}^- \longrightarrow C_6H_5CONH_2$$
$$(M + m/e\ 119)$$

(161) and ammonia (converted to benzamide), which were characterized by the fragmentations shown.

A relatively simple exercise gave the expected result when the nitrogen

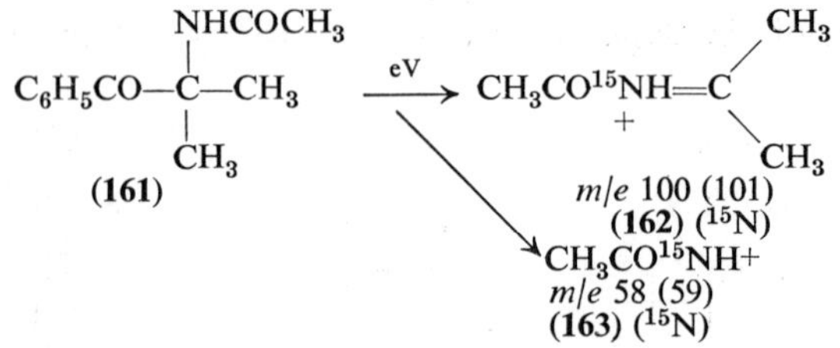

$$\underset{(161)}{C_6H_5CO-\underset{\underset{CH_3}{|}}{\overset{\overset{NHCOCH_3}{|}}{C}}-CH_3} \xrightarrow{eV} CH_3CO^{15}NH=\underset{CH_3}{\overset{CH_3}{C}}$$

$$\overset{+}{\underset{\substack{m/e\ 100\ (101) \\ (162)\ (^{15}N)}}{}}$$

$$CH_3CO^{15}NH^+$$
$$m/e\ 58\ (59)$$
$$(163)\ (^{15}N)$$

of an isothiazole product (165) was shown to derive from ammonia rather than from the isothiazolium precursor (164) (113).

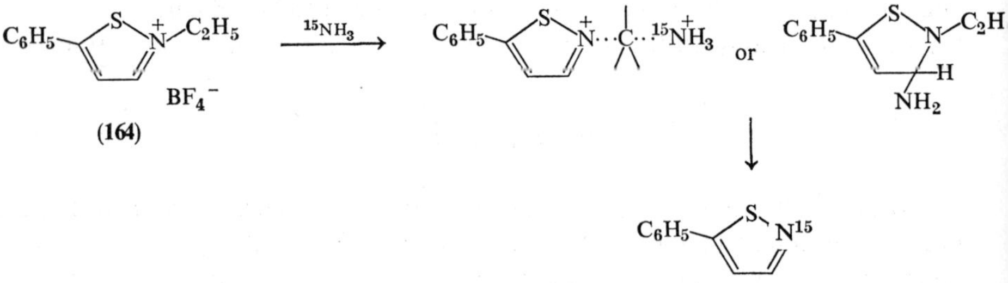

$$\underset{(164)}{C_6H_5-\underset{BF_4^-}{\overset{S}{\diagdown}}\overset{+}{N}\diagdown^{C_2H_5}} \xrightarrow{^{15}NH_3} C_6H_5-\overset{S}{\diagdown}\overset{+}{N}\cdots\overset{|}{C}\cdots^{15}\overset{+}{N}H_3 \quad or \quad C_6H_5-\overset{S}{\diagdown}\underset{\underset{NH_2}{|}}{\overset{N-C_2H}{\diagdown}}-H$$

$$\downarrow$$

$$\underset{(165)}{C_6H_5-\overset{S}{\diagdown}N^{15}}$$

C. Nitrogen-15 and Deuterium

A rather different illustration of the use of ^{15}N is provided by the work of Axenrod and Milne (94), who demonstrated that the nitrosamine rearrangement of 166, to the amidoxime 167, proceeded intermolecularly rather than intramolecularly. This was accomplished by employing equal quantities of D_4- and ^{15}N-labeled nitrosamines and showing the products to consist of

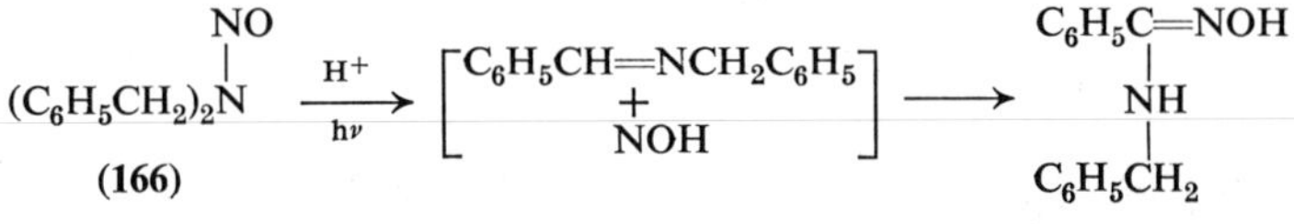

$$\underset{(166)}{(C_6H_5CH_2)_2\overset{\overset{NO}{|}}{N}} \xrightarrow[h\nu]{H^+} \left[\overset{C_6H_5CH=NCH_2C_6H_5}{\underset{NOH}{+}} \right] \longrightarrow \underset{(167)}{\underset{C_6H_5CH_2}{\overset{C_6H_5C=NOH}{\underset{|}{N}H}}}$$

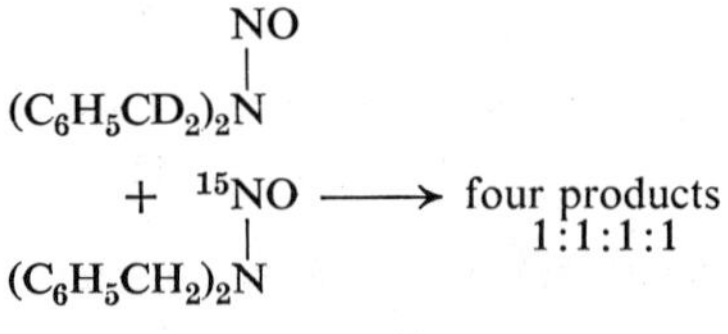

$$\underset{(166)\ (D_4)\ +\ (167)\ (^{15}N)}{\underset{1:1}{\begin{array}{c} (C_6H_5CD_2)_2\overset{\overset{NO}{|}}{N} \\ +\ ^{15}NO \longrightarrow \underset{1:1:1:1}{four\ products} \\ (C_6H_5CH_2)_2N \end{array}}}$$

equimolar amounts of unlabeled, mono-, tetra- and penta-labeled compounds.

D. Carbon-13

Relatively few organic mechanistic studies have involved the use of ^{13}C (in contrast to mass-spectral mechanistic studies, see above). However, the mercuric ion oxidation of propene-1-^{13}C gave some very interesting results (114). Although the minor products, acetone (10%) and propionaldehyde (2%), retained their isotopic identity, the main product, acrolein (88%), was shown to contain equal amounts of labeling at the two terminal carbon atoms, requiring a symmetrical intermediate like that shown.

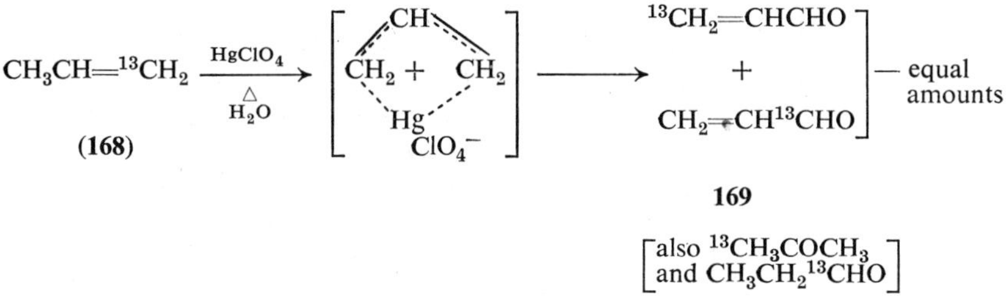

$$CH_3CH{=}^{13}CH_2 \xrightarrow[\substack{\triangle \\ H_2O}]{HgClO_4} \left[\begin{array}{c} CH \\ CH_2 + CH_2 \\ Hg \\ ClO_4^- \end{array} \right] \longrightarrow \left[\begin{array}{c} {}^{13}CH_2{=}CHCHO \\ + \\ CH_2{=}CH^{13}CHO \end{array} \right] \begin{array}{l} - \text{ equal} \\ \text{ amounts} \end{array}$$

(168)

169

$$\left[\begin{array}{l} \text{also } {}^{13}CH_3COCH_3 \\ \text{and } CH_3CH_2{}^{13}CHO \end{array} \right]$$

E. Oxygen-18 and Deuterium

In a study designed to distinguish intermolecular from intramolecular methyl transfer in the isomerization of 170 to 171, an equimolar mixture of

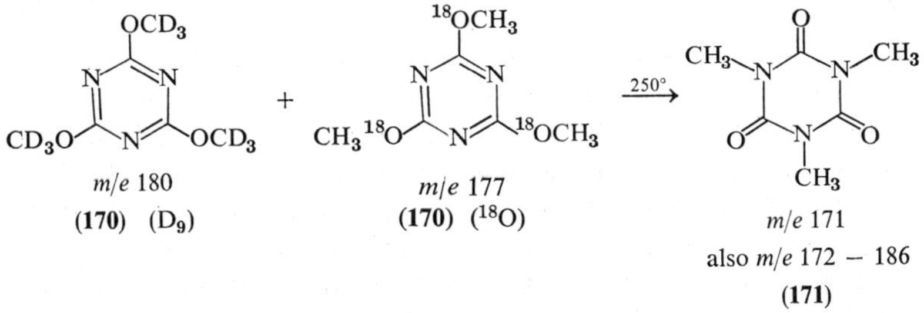

m/e 180
(170) (D$_9$)

m/e 177
(170) (^{18}O)

m/e 171
also m/e 172 − 186
(171)

the ^{18}O and CD$_3$-labeled compounds was heated at 250° (115). The trimethyl amide isolated gave molecular ions from m/e 171–186, whose intensities indicated a statistical distribution, arising from intermolecular transfer.

F. Deuterium

Another very intriguing use of mass spectrometry was that employed in establishing the intermediacy (or lack thereof) of a metallocyclobutadiene (or equivalent) in the metal-catalyzed conversion of butyne to hexamethyl-benzene **(174)** (116). Only from such a symmetrical intermediate could

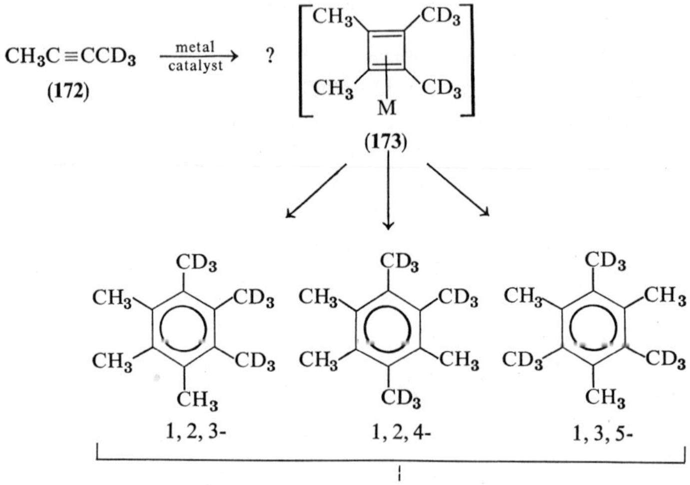

$$CH_3C \equiv CCD_3 \xrightarrow[\text{catalyst}]{\text{metal}} ?$$
(172)

(173)

1, 2, 3-   1, 2, 4-   1, 3, 5-

(174) **(D₉)**

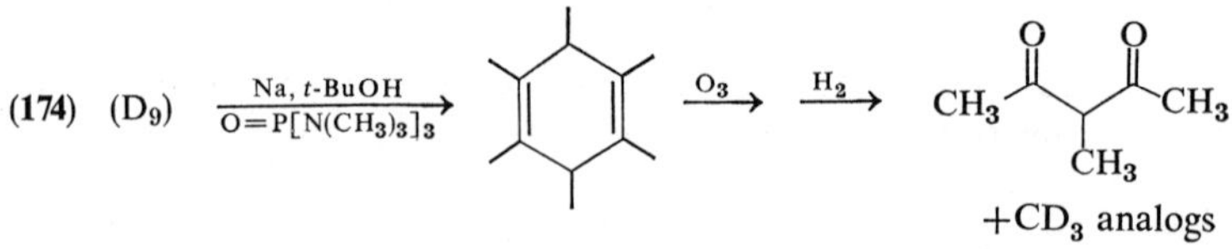

1,2,3-tris(trideuteromethyl)-4,5,6-trimethylbenzene **(174)** arise in addition to the 1,3,4- and 1,3,5-isomers. A "concerted" mechanism would give only the latter two isomers. While some catalysts [$(C_5H_5CN)_2Pd(II)Cl_2$; $AlCl_3$] gave the 1,2,3-isomer in reasonable amounts, most [$(C_6H_5)_3Cr\cdot3THF$; $\{C_6H_3(CH_3)_3\}_2Co(II)$;$Co_2(CO)_8$;$(CH_2CHCN)_2Ni(O)$;and $TiCl_4\cdot(iso\text{-}C_4H_9)_3Al$] did not. Since ordinary mass-spectral analysis cannot distinguish between the isomers, it was necessary to employ the degradation shown, to give 3-methyl-2,4-pentanedione with varying degrees of deuteration.

Deuterium labeling has been of some help in studying photochemical reactions. Among these studies is the report by Haller and Srinivasan (117) that the tricyclooctane **176** is formed without loss of deuterium from 1,5-cyclooctadiene-$^2H_{12}$ **(175)**, thus excluding an earlier-proposed radical mechanism.

The primary process in formaldehyde photolysis was reported to be $CH_2O \xrightarrow{h\upsilon} H_2 + CO$, since irradiation of formaldehyde-D_2 in the presence of unlabeled formaldehyde or propene gave little or no HD ($H_2 + D_2$ in the

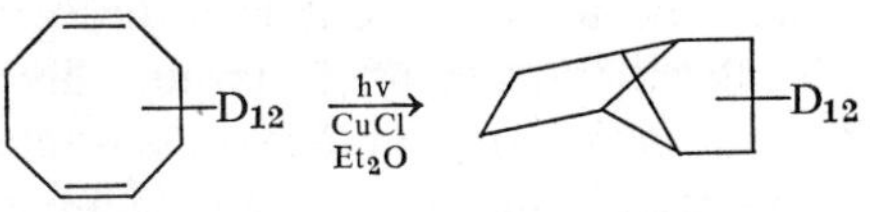

(175)　　　　　　　　　　(176)

former case, D_2 in the latter) (118). Of course, the complex process of hydrogenation continues to employ (119) deuterium labeling as an invaluable tool.

Many reactions have been studied at high temperatures by means of deuterium labeling, several by Fields and Meyerson (120–122). All give a complex mixture of products with multiple labeling by deuterium. In most cases a benzyne intermediate has been shown to be involved.

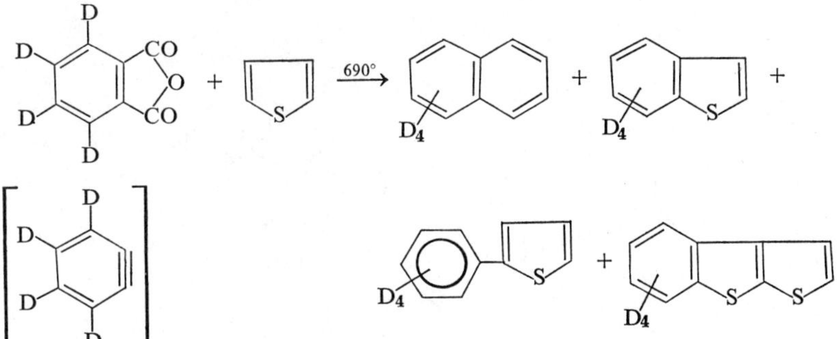

(177)

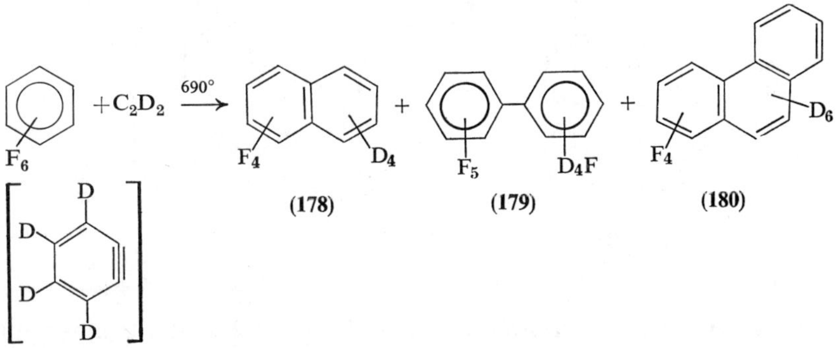

(178)　　　(179)　　　(180)

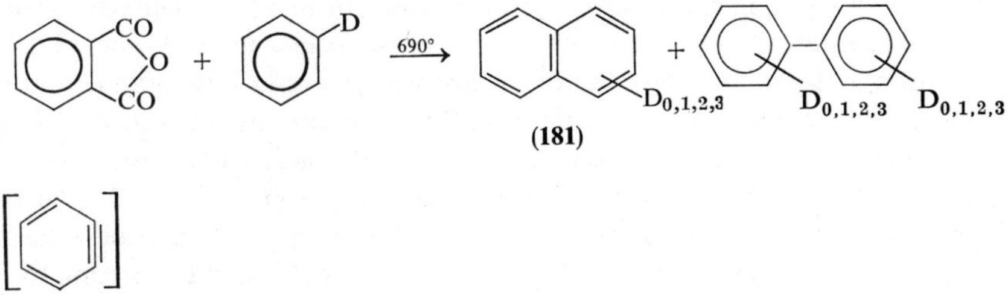

(181)

Similar studies have been carried out by other groups. Three, which apparently involve localized transfer of deuterium, are shown. In the first (123), deuterium from the *ortho* position of styrene is retained in the product (183), presumably by intramolecular transfer to the 9-position. In the second

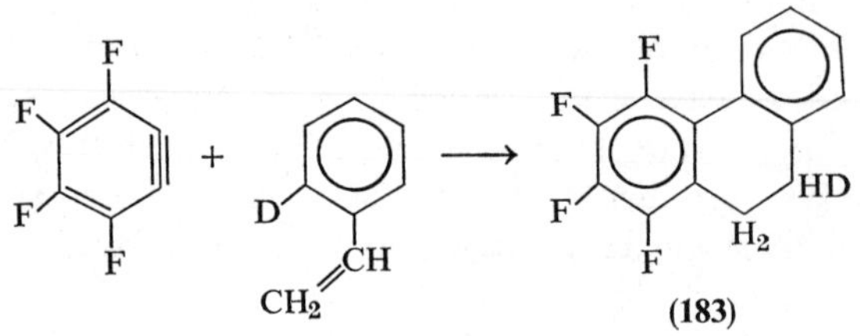

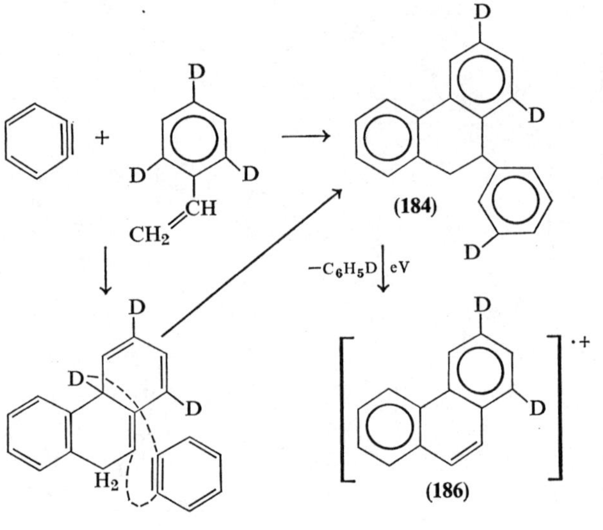

(123), deuterium from an *ortho* position of styrene-2,4,6-D$_3$ is apparently transferred to a phenyl group (*ortho* position, formed from benzyne), since all three deuterium atoms are retained in the molecular ion of 9-phenyl-9,10-dihydrophenanthrene (184), but one is lost with benzene in the fragmentation to give the phenanthrene molecular ion. Both reactions are in accord with the proposed intermediates. In the third (124), the aldehyde deuterium was exchanged with the hydrogens of the *o*-isopropenyl group of 187 to give a random distribution of label. Apparently the same reaction occurs under electron bombardment, since the M − 2 peak (which should be strong for a deuterioaldehyde) is weak, while the M − 1 peak is strong. The intermediate **188** has been proposed for the reaction.

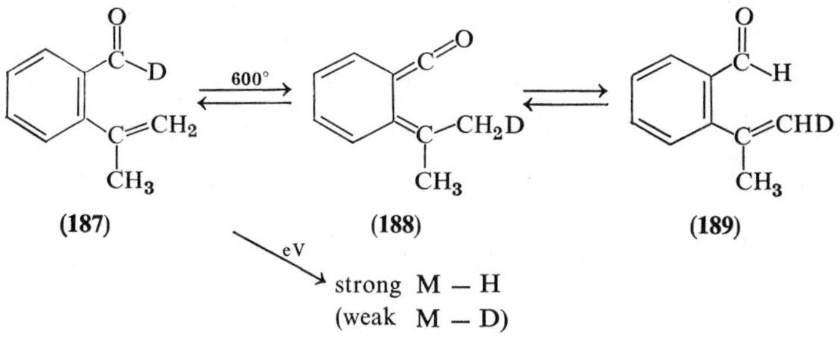

VI. STRUCTURE STUDIES

The two preceding sections have described the use of stable isotopes in studies of mass-spectral fragmentations and in studies of organic reaction mechanisms. The purpose of this section is to review some uses of these isotopes for the purpose of establishing the structure of organic compounds by mass spectrometry.

A. Oxygen-18

The introduction of ^{18}O into ketones by equilibration with $H_2^{18}O$ in the presence of acid provides a convenient means of labeling such compounds for mass spectral studies (125). The technique could certainly lend itself to

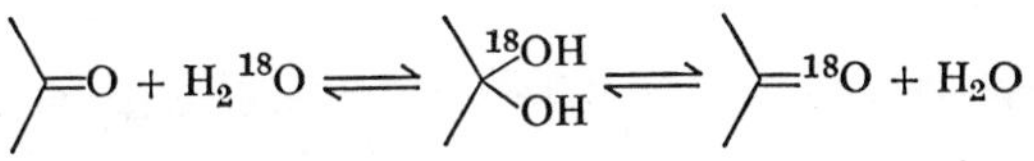

mass-spectral solution of structure problems. The total incorporation of ^{18}O into steroids would reveal the number of exchangeable groups; while the location of the label might be indicated by partial incorporation, since the rate of incorporation depends on the position of the keto group.

Similarly, employing this technique Bolker and Kung found that when **190** was heated for 3 min at 90–95° in 1.0 N HCl in $H_2^{18}O$, the N—OH group was stable, but the keto oxygen underwent complete exchange (126). The

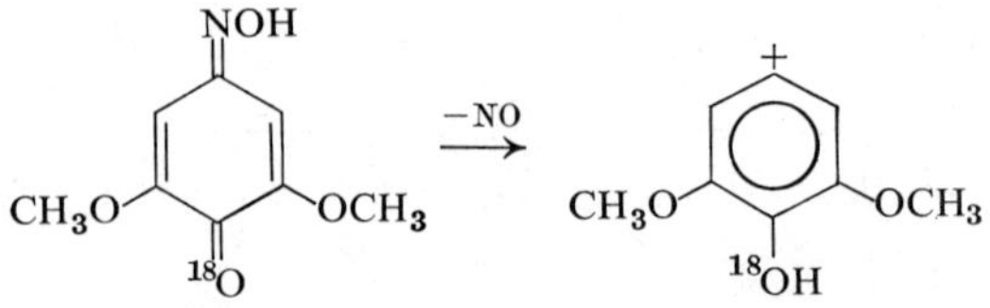

(190)

M·+[m/e 183 (185)] m/e 153 (155)

mass spectrum of **190** had been characterized previously using those of its acetate and its deuterium substituted (OD) derivative.

B. Nitrogen-15

Optically active *trans*-cyclooctene (**191**) was obtained by the Hofmann degradation of (-)-N-n-butyl-N-isobutyl-N-methylcyclooctylammonium salts (**192**). The hydroxide gave a 1.4% optical yield, while the perchlorate

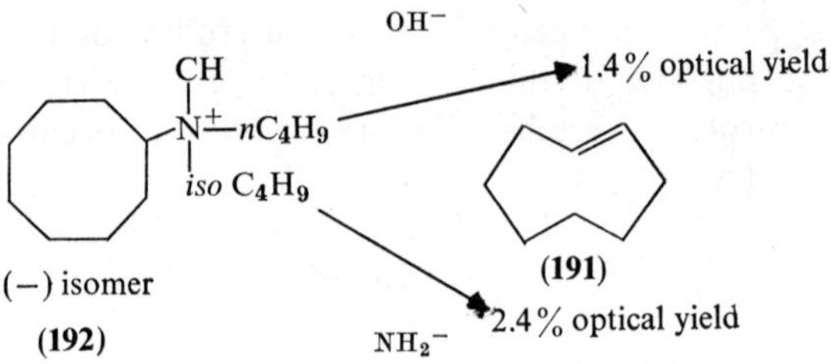

(−) isomer

(192)

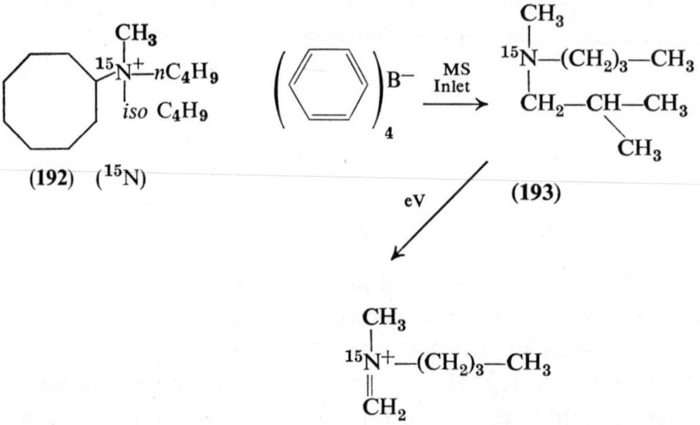

(191)

(192) (¹⁵N)

(193)

(194)

m/e 100 (¹⁵N; 101)

treated with potassium amide gave a 2.4% optical yield (128). Employing ^{15}N as the isotopic label, Cope showed that the optical purity of the resolved quaternary ammonium cation was 99.6%, as determined by the isotopic dilution method with analysis by mass spectrometry. The tetraphenylboron salt was used for the mass-spectral studies, with the ion at m/e 100 (^{15}N, 101) (**194**) providing the measure of ^{15}N incorporation.

Mass spectrometry was employed in assigning a structure to the product of 2-amino-Δ^2-thiazoline (**195**) with potassium cyanate, which could have been 2-ureido-Δ^2-thiazoline (**196**) or 2-imino-3-carbamoylthiazolidine (**197**) (129). Employing the 2-^{15}N-derivative and desulfurizing the product gave urea-^{15}N (**198**), establishing the structure of the product as **196**.

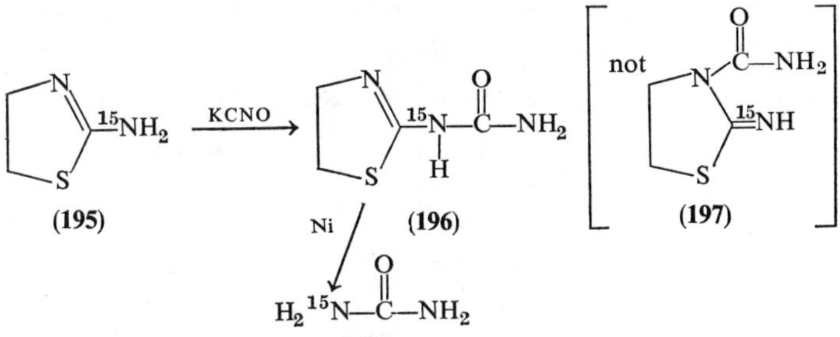

Similarly, Milne determined the structure of the adduct of **195** with phenyl isocyanate as **199** (130). The molecular ion of **201** was measured by high-resolution mass spectrometry as m/e 137.061; $C_7H_8{}^{14}N^{15}NO$ requires m/e 137.061.

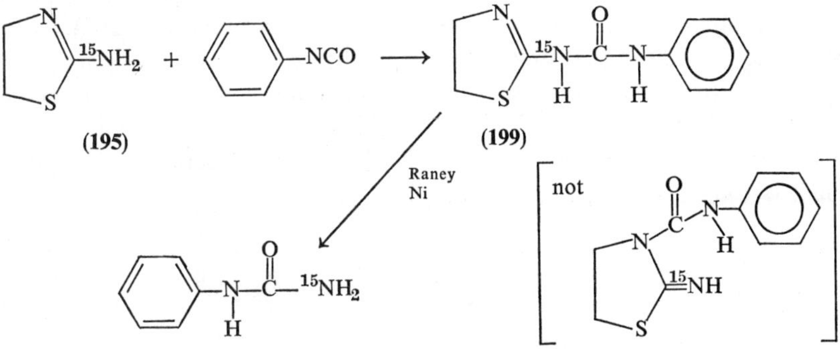

C. Deuterium

The use of deuterium, and particularly CD_3 groups, for labeling the N-terminal portion of peptides has been discussed earlier (9) and is discussed in some detail in Chapter 9. A somewhat different mass-spectral sequencing of two peptides containing an N-terminal pyroglutamic acid residue was described by Thomas *et al.* (131). The peptides, after O,N-permethylation (with CD_3I and NaH in DMF), gave excellent mass spectra. The heptapeptide of the zymogen of phospholipase A demonstrated the following sequence:

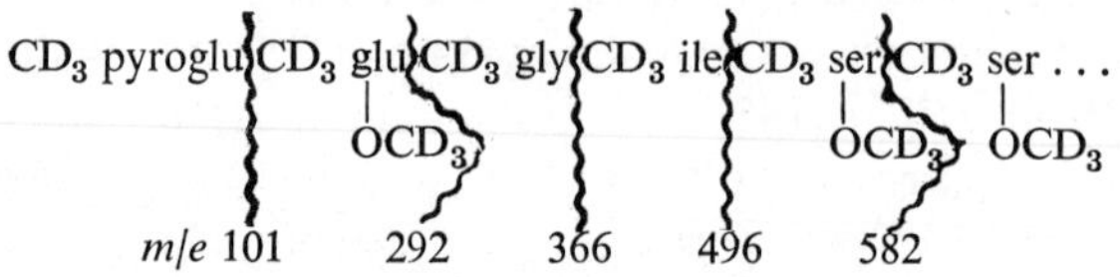

$$CD_3 \text{ pyroglu} \mid CD_3 \text{ glu} \mid CD_3 \text{ gly} \mid CD_3 \text{ ile} \mid CD_3 \text{ ser} \mid CD_3 \text{ ser} \ldots$$

$$\overset{|}{O}CD_3 \qquad\qquad\qquad \overset{|}{O}CD_3 \quad \overset{|}{O}CD_3$$

$$m/e\ 101 \qquad 292 \qquad 366 \qquad 496 \qquad 582$$

The docosapeptide from the λ-chain of pig immunoglobulin gave a mass spectrum (without CD_3 labeling) consistent with either of the following two sequences (indicating two components):

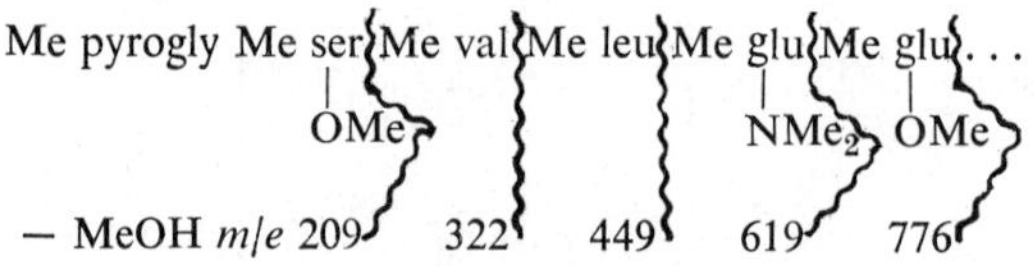

$$\text{Me pyrogly Me ser} \mid \text{Me val} \mid \text{Me leu} \mid \text{Me glu} \mid \text{Me glu} \ldots$$

$$\overset{|}{O}Me \qquad\qquad\qquad \overset{|}{N}Me_2 \quad \overset{|}{O}Me$$

$$- \text{ MeOH } m/e\ 209 \qquad 322 \qquad 449 \qquad 619 \qquad 776 \tag{A}$$

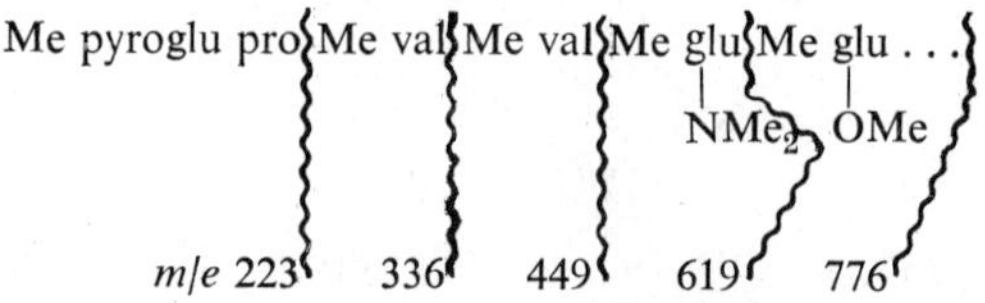

$$\text{Me pyroglu pro} \mid \text{Me val} \mid \text{Me val} \mid \text{Me glu} \mid \text{Me glu} \ldots$$

$$\overset{|}{N}Me_2 \quad \overset{|}{O}Me$$

$$m/e\ 223 \qquad 336 \qquad 449 \qquad 619 \qquad 776 \tag{B}$$

A study that illustrates the use of mass spectrometry for determining the number of exchangeable hydrogens with deuterium as an aid to structure proof is that of Forward and Whiting (132). Four structures were possible for the photodimer of 2-benzyl-5-benzylidenecyclopentanone (**202**). Mass spectrometry showed that only two hydrogens exchanged in the photodimer when it was heated at reflux for 12 hours in a solution of sodium in ethanol-OD. This information, coupled with the other mass-spectral data and the nuclear magnetic resonance studies, established the structure of the photo-dimer as **203**.

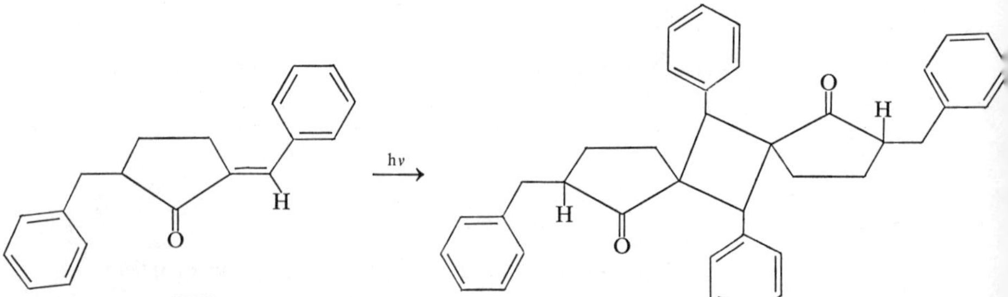

(202)   (203)

Jones and Marples employed mass-spectral evidence in assigning the structure of **205**, arising from treatment of the 6β-acetoxy-$8\beta,9\beta$-epoxy-11-ketone **204** with methanolic potassium hydroxide (133).

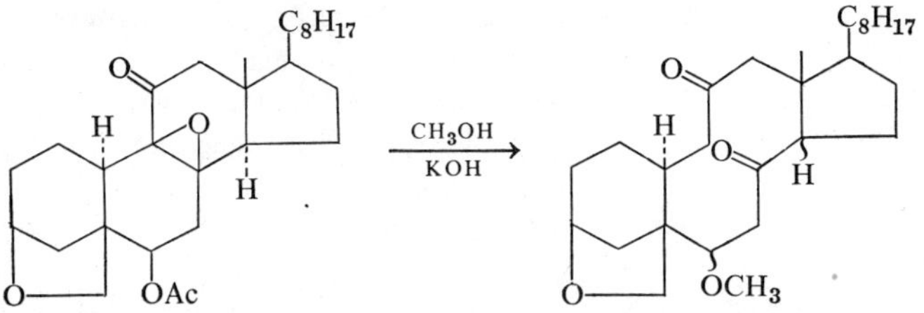

(204)   (205)

Mass-spectral analysis of the product resulting from the treatment of **205** with C_2H_5OD-NaOD (72 hrs at 60°) showed ions at m/e 460, 461, 462, 464, 465 corresponding to the d_0 (7%), d_1 (25%), d_2 (10%), d_4 (9%), and d_5 (50%) species. Although the expected d_6 and d_7 species were absent, presumably since two of the hydrogen atoms of **205** are exchanged with difficulty, the incorporation of more than four deuterium atoms indicated the presence of a diketone, apparently arising from a complex reduction.

REFERENCES

1. H. Budzikiewicz, C. Djerassi, and D. H. Williams, *Mass Spectrometry of Organic Compounds*, Holden-Day, Inc., San Francisco, Calif., 1967.

2. H. Budzikiewicz, C. Djerassi, and D. H. Williams, *Structure Elucidation of Natural Products by Mass Spectrometry, Vol. I: Alkaloids*, Holden-Day, Inc., San Francisco, Calif., 1964.

3. Ibid., *Vol. II: Steroids, Terpenoids, Sugars, and Miscellaneous Classes*, Holden-Day, Inc., San Francisco, Calif., 1964.

4. J. H. Beynon, R. A. Saunders, and A. E. Williams, *The Mass Spectra of Organic Molecules*, Elsevier Publishing Co., Amsterdam, The Netherlands, 1968.

5. R. Viallard, *Bull. Soc. Chim. France*, 3695 (1966).

6. M. Corval and M. R. Viallard, *Bull. Soc. Chim. France*, 3710 (1966).

7. S. Meyerson and E. K. Fields, *Science*, **166**, 325 (1969).

8. G. E. Van Lear and F. W. McLafferty, *Ann. Rev. Biochem.* **38**, 289 (1969).

9. K. L. Rinehart, Jr. and T. H. Kinstle, *Ann. Rev. Phys. Chem.*, **19**, 301 (1968).

10. H. D. Beckey, *Z. Naturforsch.*, **14a**, 712 (1959).

11. Z. Korostyshevskii and I. V. Goldenfeld, *Dokl. Akad. Nauk. SSSR.*, **172**, 1364 (1967).

12. J. H. Beynon, *Mass Spectrometry and its Applications to Organic Chemistry*,'' Elsevier Publishing Co., Amsterdam, The Netherlands, 1960.

13. J. H. Beynon and A. E. Williams, *Mass and Abundance Tables for Use in Mass Spectrometry*, Elsevier Publishing Co., Amsterdam, The Netherlands, 1963.

14. R. A. Saunders and A. E. Williams, in F. W. McLafferty, (Ed.), *Mass Spectrometry of Organic Ions*, Academic Press, New York, 1963, Chapter 8.

15. H. Knöppel and W. Beyrich, *Tetrahedron Lett.*, 291 (1968).

16. (a) M. Senn, W. J. Richter and A. L. Burlingame, *J. Amer. Chem. Soc.*, **87**, 680 (1965); (b) W. J. Richter, M. Senn and A. L. Burlingame, *Tetrahedron Lett.*, 1235 (1965).

17. M. M. Bursey and F. W. McLafferty, *Carbonium Ions*, Interscience, New York, 1968.

18. J. L. Sommerville, (Ed.), *The Isotope Index*, Scientific Equipment Co., Indianapolis, 1967.

19. H. Nakano, H. Sato, and B. Tamaoki, *Biochem. Biophys. Res. Commun.*, **22**, 425 (1966).

20. H. Nakano, H. Inano, H. Sato, M. Shikita, and B. Tamaoki, *Biochim. Biophys. Acta*, **137**, 335 (1967).

21. H. Nakano, C. Takemoto, H. Sato, and B. Tamaoki, *Biochim. Biophys. Acta*, **152**, 186 (1968).

22. H. Takemoto, H. Nakano, H. Sato, and B. Tamaoki, *Biochim. Biophys. Acta*, **152**, 749 (1968).

23. H. Nakano, H. Sato, and B. Tamaoki, *Biochim. Biophys. Acta*, **164**, 585 (1968).

24. E. E. van Tamelen, J. D. Willett, and R. B. Clayton, *J. Amer. Chem. Soc.*, **89**, 3371 (1967).

25. J. D. Willett, K. B. Sharpless, K. E. Lord, E. E. van Tamelen, and R. B. Clayton, *J. Biol. Chem.*, **242**, 4182 (1967).

26. R. Ryhage and B. Samuelsson, *Biochem. Biophys. Res. Commun.*, **19**, 279 (1965).

27. N. Itada, *Biochem. Biophys. Res. Commun.*, **20**, 149 (1965).

28. S. R. Thorpe and C. C. Sweeley, *Biochemistry*, **6**, 887 (1967).

29. J. Daly, M. Levitt, G. Guroff, and S. Udenfriend, *Arch. Biochem. Biophys.*, **126**, 593 (1968).

30. E. Heinz, A. P. Tulloch, and J. F. T. Spencer, *J. Biol. Chem.*, **244**, 882 (1969).

31. R. Caprioli and D. Rittenberg, *Biochemistry*, **8**, 3375 (1969).

32. H. Follmann and H. P. C. Hogenkamp, in press.

33. P. D. Shaw and J. A. McCloskey, *Biochemistry*, **6**, 2247 (1967).

34. H. P. Broquist, A. J. Aspen, S. D. Aust, J. J. Snyder, R. A. Gardiner, and K. L. Rinehart, Jr., in press.

35. (a) A. K. Bose, K. G. Das, P. T. Funke, I. Kugajevsky, O. P. Shukla, K. S. Khanchandani, and R. J. Suhadolnik, *J. Amer. Chem. Soc.*, **90**, 1038 (1968); (b) A. K. Bose, K. S. Khanchandani, R. Tavares, and P. Funke, 5th International Symposium on the Chemistry of Natural Products, London, July 8-13, 1968; Abstracts, p. 119.

36. G. R. Waller, R. Ryhage, and S. Meyerson, *Anal. Biochem.*, **16**, 277 (1966).

37. G. Popják, D. S. Goodman, J. W. Cornforth, and R. Ryhage, *J. Biol. Chem.*, **236**, 1934 (1961).

38. G. Popják, J. W. Cornforth, R. H. Cornforth, R. Ryhage, and D. S. Goodman, *J. Biol. Chem.*, **237**, 56 (1962).

39. E. S. Hall, F. McCapra, T. Money, K. Fukumoto, J. R. Hanson, B. S. Mootoo, G. T. Phillips, and A. I. Scott, *Chem. Commun.*, 348 (1966).

40. G. J. Schroepfer, Jr., *J. Biol. Chem.*, **241**, 5441 (1966).

41. G. J. Schroepfer Jr., W. H. Lee, R. Kammereck, and J. A. McCloskey, 154th Natl. Mtg. Amer. Chem. Soc., Chicago, Ill., Sept., 1967, Abstract No., 140C.

42. (a) W. K. Saur, H. L. Crespi, E. A. Halevi, and J. J. Katz, *Biochemistry*, 7, 3529 (1968); (b) W. K. Saur, H. L. Crespi, L. Harkness, G. Norman, and J. J. Katz, *Anal. Biochem.*, 22, 424 (1968).

43. M. Lenfant, E. Zissman, and E. Lederer, *Tetrahedron Lett.*, 1049 (1967).

44. L. M. Jackman, I. G. O'Brien, G. B. Cox, and F. Gibson, *Biochim. Biophys. Acta.*, 141, 1 (1967).

45. A. D. Argoudelis, T. E. Eble, J. A. Fox, and D. J. Mason, *Biochemistry*, 8, 3408 (1969).

46. P. Goldman and G. W. A. Milne, *J. Biol. Chem.*, 241, 5557 (1966).

47. P. Goldman, G. W. A. Milne, and D. B. Keister, *J. Biol. Chem.*, 243, 428 (1968).

48. J. L. Holtzman, J. R. Gillette, and G. W. A. Milne, *J. Biol. Chem.*, 242, 4386 (1967).

49. J. Holtzman, J. R. Gillette, and G. W. A. Milne, *J. Amer. Chem. Soc.*, 89, 6341 (1967).

50. L. G. Sparrow, P. P. K. Ho, T. K. Sundaram, D. Zach, E. J. Nyns, and E. E. Snell, *J. Biol. Chem.*, 244, 2590 (1969).

51. H. Eriksson, J. A. Gustafsson, and J. Sjövall, *European J. Biochem.*, 9, 550 (1969).

52. G. Spiteller, *Massenspektrometrische Strukturanalyse Organischer Verbindungen*, Verlag Chemie, Weinheim/Bergstrasse, 1964.

53. (a) L. Friedman, *Ann. Rev. Phys. Chem.*, 19, 273 (1968); (b) V. Cermak, A. Dalgarro, E. E. Ferguson, L. Friedman, and E. W. McDaniel, *Ion-Molecule Reactions*, Wiley, New York, 1969.

54. K. L. Rinehart, Jr., R. Ryhage, and H. B. Renfroe, 145th Natl. Mtg. Amer. Chem. Soc., New York, Sept. 8–13, 1963; Paper No. Q27.

55. R. A. W. Johnstone, B. J. Millard and D. S. Millington, *Chem. Commun.*, 600 (1966).

56. J. L. Cotter and R. A. Dine-Hart, *Chem. Commun.*, 809 (1966).

57. J. Bonham, E. McLeister, and P. Beak, *J. Org. Chem.*, 32, 639 (1967).

58. M. G. Moshonas and E. D. Lund, *J. Agr. Food Chem.*, 17, 802 (1969).

59. P. Brown and C. Djerassi, *J. Amer. Chem. Soc.*, 89, 2711 (1967).

60. B. G. Pring and N. E. Stjernström, *Acta Chem. Scand.*, 22, 549 (1968).

61. T. H. Kinstle, O. L. Chapman, and M. Sung, *J. Amer. Chem. Soc.*, 90, 1227 (1968).

62. R. F. Schimbor, Ph.D. thesis, University of Illinois, 1966.

63. K. L. Rinehart, Jr., R. F. Schimbor, and T. H. Kinstle, *Antimicrobial Agents and Chemotherapy—1965*, 119 (1966).

64. A. K. Bose, I. Kugajevsky, P. T. Funke, and K. G. Das, *Tetrahedron Lett.*, 3065 (1965).

65. V. N. Bochkarev, N. S. Ovchinnikova, N. S. Wul'fson, E. M. Kleiner, and A. S. Khokhlov, *Dokl. Akad. Nauk, SSSR.*, 172, 1079 (1967).

66. H. M. Grubb and S. Meyerson, in F. W. McLafferty, (Ed.), *Mass Spectrometry of Organic Ions*, Academic Press Inc., New York, N.Y., 1963, Chapter 10.

67. S. Meyerson, *Record Chem. Progr.*, 26, 257 (1965).

68. S. Meyerson and P. N. Rylander, *J. Chem. Phys.*, 27, 901 (1957).

69. K. L. Rinehart, Jr. and A. C. Buchholz, 154th Natl. Mtg., Amer. Chem. Soc., Chicago, Ill., Sept. 11–15, 1967, Paper No. S85.

70. K. L. Rinehart, Jr., A. C. Buchholz, G. E. Van Lear, and H. L. Cantrill, *J. Amer. Chem. Soc.*, 90, 2983 (1968).

71. A. Siegel and K. L. Rinehart, Jr., 158th Natl. Mtg., Amer. Chem. Soc., New York, N.Y., Sept. 7–12, 1969, Paper No. 066.

72. A. V. Robertson, M. Marx, and C. Djerassi, *Chem. Commun.*, 414 (1968).

73. K. L. Rinehart, Jr. and A. C. Buchholz, *J. Amer. Chem. Soc.*, **90**, 1973 (1968).

74. A. V. Robertson and C. Djerassi, *J. Amer. Chem. Soc.*, **90**, 6992 (1968).

75. P. D. Woodgate and C. Djerassi, *Tetrahedron Lett.*, in press.

76. M. Marx and C. Djerassi, *J. Amer. Chem. Soc.*, **90**, 678 (1968).

77. D. G. I. Kingston, 158th Natl. Mtg., Amer. Chem. Soc., New York, Sept. 7–12, 1969, Paper No. 067.

78. N. G. Foster and R. W. Higgins, *Org. Mass Spectrom.*, **1**, 191 (1968).

79. K. D. Berlin and R. D. Shupe, *Org. Mass Spectrom.*, **2**, 447 (1969).

80. R. A. W. Johnstone and B. J. Millard, *Z. Naturforsch.*, **21a**, 604 (1966).

81. R. A. W. Johnstone and S. D. Ward, *Adv. Mass Spect.*, **4**, 211 (1968).

82. R. A. W. Johnstone and B. J. Millard, *J. Chem. Soc. (C)*, 1955 (1966).

83. G. G. Meisels, J. Y. Park, and B. G. Giessner, *J. Amer. Chem. Soc.*, **91**, 1555 (1969).

84. H. H. Voge, C. D. Wagner, and D. D. Stevenson, *J. Catalysis*, **2**, 58 (1963).

85. R. A. W. Johnstone, B. J. Millard, F. M. Dean, and A. W. Hill, *J. Chem. Soc. (C)*, 1712 (1966).

86. F. M. Dean, J. Goodchild, R. A. W. Johnstone, and B. J. Millard, *J. Chem. Soc. (C)* 2232 (1967).

87. T. H. Kinstle and J. G. Stam, *Chem. Commun.*, 185 (1968).

88. A. K. Bose, I. Kugajevsky, P. T. Funke, and K. G. Das, *Tetrahedron Lett.*, 1725 (1964).

89. M. S. Manhas and R. T. So, *15th Annual Conference on Mass Spectrometry and Allied Topics*, ASTM, Committee E-14, 1967, Denver, Colorado, Paper 121.

90. E. S. Gould, *Mechanism and Structure in Organic Chemistry*, Holt, Rinehart and Winston, New York, 1959.

91. J. Hine, *Physical Organic Chemistry*, McGraw-Hill, New York, 1962.

92. R. Stewart, *The Investigation of Organic Reactions*, Prentice-Hall, Englewood Cliffs, N.J., 1966.

93. R. Breslow, *Organic Reaction Mechanisms*, 2nd ed., Benjamin, New York, 1969.

94. T. Axenrod and G. W. A. Milne, *Tetrahedron*, **24**, 5775 (1968).

95. R. H. Shapiro, T. E. McEntee, Jr., and D. L. Coffen, *Tetrahedron*, **24**, 2809 (1968).

96. C. Lifshitz and M. Shapiro, *J. Chem. Phys.*, **45**, 4242 (1966).

97. C. Lifshitz and M. Shapiro, *J. Chem. Phys.*, **46**, 4912 (1967).

98. J. A. McCloskey, A. M. Lawson, and F. A. J. M. Leemans, *Chem. Commun.*, 285 (1967).

99. C. Djerassi and B. Zeeh, *Chem. Ind. (London)*, 358 (1967).

100. J. A. Osborn, F. H. Jardine, J. F. Young, and G. Wilkinson, *J. Chem. Soc. (A)*, 1711 (1966).

101. H.-J. Hansen, B. Sutter, and H. Schmid, *Helv. Chim. Acta*, **51**, 828 (1968).

102. A. Fry and M. Lusser, *J. Org. Chem.*, **31**, 3422 (1966).

103. G. Illuminati and F. Monacelli, *J. Inorg. Nucl. Chem.*, **29**, 1265 (1967).

104. E. N. Gur'yanova, P. I. Saukov, A. I. Kutepova, V. V. Grigor'ev, and N. I. Grishko, *Zh. Org. Khim.*, **3**, 2199 (1967).

105. H. Richey, Jr. and J. M. Richey, *J. Amer. Chem. Soc.*, **88**, 4971 (1966).

106. H. E. Audier, J. F. Dupir, and J. Julien, *Bull. Soc. Chim. France*, 2812 (1966).

107. K. G. Rutherford and R. M. Ottenbrite, *Can. J. Chem.*, **45**, 679 (1967).

108. S. Fliszár, J. Carles, and J. Renard, *J. Amer. Chem. Soc.*, **90**, 1364 (1968).

109. W. B. DeMore and O. F. Raper, *J. Chem. Phys.*, **46**, 2500 (1967).

110. M. Neiman, A. Blumenfeld, and B. Kovarskaya, *J. Polymer Sci.*, **4**, 2901 (1966).

111. B. S. Shasha, W. M. Doane, C. R. Russell, and C. E. Rist, *J. Org. Chem.*, **34**, 1642 (1969).

112. N. J. Leonard and B. Zwanenburg, *J. Amer. Chem. Soc.*, **89**, 4456 (1967).

113. J. M. Landesberg and R. A. Olofson, *Tetrahedron*, **22**, 2135 (1966).

114. J.-C. Strini and J. Metzger, *Bull. Soc. Chim. France*, 3150 (1966).

115. L. Paoloni, G. Nencini, M. L. Tosato, and T. Sabatori, *Gazz. Chim. Ital.*, **97**, 635 (1967).

116. G. M. Whitesides and W. J. Ehmann, *J. Amer. Chem. Soc.*, **91**, 3800 (1969).

117. I. Haller and R. Srinivasan, *J. Amer. Chem. Soc.*, **88**, 5084 (1966).

118. B. A. DeGraff and J. G. Calvert, *J. Amer. Chem. Soc.*, **89**, 2247 (1967).

119. R. L. Burwell, Jr. and K. Schrage, *Faraday Soc. Disc.*, **41**, 215 (1966).

120. E. K. Fields and S. Meyerson, *J. Org. Chem.*, **34**, 2475 (1969).

121. E. K. Fields and S. Meyerson, *Tetrahedron Lett.*, 571 (1967).

122. E. K. Fields and S. Meyerson, *J. Amer. Chem. Soc.*, **88**, 21 (1966).

123. R. Harrison, H. Heaney, J. M. Jablonski, K. G. Mason, and J. M. Sketchley, *J. Chem. Soc.* (*C*), 1684 (1969).

124. R. F. C. Brown and M. Butcher, *Aust. J. Chem.*, **22**, 1457 (1969).

125. A. M. Lawson, F. A. J. M. Leemans and J. A. McCloskey, *15th Annual Conference on Mass Spectrometry and Allied Topics,'ASTM, Committee E-14*, 1967, Denver, Colorado, Paper 116.

126. H. J. Bolker and F. L. Kung, *Can. J. Chem.*, **47**, 2109 (1969).

127. P. Capella and C. M. Zorzut, *Anal. Chem.*, **40**, 1458 (1968).

128. A. C. Cope, W. R. Funke, and F. N. Jones, *J. Amer. Chem. Soc.*, **88**, 4693 (1966).

129. D. L. Klayman, A. Senning, and G. W. A. Milne, *Acta. Chem. Scand.*, **21**, 217 (1967).

130. D. L. Klayman, J. J. Maul, and G. W. A. Milne, *Tetrahedron Lett.*, 281 (1967).

131. G. H. de Haas, F. Franek, B. Keil, D. W. Thomas, and E. Lederer, *Febs. Lett.*, **4**, 25 (1969).

132. G. C. Forward and D. A. Whiting, *J. Chem. Soc.* (*C*), 1868 (1969).

133. J. G. L. Jones and B. A. Marples, *Chem. Commun.*, 872 (1969).

Mass Spectrometry in Peptide Chemistry

M. M. SHEMYAKIN, YU A. OVCHINNIKOV,
AND A. A. KIRYUSHKIN

*Institute for Chemistry of Natural Products, USSR Academy
of Sciences, Moscow, USSR*

I. Introduction 289
II. Earlier Work on Mass-Spectrometric Approaches to Amino Acid Sequence
Determination in Peptides 290
III. Peptide Chain Fragmentation Patterns in *N*-Acylpeptide Esters 292
IV. Fragmentation of *N*-Acylpeptide Esters as a Function of Constituent Amino
Acid Residues 295
    A. Aliphatic Amino Acids 295
    B. Proline 297
    C. Hydroxyamino Acids 297
    D. Monoamino Dicarboxylic Acids 297
    E. Aromatic and Heterocyclic Amino Acids 299
    F. Methionine 301
    G. Cystine 302
    H. Cysteine 303
    I. Lysine and Ornithine 304
    J. Arginine 304
V. Cyclic Peptides and Depsipeptides 307
VI. Application of Mass Spectrometry to the Solution of Structural and Analytical
Problems in Peptide and Protein Chemistry 311
References 320

I. INTRODUCTION

The importance of mass spectrometry in peptide chemistry depends mainly on the possibility that it may be successfully applied to the determination of the sequence of amino acid residues in peptides, particularly those peptides obtained by selective cleavage of proteins. Chemical or enzymatic cleavage of protein chains into polypeptide fragments and the separation of the latter now presents no fundamental difficulties. However, a rapid and straightforward determination of amino acid sequence in the peptides thus obtained is not always possible, the moreso since in most cases only micro amounts of material are available. All the methods used to this end are

289

extremely laborious and usually permit determination of only a partial sequence. However, as the problem of primary structure determination in proteins is so crucial and the cases encountered are so diverse, it would be advantageous to have several methods of determination to check and supplement one another. In recent years mass spectrometry has been developed as a method of structure elucidation of the peptide products of protein cleavage, that could compete successfully or, moreover, be combined with other methods of protein analysis, including Edman degradation.

This review* is written with the aim of discussing the literature on mass spectra of peptides from the viewpoint of its practical value, and attention is mainly directed to the work that has led to the introduction and successful development of methods of mass-spectrometric determination of amino acid sequence in peptides.

II. EARLIER WORK ON MASS-SPECTROMETRIC APPROACHES TO AMINO ACID SEQUENCE DETERMINATION IN PEPTIDES

There have been some reports (1, 5, 6) on mass spectrometry of several simple di- and tripeptides, but in general, free peptides are not amenable to mass-spectrometric investigation, as they are too involatile and thermally unstable. Thus dipeptides, upon heating, give diketopiperazines fairly easily, and the resulting mass spectra naturally do not permit any conclusions with regard to the amino acid sequence in the initial peptide (6).

The first attempts to work out some practical procedures to allow the application of mass spectrometry to amino acid sequence determination in peptides were made by Biemann in the late 1950s and early 1960s (1, 7–9) [see also (10)]. Biemann suggested reduction of N-acetylpeptide esters with lithium aluminum hydride (or deuteride) to the corresponding polyamino-alcohols (1), which are highly volatile; the decomposition under electron impact of these products is very specific, involving mainly C—C bond splitting, as follows:

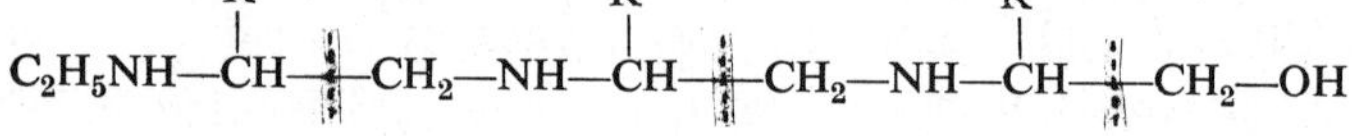

$$C_2H_5NH-\overset{\displaystyle R}{\underset{\displaystyle |}{CH}}\text{--}CH_2-NH-\overset{\displaystyle R}{\underset{\displaystyle |}{CH}}\text{--}CH_2-NH-\overset{\displaystyle R}{\underset{\displaystyle |}{CH}}\text{--}CH_2-OH$$

(1)

Identification in the mass spectra of polyaminoalcohols (1) of the resulting fragments provides the necessary information on amino acid residue sequence in the original peptide. However, preparative reduction of

---

* For earlier reviews on the mass spectrometry of peptides, see Ref. 1–4.

peptides of more than 4 or 5 amino acid residues (even if all of these are aliphatic) on a small scale with good yields is difficult. Biemann therefore suggested rather a complicated scheme of polypeptide analysis comprising the following steps: (1) nonspecific hydrolysis of the compound under investigation to a mixture of lower peptides; (2) acetylation and esterification of the shorter peptides to convert them into compounds soluble in organic solvents; (3) reduction of the mixture of the N-acylpeptide esters with $LiAlH_4$ or (for labeling) with $LiAlD_4$; and (4) gas-chromatographic separation of the resulting polyaminoalcohols and mass-spectrometric analysis of separate fractions. Whenever the side chains of amino acid residues contained a hydroxyl group, or acquired one upon reduction of the N-acylpeptide ester (serine, aspartic acid, etc.), the polyaminopolyols had to be converted with $SOCl_2$ into polyamino chlorides and subsequently reduced with $LiAlD_4$ to polyamines. The intermediate aminochlorides, however, turned out to be unstable and liable to undergo diverse inter- and intramolecular reactions, complicating the interpretation of the mass spectra. Histidine- and arginine-containing peptides were even more troublesome. Because of all these difficulties, Biemann's technique has not been applied to the structure elucidation of natural peptides, though it was considered a most promising method until the mid-1960s, when it was shown that intermediate N-acylpeptide esters can be used directly for mass spectrometry.

It was found in 1958 by Andersson (11) and in 1961 by Stenhagen (12) that methyl esters of trifluoroacetyl peptides give reproducible mass spectra that reflect the amino acid sequence of the peptides. A little later, Weygand (13) came to the same conclusion, having shown that mass spectra of N-trifluoroacetyl peptide esters exhibit peaks corresponding to species such as $[CF_3CO-NHCHR]^+$ and $[H_2N-CHR-COOMe]^+$, which are very informative as to the amino acid sequence. Similar ions were also found in the mass spectra of acetylpeptides and their esters (14–16).

It was now obvious that for peptides—at any rate, for those constructed of a smaller number of simple (e.g., aliphatic) amino acids—the amino acid sequence could be deduced from the mass spectra of their N-acylated esters. However, the number of investigated peptides was still too small, and the studies performed were too out of touch with the actual problems of analytical peptide and protein chemistry, for an accurate assessment of the true possibilities of mass spectrometry in this field. Therefore detailed investigations of the mass-spectrometric behavior of N-acylpeptide esters were undertaken by several research groups. Fragmentation routes common to all N-acylpeptide esters irrespective of their amino acid composition were investigated and in addition, some peculiar fragmentations due to the presence of certain amino acid residues were studied. Simultaneously, effective methods for the conversion of free peptides into compounds suitable for mass spectrometry were worked out, rational application of mass spectrometry to structural problems of peptide and protein chemistry was suggested, the scope of the methods was determined, and their extension to larger peptides was studied. The course of this work and the main results have been reported in a series of review papers (17–23).

III. PEPTIDE CHAIN FRAGMENTATION PATTERNS IN N-ACYLPEPTIDE ESTERS

The general pattern of peptide chain fragmentation for *N*-acylpeptide esters found to be common to peptides containing most amino acids was elucidated by 1965–1966 (17–26).

Usually the primary process in the decomposition of the *N*-acylpeptide ester molecular ion (2) is the splitting of one of the amide bonds (or of the C-terminal ester bond) with localization of the positive charge at the *amino acid fragment* (3), which carries the *N*-acyl protecting group. The fragment (3) decomposes further, eliminating the amino acid residue either in one step (3 → 5) or in two steps (3 → 4 → 5) via an intermediate *aldimine fragment*

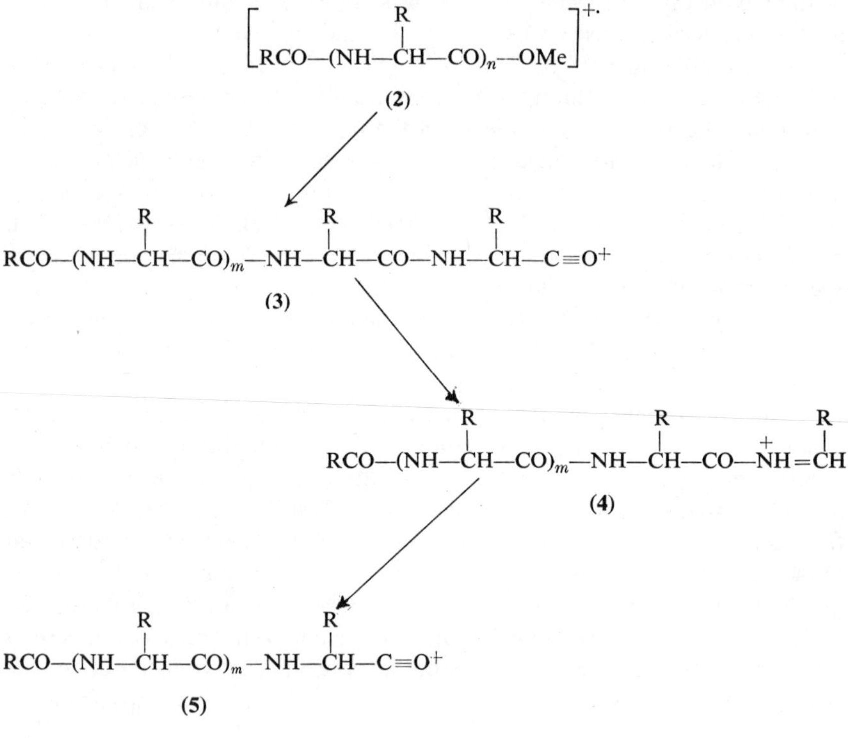

$$\left[RCO-(NH-\overset{\overset{\displaystyle R}{|}}{CH}-CO)_n-OMe \right]^{+\cdot}$$

(2)

$$RCO-(NH-\overset{\overset{\displaystyle R}{|}}{CH}-CO)_m-NH-\overset{\overset{\displaystyle R}{|}}{CH}-CO-NH-\overset{\overset{\displaystyle R}{|}}{CH}-C\equiv O^+$$

(3)

$$RCO-(NH-\overset{\overset{\displaystyle R}{|}}{CH}-CO)_m-NH-\overset{\overset{\displaystyle R}{|}}{CH}-CO-\overset{+}{NH}=\overset{\overset{\displaystyle R}{|}}{CH}$$

(4)

$$RCO-(NH-\overset{\overset{\displaystyle R}{|}}{CH}-CO)_m-NH-\overset{\overset{\displaystyle R}{|}}{CH}-C\equiv O^+$$

(5)

(**4**); amino acid fragments such as **5** decompose further until the ion RCO$^+$ is formed. Sometimes the amino acid fragment loses two or more amino acid residues in a single step.

This so-called "*amino acid type of fragmentation*" (24) results in cleavage of all the amide bonds in the *N*-acylpeptide ester. The different m/e values of amino acid and/or aldimine fragments corresponding to the splitting of the respective amide bonds readily give the amino acid sequence in the starting peptide.

The relative intensities of the amino acid and aldimine ions are affected by a series of factors such as peptide chain length, the nature of *N*-acyl group and the C-terminal residue of the fragment, and the like. The longer the peptide chain, the more abundant the amino-acid fragment ions, whereas with di- and tripeptides the aldimine ions (**4**) are often more intense than the corresponding amino acid ions (**3**). With trifluoroacetyl peptides the aldimine ions are usually far more intense than the spectra of peptides acylated with aliphatic acids; thus methyl esters of *N*-trifluoroacetyl peptides containing two or three amino acid residues form aldimine fragments from the molecular ion in a one-step process. Intensities of aldimine fragment ions usually are fairly high when the C-terminal amino acid residue has a branched side chain, while aldimine fragment ions having glycine as a C-terminus are either quite weak or altogether absent.

Peptide chain cleavage in the manner described above is sometimes accompanied by hydrogen atom migration, which is usually caused by a rearrangement process of some sort. For example, C_α-$C_{carb.}$ bond splitting can proceed as follows (17), resulting in an ion (**6**) differing from the aldimine fragment by 1 amu:

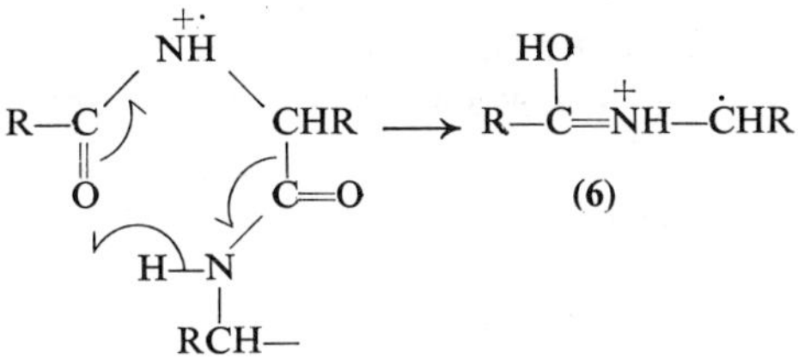

$$R-C(=O)\cdots \overset{+\cdot}{NH}\cdots CHR \cdots C=O \cdots H-N \cdots R\dot{C}H- \longrightarrow R-\overset{HO}{\underset{}{C}}=\overset{+}{NH}-\dot{C}HR \quad (\mathbf{6})$$

Aldimine fragments can also decompose to amine fragments of type **7**:

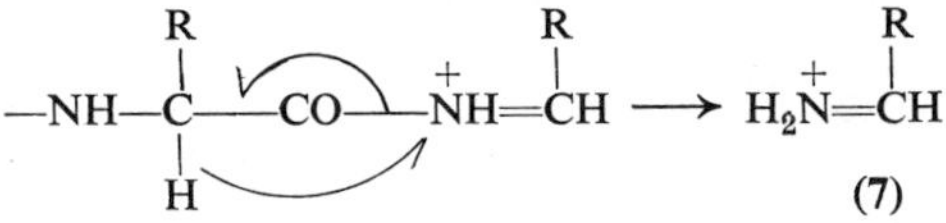

$$-NH-\underset{H}{\overset{R}{C}}-CO-\overset{+}{NH}=\overset{R}{\underset{}{CH}} \longrightarrow H_2\overset{+}{N}=\overset{R}{\underset{}{CH}} \quad (\mathbf{7})$$

It seems that it is this process that accounts for the very intense ions of lower m/e, which correspond to amine fragments of the constituent amino acid residues.

Of the other relatively minor fragmentation patterns there should be mentioned the formation of ions of the type $[H_2N\text{-}CHR\text{-}COOR]^+$, characteristic of the C-terminal amino acids. These ions are more intense for phenyl esters (27). The identification of these ions is facilitated by comparison of the mass spectra of methyl and trideuteromethyl esters of N-acylpeptides (28). On fragmentation of trifluoroacetyl and heptafluorobutyryl derivatives of peptide methyl esters, ions of type $[CO\text{-}NH\text{-}CHR\text{-}COOMe]^+$ and $[CO\text{-}NH\text{-}CHR\text{-}CO\text{-}NH\text{-}CHR\text{-}COOMe]^+$ are also formed (27, 29).

Specific fragmentations due to the nature of the side chains of constituent amino acid residues will be discussed below.

Another process common to all peptides is dehydration, which affects the amide groups. Mass spectra of most peptide esters exhibit peaks at m/e (M − 18). There are cases in which these peaks are very intense, and in such cases the amino acid and aldimine fragment peaks (4 and 5) are accompanied by satellite peaks 18 amu lower in mass. The process appears to proceed as follows:

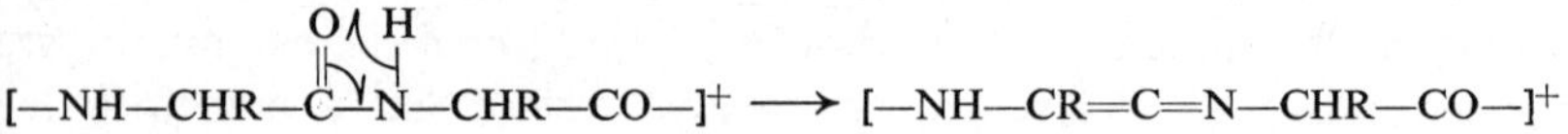

$$[-NH-CHR-\overset{O}{\overset{\|}{C}}-\overset{H}{\overset{|}{N}}-CHR-CO-]^+ \longrightarrow [-NH-CR=C=N-CHR-CO-]^+$$

It is conceivable however, that other types of participation may be involved in this elimination. Dehydration is more pronounced for peptides of low volatility, and this suggests the thermal character of the process; however, the presence of metastable ions shows the reaction to be partially induced by electron impact. Dehydration is also favored by glycine and aromatic and heterocyclic amino acid residues (23, 30).

Finally, it should be noted that N-acylpeptide esters such as 8, acylated by aliphatic acids of sufficiently high molecular weight under electron impact, can split off a portion of their N-acyl group to give ions of N-acetylpeptide esters of type 9 (31):

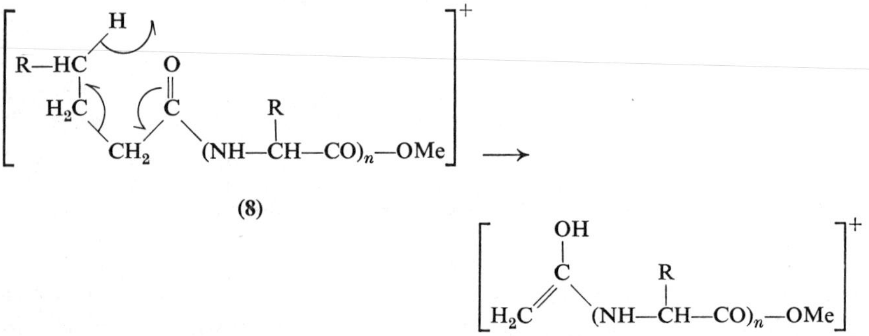

$$\left[\begin{array}{c} R-HC \\ H_2C \\ \quad CH_2 \end{array} \begin{array}{c} H \\ O \\ C \\ (NH-CH-CO)_n-OMe \end{array} \right]^+$$

(8)

$$\longrightarrow$$

$$\left[\begin{array}{c} OH \\ C \\ H_2C \end{array} \begin{array}{c} R \\ (NH-CH-CO)_n-OMe \end{array} \right]^+$$

(9)

Ions such as **9** are most abundant in the spectra of highly volatile peptides of low molecular weight.

Peaks of type **9** and those of $(M-H_2O^+)$ are useful in the identification of the molecular ion.

IV. FRAGMENTATION OF N-ACYLPEPTIDE ESTERS AS A FUNCTION OF CONSTITUENT AMINO ACID RESIDUES

The fragmentation of the peptide chain of N-acylpeptide esters described above proceeds simultaneously with other processes characteristic of the nature of the side chains of the constituent amino acid residues. These processes include loss of the side chain or of a part of it, specific cleavage of the peptide chain, and intermolecular alkylation.

A. Aliphatic Amino Acids

There has been no mention in the literature of any specific fragmentation pattern that could be ascribed to glycine and alanine residues in the peptide chain; somewhat peculiar, perhaps, is the low abundance mentioned above of aldimine fragments with C-terminal glycine. Elimination of a methyl group from the alanine residue should be possible under mass-spectrometric conditions, but apparently does not occur to any noticeable extent.

On the contrary, mass spectra of valine-containing peptides have been repeatedly reported (e.g., 1, 3, 4, 17, 18, 21, 23) to exhibit ions resulting from elimination of the valine side chain either as an olefin (**10** → **11**) or as a radical (**10** → **12**). Analogous elimination of the side chain was observed

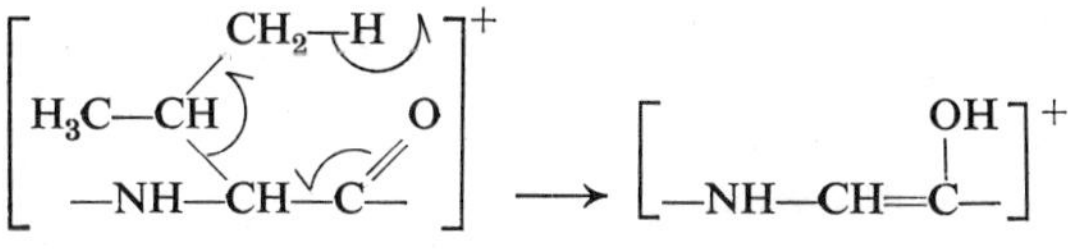

$$\left[\begin{array}{c} \text{CH}_2\text{—H} \\ \text{H}_3\text{C—CH} \qquad \text{O} \\ \text{—NH—CH—C—} \end{array}\right]^+ \longrightarrow \left[\begin{array}{c} \text{OH} \\ \text{—NH—CH=C—} \end{array}\right]^+$$

(10)   (11)

$$\left[\begin{array}{c} \text{C}_3\text{H}_7 \\ \text{—NH—CH—CO—} \end{array}\right]^+ \longrightarrow [\text{—NH—}\overset{\centerdot}{\text{C}}\text{H—CO—}]^+$$

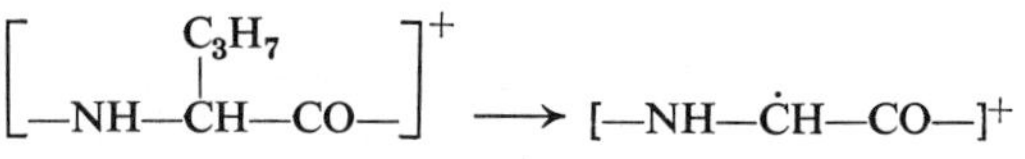

(10)   (12)

also for leucine- and isoleucine-containing peptides; however, these are known alternatively to lose only parts of their side chains. Leucine in particular very commonly loses either $C_3H_7^-$ or propylene.

In connection with this, the possibility of distinguishing between leucine and isoleucine residues in peptides may be briefly discussed. Some differences in the mass-spectrometric behavior of leucine and isoleucine derivatives have been reported in the literature (e.g., see, Ref. 3, 32) and attempts were made to draw some conclusions regarding the nature of the isomeric leucines in dipeptides. Thus Weygand (17) reported that methyl esters of trifluoroacetyl-peptides with an N-terminal isoleucine give intense peaks at m/e 168 and 154, whereas the analogous peptides with N-terminal leucine gave instead a peak at m/e 140. The corresponding ion formation could be visualized in the following way:

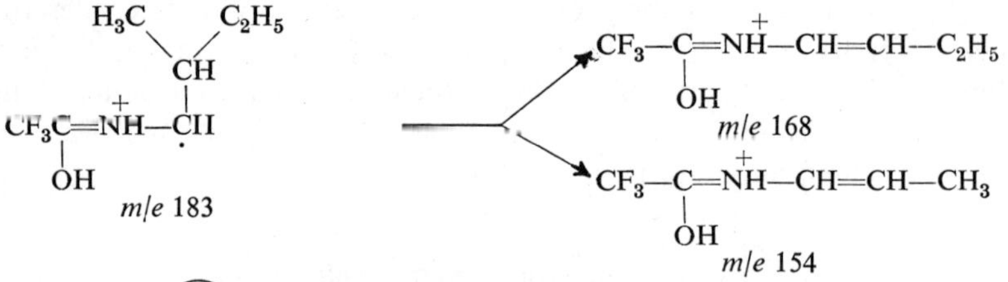

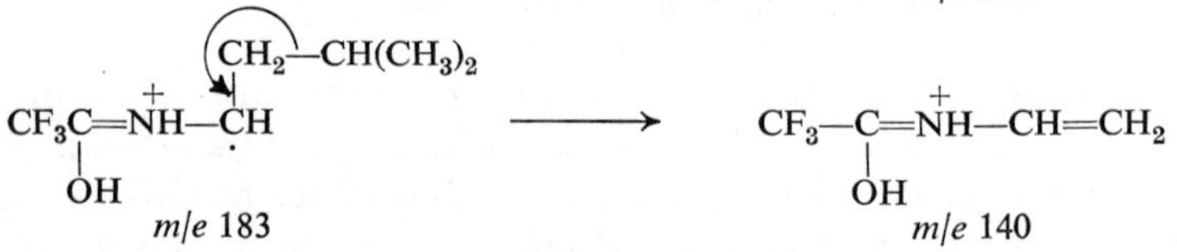

It was also shown (33) that unlike the corresponding leucine-containing peptides, methyl esters of trifluoroacetylpeptides with C-terminal isoleucine produce ions of m/e (M—127) (**13**):

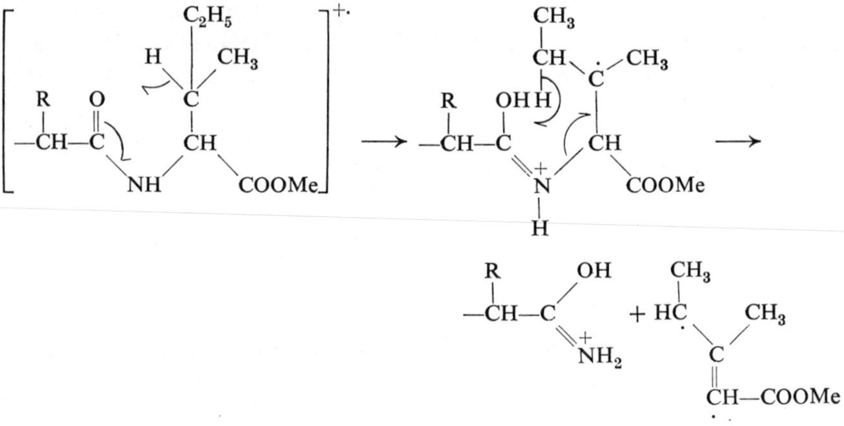

In general, however, the mass-spectrometric identification of the isomeric leucines in peptides is difficult (3, 23), and in natural peptides of complex structure leucine and isoleucine usually cannot be distinguished mass spectrometrically (34), although for peptides containing residues of only one or other of the isomers in question, the choice can be made on the basis of the amino acid analysis quite easily and unequivocally.

B. Proline

Proline residues do not affect the mass spectra of N-acylpeptide esters to any considerable degree. Decomposition of the heterocyclic moiety of this amino acid appears to take place only in the aldimine fragments containing a C-terminal proline (14), which are capable of eliminating a part of the pyrroline ring to give ions (15) and (16) of low abundance (18, 20, 21, 23).

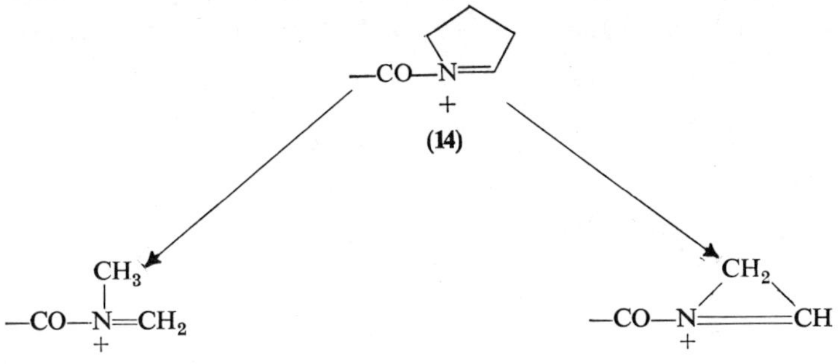

$$-CO-N\overset{+}{=} \text{(pyrroline ring)}$$

(14)

$$-CO-\overset{+}{N}=CH_2 \qquad \qquad -CO-\overset{+}{N}\overset{CH_2}{=\!\!=\!\!=}CH$$
$$CH_3$$

(15)   (16)

C. Hydroxyamino Acids

It has been shown repeatedly that fragmentation of peptides containing residues of hydroxyamino acids (serine, threonine, hydroxyproline, and homoserine) is accompanied by dehydration of the side chain (1, 15, 17, 18, 23, 35). Serine and threonine residues (17) are characterized also by loss of the whole side chain with hydrogen atom migration (17 → 18).

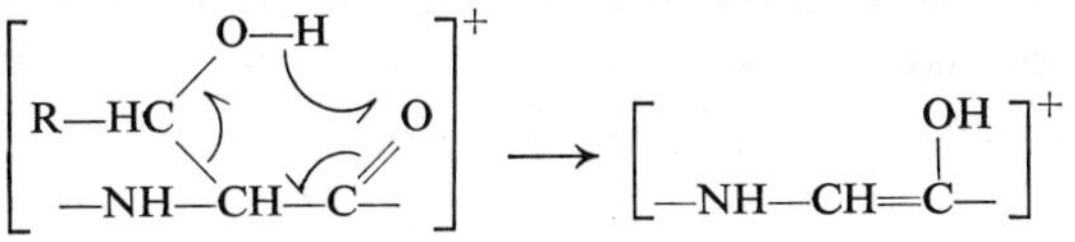

$$\left[\begin{array}{c} O-H \\ R-HC \quad \quad O \\ -NH-CH-C- \end{array} \right]^{+} \longrightarrow \left[\begin{array}{c} OH \\ -NH-CH=C- \end{array} \right]^{+}$$

D. Monoamino Dicarboxylic Acids

Fragmentation of N-acylpeptide esters containing esterified residues of monoamino dicarboxylic acids (aspartic and glutamic acids) and their ω-amides (asparagine and glutamine) have been investigated quite elaborately

(23, 31, 35–38). It should be noted that peptides of this composition undergo mainly the amino acid type of fragmentation, though *a priori*, one would expect the ω-carbalkoxy or ω-amide group from the dicarboxylic acid residue to compete with the C-terminal carbalkoxy group for the localization of the positive charge, thus giving rise to noncharacteristic ions that do not represent the amino acid sequence of the peptide under investigation. However, if the positive charge is localized at the ω-substituent of the residue, this does not appear to initiate any novel fragmentation patterns in the molecule, the hydrocarbon side chain of amino dicarboxylic acid presumably preventing the positive charge from migrating from the ω-substituent to the peptide chain.

In addition to the amino acid type of fragmentation, compounds (19) and (21) also suffer conversion of the ω-carbalkoxy group or the ω-carboxamide group into a ketene (20). Loss of NH_3 from the latter is usually a

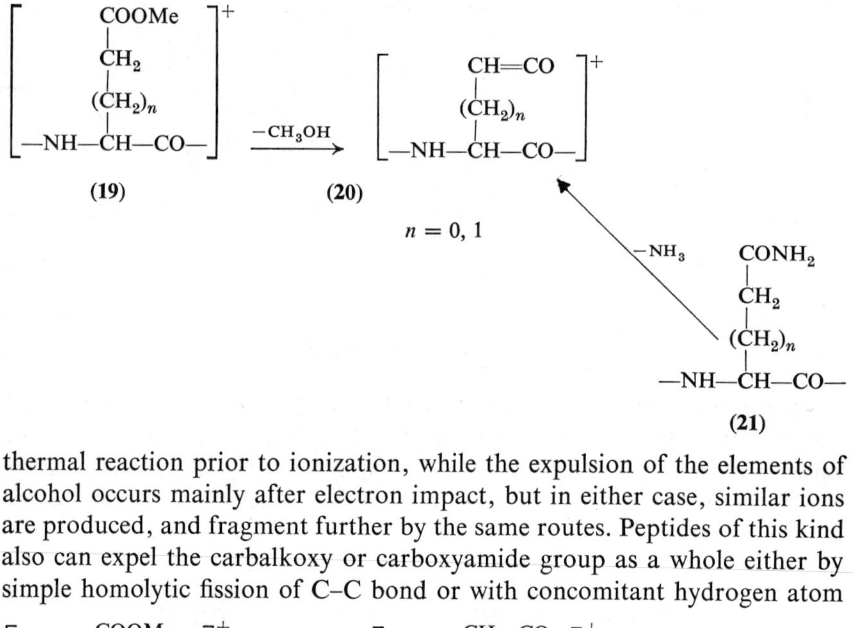

$$
\left[\begin{array}{c} \text{COOMe} \\ | \\ \text{CH}_2 \\ | \\ \text{(CH}_2)_n \\ | \\ \text{—NH—CH—CO—} \end{array} \right]^{+}
\xrightarrow{-\text{CH}_3\text{OH}}
\left[\begin{array}{c} \text{CH}=\text{CO} \\ | \\ \text{(CH}_2)_n \\ | \\ \text{—NH—CH—CO—} \end{array} \right]^{+}
$$

(19)   (20)

$$n = 0, 1$$

$$
\xrightarrow{-\text{NH}_3}
\left[\begin{array}{c} \text{CONH}_2 \\ | \\ \text{CH}_2 \\ | \\ \text{(CH}_2)_n \\ | \\ \text{—NH—CH—CO—} \end{array} \right]
$$

(21)

thermal reaction prior to ionization, while the expulsion of the elements of alcohol occurs mainly after electron impact, but in either case, similar ions are produced, and fragment further by the same routes. Peptides of this kind also can expel the carbalkoxy or carboxyamide group as a whole either by simple homolytic fission of C–C bond or with concomitant hydrogen atom

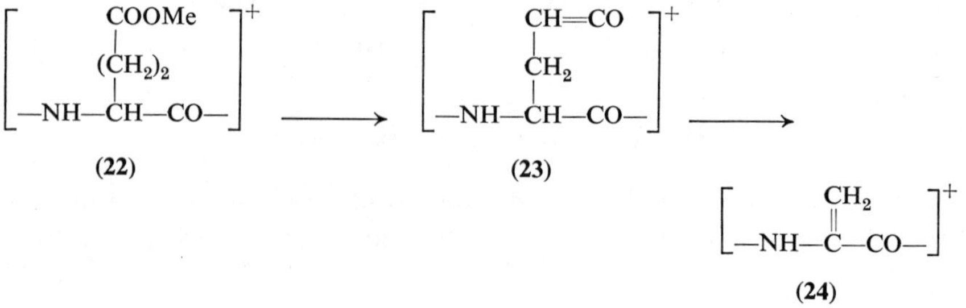

$$
\left[\begin{array}{c} \text{COOMe} \\ | \\ \text{(CH}_2)_2 \\ | \\ \text{—NH—CH—CO—} \end{array} \right]^{+}
\longrightarrow
\left[\begin{array}{c} \text{CH}=\text{CO} \\ | \\ \text{CH}_2 \\ | \\ \text{—NH—CH—CO—} \end{array} \right]^{+}
\longrightarrow
$$

(22)   (23)

$$
\left[\begin{array}{c} \text{CH}_2 \\ \| \\ \text{—NH—C—CO—} \end{array} \right]^{+}
$$

(24)

migration. In addition, the peptides with monoamino dicarboxylic acids are sometimes observed to eliminate the amino acid chain either entirely or partially; cases have been reported in which the stepwise character of the process $(22 \rightarrow 23 \rightarrow 24)$ was supported by the corresponding metastable peaks (23, 35).

Differences in the mass-spectrometric behavior of α- and ω-peptides of monoamino dicarboxylic acids [(25) and (26), respectively] make it possible

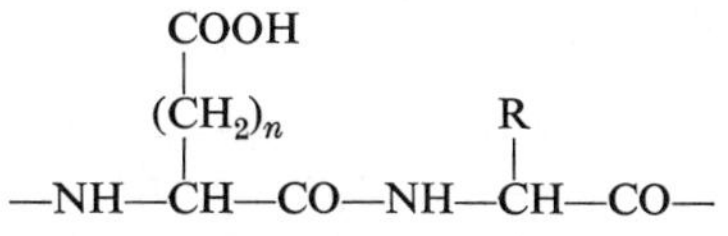

(25) 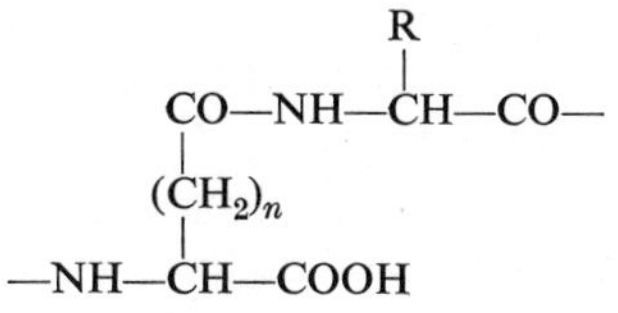   (26)

to identify the type of amide bond in these peptides from their mass spectra (31, 39). Esters of N-acyl derivatives of α-peptides containing a glutamic acid residue, and thus a γ-carbomethoxy group, can, in addition to the fragmentation described above, also lose elements of water from the molecular ion and other fragments containing the glutamic acid γ-methyl ester residue. Elimination of this kind can be envisaged as $(27 \rightarrow 28 \rightarrow 29)$. This reaction of course

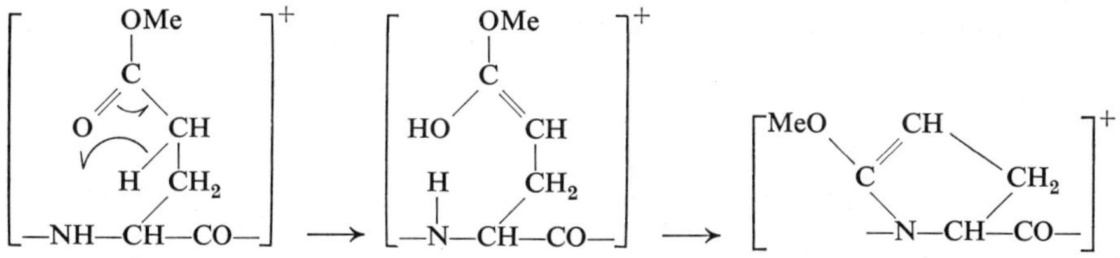

(27)   (28)   (29)

would not proceed if the γ-carboxyl of the glutamic acid is involved in a peptide bond. On the other hand (31), mass spectra of γ-peptides of glutamic acid, as well as those of β-peptides of aspartic acid, usually do not show peaks corresponding to aldimine fragments with the C-terminal residue of this amino acid.

E. Aromatic and Heterocyclic Amino Acids

Peptides containing residues of aromatic amino acids (phenylalanine and tyrosine) and heterocyclic amino acids (histidine and tryptophan) undergo, in addition to the amino acid type of fragmentation, a series of specific conversions (17, 18, 21, 30, 40–42). First, the side chain of these amino acids can be eliminated as the cation $ArCH_2^+$. Peaks corresponding to the ions $ArCH_2^+$ (m/e 130, 107, 91, and 81 for tryptophan, tyrosine, phenylalanine, and histidine respectively) are present in the mass spectra of all the peptides containing these amino acids and, as a rule, are the most prominent peaks in the spectrum. Naturally, such ions do not provide any information on the amino acid sequence in peptides, but they can serve as additional evidence as to the presence of these amino acids in the peptide. This is particularly useful for tryptophan, determination of which in peptides is quite troublesome.

Alternatively, the side chain of aromatic and heterocyclic amino acids can be eliminated as the radical $ArCH_2^{\cdot}$, with further decomposition of the resulting ion according to the amino acid type of fragmentation. In the case of phenylalanine or histidine, elimination of $ArCH_2^{\cdot}$ is a result of homolytic cleavage of C_α—C_β bond, while in the molecular ions of tryptophan- or tyrosine-containing peptides, it is accompanied by migration of a hydrogen atom to the charged fragment; therefore, the mass spectra of derivatives of tryptophan- and tyrosine-containing peptides show very intense peaks at m/e (M − 129) and (M − 106), respectively. It has been shown (17) that in tyrosine residues the migrating hydrogen atom is that of the phenolic hydroxyl.

Under mass-spectrometric conditions, rupture of the N—C_α bond of the aromatic (heterocyclic) amino acid residue often occurs with concomitant migration of a hydrogen atom from the methylene group (**30** → **31**). The

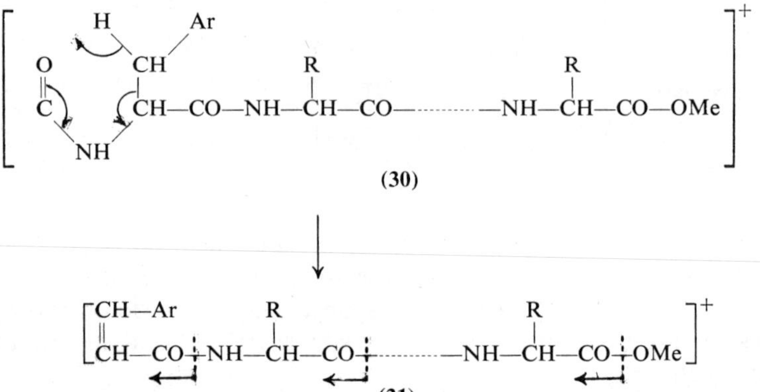

(30)

(31)

resulting ion (**31**) then fragments according to the amino acid fragmentation pattern. These three processes are most pronounced in tryptophan-containing peptides and much less so with peptides containing histidine and phenylalanine.

Peptides such as **32** are also capable of N–C$_\alpha$ bond cleavage at tyrosine residues, with migration of two hydrogen atoms, resulting in ions such as (**33**), which then eliminate NH_3 to produce amino acid fragments (**34**) (30, 43).

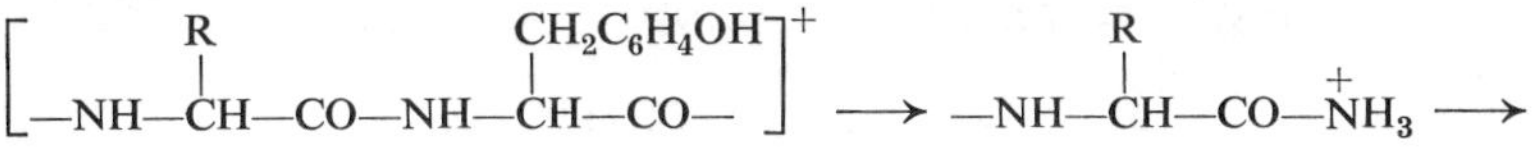

$$\left[-NH-\underset{\underset{\textbf{(32)}}{}}{\overset{\overset{R}{|}}{CH}}-CO-NH-\underset{}{\overset{\overset{CH_2C_6H_4OH}{|}}{CH}}-CO- \right]^+ \longrightarrow -NH-\underset{\underset{\textbf{(33)}}{}}{\overset{\overset{R}{|}}{CH}}-CO-\overset{+}{N}H_3 \longrightarrow$$

$$-NH-\underset{}{\overset{\overset{R}{|}}{CH}}-C\equiv O^+$$

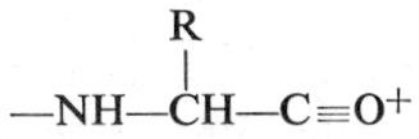

(34)

Mass spectra of histidine-containing N-acylpeptide esters show satellite peaks 14 amu higher than those corresponding to the molecular ion and fragment ions containing histidine (18–21, 23, 30, 40, 44, 45). These arise by thermal intermolecular N-methylation of the imidazole ring, the C-terminal carbomethoxy group being the main source of the methyl groups (42) as follows:

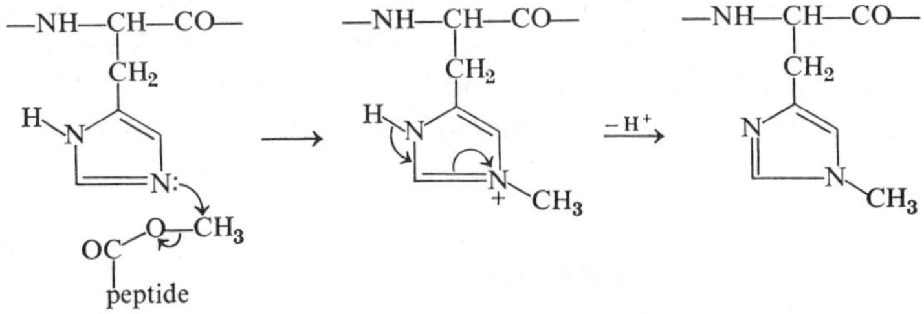

Intermolecular methylation is more prevalent when the temperature of the ion source is over 200°. An analogous but less important process has been observed for derivatives of tryptophan-containing peptides (45).

Peptides that contain residues of hydroxyamino acids or other aromatic (heterocyclic) amino acids in addition to histidine often give molecular ions and amino-acid fragmentation-pattern ions of very low, or zero, abundance and very intense peaks 4 amu lower in mass. This is due to intermolecular N-methylation and dehydration, two processes that are particularly typical of involatile peptides containing more than one aromatic (heterocyclic) amino acid residue. These conclusions were confirmed by accurate mass measurements of the appropriate ions and by observation of the corresponding metastable peaks (45).

F. Methionine

In the course of amino acid fragmentation of the N-acylpeptide esters, methionine residues (**35**) are very apt to lose their entire side chain, giving rise to very abundant radical ions (36), sometimes with concomitant migration

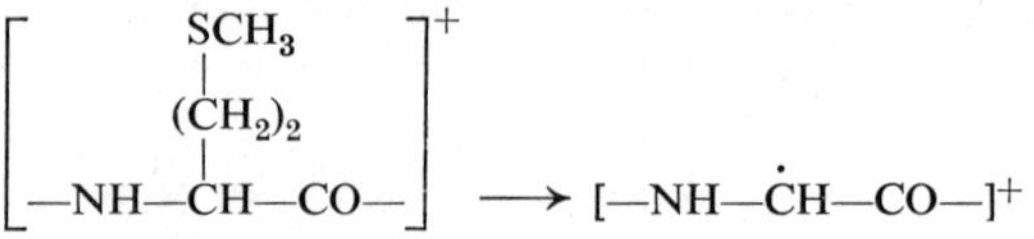

$$\left[\begin{array}{c} SCH_3 \\ | \\ (CH_2)_2 \\ | \\ -NH-CH-CO- \end{array}\right]^+ \longrightarrow [-NH-\overset{\cdot}{C}H-CO-]^+$$

(35)   (36)

of the hydrogen atom to the charged fragment. Simultaneous fissions of the C_β–C_γ and C_γ–S bonds in the methionine side chain have also been observed, but the corresponding peaks usually are of low intensity (18, 20, 21, 23, 46). Methionine sulfoxide residues, however, exhibit C_γ–S cleavage much more readily, while methionine sulfone, on the contrary, suffers practically no cleavage of its side chain (47).

G. Cystine

Cystine-containing N-acylpeptide esters (37) usually do not give molecular ions, and in their mass spectra the peaks of highest m/e are those of m/e $(M/2 + 1)$ (43, 46, 48). The reason for this lies in the extreme ease of S–S bond fission, leading to cysteine peptides with unprotected SH groups (38). Simultaneously, C_β–S and C_α–C_β bonds may also be broken, yielding ions (39), and (40) that also can arise from ions of the type 38. Ions 38, 39, and 40

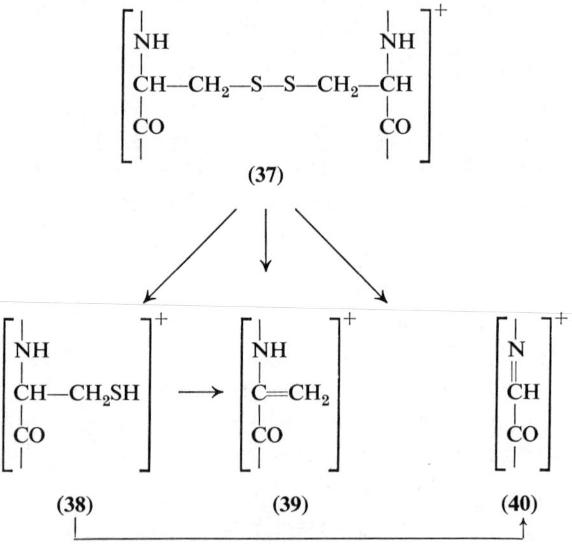

$$\left[\begin{array}{c} | \\ NH \qquad\qquad NH \\ | \qquad\qquad\quad | \\ CH-CH_2-S-S-CH_2-CH \\ | \qquad\qquad\qquad\quad | \\ CO \qquad\qquad CO \\ | \qquad\qquad\quad | \end{array}\right]^+$$

(37)

$$\left[\begin{array}{c} | \\ NH \\ | \\ CH-CH_2SH \\ | \\ CO \\ | \end{array}\right]^+ \longrightarrow \left[\begin{array}{c} | \\ NH \\ | \\ C=CH_2 \\ | \\ CO \\ | \end{array}\right]^+ \qquad \left[\begin{array}{c} | \\ N \\ \| \\ CH \\ | \\ CO \\ | \end{array}\right]^+$$

(38)   (39)   (40)

then undergo an amino acid type of fragmentation, and this makes it possible to carry out amino acid sequence determination in cystine-containing peptides. The mass spectrum of an asymmetrical cystine-containing peptide is the sum of the mass spectra of the two component peptides; interpretation of such a spectrum on the basis of the amino acid fragmentation pattern is possible, though it may sometimes be difficult (cf. Biemann *et al.* (49) on the analysis of the mass spectra of peptide mixtures).

H. Cysteine

Fragmentation patterns of N-acylpeptide esters containing S-carboxymethylcysteine (43, 46, 48), S-carboxamidomethyl-cysteine (43, 46), S-β-aminoethylcysteine (43, 46, 48), S-benzylcysteine (50), and cysteic acid (43, 46) have been studied.

For the mass spectrometric determination of the amino acid sequence in cysteine-containing peptides, the S-carboxymethyl and S-carboxamidomethyl derivatives of cysteine are most convenient. (It might be noted that in protein chemistry, carboxymethylation is regarded as one of the most reliable protective procedures for the protection of SH groups). Esters of N-acyl-S-carboxymethylcysteine-containing peptides (41) exhibit amino acid type of fragmentation and in addition can expel the carbomethoxy group or methylmercaptoacetate, yielding ions (42) and (43) respectively; apart from this,

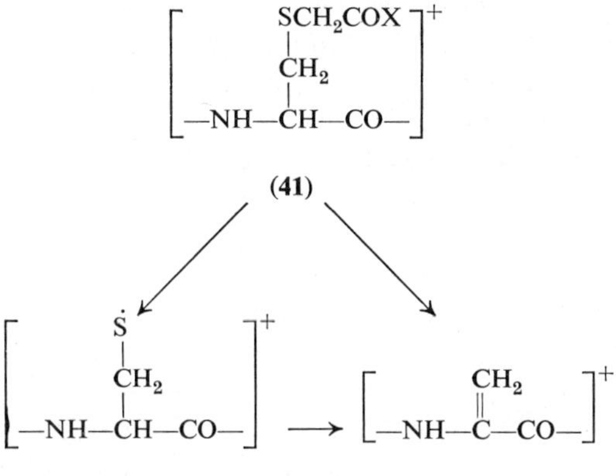

$$\left[\begin{array}{c} \text{SCH}_2\text{COX} \\ | \\ \text{CH}_2 \\ | \\ -\text{NH}-\text{CH}-\text{CO}- \end{array}\right]^+$$

(41)

$$\left[\begin{array}{c} \overset{\cdot}{\text{S}} \\ | \\ \text{CH}_2 \\ | \\ -\text{NH}-\text{CH}-\text{CO}- \end{array}\right]^+ \longrightarrow \left[\begin{array}{c} \text{CH}_2 \\ \| \\ -\text{NH}-\text{C}-\text{CO}- \end{array}\right]^+$$

(42)   **(43)**

$$X = \text{OMe, NH}_2$$

N—C_α bond fission of the carboxymethylcysteine residue in a process analogous to that discussed above for the derivatives of peptides containing aromatic and heterocyclic amino acid residues has been observed.

Methyl esters of S-β-aminoethylcysteine-containing N^α, N^ω-di-acyl-peptides (44), under mass spectrometric conditions are very readily converted (apparently by a thermal process) into dehydroalanine derivatives (45). Since the latter undergo fragmentation of the amino acid type, the sequence determination of N-β-aminoethylcysteine-containing peptides presents no difficulties.

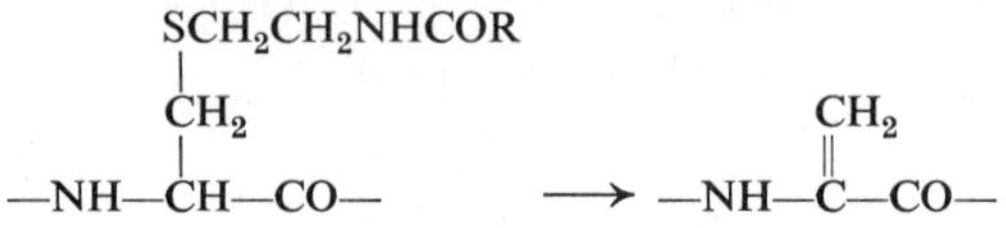

$$\begin{array}{cc}
\begin{array}{c}
\text{SCH}_2\text{CH}_2\text{NHCOR} \\
| \\
\text{CH}_2 \\
| \\
-\text{NH}-\text{CH}-\text{CO}- \\
\\
\textbf{(44)}
\end{array}
& \longrightarrow
\begin{array}{c}
\\
\text{CH}_2 \\
\| \\
-\text{NH}-\text{C}-\text{CO}- \\
\\
\textbf{(45)}
\end{array}
\end{array}$$

Methyl esters of N-trifluoroacetyl derivatives of S-benzylcysteine-containing peptides mainly follow the same fragmentation routes as the corresponding S-carboxymethyl derivatives.

N-Acylpeptide esters containing cysteic acid residues are apparently thermally unstable and suffer random fission of their peptide chain, giving uninterpretable mass spectra.

I. Lysine and Ornithine

N-Acylpeptide esters containing residues of the diamino monocarboxylic acids lysine and ornithine, also give the amino acid type of fragmentation if the ω-amino group is protected by acylation (e.g., see Ref. 18, 22, 23). Specific to acylated lysine- and ornithine-containing peptides is the cleavage of the C–C and C–N bonds in the side chain carrying the ω-amide group (51, 52); this process is often accompanied by hydrogen atom migration and follows the decomposition pattern of secondary amides (53).

J. Arginine

The situation for arginine-containing peptides is somewhat peculiar. N^α-Acylpeptide esters containing arginine residues with an unprotected or diacylated guanidino group are unsuitable for mass spectrometry, owing partly to their thermal instability and partly to the tendency of the positive charge to be localized at the guanidino group, resulting in mass spectra that show no peaks corresponding to the amino acid fragmentation pattern (51, 52, 54). Many earlier attempts to obtain interpretable mass spectra for free arginine (55) or for its derivatives [ethyl ester (1), N^α-acetylarginine (55), methyl ester of N^α-2,4-dinitrophenylarginine (56), etc.] proved to be unsuccessful.

In a quest for derivatives that could be used for mass-spectrometric purposes, a number of methods of modification of arginine residues were tested. Satisfactory results were obtained by converting the arginine residues in peptides into derivatives of ornithine by treatment with aqueous hydrazine; the esters of diacyl derivatives of the resulting ornithine-containing peptides, as discussed above, were found to be suitable for mass-spectrometric determination of the amino acid sequence (51, 52, 54).

Another way of converting arginine-containing peptides into compounds amenable to mass spectrometry involves condensation of the arginine guanidino group of the N^α-acylpeptide (46) with β-dicarbonyl compounds (1,1,3,3-tetraalkoxypropane or acetylacetone) under acid conditions or with α-dicarbonyl compounds (e.g., 1,2-cyclohexanedione) under mildly alkaline conditions (this is important for peptides containing tryptophan, which are unstable in acidic media).* The resulting N^α-acylpeptide esters (obtained directly by the first route and in the second route after subsequent esterification) contain N^δ-pyrimidylornithine (47, R = H), N^δ-dimethyl pyrimidylornithine (47, R = CH$_3$), and N^δ-substituted ornithine (48) respectively (51, 52, 54, 54a).

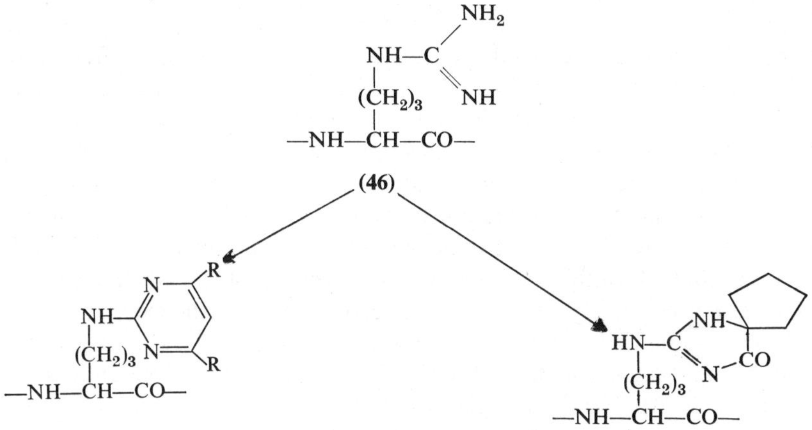

(47);R = H or CH$_3$ (48)

Methyl esters of N^α-acylpeptides containing the N^δ-pyrimidylornithine (or dimethylpyrimidylornithine) residue (47) are thermally quite stable; under electron impact they undergo amino acid fragmentation mainly, making it possible to use mass spectrometry for amino acid sequence determination in peptides of this kind. Stepwise decomposition of the N^δ-pyrimidylornithine residue results in characteristic ions at m/e 193, 177 (176), 163, 148, 136, 122 (123), 108, 95 (96), and 79. An analogous group of ions 28 amu to higher mass is found in the mass spectra of peptides with N^δ-dimethylpyrimidylornithine residues. Peaks at m/e 193 and 163 are due to the

---

* Similar reactions have been recommended for modification of arginine residues in proteins (57, 58) and for the preparation of an arginine derivative amenable to gas-liquid chromatography (59).

ions **49** and **51**; these are further converted into ions of m/e 177 (**50**) and 148 (**52**) respectively. Ion-radicals at m/e 136, 122, and 108 are formed by

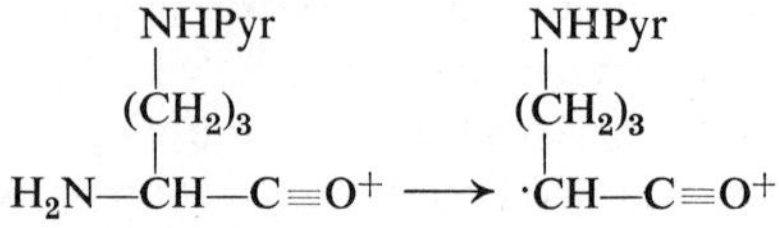

$$
\begin{array}{cc}
\text{NHPyr} & \text{NHPyr} \\
| & | \\
(\text{CH}_2)_3 & (\text{CH}_2)_3 \\
| & | \\
\text{H}_2\text{N—CH—C}\equiv\text{O}^+ \longrightarrow & \cdot\text{CH—C}\equiv\text{O}^+ \\
m/e\ 193 & m/e\ 177 \\
(\mathbf{49}) & (\mathbf{50})
\end{array}
$$

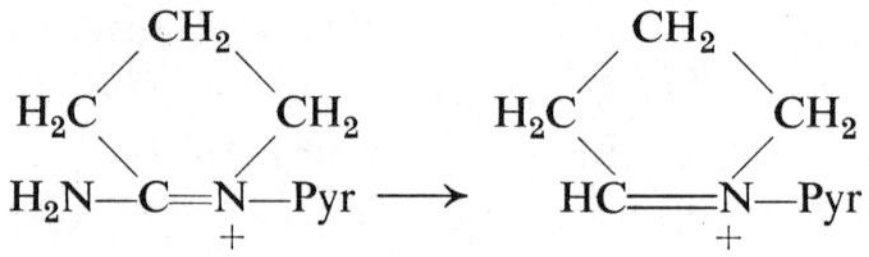

$$
\begin{array}{cc}
\quad\quad\text{CH}_2 & \quad\quad\text{CH}_2 \\
\diagup\quad\quad\diagdown & \diagup\quad\quad\diagdown \\
\text{H}_2\text{C}\quad\quad\quad\text{CH}_2 & \text{H}_2\text{C}\quad\quad\quad\text{CH}_2 \\
\diagdown\quad\quad\diagup & \diagdown\quad\quad\diagup \\
\text{H}_2\text{N—C}{=}\text{N—Pyr} \longrightarrow & \text{HC}{=}\text{N—Pyr} \\
\quad\quad + & \quad\quad + \\
m/e\ 163 & m/e\ 148 \\
(\mathbf{51}) & (\mathbf{52})
\end{array}
$$

homolytic fission of C–C bonds in the side chain; this fission is often accompanied by hydrogen atom migration, especially during cleavage of C_β–C_γ bond. Ions at m/e 95 and 96 arise from C_δ–N bond cleavage with concomitant transfer of one or two hydrogen atoms to the charged fragment, and the peak at m/e 79 corresponds to the pyrimidyl ion less a hydrogen atom (51, 52).

Another characteristic feature of derivatives of N^δ-pyrimidylornithine-containing peptides is cleavage under electron impact of the N–C_α bond at the N^δ-pyrimidylornithine residue with localization of the positive charge at the heterocyclic moiety; the process is accompanied by the expulsion of two hydrogen atoms from the charged fragment. If the N^δ-pyrimidylornithine residue is at the C-terminus of the peptide chain, there will be an abundant ion at m/e 206, probably with the structure **53**.

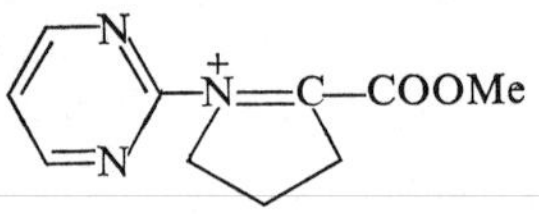

$$
\text{Pyr}{-}\overset{+}{\text{N}}{=}\text{C—COOMe}
$$

(**53**)

The secondary amino group in the N^δ-pyrimidylornithine residue is susceptible to intermolecular methylation under mass-spectrometric conditions, giving rise to satellite peaks 14 amu higher than the molecular ions and the ions due to the amino acid type of fragmentation, very much in the same way as with histidine- and tryptophan-containing peptides. When a N_δ-pyrimidylornithine-containing peptide has a histidine or another N^δ-pyrimidyl-ornithine residue in it, its molecular ion (and frequently the amino acid fragmentation peaks) will be accompanied by peaks 14 *and* 28 amu higher in mass. On the other hand, the mass spectra of peptides with two N^δ-pyrimidyl-ornithine residues also show characteristic peaks corresponding to the molecular ion (and the amino acid fragmentation ions) minus a portion of the N^δ-pyrimidylornithine side chain:

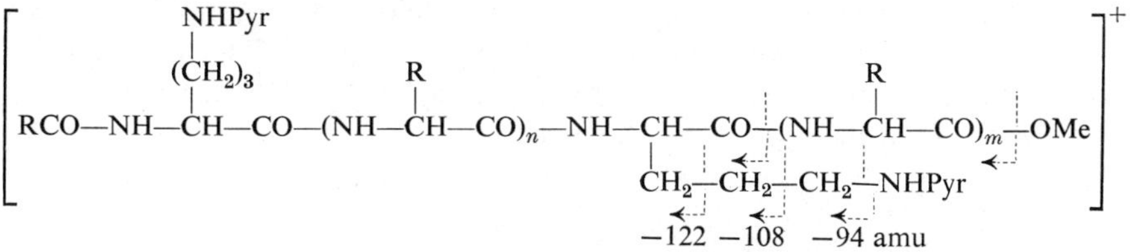

$$
\left[
\begin{array}{c}
\text{NHPyr} \\
| \\
(\text{CH}_2)_3 \qquad\qquad \text{R} \qquad\qquad\qquad\qquad \text{R} \\
| \qquad\qquad\qquad | \qquad\qquad\qquad\qquad | \\
\text{RCO—NH—CH—CO—(NH—CH—CO)}_n\text{—NH—CH—CO}\!-\!(\text{NH—CH—CO)}_m\!-\!\text{OMe} \\
\text{CH}_2\!-\!\text{CH}_2\!-\!\text{CH}_2\!-\!\text{NHPyr} \\
-122\ -108\ -94\ \text{amu}
\end{array}
\right]^+
$$

The derivatives of imidazolinonylornithine-containing peptides (48) are less volatile than the corresponding N^δ-pyrimidylornithine-containing peptides, but the amino acid fragmentation pattern in their mass spectra is clear enough, and all the main fragmentation features observed for N^δ-pyrimidylornithine-containing peptides can also be found here. As in the case of pyrimidylornithine derivatives, the mass spectra of imidazolinonyl-ornithine-containing peptides have a group of peaks at m/e 221, 206, 194, 180, 166, 152, and 138, corresponding to ions formed by stepwise decom-position of the N^δ-imidazolinonylornithine residue. Intermolecular thermal methylation and dehydration are very typical for N^δ-imidazolinonylornithine-containing peptides; the superposition of the two processes results in intense peaks appearing as satellites of the molecular ion and amino acid fragmenta-tion ions 4 amu to lower mass. The cyclopentane ring can also decompose, as shown below (51, 52):

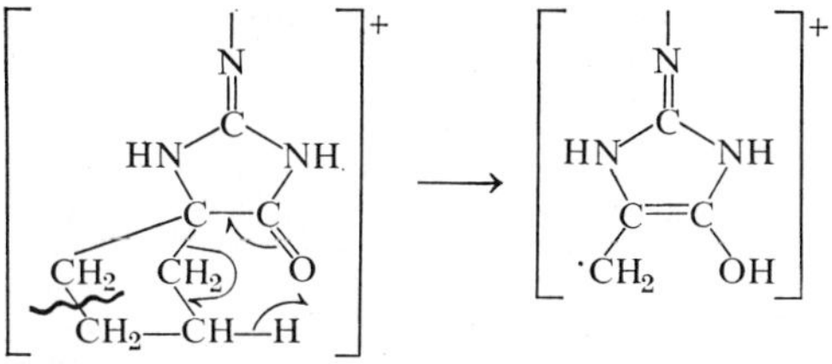

$$
\left[
\begin{array}{c}
| \\
\text{N} \\
\| \\
\text{C} \\
\text{HN}^{\diagup}\ ^{\diagdown}\text{NH} \\
| \qquad | \\
\text{C—C} \\
\diagup\ \ \ \ \diagup\ \ \diagdown \\
\text{CH}_2\ \ \text{CH}_2\ \ \ \text{O} \\
\diagdown\ \diagup \\
\text{CH}_2\text{—CH—H}
\end{array}
\right]^+
\longrightarrow
\left[
\begin{array}{c}
| \\
\text{N} \\
\| \\
\text{C} \\
\text{HN}^{\diagup}\ ^{\diagdown}\text{NH} \\
| \qquad | \\
\text{C=C} \\
| \qquad | \\
\cdot\text{CH}_2\ \ \ \text{OH}
\end{array}
\right]^+
$$

All this considerably complicates the mass spectra of N^δ-imidazolinonyl-ornithine-containing peptides (48); therefore, when possible, it is more expedient when dealing with arginine-containing peptides to convert these into the N^δ-acyl- or N^δ-pyrimidylornithine-containing derivatives.

V. CYCLIC PEPTIDES AND DEPSIPEPTIDES

The fragmentation of cyclic peptides and depsipeptides is predicated upon the point of cleavage of the ring and then by the fragmentation behavior of the resulting linear fragment.

It has been suggested (15) that the main mode of ring opening in cyclo-peptides involves expulsion of elements of one of amide bonds (CONH). However, it was subsequently shown that although elimination of CONH does occur, the majority of the fragment ions in the mass spectra of cyclo-peptides are formed by different mechanisms (60, 61). The residue of any amino acid can be eliminated as a whole or as an amine fragment, and the resulting ion-radical (e.g., **54**) decomposes further by C–C and C–N bond rupture so that the positive charge remains localized on the —CHR—C≡O+ group:

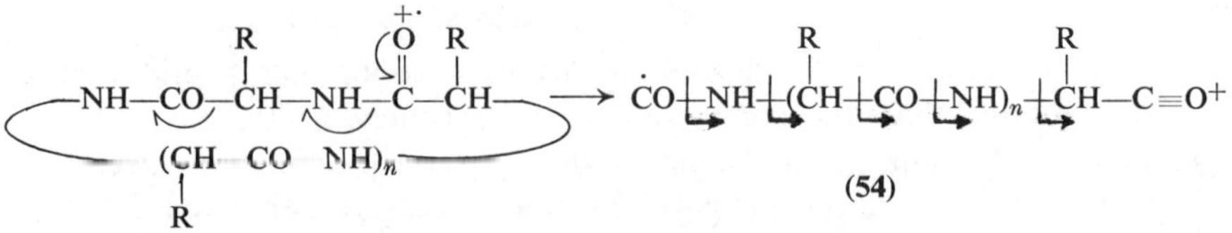

$$-NH-CO-CH-NH-\overset{\overset{+\cdot}{O}}{\underset{\underset{R}{(CH-CO-NH)_n}}{C}}-CH- \longrightarrow \dot{C}O|NH|\overset{R}{CH}|CO|NH)_n|\overset{R}{CH}-C\equiv O^+$$

(54)

If the cyclopeptide under investigation contains a residue of an aromatic amino acid (**55**), the latter, under electron impact, will suffer N–C$_\alpha$ bond cleavage, and the resulting linear fragment (**56**) then undergoes the amino acid fragmentation as it would with the linear peptides:

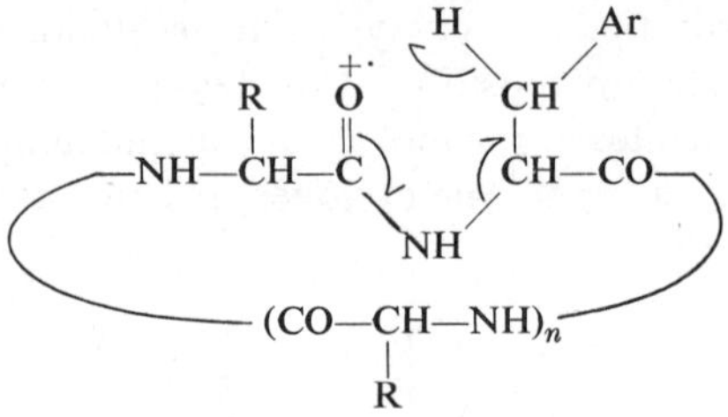

$$-NH-\overset{R}{CH}-\overset{\overset{+\cdot}{O}}{C} \quad \overset{H \quad Ar}{\underset{CH-CO-}{CH}}$$

(55)

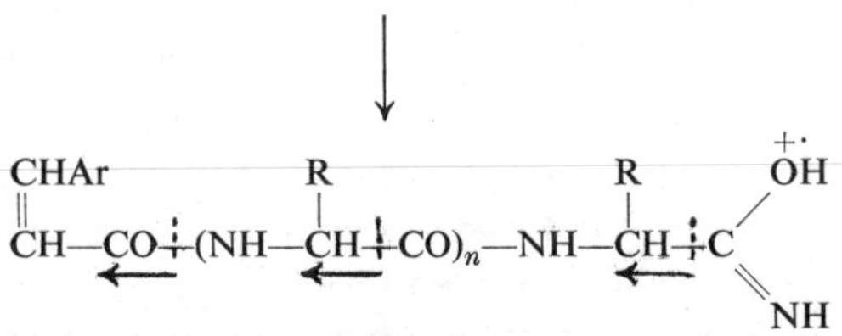

(56)

$$\overset{CHAr}{\underset{\leftarrow}{CH}}-CO\dot{+}(NH-\overset{R}{\underset{\leftarrow}{CH}}CO)_n-NH-\overset{R}{\underset{\leftarrow}{CH}}\dot{+}C\overset{\overset{+\cdot}{OH}}{\underset{NH}{}}$$

The fragmentation patterns of the linear peptides described above as due to the nature of the constituent amino acids are true for the cyclic peptides.

The roughly equal probability of ring opening at any amino acid residue and of further fragmentation by a number of routes makes the interpretation of the mass spectra of cyclic peptides more difficult than those of linear peptides. Indeed, all the work on the structure elucidation of cyclic peptides shows that their mass spectra are actually used only for molecular-weight determination (33, 62–64).

A great number of papers are devoted to the investigation of cyclic depsipeptide fragmentation. Mass-spectrometric behavior of cyclic depsipeptides depends on the ring size and to a lesser extent on the nature of constituent hydroxy and amino acids. In cyclotetradepsipeptides with regularly alternating α-amino and α-hydroxy acid residues, the first fragmentation usually involves elimination of the ester grouping; the linear ion formed then decomposes further by consecutive expulsion of amino and hydroxy acid residues in a manner generally similar to the amino acid fragmentation pattern of linear N-acylpeptide esters (65, 66). If the cyclotetradepsipeptide contains β-hydroxy and α-amino acid residues (e.g., the antibiotic serratamolide) different ester and amide bonds may be cleaved simultaneously, and the molecular ion may thus give rise to different breakdown products (67). Every decomposition route is accompanied by side chain expulsion.

Mass spectrometry has been successfully used for the structure elucidation of the natural cyclotetradepsipeptides angolide (57) (68) and isarolides A, B, and C (58, 59, and 60 respectively) (69).

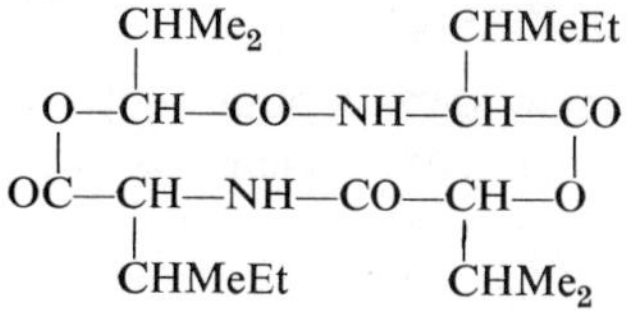

(57)

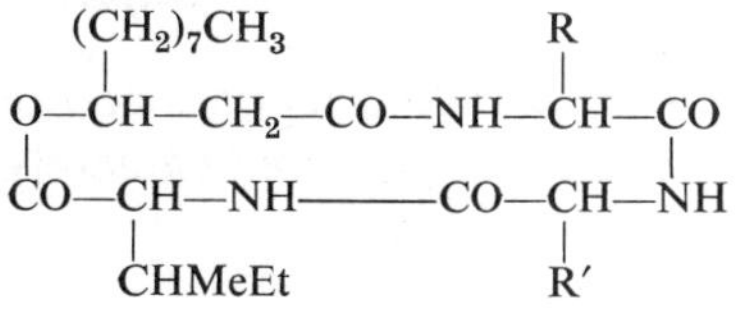

(58) R = CHMe$_2$; R′ = CH$_2$Ph
(59) R = CH$_2$Ph; R′ = CHMe$_2$
(60) R = R′ = CH$_2$Ph

The characteristic fragmentation of cyclic hexadepsipeptides with several hydroxy acid residues involves consecutive expulsion of hydroxy and amino acid residues, the ring opening of the molecular ion proceeding either by direct elimination of a hydroxy or an amino acid residue or by expulsion of a side chain from an amino acid residue with simultaneous N–C bond cleavage in the ring, as is shown here for sporidesmolide I (61) (24, 65, 66, 70–72):

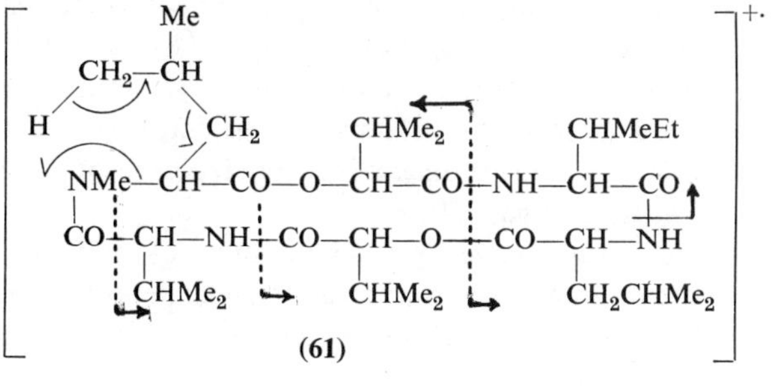

(61)

The structure of the natural cyclic hexadepsipeptide, sporidesmolide III (**62**), was deduced from its mass spectrum (72). Mass-spectral studies of the mixture of enniatin antibiotics have shown some *Fusarium* strains to produce two antibiotics enniatin A_1 and enniatin B_1 (**63** and **64**, respectively) besides the known enniatins A and B (73).

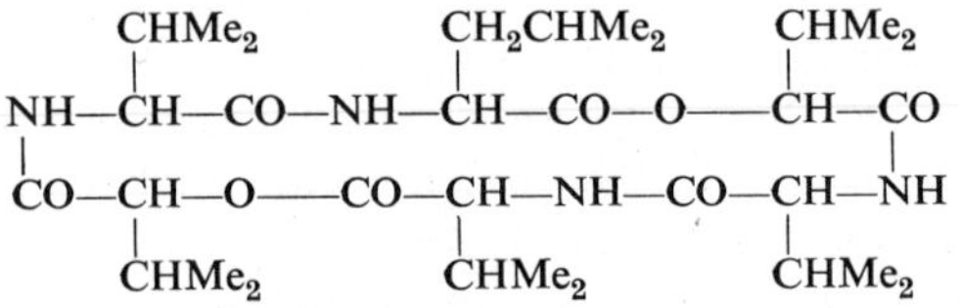

$$
\begin{array}{ccc}
\overset{\displaystyle \text{CHMe}_2}{|} & \overset{\displaystyle \text{CH}_2\text{CHMe}_2}{|} & \overset{\displaystyle \text{CHMe}_2}{|} \\
\text{NH—CH—CO—NH—CH—CO—O——CH—CO} \\
| \qquad\qquad\qquad\qquad\qquad\qquad\qquad\qquad | \\
\text{CO—CH—O——CO—CH—NH—CO—CH—NH} \\
\underset{\displaystyle \text{CHMe}_2}{|} \qquad\quad \underset{\displaystyle \text{CHMe}_2}{|} \qquad\quad \underset{\displaystyle \text{CHMe}_2}{|}
\end{array}
$$

(62)

$$
\begin{array}{ccc}
\overset{\displaystyle \text{CHMeEt}}{|} & \overset{\displaystyle \text{CHMe}_2}{|} & \overset{\displaystyle \text{R}}{|} \\
\text{NMe—CH—CO—O—CH—CO—NMe—CH—CO} \\
| \qquad\qquad\qquad\qquad\qquad\qquad\qquad\qquad | \\
\text{CO—CH—O——CO—CH—NMe——CO—CH—O} \\
\underset{\displaystyle \text{CHMe}_2}{|} \qquad\quad \underset{\displaystyle \text{R}'}{|} \qquad\quad \underset{\displaystyle \text{CHMe}_2}{|}
\end{array}
$$

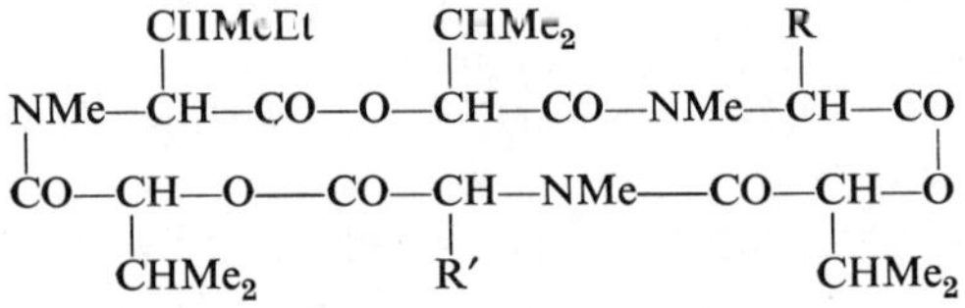

(63) R = $CHMe_2$; R′ = CHMeEt
(64) R = R′ = $CHMe_2$

Cyclic hexadepsipeptides and higher cyclic depsipeptides with a single ester bond decompose predominantly by expulsion of the elements of the ester bond, as shown below:

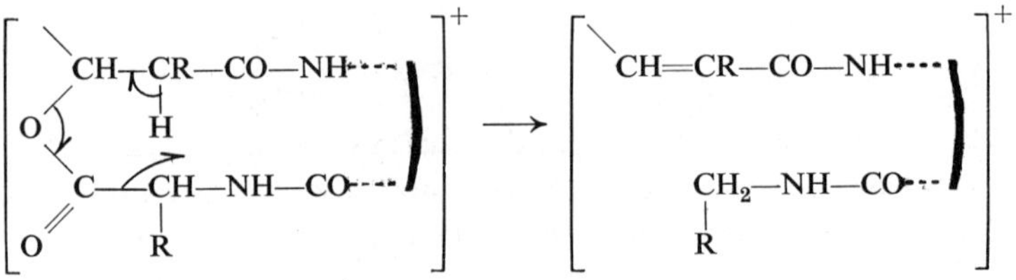

(65)

The resulting ion (**65**) decomposes further by consecutive elimination of amino acid residues, thus allowing mass-spectrometric determination of the amino acid sequence in compounds of this type (74–78). Mass spectrometry has actually been successfully used to confirm the structures of peptidolipin NA (**66**) (74, 75), antibiotics from the group of staphylomycin S (**69**), and etamycin (**70**) (76, 77), and to determine the structure of Val[6]- and Abu[1]-peptidolipins NA (**67** and **68**) (79, 80).

It should be noted, however, that compounds of this type are also capable of fragmentation reactions typical of cyclic peptides, thus sometimes complicating the mass-spectrum interpretation considerably. The structure

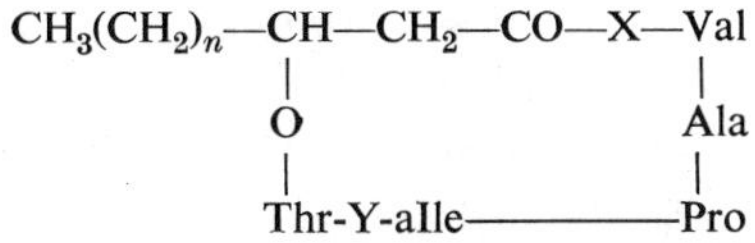

CH$_3$(CH$_2$)$_n$—CH—CH$_2$—CO—X—Val
| |
O   Ala
| |
Thr—Y-aIle————————Pro

(66) X = Thr; Y = Ala; n = 16
(67) X = Thr; Y = Val; n = 16, 17, 18
(68) X = Abu; Y = Ala; n = 16, 17, 18

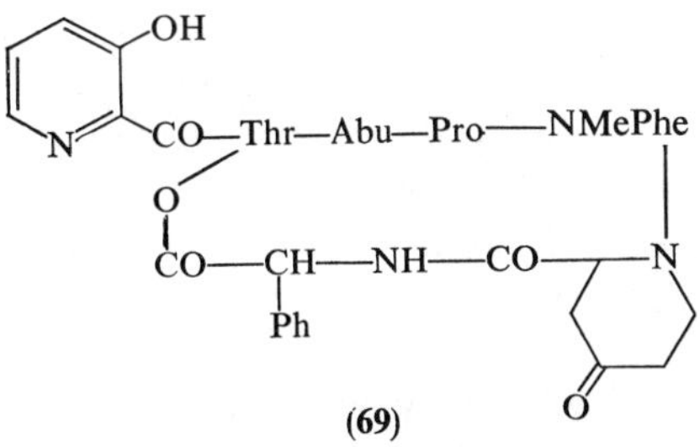

(69)

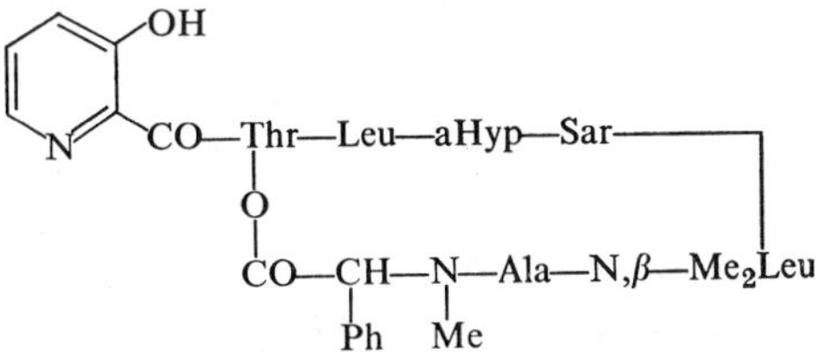

(70)

of isariin (71), for example, was deduced not from the mass spectrum of the cyclodepsipeptide itself, but from that of the ester of the corresponding linear hydroxyacylpeptide, isariinic acid, obtained by alkaline hydrolysis of isariin

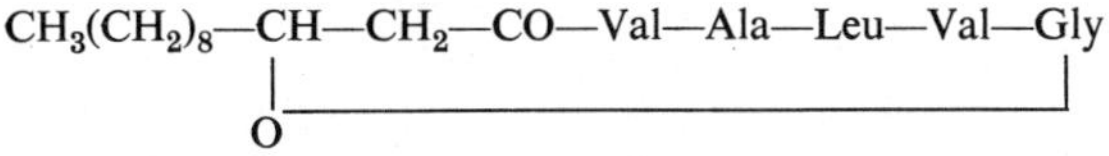

CH$_3$(CH$_2$)$_8$—CH—CH$_2$—CO—Val—Ala—Leu—Val—Gly
|
O

(71)

with subsequent esterification (81, 82). Mass-spectrometric methods were also used to establish the structure of the depsipeptide antibiotics neoantimycin (83), monamycin (83a) and surfactin (83b).

VI. APPLICATION OF MASS SPECTROMETRY TO THE SOLUTION OF STRUCTURAL AND ANALYTICAL PROBLEMS IN PEPTIDE AND PROTEIN CHEMISTRY

Recently some rational operative schemes for the mass spectrometric determination of the amino acid sequence in peptides have been worked out mainly for peptides resulting from partial hydrolysis of proteins (23, 33, 84–87).

The first step for establishing the structure for any peptide is the quantitative amino acid analysis of the substance under investigation. With modern amino acid analyzers it is quite adequately accomplished using a hundredth of a micromole of the compound. The knowledge of the amino acid composition of the peptide is necessary, for this, together with the molecular weight of the substance, determines its volatility and consequently the feasibility of the mass-spectrometric approach to the investigation of the peptide in question. Further, if the amino acid analysis shows that the peptide can be investigated mass spectrometrically, then the same amino acid composition may be used to determine the kind of the derivative most suitable for mass spectrometry and the techniques for its preparation.

The interpretation of mass spectra is considerably facilitated by the knowledge of amino acid composition. Earlier this point was rather underestimated (44, 49, 88–90), since the majority of work on amino acid sequence determination in peptides was carried out on synthetic or natural model compounds of known composition. As is discussed above, amino acid residues decompose by a number of alternate and quite peculiar fragmentation routes, obscuring the picture of the peptide mass spectrum.

Thus attempts (44, 49, 88–90) to determine simultaneously both amino acid composition and amino acid sequence from the mass spectrum were generally not very successful.

The second stage of peptide-structure elucidation by mass spectrometry involves conversion of the free peptide into a derivative amenable to mass spectrometry. Usually it is expedient to approach this by acylation of all the amino groups of the free peptide (72). N-Hydroxysuccinimide esters of carboxylic acids (73) are the best acylating agents, allowing the reaction to be carried out in aqueous or an aqueous dioxane solution of $NaHCO_3$ at room temperature (91). Under these conditions, acylation of α-amino groups of all the amino acids as well as of ω-amino groups of lysine and ornithine proceeds smoothly; the guanidino group in arginine residues remains intact. A peptide containing no arginine can be esterified immediately after acylation. The best esterification agent, which gives pure acylpeptide esters (75),

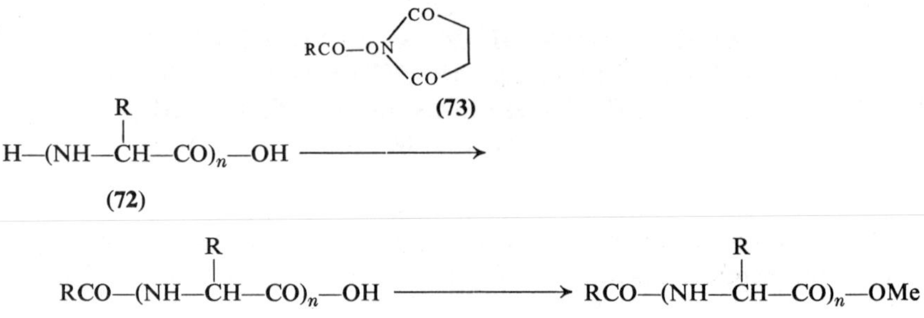

$$RCO-ON \underset{CO}{\overset{CO}{\diagup}} \diagdown$$

(73)

$$H-(NH-\underset{\underset{R}{|}}{CH}-CO)_n-OH \longrightarrow$$

(72)

$$RCO-(NH-\underset{\underset{R}{|}}{CH}-CO)_n-OH \longrightarrow RCO-(NH-\underset{\underset{R}{|}}{CH}-CO)_n-OMe$$

(74)   (75)

is methanol containing catalytic amounts of sulfuryl chloride (23, 84). At this stage, other methylating agents such as diazomethane can also be used (91).

Peptides can sometimes be acylated with acetic anhydride in methanol and then esterified (34, 92). However, under these conditions peptides frequently do not undergo acylation, and side products are common (93).

It has also been recommended (44) that the free peptide be esterified with methanolic HCl before acetylation, but esterification of free peptides under the described conditions is sometimes accompanied by peptide bond cleavage (93, 94). For peptides soluble in methanol (e.g., those containing a S-benzylcysteine residue), esterification with methanolic diazomethane and subsequent acylation has been suggested (95, 96).

N^{α}-Acyl derivatives of arginine-containing peptides that have no tryptophan residue can be condensed with 1,1,3,3-tetramethoxypropane in methanol saturated with hydrogen chloride, the arginine residue giving the residue of N^{δ}-pyrimidylornithine (**47**); under these conditions the carboxyl groups are simultaneously esterified. If the peptide has no amide bonds such as Arg-Gly, which is particularly sensitive to alkaline agents, it can be readily converted to the corresponding ornithine-containing peptide by boiling with 20% aqueous hydrazine for 20 to 30 min, and subsequent acylation and methylation by the technique described above; this applies to peptides containing both arginine and tryptophan (51, 52, 54).

The third stage of mass-spectrometric determination of the amino acid sequence in peptides is the mass-spectrometric measurement and the interpretation of the mass spectrum of the N-acylpeptide ester. For interpretation of the spectrum one must consider the general as well as the specific patterns of fragmentation of N-acylpeptide esters. Mass-spectrometric determination of the amino acid sequence is based on the amino acid type of fragmentation; indeed, identification of all the amino acid and/or aldimine fragments in the mass spectrum gives all the information necessary for the structure elucidation of the peptide. However, as the peptide can have other easily cleaved groupings and/or be capable of intermolecular alkylation, its mass spectrum may not always exhibit all the peaks corresponding to the rupture of all the amide bonds (20, 21, 23, 84, 85). Moreover, the specific conversions of the type described above involving partial or complete splitting of the side chains cause transformations of certain amino acid residues into other ones. Thus leucine expelling propylene or butylene yields alanine or glycine respectively; any aromatic or heterocyclic amino acid eliminating $ArCH_2'$ with concomitant migration of a hydrogen atom to the charged fragment would also give glycine. This is ample evidence that unfortunately a determination of the amino acid sequence solely from the amino acid type fragmentation peaks is not generally valid.

Rational application of mass spectrometry led to the structure elucidation of some peptides obtained from the products of partial hydrolysis of pig pepsin (e.g., peptide **76**) and cytoplasmic aspartate-transaminase (e.g., **77** and **78**) (84, 85). The structures of the natural peptides—fortuitin (**79**) (25) and a lipopeptide (**80**) produced by *M. johnei* —(97) were also established.

Many scientists have attempted to work out methods of simplifying the interpretation of the mass spectra. Thus Lederer (26, 31) suggested recording the mass spectra of two derivatives for each peptide differing by the nature

Ile—Leu—Gly—Asp—Val—Phe
(76)

Val—Glu—Glu—Arg
(77)

Leu—Glu—Ala—Leu—Lys
(78)

$$CH_3(CH_2)_nCO—Val—MeLeu—Val—Val—MeLeu—\overset{\overset{\displaystyle Ac}{|}}{Thr}—\overset{\overset{\displaystyle Ac}{|}}{Thr}—Ala—Pro—OMe$$
n = 18, 20
(79)

CH₃(CH₂)ₙCO—Phe—Ile—Ile—Phe—Ala—OMe
n = 14, 16, 18, 20
(80)

of N-acyl groups (e.g., $C_{17}H_{35}CO + C_{18}H_{37}CO$ or $CH_3CO + CD_3CO$). Then all the ions arising from the fragmentation of the amino acid type will give pairs of peaks differing for the cases cited by 14 and 3 amu respectively. McLafferty (89) suggested acyl groups containing a bromine or chlorine atom for the same purpose; fragments bearing those will yield characteristic groups of peaks due to the specific isotopic composition of these elements. One of the most reliable methods, however, is high-resolution mass spectrometry, at an adequate resolving power allowing exact mass measurement and subsequent calculation of the empirical formulas of the fragments. Availability of the exact masses and empirical formulas of all the fragments [the method of element maps (98–100)] permits the automated determination of the amino acid sequence in peptides (44, 49, 88–90), though it should be borne in mind that since the programs are based exclusively on the amino acid fragmentation pattern (i.e., they use for identification amino acid and aldimine fragments only) they need not give a correct, unambiguous answer. However, it is noteworthy that the majority of mass spectra can be deciphered at low resolving power (23, 86); for doubtful cases, of course, it is expedient to check the data obtained, by high-resolution mass spectrometry.

Metastable peaks are of great help in deciphering the mass spectra, since they permit one to follow the stepwise decomposition of molecular ions of N-acylpeptide esters by amino acid type of fragmentation proceeding (to some degree at least) by the successive elimination of amino acid residues beginning with the C-terminus (19, 30, 35, 43, 52). In this respect the semiautomatic registration and calculation technique worked out by Barber (101) and based on the possibility (102) established earlier of identifying the parent ion for every daughter formed in the field-free region between two sectors of the MS902 mass spectrometer is of great promise. Computer analysis of high-resolution mass spectra and metastable mass spectra allows to determine the structure of each peptide in a mixture (102a).

Yet another approach to mass-spectrometric analysis of peptides was advanced by Ya M. Varshavskii (103, 104), who suggested for this purpose the use of monochromatic photons, a technique that might lead perhaps to regulated peptide fragmentation and thus facilitate the interpretation of mass spectra. It appears from some preliminary data that other types of ionization (chemical ionization, field ionization) might prove useful for structure elucidation of peptides, particularly thermally labile and involatile compounds (105, 106, 106a). Further discussion of these techniques is to be found in Chapter 7.

In recent years considerable efforts have been made to establish the limits of the mass-spectrometric method and to find the ways to extend them to include larger peptides. It is clear that the most serious limitation is the degree of volatility of the peptide derivative, and this depends on the nature and the number of the constituent amino acids. Consequently, satisfactory mass spectra have been reported for certain nonapeptides (18, 25, 31), while some pentapeptides give incomplete mass spectra bereft of molecular ions (23, 30). Different amino acids affect the peptide volatility differently. Heterocyclic and aromatic amino acid residues reduce volatility more than all others; of these histidine and tyrosine are the most effective. Generally, if a peptide has three or four such residues, its mass spectrum is unobtainable. Other amino acids, especially cysteine and its S-derivatives, enhance the thermal instability of the peptide, preventing mass spectrometry of moderately high-molecular-weight peptides of low volatility.

The volatility of N-acylpeptide esters is also affected by the nature of N-acyl group. In addition to information on peptide esters acylated with aliphatic acids and trifluoroacetic acid, there is much mass-spectral data on benzoyl and pentadeuterobenzoyl (27, 107), ethoxycarbonyl (91, 93, 108), benzyloxycarbonyl (27, 93, 109, 110), phthaloyl (27, 93, 111, 112), heptafluorobutyryl (29), and other (27, 93, 113–116) peptide derivatives. Investigations of the effect of different N-acyl groups on the peptide derivative volatility and on the relative abundances of the amino acid and aldimine fragments show that the peptides acylated with low-molecular-weight aliphatic acids (from acetic acid to decanoic acid) are most suitable for mass spectrometry (21, 23, 93). Other peptide derivatives were either too involatile (e.g., 2,4-dinitrophenyl or phthaloyl derivatives) or too apt to decompose thermally (e.g., benzyloxycarbonyl derivatives). Earlier it was assumed (17, 29) that trifluoroacetyl and heptafluorobutyryl derivatives of peptide methyl esters have enhanced volatility, but recently it was shown (93) that they have no advantage over acetyl derivatives if a direct inlet system is used.

To increase volatility Lederer suggested exhaustive N-methylation of amide bonds in N-acylpeptides (117). For this purpose CH_3I in dimethylformamide in the presence of silver oxide (54a, 92, 117–119), sodium hydride in dimethylformamide (34, 87), and sodium derivatives of dimethylsulfoxide (120, 121) or of dimethylacetamide (122) in corresponding solvents were used. In this way N-acylpeptide esters containing N-methylamino acid residues are obtained. Not only are these much more volatile than the corresponding desmethyl derivatives but their spectra usually are easier to interpret. The possibilities of the method were demonstrated by the amino acid sequence elucidation in the octadecapeptide (81) and the docosapeptide (82) obtained from the hydrolysis products of the immunoglobulin λ-chain and of the heptapeptide (83) from the zymogen of phospholypase A (87). N-Permethyla-

Ala—Thr—Leu—Thr—Ile—Thr—Gly—Ala—Gln—Ala— . . .
(81)
Pyroglu—Thr—Val—Leu—Gln—Glu— . . .
(82)
Pyroglu—Glu—Gly—Ile—Ser—Ser—Arg
(83)

tion has also been used in the amino acid sequence determination of the natural lipopeptides mycoside C_{b1} (84) (120), stendomycin (85) (123), esperine (86), (124) and viscosic acid (87) (47).

Comparative studies of different N-permethylation techniques showed (125) that the use of silver oxide frequently resulted in undesirable side reactions, particularly for peptides containing aspartic and glutamic acid residues, methionine, tryptophan, and also some others (cf. 30, 43, 126), while sodium hydride in dimethylformamide led to C-methylation of aspartic and glutamic acid residues and glycine. Sodium derivatives of dimethylsulfoxide and dimethylacetamide were much less harmful, but even those could entail complications. For example, it was established (23, 84) that N-acylpeptides containing histidine and tryptophan residues, after methylation, give complicated mass spectra that are difficult to decode because, due to the incomplete N-methylation of the heterocyclic rings, fragmentation leads to ions differing by one or more CH_2 groups. Consequently, differentiation between glycine and alanine, serine and threonine, valine and leucine (isoleucine), and so on is difficult. The same problem can be caused by C-methylation of histidine-, tryptophan- and glycine-containing peptides, which often accompanies their N-methylation. Incomplete methylation of the ω-amino groups of lysine- and ornithine-containing peptides (84, 92) is also troublesome. Apart from this, N-permethylation may sometimes facilitate specific fragmentations that require special investigation. Thus the residues of all the sulfur-containing amino acids (methionine, cystine, cysteine), after treatment with methyl iodide, are much more apt to break at the C–S

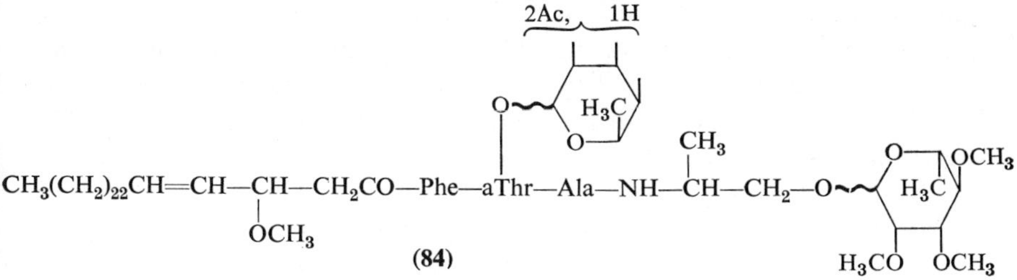

2Ac, 1H

$CH_3(CH_2)_{22}CH=CH-CH-CH_2CO-Phe-aThr-Ala-NH-CH-CH_2-O$

OCH_3

(84)

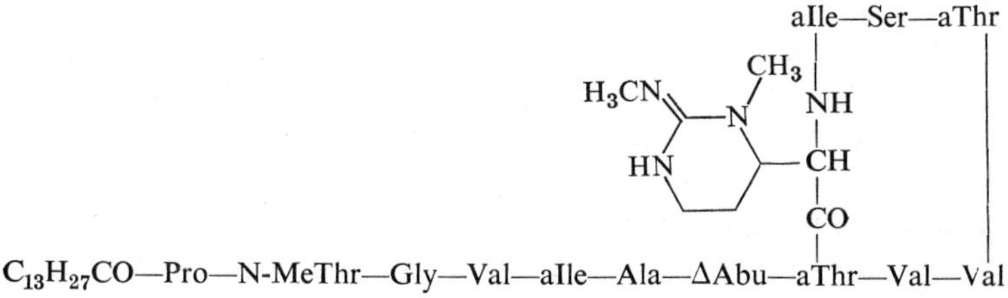

aIle—Ser—aThr

$C_{13}H_{27}CO-Pro-N-MeThr-Gly-Val-aIle-Ala-\Delta Abu-aThr-Val-Val$

(85)

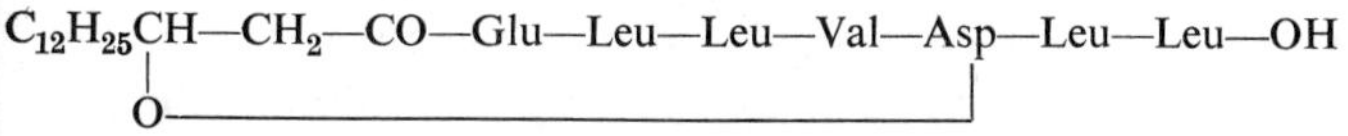

$C_{12}H_{25}CH-CH_2-CO-Glu-Leu-Leu-Val-Asp-Leu-Leu-OH$

O

(86)

$CH_3(CH_2)_6CH-CH_2CO-Leu-Glu-Thr-Val-Leu-Ser-Leu-Ser-Leu-OH*$

OH

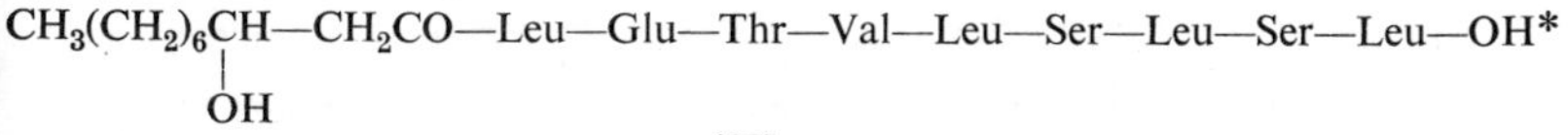

(87)

*One of Leu residues is actually Ile.

bonds, apparently with the formation of sulfonium salts that undergo further decomposition under mass-spectrometric conditions (43, 127). It has been shown recently that cysteine-containing peptides give satisfactory mass spectra after methylation with the calculated amount of the sodium derivative of dimethylsulfoxide (127a).

The volatility of peptides containing methionine, cystine, and cysteine (or any of the S-substituted derivatives of the latter) can be considerably enhanced by desulfurization; it is more expedient to perform this procedure before N-permethylation of sulfur-containing N-acylpeptides (92, 127, 128). Desulfurization by boiling a solution of N-acylpeptide ester in an inert solvent over Raney nickel (92) is accompanied by a whole set of undesirable side reactions (reduction of tyrosine and tryptophan residues, incidental fission of amide bonds, and considerable sorption of peptides on the catalyst), while the use of Raney nickel in dimethylformamide at 20° (43, 127, 128) results in the quantitative conversion of methionine residues to α-amino-butyric acid, cystine and cysteine furnishing alanine residues while the other amino acids are unchanged. Mass spectra of peptides after desulfurization generally are more easily analyzed than are those of sulfur-containing peptides. When a peptide contains in addition to cystine (cysteine) an alanine residue, desulfurization should be carried out over a catalyst obtained by leaching of a Ni/Al alloy with NaOD in D_2O; then cystine (cysteine) yields deuteroalanine (22, 127).

In the case of very involatile peptides resulting from protein hydrolysis, which are not at all amenable to mass spectrometry, it has been found advantageous to subject them to a combination of mass spectrometry and Edman degradation (with phenylisothiocyanate), allowing stepwise elimination of several amino acid residues from the N-terminus. The shortened peptide obtained by Edman procedure and converted according to the above described techniques into an N-acylpeptide ester can undergo mass spectrometry without any purification (23, 84, 129). Edman degradation is known to allow identification of a limited number of amino acid residues only, since the progressive assimilation of the side products ultimately impairs the identification of the phenylthiohydantoins and leads to dubious data as to the amino acid composition of the peptide. Those side products, however, were found not to distort to any considerable degree the character of the mass spectrum. In the case of peptides whose N-terminal amino acid(e.g., aromatic or cysteine) complicates the mass spectrum greatly, removal of this amino acid residue by the Edman technique simplifies the character of the spectrum and allows a direct answer as to the amino acid sequence in the peptide. Thus removal of the five amino acid residues from the N-terminus of bradikinin (88) by the Edman technique yielded (129) a tetrapeptide (89) that was further converted into a derivative (90) suitable for mass spectrometry. Identification of phenylthiohydantoins and decoding of the mass spectrum of the acylpeptide ester (90) led to the structure elucidation of the nonapeptide (88). The structure of the decapeptide (91) was also established

$$\text{Arg—Pro—Pro—Gly—Phe—Ser—Pro—Phe—Arg}$$
$$\textbf{(88)}$$
$$\downarrow$$
$$\text{Ser—Pro—Phe—Arg}$$
$$\textbf{(89)}$$
$$\downarrow$$
$$\text{Dec—Ser—Pro—Phe—Orn(Pyr)—OMe}$$
$$\textbf{(90)}$$
$$\text{Phe—Pro—Pro—Phe—Phe—Val—Pro—Pro—Ala—Phe}$$
$$\textbf{(91)}$$

by mass spectrometry preceded by the Edman technique used to remove two terminal amino acids.

Combined application of mass spectrometry and Edman degradation is somewhat difficult only for lysine-containing peptides, as under Edman reaction conditions the ω-amino group of the lysyl residue is converted into a phenylthiocarbamoyl (PTC) group, which complicates the interpretation of the mass spectrum. However, it has been shown (84, 129) that the PTC group is readily removed by hydrazinolysis under the conditions described for the conversion of arginine into ornithine, to give a free peptide that, after acylation and esterification, can be subjected to mass spectrometry without previous purification.

To attain the objectives mentioned above, it may also prove useful to release one or more amino acid residues from the C-terminus of the peptide by incubation with carboxypeptidase. If the peptide under investigation is obtained by tryptic hydrolysis of a protein and therefore contains an arginine (lysine) at its C-terminus, then the latter will be readily and selectively removed by carboxypeptidase B (84). The shortened peptide can then be easily isolated in a condition sufficiently pure for mass spectrometry, as was shown for **92** and **93**.

Leu—Gly—Ala—Leu—Lys
(92)

Phe—Trp—Ala—Gly—Leu—Gly—Arg
(93)

As another way to extend the scope of mass spectrometric determination of peptide structure, one should also mention a combination of gas chromatography and mass spectrometry. In this case the peptide under investigation is nonspecifically hydrolyzed to yield a mixture of short peptides that are then converted into methyl esters of the N-trifluoroacetyl derivatives. The separation of the mixture is performed by means of gas chromatography, and the mass spectrum of each fraction is measured. By this method the structures of some natural cyclopeptides were elucidated (33, 63, 64) (cf. 95, 96). However, this technique is difficult to apply to peptides of any length, the restricting factor here being the resolving power of the gas chromatograph. On the other hand, if the starting peptide contains some amide bonds particularly susceptible to hydrolysis, it will result in the absence of a number of overlapping peptides necessary to reconstruct the original peptide. As a shortcoming of the method, one must consider some difficulties in the deciphering of the mass spectra as a result of the fact that the amino acid composition of the separate fractions cannot be determined beforehand.

There have been suggestions that mass spectrometry be employed for the identification of the various amino acid derivatives obtained by degradation of peptides and proteins by classical methods [phenyl- and methylthiohydantoins (130–135), 2,4-dinitrophenyl (56) and dansyl (136) derivatives, and 2-phenyl-5-thiazolones (137)]. Sometimes mass spectrometry can be used to check the purity of synthetic peptides; recently it was shown in this way that cleavage of *tert*-butyloxycarbonyl and *tert*-butyl protective groups in tryptophan-containing peptides is accompanied by butylation of the indole ring of the tryptophan residue (138). However, the most important and also most promising field of application for mass spectrometry is the amino acid sequence determination in peptides resulting from the partial hydrolysis of proteins. Here the mass-spectrometric method can considerably simplify and speed up the procedure of the protein primary structure elucidation, the more so since the work can be carried out with micro amounts of the substance. The possibilities of the method are clear from the data cited above on its successful application to the structure elucidation of pig pepsin, cytoplasmic aspartate transaminase (84), immunoglobulins (34, 87), and the like. One can well expect mass spectrometry to find extensive use in the future in protein and peptide chemistry.

REFERENCES

1. K. Biemann, *Mass Spectrometry*, McGraw-Hill, New York, 1962, Chapter 7.
2. H. Budzikiewicz, C. Djerassi, and D. H. Williams, *Structure Elucidation of Natural Products by Mass Spectrometry*, Holden-Day, San Francisco, Calif., 1964, Vol. II, pp. 183–202.
3. K. Heyns and H.-F. Grützmacher, *Fortschr. Chem. Forschung*, **6**, 536 (1966).
4. J. H. Jones, *Quart. Rev. (London)*, **22**, 302 (1968).
5. G. A. Junk and H. J. Svec, *Anal. Biochem.*, **6**, 199 (1963).
6. H. J. Svec and G. A. Junk, *J. Amer. Chem. Soc.*, **86**, 2278 (1964).
7. K. Biemann, *Chimia*, **14**, 393 (1960).
8. K. Biemann, F. Gapp, and J. Seibl, *J. Amer. Chem. Soc.*, **81**, 2274 (1959).
9. K. Biemann and W. Vetter, *Biochem. Biophys. Res. Commun.*, **3**, 578 (1960).
10. V. G. Manusadzhyan and Ya. M. Varshavskii, *Izv. Akad. Nauk Arm. SSR, Khim. Nauki*, **17**, 156 (1964).
11. C.-O. Andersson, *Acta Chem. Scand.*, **12**, 1353 (1958).
12. E. Stenhagen, *Z. Anal. Chem.*, **181**, 462 (1961).
13. F. Weygand, A. Prox, W. König, and H. H. Fessel, *Angew. Chem.*, **75**, 724 (1963).
14. K. Heyns and H.-F. Grützmacher, *Tetrahedron Lett.*, 1761 (1963).
15. K. Heyns and H.-F. Grützmacher, *Justus Liebig's Ann. Chem.*, **669**, 189 (1963).
16. V. G. Manusadzhyan, A. M. Zyakoon, A. V. Chuvilin, and Ya. M. Varshavskii, *Izv. Akad. Nauk Arm. SSR, Khim. Nauki*, **17**, 143 (1964).
17. F. Weygand, A. Prox, H. H. Fessel, and K. K. Sun, *Z. Naturforsch.*, **20b**, 1169 (1965).
18. M. M. Shemyakin, Yu. A. Ovchinnikov, A. A. Kiryushkin, E. I. Vinogradova, A. I. Miroshnikov, Yu. B. Alakhov, V. M. Lipkin, Yu. B. Shvetsov, N. S. Wul'fson, B. V. Rosinov, V. N. Bochkarev, and V. M. Burikov, *Nature*, **211**, 361 (1956).
19. E. Lederer and B. C. Das, in *Peptides, Proc. 8th European Peptide Symp.*, North-Holland, Amsterdam, The Netherlands, 1967, p. 131.
20. M. M. Shemyakin, Yu. A. Ovchinnikov, and A. A. Kiryushkin, in *Peptides, Proc. 8th European Peptide Symp.*, North-Holland, Amsterdam, 1967, p. 155.

21. Yu. A. Ovchinnikov, A. A. Kiryushkin, E. I. Vinogradova, B. V. Rosinov, and M. M. Shemyakin, *Biokhimia*, **32**, 427 (1967).

22. E. Lederer, *Pure Appl. Chem.*, **17**, 489 (1968).

23. M. M. Shemyakin, *Pure Appl. Chem.*, **17**, 313 (1968).

24. N. S. Wul'fson, V. A. Puchkov, B. V. Rosinov, Yu. V. Denisov, V. N. Bochkarev, M. M. Shemyakin, Yu. A. Ovchinnikov, A. A. Kiryushkin, E. I. Vinogradova, and M. Yu. Feigina, *Tetrahedron Lett.*, 2805 (1965).

25. M. Barber, P. Jolles, E. Vilkas, and E. Lederer, *Biochem. Biophys. Res. Commun.*, **18**, 469 (1965).

26. E. Bricas, J. van Heijenoort, M. Barber, W. A. Wolstenholme, B. C. Das, and E. Lederer, *Biochemistry*, **4**, 2254 (1965).

27. A. Prox and K. K. Sun, *Z. Naturforsch.*, **21b**, 1028 (1966).

28. J. P. Flikweert, W. Heerma, Th J. Penders, G. Dijkstra, and J. F. Arens, *Rec. Trav. Chim. Pays Bas*, **86**, 293 (1967).

29. B. A. Andersson, *Acta Chem. Scand.*, **21**, 2906 (1967).

30. V. M. Lipkin, Mass Spectrometric Determination of the Amino Acid Sequence in Peptides Containing Residues of Aromatic and Heterocyclic Amino Acids, thesis, Moscow, 1968.

31. J. van Heijenoort, E. Bricas, B. C. Das, E. Lederer, and W. A. Wolstenholme, *Tetrahedron*, **23**, 3403 (1967).

32. K. Biemann, J. Seibl, and F. Gapp, *J. Amer. Chem. Soc.*, **83**, 3795 (1961).

33. A. Prox and F. Weygand, in *Peptides Proc. 8th European Peptide Symp.*, North-Holland, Amsterdam, 1967, p. 158.

34. F. Franek, B. Keil, D. W. Thomas, and E. Lederer, *FEBS Lett.*, **2**, 309 (1969).

35. A. I. Miroshnikov, Mass Spectrometric Determination of the Amino Acid Sequence in Peptides Containing Residues of Monoamino Dicarboxylic Acids and Their ω-Amides, thesis, Moscow, 1968.

36. N. S. Wul'fson, V. N. Bochkarev, B. V. Rosinov, M. M. Shemyakin, Yu. A. Ovchinnikov, A. A. Kiryushkin, and A. I. Miroshnikov, *Tetrahedron Lett.*, 39 (1966).

37. M. M. Shemyakin, Yu. A. Ovchinnikov, A. A. Kiryushkin, A. I. Miroshnikov, and B. V. Rosinov, *Zh. Obshch. Khim.*, **40**, 407 (1970).

38. M. M. Shemyakin, Yu. A. Ovchinnikov, A. A. Kiryushkin, A. I. Miroshnikov, and B. V. Rosinov, *Zh. Obshch. Khim.*, **40**, 443 (1970).

39. A. A. Kiryushkin, A. I. Miroshnikov, Yu. A. Ovchinnikov, B. V. Rosinov, and M. M. Shemyakin, *Biochem. Biophys. Res. Commun.*, **24**, 943 (1966).

40. M. M. Shemyakin, Yu. A. Ovchinnikov, A. A. Kiryushkin, E. I. Vinogradova, Yu. B. Alakhov, V. M. Lipkin, and B. V. Rosinov, *Zh. Obshch. Khim.*, **38**, 798 (1968).

41. H.-F. Grützmacher and K. Heyns, *Justus Liebig's Ann. Chem.*, **698**, 24 (1966).

42 P. Pfaender, *Justus Liebig's Ann. Chem.*, **707**, 209 (1967).

43. V. A. Gorlenko, Mass Spectrometric Determination of the Amino Acid Sequence in Cysteine- and Cystine-containing Peptides, thesis, Moscow, 1969.

44. M. Senn, R. Venkataraghavan, and F. W. McLafferty, *J. Amer. Chem. Soc.*, **88**, 5593 (1966).

45. G. W. A. Milne, A. A. Kiryushkin, Yu. B. Alakhov, V. M. Lipkin, and Yu. A. Ovchinnikov, *Tetrahedron*, **26**, 299 (1970).

46. Yu. A. Ovchinnikov, A. A. Kiryushkin, V. A. Gorlenko, and B. V. Rosinov, *Zh. Obshch. Khim.*, in press.

47. A. A. Kiryushkin, unpublished data.

48. A. A. Kiryushkin, V. A. Gorlenko, Ts. E. Agadzhanyan, B. V. Rosinov, Yu. A. Ovchinnikov, and M. M. Shemyakin, *Experimentia*, **24**, 883 (1968).

49. K. Biemann, C. Cone, B. R. Webster, and G. P. Arsenault, *J. Amer. Chem. Soc.*, **88**, 5598 (1966).

50. E. Bayer, G. Jung, and W. König, *Z. Naturforsch.*, **22b**, 924 (1967).

51. M. M. Shemyakin, E. I. Vinogradova, Yu. A. Ovchinnikov, A. A. Kiryushkin, M. Yu. Feigina, N. A. Aldanova, Yu. B. Alakhov, V. M. Lipkin, B. V. Rosinov, and L. A. Fonina, *Tetrahedron*, **25**, 5785 (1969).

52. Yu. B. Alakhov, Mass Spectrometric Determination of the Amino Acid Sequence in Arginine-containing Peptides, thesis, Moscow, 1968.

53. J. A. Gilpin, *Anal. Chem.*, **31**, 935 (1959).

54. M. M. Shemyakin, Yu. A. Ovchinnikov, E. I. Vinogradova, M. Yu. Feigina, A. A. Kiryushkin, N. A. Aldanova, Yu. B. Alakhov, V. M. Lipkin, and B. V. Rosinov, *Experientia*, **23**, 428 (1967).

54a. J. Lenard and P. M. Gallop, *Anal. Biochem.*, **29**, 203 (1969).

55. K. Heyns and H.-F. Grützmacher, *Justus Liebig's Ann. Chem.*, **667**, 194 (1963).

56. Th. J. Penders, H. Copier, W. Heerma, G. Dijkstra, and J. F. Arens, *Rec. Trav. Chim. Pays Bas*, **85**, 216 (1966).

57. T. M. King, *Biochemistry*, **5**, 3454 (1966).

58. K. Toi, E. Binum, E. Norris, and H. A. Itano, *J. Biol. Chem.*, **240**, 3455 (1965).

59. H. Vetter-Diechtl, W. Vetter, W. Richter, and K. Biemann, *Experientia*, **24**, 341 (1968).

60. B. J. Millard, *Tetrahedron Lett.*, 3041 (1965).

61. B. V. Rosinov, V. M. Burikov, V. V. Shilin, and A. A. Kiryushkin, *Zh. Obshch. Khim.*, **38**, 2690 (1968).

62. S. Takeuchi, M. Senn, R. W. Curtis, and F. W. McLafferty, *Phytochemistry*, **6**, 287 (1967).

63. Th. Wieland, G. Lüben, H. Ottenheym, J. Faesel, J. X. de Vries, W. Konz, A. Prox, and J. Schmid, *Angew, Chem.*, **80**, 209 (1968).

64. A. Prox, J. Schmid and H. Ottenheym, *Justus Liebig's Ann. Chem.*, **722**, 179 (1969).

65. N. S. Wul'fson, V. A. Puchkov, V. N. Bochkarev, B. V. Rosinov, A. M. Zyakoon, M. M. Shemyakin, Yu. A. Ovchinnikov, V. T. Ivanov, A. A. Kiryushkin, E. I. Vinogradova, M. Yu. Feigina, and N. A. Aldanova, *Tetrahedron Lett.*, 951 (1964).

66. V. N. Bochkarev, V. A. Puchkov, N. S. Wul'fson, M. M. Shemyakin, Yu. A. Ovchinnikov, A. A. Kiryushkin, V. T. Ivanov, E. I. Vinogradova, and N. A. Aldanova, *Khim. Pryrod. Soed.*, 52 (1965).

67. C. H. Hassall and O. Thomas, *Tetrahedron Lett.*, 4485 (1965).

68. C. G. Macdonald and J. S. Shannon, *Tetrahedron Lett.*, 3113 (1964).

69. L. H. Briggs, B. J. Fergus, and J. S. Shannon, *Tetrahedron Suppl.* **8**, 269 (1966).

70. N. S. Wul'fson, V. A. Puchkov, B. V. Rosinov, A. M. Zyakoon, M. M. Shemyakin, Yu. A. Ovchinnikov, A. A. Kiryushkin, and V. T. Ivanov, *Tetrahedron Lett.*, 2793 (1965).

71. C. G. Macdonald and J. S. Shannon, *Tetrahedron Lett.*, 2087 (1964).

72. D. W. Russell, C. G. Macdonald, and J. S. Shannon, *Tetrahedron Lett.*, 2759 (1964).

73. A. A. Kiryushkin, Yu. A. Ovchinnikov, B. V. Rosinov, and N. S. Wul'fson, *Khim. Pryrod. Soed.*, 145 (1968).

74. M. Barber, W. A. Wolstenholme, M. Guinand, G. Michel, B. C. Das, and E. Lederer, *Tetrahedron Lett.*, 1331 (1965).

75. M. Guinand and G. Michael, *Biochim. Biophys. Acta*, **125**, 75 (1966).

76. A. A. Kiryushkin, V. M. Burikov, and B. V. Rosinov, *Tetrahedron Lett.*, 2675 (1967).

77. B. V. Rosinov, V. M. Burikov, I. A. Bogdanova, and A. A. Kiryushkin, *Zh. Obshch. Khim.*, **39**, 891 (1969).

78. D. G. I. Kingston, Lord Todd, and D. H. Williams, *J. Chem. Soc.*, (C) 1669 (1966).

79. M. Guinand, G. Michel, B. C. Das, and E. Lederer, *Vietnam. Chim. Acta*, 37 (1966).

80. M. Guinand, M. J. Vacheron, G. Michel, B. C. Das, and E. Lederer, *Tetrahedron Suppl.*, **7**, 221 (1966).

81. W. A. Wolstenholme and L. C. Vining, *Tetrahedron Lett.*, 2785 (1966).

82. A. A. Kiryushkin, Yu. A. Ovchinnikov, B. V. Rosinov, and N. S. Wul'fson, *Khim. Pryrod. Soed.*, 149 (1966).

83. L. Caglioti, D. Misiti, R. Mondelli, A. Selva, F. Argamone, and G. Cassinelli, *Tetrahedron*, **25**, 2193 (1969).

83a. K. Bevan, J. S. Davies, M. J. Hall, C. H. Hassall, R. B. Mortin, D. A. S. Phillips, Y. Ogihara, and W. A. Thomas, *Experientia*, **26**, 122 (1970).

83b. A. Kakinuma, A. Ducluda, T. Shima, H. Sugino, M. Isono, G. Tamura, and K. Arima, *Agr. Biol. Chem.*, **33**, 1669 (1969).

84. M. M. Shemyakin, Yu. A. Ovchinnikov, E. I. Vinogradova, A. A. Kiryushkin, M. Yu. Feigina, N. A. Aldanova, Yu. B. Alakhov, V. M. Lipkin, A. I. Miroshnikov, B. V. Rosinov, and S. A. Kazaryan, *FEBS Lett.*, in press.

85. V. M. Lipkin, Yu. B. Alakhov, N. A. Aldanova, M. Yu. Feigina, A. A. Kiryushkin, A. I. Miroshnikov, Yu. A. Ovchinnikov, B. V. Rosinov, M. M. Shemyakin, and E. I. Vinogradova, Report to the 10th European Peptide Symposium, Abano-Terme, Italy, 1969.

86. H. R. Morris, A. J. Geddes, and G. N. Graham, *Biochem. J.*, **111**, 38p (1969).

87. G. H. de Haas, F. Franek, B. Keil, D. W. Thomas, and E. Lederer, *FEBS Lett.*, **4**, 25 (1969).

88. K. Biemann, C. Cone, and B. R. Webster, *J. Amer. Chem. Soc.*, **88**, 2597 (1966).

89. M. Senn and F. W. McLafferty, *Biochem. Biophys. Res. Commun.*, **23**, 381 (1966).

90. M. Barber, P. Powers, M. J. Wallington, and W. A. Wolstenholme, *Nature*, **212**, 784 (1966).

91. A. A. Kiryushkin, Yu. A. Ovchinnikov, M. M. Shemyakin, V. N. Bochkarev, B. V. Rosinov, and N. S. Wul'fson, *Tetrahedron Lett.*, 33 (1966).

92. D. W. Thomas, B. C. Das, S. D. Gero, and E. Lederer, *Biochem. Biophys. Res. Commun.*, **32**, 519 (1968).

93. R. T. Aplin, I. Eland, and J. H. Jones, *Org. Mass Spectrom.*, **2**, 795 (1969).

94. H. Hörman, W. Grassmann, E. Wünsch, and H. Preller, *Chem. Ber.*, **89**, 933 (1956).

95. E. Bayer, E. Hagenmayer, W. König, H. Pauschmann, and W. Sautter, *Z. Anal. Chem.*, **243**, 670 (1968).

96. E. Bayer and W. A. Koenig, *J. Gas Chrom.*, **7**, 95 (1969).

97. G. Laneelle, J. Asselineau, W. A. Wolstenholme, and E. Lederer, *Bull. Soc. Chim. France*, 2133 (1965).

98. K. Biemann, P. Bommer, and D. M. Desiderio, *Tetrahedron Lett.*, 1725 (1964).

99. C. Merritt, P. Issenberg, M. L. Bazinet, B. N. Green, T. O. Merron, and J. G. Murray, *Anal. Chem.*, **37**, 1037 (1965).

100. W. J. McMurray, B. N. Green, and S. R. Lipsky, *Anal. Chem.*, **38**, 1194 (1966).

101. M. Barber, W. A. Wolstenholme, and K. R. Jennings, *Nature*, **214**, 664 (1967).

102. K. R. Jennings, *Chem. Commun.*, 283 (1966).

102a. F. W. McLafferty, R. Venkataraghavan, and P. Irving, *Biochem. Biophys. Res. Commun.*, **39**, 274 (1970).

103. V. M. Orlow, M. N. Akopyan, and Ya. M. Varshavskii, *Dokl. Akad. Nauk SSSR*, **166**, 1223 (1966).

104. V. M. Orlow and Ya. M. Varshavskii, *Dokl. Akad. Nauk SSSR*, **176**, 602 (1967).

105. H. M. Fales, G. W. A. Milne, and M. L. Vestal, *J. Amer. Chem. Soc.*, **91**, 3682 (1969).

106. E. M. Chait, T. W. Shannon, W. O. Perry, G. E. Van Lear, and F. W. McLafferty, *Int. J. Mass Spectrom. and Ion Phys.*, **2**, 141 (1969).

106a. P. Brown and G. R. Pettit, *Org. Mass Spectrom.*, **3**, 67 (1970).

107. J. P. Kamerling, W. Heerma, Th. J. Penders, and J. F. G. Vliegenthart, *Org. Mass Spectrom.*, **1**, 345 (1968).

108. J. P. Kamerling, W. Heerma, and J. F. G. Vliegenthart, *Org. Mass Spectrom.*, **1**, 351 (1968).

109. R. T. Aplin, J. H. Jones, and B. Liberek, *Chem. Commun.*, 794 (1966).

110. R. T. Aplin, J. H. Jones, and B. Liberek, *J. Chem. Soc.*, (C) 1001 (1968).

111. R. T. Aplin and J. H. Jones, *Chem. Commun.*, 261 (1967).

112. R. T. Aplin and J. H. Jones, *J. Chem. Soc.*, (C) 1770 (1968).

113. Th. J. Penders and J. F. Arens, *Experientia*, **22**, 722 (1966).

114. Th. J. Penders, W. Heerma, H. Copier, G. Dijkstra, and J. F. Arens, *Rec. Trav. Chim. Pays Bas*, **85**, 879 (1966).

115. V. Bacon, E. Jellum, W. Patton, W. Pereira, and B. Halpern, *Biochem. Biophys. Res. Commun.*, **37**, 878 (1969).

116. K. M. Baker, M. A. Shaw, and D. H. Williams, *Chem. Commun.*, 1109 (1969).

117. B. C. Das, S. D. Gero, and E. Lederer, *Biochem. Biophys. Res. Commun.*, **29**, 211 (1967).

118. B. C. Das, S. D. Gero, and E. Lederer, *Nature*, **217**, 547 (1968).

119. D. W. Thomas, B. C. Das, S. D. Gero, and E. Lederer, *Biochem. Biophys. Res. Commun.*, **32**, 199 (1968).

120. E. Vilkas and E. Lederer, *Tetrahedron Lett.*, 3089 (1968).

121. D. W. Thomas, *Biochem. Biophys. Res. Commun.*, **33**, 483 (1968).

122. K. L. Agarwal, G. W. Kenner, and R. C. Sheppard, *J. Amer. Chem. Soc.*, **91**, 3096 (1969).

123. D. W. Thomas, E. Lederer, M. Bodanszky, J. Izdebski, and I. Muramatsu, *Nature*, **220**, 580 (1968).

124. D. W. Thomas and T. Ito, *Tetrahedron*, **25**, 1985 (1969).

125. D. W. Thomas, *FEBS Lett.*, **5**, 53 (1969).

126. K. L. Agarwal, R. A. W. Johnstone, G. W. Kenner, D. S. Millington, and R. C. Sheppard, *Nature*, **219**, 489 (1968).

127. Yu. A. Ovchinnikov, A. A. Kiryushkin, V. A. Gorlenko, and B. V. Rosinov, *Zh. Obshch. Khim.*, in press.

127a. M. L. Polan, W. J. McMurray, S. R. Lipsky, and S. Lande, *Biochem. Biophys. Res. Commun.*, **38**, 1127 (1970).

128. A. A. Kiryushkin, V. A. Gorlenko, B. V. Rosinov, Yu. A. Ovchinnikov, and M. M. Shemyakin, *Experientia*, **25**, 913 (1969).

129. N. A. Aldanova, E. I. Vinogradova, S. A. Kazaryan, B. V. Rosinov, and M. M. Shemyakin, *Biokhimia*, in press.

130. N. S. Wul'fson, V. M. Stepanov, V. A. Puchkov, and A. M. Zyakoon, *Izv. Akad. Nauk SSSR, Otdel, Khim. Nauk*, 1524 (1963).

131. V. A. Puchkov, V. M. Stepanov, N. S. Wul'fson, and A. M. Zyakoon, *Dokl. Akad. Nauk SSSR*, **157**, 1160 (1963).

132. V. M. Stepanov, N. S. Wul'fson, V. A. Puchkov, and A. M. Zyakoon, *Zh. Obshch. Khim.*, **34**, 3771 (1964).

133. A. M. Zyakoon, V. A. Puchkov, V. M. Stepanov, and N. S. Wul'fson, *Izv. Akad. Nauk SSSR, Ser. Khim.*, 1713 (1967).

134. A. M. Zyakoon, V. M. Stepanov, N. S. Wul'fson, and V. A. Puchkov, *Izv. Akad. Nauk SSSR, Ser. Khim.*, 2410 (1967).

135. F. F. Richards, W. T. Barnes, R. E. Lovins, R. Salomone, and M. Waterfield, *Nature*, **221**, 1241 (1969).

136. G. Marino and V. Buonocore, *Biochem., J.* **110**, 603 (1968).

137. G. C. Barrett and J. R. Chapman, *Chem. Commun.*, 335 (1968).

138. Yu. B. Alakhov, A. A. Kiryushkin, V. M. Lipkin, and G. W. A. Milne, *Chem. Commun.*, 406 (1970).

The Application of Mass Spectrometry to Problems in Medicine and Biochemistry

G. W. A. MILNE

Laboratory of Chemistry, National Heart and Lung Institute, National Institutes of Health, Bethesda, Maryland

I.	Introduction	327
II.	Respiratory and Blood-Gas Analysis	328
III.	Analysis for Trace Metals	330
IV.	Detection and Assay of Drugs	332
V.	Analysis of Steroids	333
VI.	Analysis of Other Lipids	342
VII.	Analysis of Phospholipids	346
VIII.	Studies in Metabolism	350
IX.	Studies Involving the Use of Isotopes	354
X.	The Use of Mass Spectroscopy in Studying the Fate of Pesticides and Insecticides	358
XI.	Identification of Biologically Significant Natural Materials	360
XII.	Conclusion	364
	References	364

I. INTRODUCTION

As the technique of mass spectroscopy moves from its birthplace in the field of physics and is adopted first by chemists, then by biochemists, and finally by clinicians, not only are the problems to which it is applied frequently more difficult, but also, barriers in communication become serious. The mass spectroscopist possessing any medical knowledge is rare, as is the physician who is competent in mass spectroscopy. For these reasons, the impact of mass spectrometry on medicine has been minimal, and there has been only a small improvement in the 20 years since Nier's lament (1) that biologists have been slow to make use of the method.

During this period, however, the middle ground between physics and medicine has been diligently cultivated by biochemists and chemists, with the result that today mass spectrometry is a standard technique in biochemical analysis, and in this sense is now being applied indirectly and often

327

directly to problems in medicine. This review seeks to describe some of the work of the last few years in the application of mass spectroscopy to medical and biochemical problems. It is no longer possible to cover all the published work in this broad area, and therefore no claim is made for completeness. Rather, it is hoped that the examples quoted here of the use of the technique will illuminate the largely unexplored possibilities of mass spectroscopy in medicine and biochemistry. Most of the work described here dates from 1965, and for a description of earlier reports, the reader is referred to several excellent reviews that are available (2–7).

Some very important areas have been omitted, since they are treated elsewhere in this book. Thus the mass spectra of peptides are dealt with in Chapter 9, "Mass Spectrometry in Peptide Chemistry" and the use of stable isotopes is considered only briefly here, Chapter 8, "Mass-Spectral Studies Employing Stable Isotopes in Chemistry and Biochemistry" being devoted to this subject.

II. RESPIRATORY AND BLOOD-GAS ANALYSIS

The compositions of expired and blood-dissolved gas are clearly of medical significance. The partial pressures of oxygen and carbon dioxide in blood and expired air are related to a number of important functions such as cardiac output and pulmonary exchange. The rates at which such partial pressures are changing are also crucial, however, and thus a complete study of lung function in a patient is a dynamic problem and cannot be done via a batch-process type of approach as used for carrying out most other clinical analysis. Not only is it desirable to have the patient "on-line" to a gas-analyzer, it also turns out to be relatively simple to do.

A brief report of the use of mass spectrometry to monitor partial pressures of oxygen and carbon dioxide in expired air appeared in 1949 (8), and a full report on the mass-spectrometric analysis of expired gases was published in 1950 by Miller et al. (1). A continuous gas stream was withdrawn from the trachea and admitted via a leak to the source of a Dempster 180° mass spectrometer. The N_2^+, O_2^+, and CO_2^+ ions were separated by changing the accelerating voltage at constant magnetic field, and in this way, a measure of these partial pressures could be obtained very rapidly in comparison to other methods. This portable spectrometer was designed for theater or bedside use and in many ways, set the pace for subsequent developments in the field. In 1957, the design was described (9) of a mass spectrometer that was fundamentally identical to that designed by Nier (1). Following some further years of development, this spectrometer reached the commercial level at the MS-4, and later the MS-10, manufactured by AEI Ltd. of England. The operation of this machine involves the simultaneous and rapid (25 cps) triggering of the scan and a CRO output scan, with the result that an apparently stationary spectrum appears on the oscilloscope and thus the second-by-second composition of expired gas may be observed. Gas-sampling tubes, internally heated to prevent water condensation, can be placed anywhere in the respiratory system between the mouth and the lung. The results obtained with this spectrometer paralleled those reported earlier by Robertson *et al.* (10), and subsequently by Siehoff *et al.* (11), who used the 21-611 Dempster 180° mass spectrometer developed by CEC Inc. of the U.S. A similar study of gas partial pressures in alveoli (sacs of the lungs) using the MS-4 has appeared (12). Errors inherent in such methods of respiratory-gas analysis have been reported (13, 14), but the method has won such acceptance on the technical level as to be included in sophisticated systems designed for overall monitoring of acutely ill patients (15). More recently, an instrument has been developed (16) for the continuous monitoring of oxygen and carbon dioxide partial pressures. This spectrometer, manufactured by the Scientific Research Instrument Corporation of the U.S., employs magnetic deflection through a 60° arc with individual collectors placed at the appropriate focal points for ions of mass 32 and 44. Extra collectors (e.g., nitrogen and argon) may be incorporated quite simply. Scanning is no longer necessary and thus a truly continuous monitor is obtained. A double electrode, implanted in the blood vessel with the cannula, can be used to monitor blood pH, which is related to the partial pressure in the blood of carbon dioxide.

An important function of this spectrometer is to measure the concentrations of gases dissolved in blood. These data, which are of considerable clinical significance, are somewhat more difficult to obtain than are those for respiratory gases, but the problem is basically one of sample handling. A pioneering study by Woldring *et al.* (17) describes the use of a cannula sealed by a membrane and inserted into a blood vessel. Dissolved gases permeate the membrane and enter the mass spectrometer via the cannula. The choice of membrane is extremely critical, as is correctly emphasized by these authors. They report that latex is generally satisfactory but that wet Teflon is not permeable (17). Other workers have had success with latex and also silastic membranes, and report (16) that under certain conditions, wet Teflon is permeable and can be used with great success. Any such cannula probably is thrombogenic—that is, causes the blood to clot on the cannula. This is hazardous to the patient, and it clearly will alter the permeability of the membrane and the accuracy of the results. This problem has been overcome by the design of a heparinized silastic membrane, which has been shown (16) to cause no clotting over a 24-hour period.

Essentially the same techniques have been used to monitor gas concentrations in brain tissue (18). A membrane of slow permeability is required for problems of this sort because intracerebral gas equilibria are not achieved rapidly. Latex is too permeable, but Teflon, with a response time of five to seven minutes, is ideal (18).

Recently, there has been interest in the use of mass spectrometry to monitor respiratory gases during manned space flights. The bulk of the spectrometer has so far precluded its use for this purpose in manned space flight, and the American space effort does not call for such experiments before 1972. However, a mass-spectrometric system has been developed for the measurement of cardiac output by analysis of the respiratory-gas composition of a subject (19). This system permits the continuous measurement by time-of-flight mass spectrometry of the partial pressures of oxygen and carbon dioxide in expired gas. The data is collected by a small analog computer, which then immediately calculates the cardiac output. A miniaturized Mattauch-Herzog mass spectrometer has been used (20) to monitor respiratory-gas partial pressures in an unpressurized aircraft with satisfactory results.

III. ANALYSIS FOR TRACE METALS

The qualitative and quantitative analysis for metals in biological systems has less direct clinical application at present than gas analysis, except perhaps in cases of metal poisoning. There is, however, considerable interest in such data, and this interest increases as the complex role of metal ions in enzyme-catalyzed reactions is elaborated. Ultimately, when a better understanding of this role is available, it should be possible to relate medical problems to metal-ion deficiencies or surfeits and make use of an appropriate therapy. Most of the metals in question occur at the ppm level, and while techniques such as colorimetry, polarography, neutron activation analysis, and flame spectrometry have all been used with some success for their estimation, mass spectrometry is probably the only technique that can be used to determine ppm levels of all metals simultaneously.

The first application of mass spectrometry to this problem was reported by Wolstenholme (21) in 1964. Electrodes were made from a mixture of lyophilized blood plasma and graphite and then used to generate ions in a spark source. The ions were analyzed in an MS-7 Mattauch-Herzog double-focusing mass spectrometer, as is necessary for ions generated in a discharge, and collected on a photoplate, each element giving a line whose density is roughly proportional to the number of ions (see Chapter 2, "Photographic Techniques in Organic High-Resolution Mass Spectrometry"). The density of the lines was estimated visually with an accuracy of $\pm 30\%$, using the $^{13}C^+$ line from the graphite as an internal standard. In this way the concentrations in lyophilized plasma of some 26 metals were estimated. The concentrations ranged from 0.05 ppm (Cs) to 25,000 ppm (K) and compared well with data from other sources.

This method was further developed by Evans and Morrison (22), who used a Nuclide GRAF-2 mass spectrometer of Mattauch-Herzog design. These workers reported in considerable detail the techniques of sample ignition, preparation of electrodes, and data reduction used to measure the concentrations in various tissues of some 30 metals. Removal of organic materials from samples by ashing is necessary, since otherwise the organic ions generated may not be resolved from metal ions, which will thus be obscured. Care must be taken to account for all the isotopic modifications of each element, and the problems caused by ill-resolved inorganic ions must be considered. Quantitation was best achieved by the independent estimation of the concentration of copper in a standard sample. The method is sensitive to about 10 ppb, with an accuracy and reproducibility of $\pm 10-25\%$, and is applicable to such tissues as blood serum, kidney tumor, lung tissue, bone, and plant leaves. As is correctly pointed out by the authors, however, the complexity of the technique currently precludes its use in routine clinical work.

An interesting application of this same technique has been reported by Hardwick and Martin (23), who sought to estimate the concentrations of various trace elements in dental tissues. A surprisingly large number of elements, 37, was detected at levels of 0.1 ppm or higher in enamel, dentine, or plaque, and further investigations of this sort should help elicit the roles of such elements as Rb, Mo, and Se in the formation or prevention of caries. The use of spark source mass spectrometry has been reported (24) for the estimation of some 25 elements in hair. With a sensitivity permitting the detection of 0.1 $\mu g/g$ of hair, this method would appear to have promising forensic possibilities.

The use of mass-spectrometric techniques to measure the equilibrium concentrations of Group I metals in human organs has been reported (25), and the quantitative analysis by spark source mass spectrometry of 25 metals in the ash of various plants has also been reported (26).

The estimation of the level in human tissues of foreign metals is a problem of considerable industrial significance, and while most such analyses, such as of lead in urine, have been handled for years with great success by normal chemical methods, such techniques are less successful at the lower levels important for more toxic metals and cannot easily distinguish between lethal and harmless isotopes. Thus Howard (27) devised a method based on mass spectrometry to measure the concentration of each of the isotopes of uranium and also the total α-activity. After appropriate preparation, the sample is evaporated onto a rhenium mass-spectrometer filament. This is used to generate ions, which are separated in a 60° magnetic sector that is scanned. The ions are collected either in a multiplier or on a vibrating reed electrometer. The relative standard deviation varied from 0.3% for ^{238}U (~97%) to 95% for ^{234}U (0.04%).

Finally, the field of forensic chemistry is not without its opportunities for mass spectrometry. The report by Weinig and Zink (28) on the use of mass spectrometry to determine the metal thallium at toxic levels in human organs, is typical of such applications.

IV. DETECTION AND ASSAY OF DRUGS

The ingestion of any chemical by a patient creates several problems that are of interest for different reasons. In cases in which accidental or deliberate poisoning has taken place, effective treatment often requires a knowledge of the physical location of the drug some hours after ingestion. Some drugs may still be found largely in the intestinal tract long after others have been completely absorbed into the blood stream. Treatment for one type would be of minimal value, if not contraindicated for the other. The distribution characteristics of many drugs are well known, and thus knowledge of the nature of the drug and of the probable time since ingestion may indicate the appropriate treatment.

The biological events initiated by the drug are also of considerable interest to biochemists and clinicians alike. In recent years, a great deal of effort has been put into the problem of identification of the metabolites *in vivo*, and this is discussed later in this chapter. The present section deals with the use of mass spectrometry in the detection and quantitation of drugs in human organs.

Prominent among the drugs that are deliberately and accidentally consumed in overdoes are barbiturates. A number of studies of the mass spectra of the six commonly available barbiturates, amobarbital (1, R = $(CH_3)_2CHCH_2CH_2$), barbital (1, R = C_2H_5), butethal (1, R = $n\text{-}C_4H_9$), pentobarbital (1, R = $CH_3CH_2CH_2CH(CH_3)$), phenobarbital (1, R = C_6H_5),

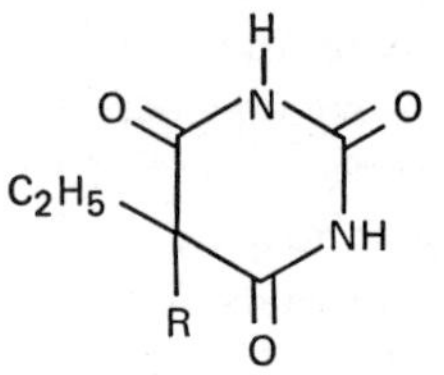

(1)

and secobarbital (**2**) have been reported (29, 30, 31). It is variously claimed that
these barbiturates do (31) and do not (30) give molecular ions, but it is clear

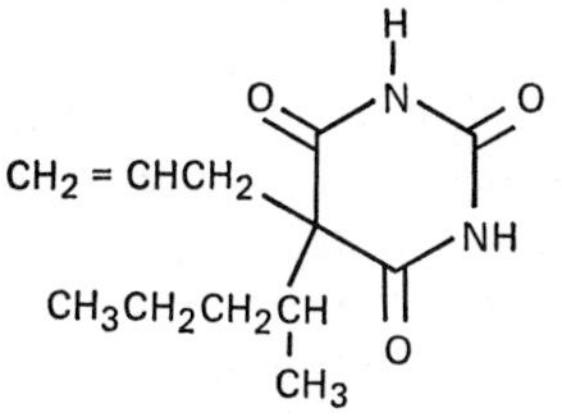

(2)

that if present, the moleculari ons are of very low abundance. In spite of this
disadvantage, the spectrum of each barbiturate is unique, and thus the method
has promise for the identification of individuals within the class, which can
itself be easily recognized by standard color reactions (32). No reports
appear to be available, however, on the mass-spectrometric determination
of barbiturates in biological materials such as stomach washings or serum.
It is possible that such determinations may be rather difficult in view of the
impurity of the samples obtained from these sources, coupled with the
reluctance of the barbiturates to give intense molecular ions. Chemical-
ionization mass spectrometry, on the other hand, as discussed in Chapter 7,
"Newer Ionization Techniques," is ideal for the identification of barbiturates
in gastric contents because they give, as the only significant ion by this technique, a quasi-molecular ion (33). This permits the specific identification of all
the major barbiturates, although amobarbital and pentobarbital, having
the same molecular weight, will still be indistinguishable from each other.

Some practically valuable work has been reported (34) on the determination of the sedative glutethimide (Doriden[R], * **3**) in stomach washings and

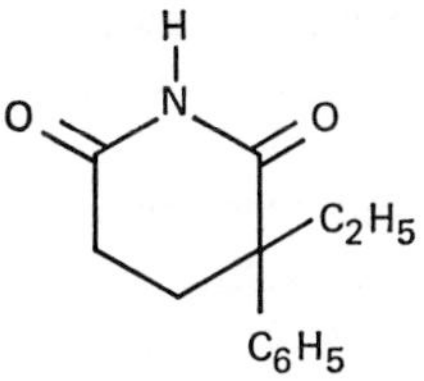

(3)

of a metabolite, α-phenylglutarimide, in urine. This drug is extremely lethal
in overdoses and is difficult to identify and quantitate. It can readily be isolated
from stomach washings by TLC, however, and can be identified easily from
its mass spectrum, which exhibits a molecular ion at m/e 217 (12%). The

* Ciba Pharmaceutical Products, Inc.

minimal amount detectable by this method is about 15 μg. A major metabolite of glutethimide, α-phenylglutarimide, can be readily detected in urine by TLC and mass spectrometry, since it also gives a molecular ion at m/e 189 (17%).

Deliberate poisoning with insecticides is relatively rare, but accidental poisoning is common and may become more common in the future. Weinig et al. (35) have shown that the "drin" insecticides, aldrin and dieldrin, may be detected relatively easily at lethal levels in human organs by mass spectrometry and even by infrared spectrophotometry. Since, as is discussed later, the mass spectra of all the major insecticides have been studied, further cases of insecticide poisoning should, thanks to this work, prove to be relatively routine problems in mass-spectrometric analysis.

V. ANALYSIS OF STEROIDS

Steroids are of great clinical importance because they are intimately involved in many biochemical processes normal to the healthy organism, such as reproduction, and, on the other hand, their presence can be deleterious as, for example, in cardiovascular disease. They are relatively accessible, compared to many mammalian-derived compounds, and also easy to handle in that they are usually stable, nonpolar, relatively low-molecular-weight compounds. It is not surprising, then, that within the scope of this chapter, the amount of published work on steroids far outweighs that on all other subjects put together and must be somewhat abridged in review.

Brooks has very correctly pointed out (36) that in the study of biologically derived steroids, mass spectrometry plays a crucial but supporting role to gas-liquid chromatography. The mass-spectrometric monitoring of gas-liquid chromatograms is certainly the most accurate way to identify compounds emerging from a gas chromatogram and so to establish retention times. Once this has been done, however, retention times alone can be used to characterize steroids or their derivatives, and in a clinical laboratory handling a large number of routine samples, gas chromatography has been proved feasible in many cases, while the problems with gas-chromatographic/ mass-spectroscopic combinations are relatively serious and range from the technical to the economic. This section will be devoted to the use of mass spectrometry in a research environment, where it is used to study particular phenomena and also to develop clinically useful analytical techniques.

Work published before 1967 in this area has been well reviewed by Adlercreutz (37), who describes the role of mass spectrometry as an important check on other methods of estimation of steroids in biological sources, and alternatively as a very powerful technique for the investigation of unknown steroids derived from these and other sources.

Many examples of the use of mass spectrometry in the first role have been reported. Sjövall *et al.* (38) used gas chromatography-mass spectroscopy to develop a gas-chromatographic method of estimating the amounts of dehydroepiandrosterone sulfate in serum. This technique was later extended to develop assay techniques for the steroids in infant feces (39). Free sterols were extracted in hexane, converted to their trimethylsilyl ethers, separated

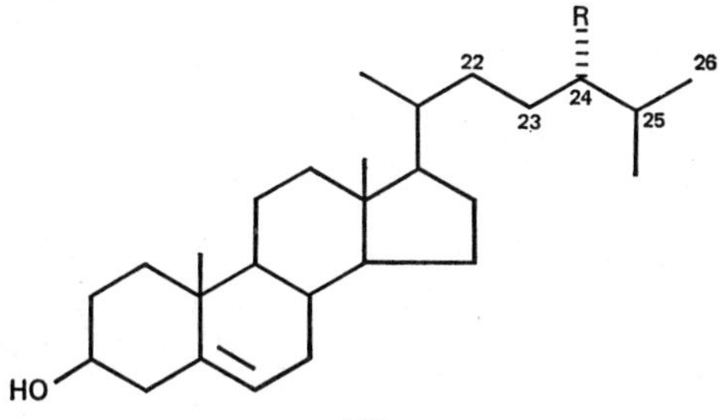

(4)

by gas chromatography, and identified by mass spectrometry as cholesterol (**4**, R = H), campesterol (**4**, R = CH$_3$), stigmasterol (**5**), and β-sitosterol

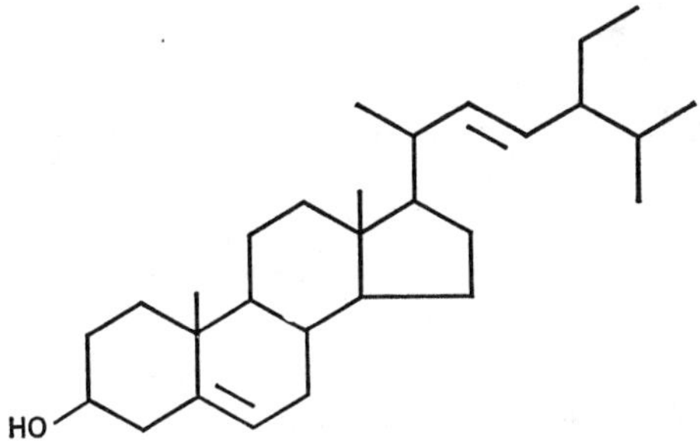

(5)

(**6**). The steroids occurring as sulfates were hydrolyzed on an ion-exchange

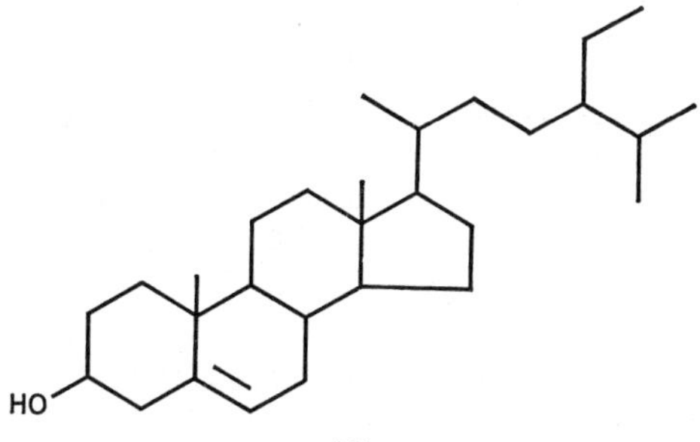

(6)

resin, eluted, and similarly identified as cholesterol (**4**, R = H), campesterol (**4**, R = CH$_3$), β-sitosterol (**6**), and 22α- and 24ε-hydroxycholesterol. The sulfate of 26-hydroxycholesterol is found in samples from infants of 1–4 months and that of stigmasterol (**5**) in feces from infants of 6–12 months. This same approach has also been used (40) in the identification and quantitation in human peripheral plasma of the monosulfates of cholesterol, dehydroepiandrosterone (**7**), epiandrosterone and androsterone (**8**), etio-

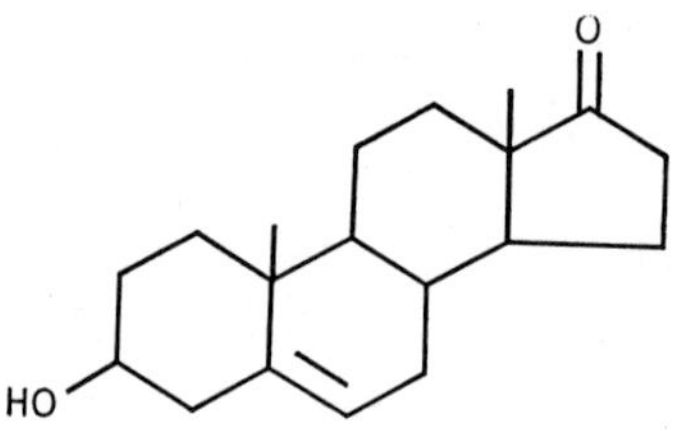

(7)

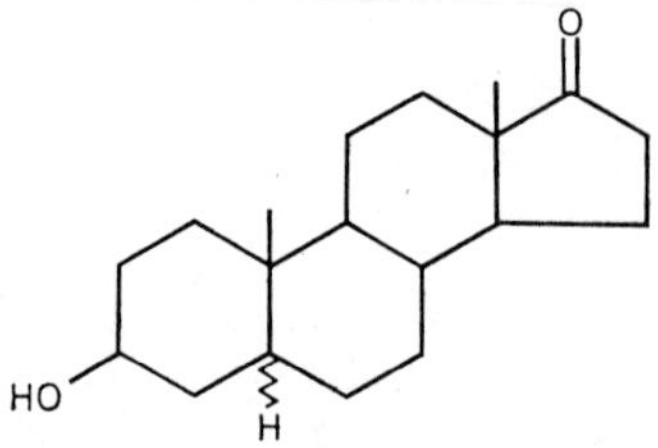

(8)

cholanolone (**9**), pregnenolone (**10**), pregn-5-ene-3β,20α-diol (**11**) androst-5-

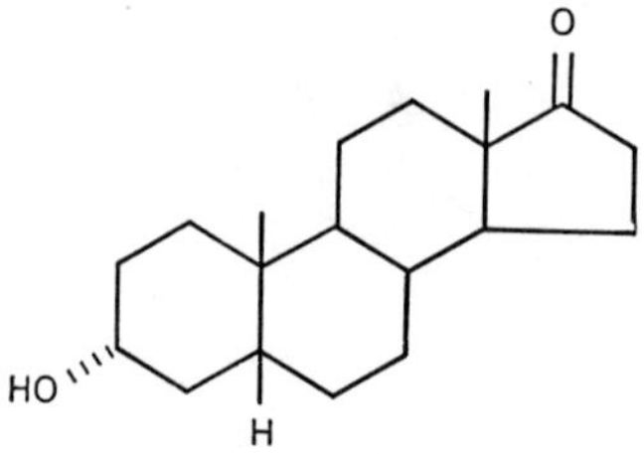

(9)

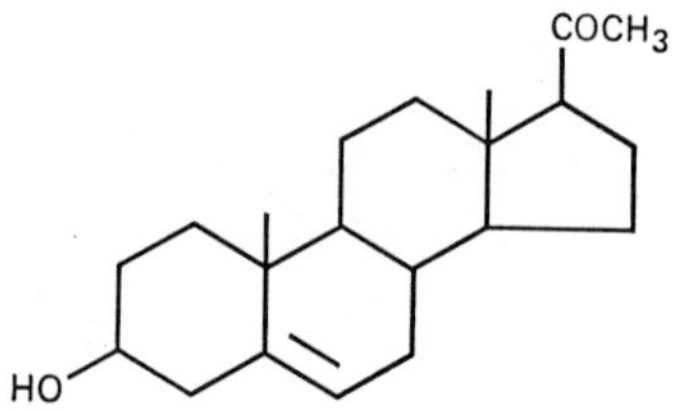

(10)

ene-3β,17β-diol (**12**), and the disulfates of androst-5-ene-3β,17β-diol, androst-5-ene-3β,17α-diol, and pregn-5-ene-3β,20α-diol (**11**). All these

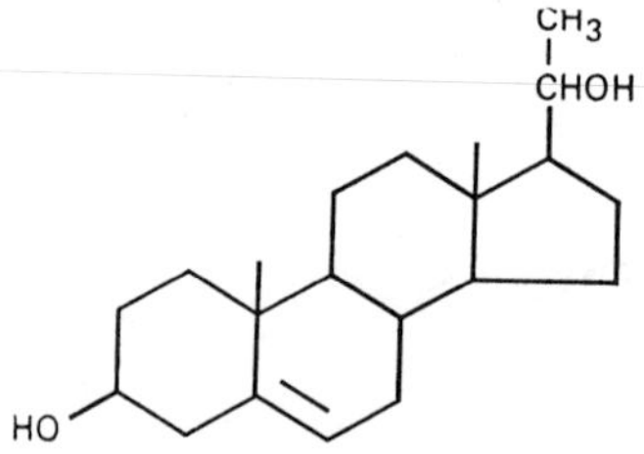

(II)

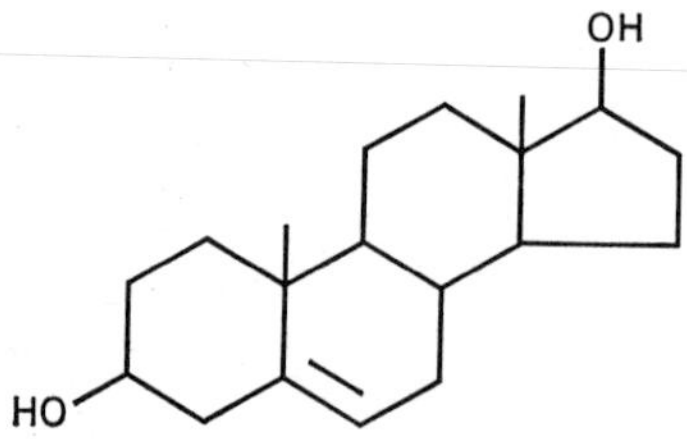

(12)

sterols may be assayed now by gas chromatography alone, as may the sterols occurring as sulfates in urine (41).

The incidence of high levels of aldosterone **(13)** in the urine of some individuals has made it necessary to develop methods for routine assays of this steroid. The compound is particularly difficult to quantitate accurately and rapidly, and earlier methods (42), while accurate and sensitive, were not practical for routine chemical screening. By using gas-chromatographic/ mass-spectroscopic techniques, a gas-chromatographic technique has been developed (43) to assay aldosterone rapidly as the γ-lactone derived by periodate oxidation. The method is sensitive enough for most clinical situations, and analyses can be carried out relatively rapidly ($\sim$20 per day) by a technician.

The standard methods of determination of estriol **(14, R = OH)**, as its tris-trimethylsilyl derivative, in the urine of pregnant and nonpregnant females (44, 45) were rechecked by gas chromatography-mass spectroscopy (46) and found to be entirely specific to estriol. In a parallel study (47), however, it was shown that gas-chromatographic/mass-spectroscopic assay of 17β-estradiol **(14, R = H)** in the urine of nonpregnant females taking the

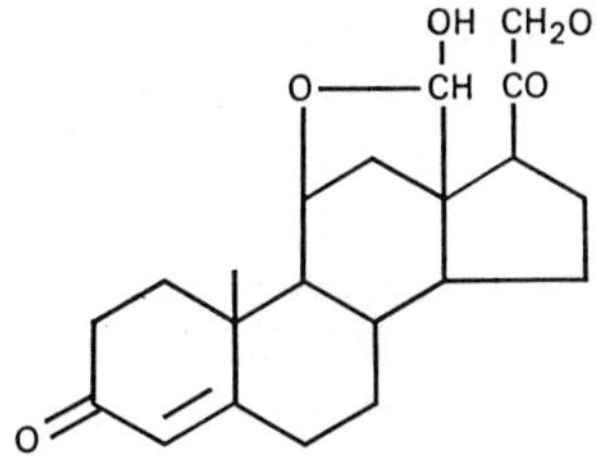

(13) 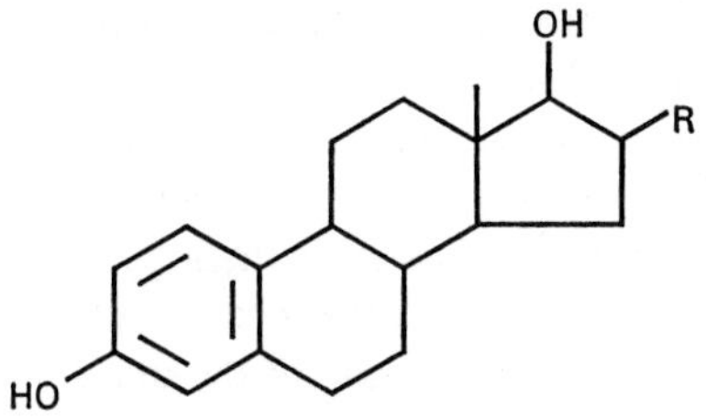   **(14)**

ovulation-promoting bis(*p*-acetoxyphenyl)cyclohexylidene-methane gave results considerably higher than those obtained by the usual colorimetric assay (45). Further investigation by gas chromatography-mass spectroscopy revealed the problem to be interference in the colorimetric method by the drug's metabolites, which were partially characterized by their mass spectra.

Using gas-chromatographic assay procedures developed by gas chromatography-mass spectroscopy as described above, the Helsinki group has accomplished a number of interesting studies on the identification of estrogen metabolites in the urine of pregnant females (48); progesterone metabolites in plasma, bile and urine of females (49); and the excess estrogens excreted by women treated with human pituitary follicle-stimulating hormone (50). Using the same approach, the Houston group has developed (51) analytical procedures to measure changes in biliary excretion of cholesterol, bile acids, and their conjugates and phospholipids. These procedures have been used to determine the normal profiles of such materials in bile and to relate changes in the profiles to drug dosage.

The second vital role of mass spectrometry in mammalian steroid biochemistry has been to aid in the identification of newly discovered metabolites. The power of mass spectrometry, particularly when used in conjunction with gas chromatography, as discussed in detail in Chapter 5, "Gas Chromatography-Mass Spectrometry" is its high sensitivity. Structure determination on microgram quantities is now routine, and various techniques promise still higher sensitivity. Thus the accelerating voltage alternator (AVA) developed for the LKB-9000 by Ryhage's group (52) monitors the effluent of a gas chromatograph not by the total ion current of the attached mass spectrometer but by the intensity of a single ionic species characteristic of the compound in question. Since the gain of the multiplier may thus be used, the overall sensitivity is considerably enhanced, and complete gas chromatographic separation of two compounds is no longer necessary in order to quantitate either or both of them. Using this technique, nanogram (10^{-9} g) level detection of bis-trimethylsilyl pregnanediol has been achieved (37), and some impressive claims of picogram (10^{-12} g) level work have been made (53).

Infant urine is now known to contain a number of steroids regarded as "unusual" in that they are not found in the urine of older subjects. Such a steroid was isolated from the urine of a one- to-three-day-old child in quantities of the order of 10^{-4} g and identified (54) by gas chromatography-mass spectroscopy of its trimethylsilyl derivative as 16β-hydroxydehydroepiandrosterone (15). General procedures have been developed (55) for the identification and estimation of all the major steroids in the urine of newborn infants. Supposedly "normal" profiles of such steroids have been measured, and deviations from such profiles may be useful as early indications of some disorder.

A similar type of finding was that of the unusual stereoisomer 5α-pregnane-3α,20α,21-triol (16), which occurs as its monosulfate in human

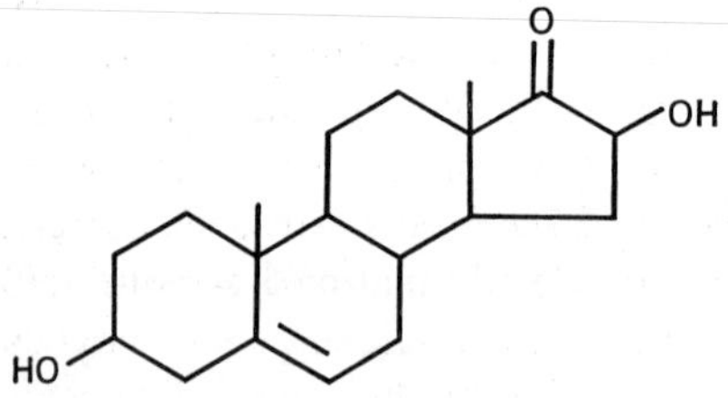

(15) 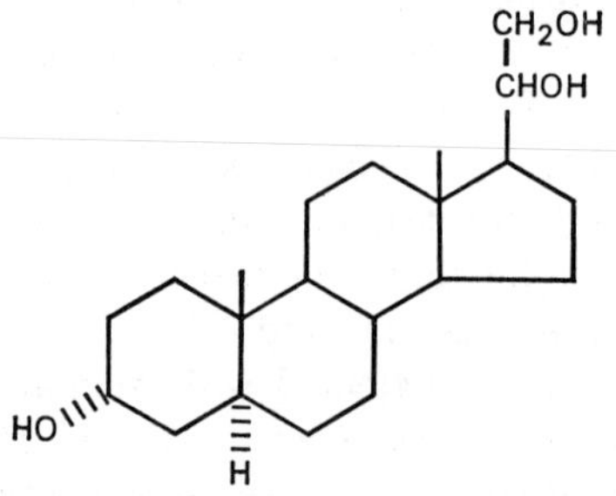   (16)

pregnancy plasma (56). Once again, the triol was separated from other sterols by gas chromatography of the mixture of trimethylsilyl derivatives and identified as a pregnane-3,20,21-triol isomer solely by mass spectrometry. The sterochemistry was settled by a series of elegant chemical steps, carried out with less than 2 μg of material and monitored by gas chromatography-mass spectroscopy.

A series of mono- and dihydroxy bile acids have been isolated (57) from human feces in mg quantities and identified by gas chromatography-mass spectroscopy as 3,12- or 3,7-dioxygenated 5β-cholanoic acids. Each of the eight possible diols was isolated, as were the 3-keto and 12-keto modifications. A previously undetected steroid, 11-dehydro-17α-hydroxyestradiol (17), has been identified (58) at the μg level in the urine of pregnant women near term, again by gas chromatography-mass spectroscopy of the trimethylsilyl derivative. In a related study (59), gas chromatography-mass spectroscopy was used to demonstrate the presence in term amniotic fluid of eight specific estrone derivatives and two androstane derivatives. Further work has been reported on the identification of estrogens in pregnancy plasma and cord plasma (60–62), and the oxygenation of estrone in the 2-position has been observed by two groups (62, 63), the metabolism of progesterone in humans (64) and in the identification (65) of Venning's (66) "sodium pregnanediol glucuronidate" as a mixture of 5β-pregnane-3α-20α-diol (18, 82%), 5α-pregnane-3α,20α-diol (18, 1%), and 3α-hydroxy-5β-

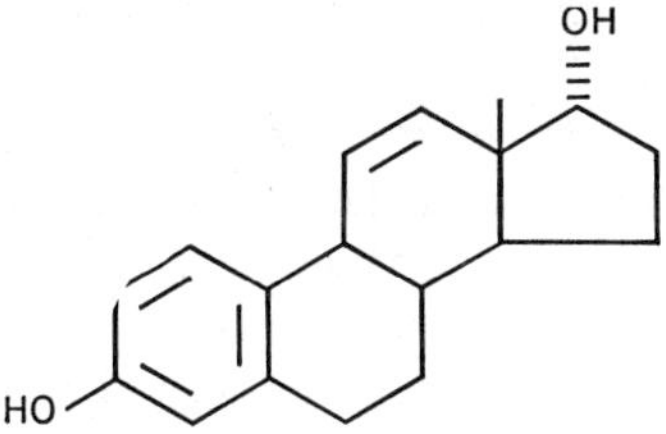

(17) 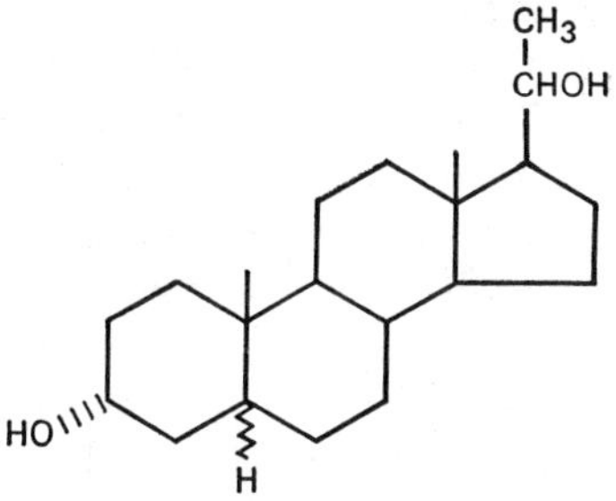   (18)

pregnane-20-one (19,17%). The analysis of bile acids and their derivatives in biological materials has been reviewed by Sjövall (67).

Known precursors of cholesterol, such as lanosterols, dimethyl cholesterols, and methyl cholesterols, have been isolated from human serum (68). The levels of such compounds, particularly of lanosterol (20), fall markedly during fasting, and their use as indicators of cholesterol biosynthesis has been proposed. Similarly, lanosterols have been observed in developing and mature brain tissue (69).

As is discussed later, a considerable amount of work involving mass spectrometry has been done, with a view to the identification of the metabolites of steroidal drugs, which are now in very wide use as oral

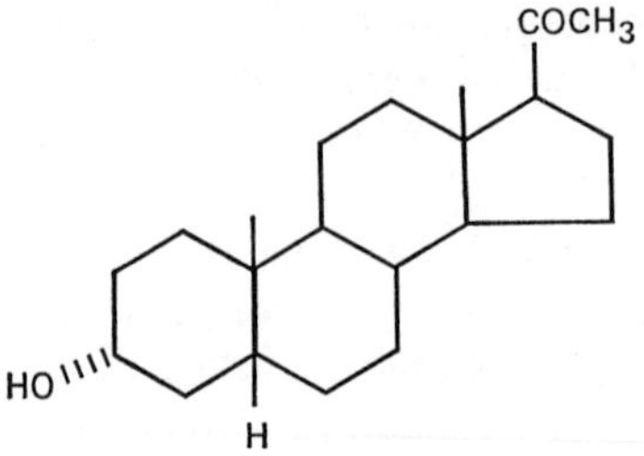

(19) 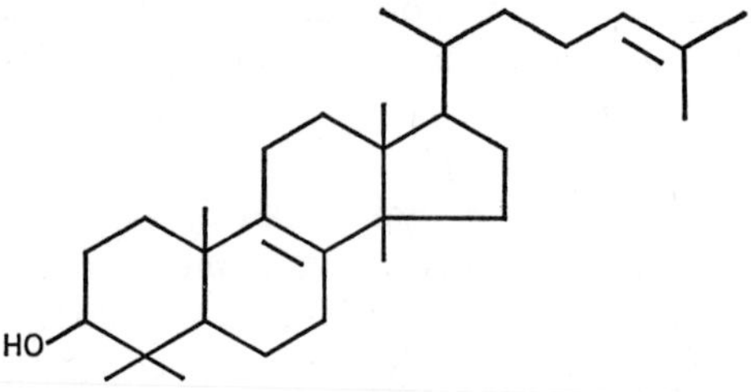   (20)

contraceptives. Typical of this work is the identification (70) of several metabolites of the potent progestogen, norgestrel, (21) in humans, and the subsequent partial delineation of the metabolic routes involved.

Without exception, the reports discussed above describe the use of mass spectrometry either to identify unknown, often new, steroids from mammalian systems or to develop assay procedures for known steroids or their derivatives. The character of the first of these problems has been changed completely and irreversibly by the application of mass spectrometry. The sensitivity of the technique has opened a new field in steroid research, in which structure determination at the microgram level has become routine. As a result, more insight has been gained into steroid metabolism, for example, and a new level of subtlety has been revealed and is being actively explored.

The impact of mass spectrometry on the second of these applications, however, is considerably more problematical. Mass spectrometry is a poor choice as a technique on which to base an assay procedure. The total throughput time of the mass spectrometer is very long, and the sample-handling techniques required do not lend themselves to automation. In contrast, a scintillation counter, having a short throughput time and being easily totally automated, can handle hundreds of samples per day. Second, sensitive as the mass spectrometer is, it is not adequately sensitive, nor is it likely ever to be so, with present technology, to study steroid hormones at the levels at which they are active. Mass spectrometers are indeed very sensitive with respect to the number of ions necessary to produce a signal, but in terms of detection of very small amounts of a steroid in an extraordinarily complex mixture such as serum, biological methods of analysis, such as radioimmunoassay, are vastly superior. Not only do they require relatively little preparation of the biological sample, but they can be and are being used routinely at the picogram (10^{-12} g) level and require little financial investment in terms of equipment beyond a scintillation counter. The complexity of the gas-chromatographic/mass-spectroscopic technique and the equipment involved is such that, sensitivity notwithstanding, it still receives vigorous competition as an analytical procedure from methods such as thin-layer and paper chromatography.

A fringe benefit, often taken for granted, of all the experience that is now available in gas-chromatographic/mass-spectroscopic work is that it has validated the early assumption that the majority of compounds, if they emerge at all from a gas chromatogram, emerge chemically unchanged.

The occurrence of steroids is not limited to the human organism, of course, and a considerable amount of work has been done on sterols from animals, plants, and microorganisms. Bovine cardiac muscle has been found (71) to contain cholesteryl alkyl ethers, particularly the hexadecyl ether. The chemistry of the polyhydroxy steroids, the ecdysones, which are the hormones thought to control the molting process in crustaceans, has involved some mass spectrometry (72), although the crucial structure determination was done by X-ray crystallography. Fecal sterols in rats have been investigated (73) and found to be mostly identical to those found in humans (39). Pigs treated with either of the two inhibitors of cholesterol biosynthesis, *trans*-1,4-bis(2-chlorobenzoylaminomethyl)cyclohexene (Ayerst, AY-9944) and 20,25-diazacholesterol (Searle, SC-12937) were found (74) to accumulate cholesta-5,7,24-trien-3β-ol in lung tissue. A new bile acid, 3α,7α-dihydroxy-5β-cholestanoic acid (22), has been identified (75) in

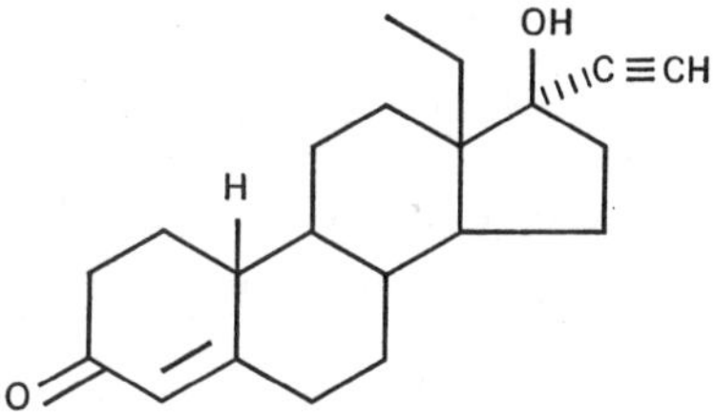

(21) 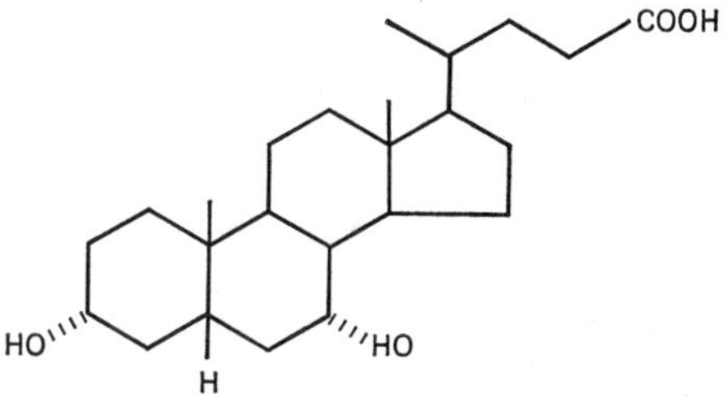   (22)

alligator bile, and a variety of sterols and bile acids have been found in guinea-pig bile (76). Studies on vitamin D metabolism have been greatly assisted by the use of mass spectrometry, and the 25-hydroxylated derivatives of vitamins D_2 and D_3 have both been identified (77, 78) in the plasma of pigs to which the vitamin had been administered. A mass-spectrometric study (79) has confirmed the earlier report (80) that the major sterol in the oyster, *Ostrea gryphea*, is 24-methylenecholesterol.

Steroids in plants have received a great deal of attention, and gas chromatography-mass spectroscopy has been used to identify fourteen sterols in oat seed (81), desmosterol (24-dehydrocholesterol), cholesterol, and six minor sterols in red algae (82), and stigmast-7-enol and stigmasta-7,24(28)-dienol in flax rust uredospores (83). Numerous sterols from pollen have been identified (84, 85) and a review has been published (86) on the use of gas chromatography-mass spectroscopy for the identification of plant sterols.

Some work has also been done using mass spectrometry to identify sterols in microorganisms. Cholesterol was isolated as the major steroid of *S. olivaceus*, which can metabolize it to, *inter alia*, androst-4-ene-3,17-dione, and androsta-1,4-diene-3,17-dione (87). It has been shown that the ethyl side chain in the C_{29} slime-mold sterols, such as stigmasta-22-en-3β-ol, is derived exclusively from methionine (88), and deuterium labeling demonstrated that all five hydrogens also come from methionine, thus precluding ethylidene derivatives as intermediates. The behavior of steroids in microorganisms is discussed further in the section on metabolism.

Among the many papers on the chemistry of steroids, some are particularly pertinent to the mass-spectrometric study of biological steroids. Much of the mass-spectrometry technique is discussed in the review by Leemans and McCloskey (4). General methods for trimethylsilylation, gas chromatography, and mass spectrometry are reported (89, 90). The gas chromatography-mass spectroscopy of the trimethylsilyl derivatives of cardiac aglycones has also been published (91), and a technique for determining the configuration at C_5 of 3,6-diketo steroids has been developed (92), based on the mass spectra of the corresponding *O*-methyl oximes (93).

VI. ANALYSIS OF OTHER LIPIDS

Lipids are ubiquitous and, like steroids, they are extremely easy to handle by modern techniques of gas-phase analysis. Some quite spectacular solutions to chemical and biochemical problems concerning lipids have been achieved, and this particular section will attempt to treat a few of these in some detail.

The most fruitful mass-spectrometric approach to problems involving biologically derived lipids is probably that of coupled gas chromatography-mass spectrometry (4, 94, 95). An early example of the power of this technique was provided by the identification (96), in a single gas-chromatographic/mass-spectroscopic run, of 3 major (C_{29}, C_{31}, C_{33}), and 15 minor *n*-alkanes in bovine feces. A series of *n*-alkanes has also been identified (97), using an identical procedure, in bovine brain.

From the literally hundreds of papers describing similar identification of known lipids in various biological sources might be quoted those reporting pristane (2,6,10,14 tetramethyl-*n*-pentadecane) in skin (98) and also in hair and ovarian demoid cysts (99); α-tocopherol (**23**) from heart muscle (100); 3-D-hydroxypalmitic acid in yeast metabolites (101); pristane, phytane (2,6,10,14 tetramethyl-*n*-hexadecane), and normal and branched-chain fatty acids in shark liver oil (102), and methyl and ethyl esters of long-chain fatty acids in ox pancreas (103).

Valuable as such reports are, the work they describe usually does not

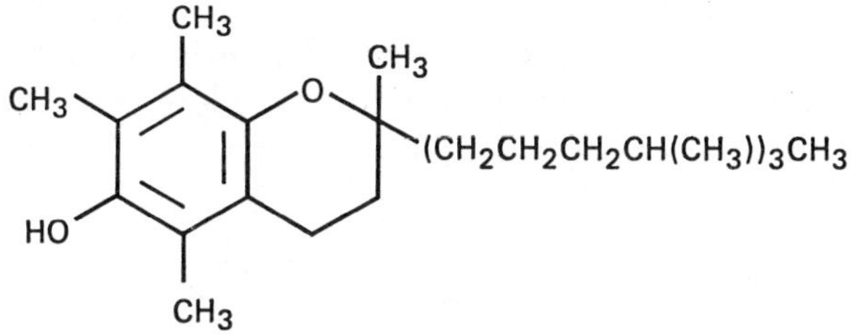

(23)

exploit the full power of the gas-chromatographic/mass spectroscopic combination. Its sensitivity and ability to cope with extremely impure materials are quite remarkable and are perhaps better illustrated by work on lipid metabolism and chemical transformation.

Double bonds in microgram levels of mono-unsaturated fatty acids may be located by osmium tetroxide or $KMnO_4$ oxidation to the glycol. This is bis-trimethylsilylated and admitted via a gas chromatograph to a mass spectrometer, where it gives a spectrum in which the two major ions are those formed by cleavage of the carbon–carbon bond of the derivatized glycol (104–108).

The waxes that are produced in the preen gland of water fowl are of vital importance, since, coated on the bird's feathers, they make possible swimming, as opposed to sinking. Odham (109, 110) has analyzed the mixture of lipids from shellduck and swan preen gland by gas chromatography-mass spectrometry and found it to consist mainly of the esters formed between C_{11} and C_{12} branched-chain acids and C_{16}, C_{17}, and C_{18} n-alkyl alcohols. All these components could be easily identified by gas chromatography alone, and in fact, mass spectrometry is used in only a confirmatory role in this work. The constituents of the chicken's presumably vestigial preen gland are more complex, however, and mass spectrometry played a larger part in their identification (111). In addition to the normal C_{10}–C_{20} fatty esters, which present no problem, there is a nonsaponifiable fraction consisting of three homologous so-called uropygiols. The gas-chromatographic behavior served to identify them as diols; their formation of acetonides showed them to be vic-diols, and the mass spectra of the mixture of diols revealed the presence of ions of the type $C_nH_{2n+1}O$, but none containing two oxygens. This led to a tentative identification of the uropygiols as C_{22}, C_{23}, and C_{24}-2,3-dihydroxy alkanes, and this was confirmed by the appropriate periodate cleavage experiments.

A lipoxidase in soybeans catalyzes the conversion of linoleic acid to its hydroperoxide and thence to 13-hydroxystearic acid. It has been shown (112) by experiments using either $^{18}O_2$ or $H_2^{18}O$, that the oxygen is incorporated from air and not from water.

The C_{20} hydrocarbon residue in chlorophyll is released from the chromo-phore by hydrolysis in the digestive tract of ruminants, and the resulting allylic alcohol, phytol, is found in most dairy produce. This alcohol is con-verted in the human liver via the olefinic acid, phytenic acid (24), to the fully

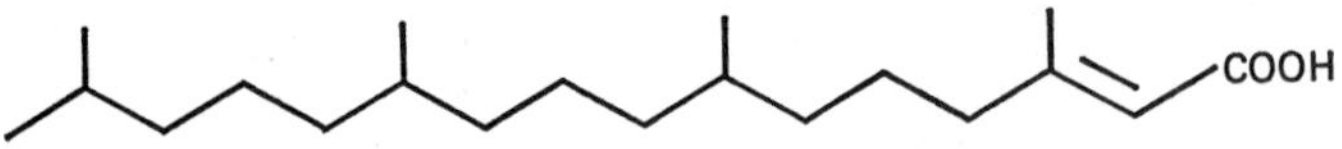

(24)

saturated phytanic acid. The phytenic acid isolated from the intestinal lymph of rats was found (113) to be a mixture of five double bond isomers, the *trans* and *cis* Δ^2 and Δ^3 and the 3-methylene compound. These were each isolated at the mg level by TLC and two-column GLC and identified by time-averaged nmr and mass spectrometry before and after hydrogenation or oxidative cleavage.

Individuals suffering from heredopathia atactica polyneuritiformis (Refsum's disease) are unable to metabolize phytanic acid. In normal sub-jects, phytanic acid was found to be converted to pristanic acid, whose first four metabolites were readily identified (114) by gas chromatography-mass spectroscopy as the expected products of successive β-oxidative cleavage, which is common for fatty acids. Phytanic acid has a β-methyl group, however, and the first step in its metabolism cannot be β-oxidation. Instead, α-hydroxyphytanate is formed (114, 115) and then oxidatively decarboxylated. Refsum's-disease patients are unable to convert phytanate to α-hydroxy-phytanate, since they apparently do not possess the necessary enzyme (116). The indicated treatment, which produces some improvement, is simply a low-phytol diet—that is, avoidance of dairy produce, meat, and the like.

Typical of the way in which mass spectrometry is often useful in bio-chemical problems is the identification (117, 118) of S-palmityl pantetheine (25) as a metabolite in rat liver plasma membrane preparations of palmityl

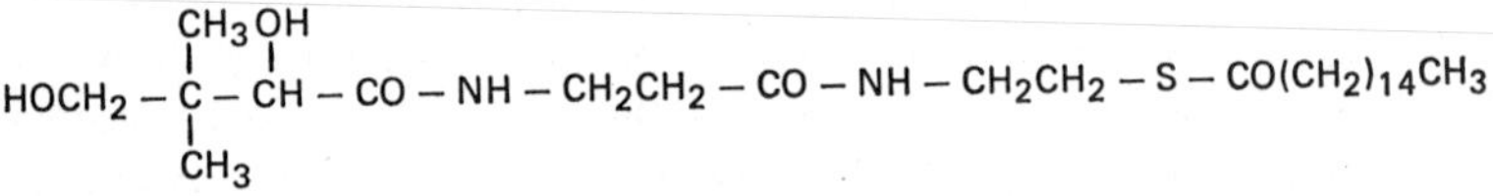

$$HOCH_2 - \underset{\underset{CH_3}{|}}{\overset{\overset{CH_3}{|}}{C}} - \underset{\underset{}{|}}{\overset{\overset{OH}{|}}{CH}} - CO - NH - CH_2CH_2 - CO - NH - CH_2CH_2 - S - CO(CH_2)_{14}CH_3$$

(25)

coenzyme-A. A submicrogram quantity of the impure metabolite gave a mass spectrum with no satisfactory molecular ion, but having fragment ions consistent with its formulation as S-palmityl pantetheine. This possibility was then easily checked by comparison with the synthetic material.

A lipid family that has attracted considerable attention recently is the prostaglandin group. Although they were discovered 30 years ago by

Goldblatt and v. Euler (119), they have been studied very little until recently, mainly because they occur in extraordinarily small quantities. The identification of the three principal prostaglandins, E_1 (26), E_2 (27), and E_3 (28),

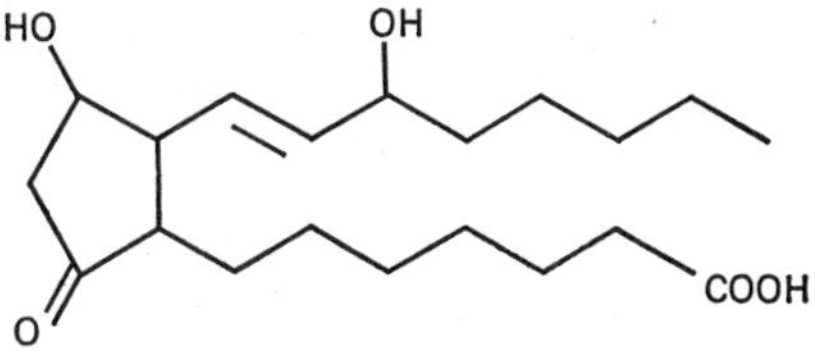

(26)

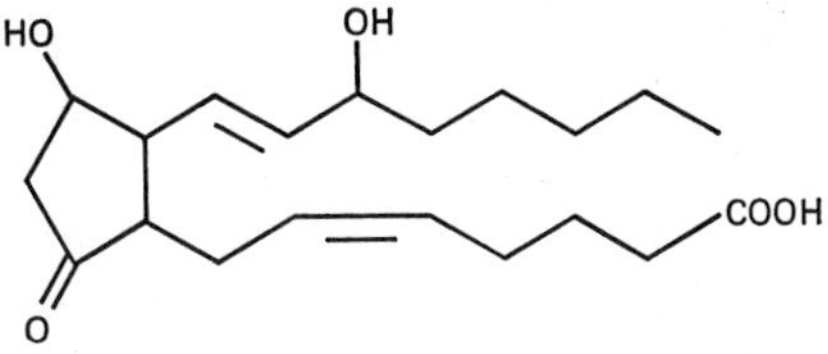

(27)

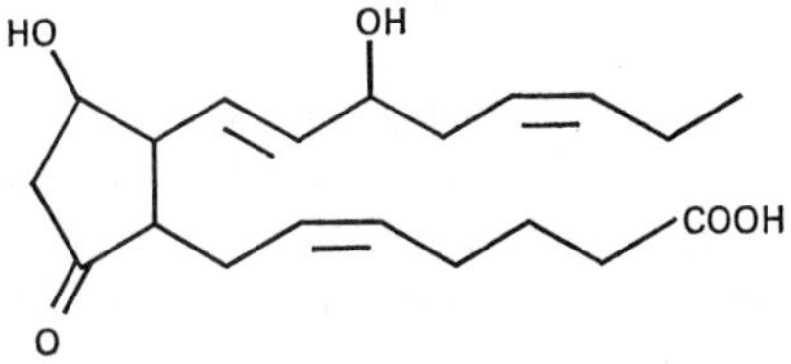

(28)

was effected by a combination of ultraviolet spectroscopy and gas chromatography-mass spectroscopy (120) entirely at the submilligram level. The identification and location of all the functional groups could be accomplished by a combination of these techniques after each chemical transformation. The structure determination of the prostaglandins is a classic of modern analytical organic chemistry (121). Following this basic work, the same group and others have brought the same techniques to bear on the problems of identifying the metabolites of the biologically potent prostaglandins (122–124), and methods for the gas-chromatographic separation and mass-spectrometric identification of all the natural prostaglandins have been developed (125).

Work on other groups of biologically important lipids, the ubiquinones and the tocopherols, has also been facilitated considerably by the advent of gas chromatography-mass spectroscopy. The mass spectra of ubiquinols and ubiquinones have been recorded (126–130) and the technique has been used (131) to detect ubiquinones in normal and parasitized duck blood. Studies of the metabolism (132) and the unusual dimerization and trimerization reactions (133) undergone *in vitro* by α-tocopherol (23) have been carried out by gas chromatography-mass spectroscopy.

The lipid diols, phthiocerol A and B, in human tubercle bacilli have each been identified by mass spectrometry (134, 135) as homologous pairs (C_{33} and C_{35}) of branched-chain methoxy-1,3-diols.

Lipids have been characterized in a variety of other sources by the use of coupled gas chromatography-mass spectroscopy. A series of C_{17}–C_{29}

mono-, di-, and triene hydrocarbons has been identified (136) in microscopic algae, and the major hydrocarbons in cotton buds have been shown (137) to be C_nH_{2n+2} normal alkanes with $22 < n < 32$. The major fatty acids in the lipid-rich alga *Botryococcus Braunii* (Kützig) have been characterized by similar techniques as palmitic acid, oleic acid, octacosenoic acid, and some α,ω-dicarboxylic acids (138). A series of mono- and dihydroxy fatty acids have been isolated (139) from apple cuticle, and in addition to hydrocarbons and fatty acids, the lipids in cabbage leaf have been found (140) to contain a symmetrical C_{29} ketone. The cuticular lipids of the larva *Teneline Molitov L.* have been identified (141) by gas chromatography-mass spectroscopy as have those in the body fat of the parasitic roundworm (142), *Ascaris lumbricoides.* The metabolism of linoleic and linolenic acids in peas has been studied using gas chromatography-mass spectroscopy (143).

As an exercise preparatory to studying lunar rock samples in search of organic molecules, several groups have studied the lipids found in geological specimens. Thus among the organic materials in the Green River Formation (Eocene) shale have been identified normal fatty acids (144), C_{12}–C_{18} dicarboxylic acids (145), methylketo acids (146), aromatic acids (147), and branched-chain fatty acids (148). The English group (149) has identified a series of ω-hydroxy and dihydroxy fatty acids in 5000-year-old freshwater lake sediment. Such work places on a very secure foundation the subsequent demonstration by mass spectrometry that the lunar surface at the Apollo 11 landing site is essentially devoid of organic material (150).

The value and potential of gas chromatography-mass spectroscopy in the lipid area is well defined by one of the leaders in the field, Ryhage, who describes (151) the separation and identification *in one day* of 52 methyl esters found in butter-fat. Here the value of the technique becomes very clear. Although the major components in this material could be easily identified in a number of ways, the great sensitivity of gas chromatography-mass spectroscopy is essential for the investigation of the minor (0.02 mole %) components, and the speed of the technique speaks for itself. In this sense, gas chromatography-mass spectroscopy is at present without competition.

VII. ANALYSIS OF PHOSPHOLIPIDS

Naturally occurring phosphates such as phospholipids and nucleotides are usually presumed to be somewhat unpromising subjects for mass spectrometry, because of their remarkable involatility and instability. Phospholipids are, however, of very major importance in mammalian biochemistry, since they appear to be intimately involved in numerous fundamental processes such as the construction of the nervous system, brain chemistry, membrane permeability, electron transport in mitochondria, and heart disease. In recent years, some effort has been made, mainly by groups in Stockholm, Michigan and Houston, to come to grips with these compounds, using gas chromatography and mass spectrometry. The purpose of this section is to review some of this work.

The free phosphate group in human plasma sphingomyelin (29; R, R' = long chain alkanes or alkenes) prevents direct assay of the material by

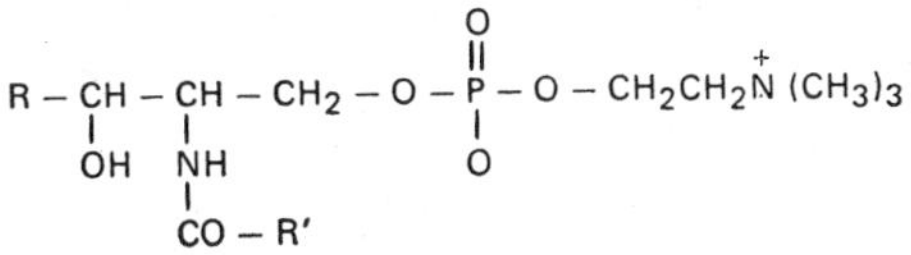

$$R-\underset{\underset{OH}{|}}{CH}-\underset{\underset{NH}{|}}{CH}-CH_2-O-\underset{\underset{O}{\overset{O}{\overset{||}{P}}}}{}-O-CH_2CH_2\overset{+}{N}\,(CH_3)_3$$

$$\underset{CO-R'}{|}$$

(29)

gas chromatography, and although various colorimetric assays have been developed, these are inevitably nonspecific with respect to the various fatty-acid residues that may be present. The Michigan group has demonstrated (152) the utility of gas chromatography-mass spectroscopy for the identification of the specific fatty-acid derivatives released from sphingomyelin by methanolysis. Such treatment of human plasma sphingomyelin led, in addition to sphing-4-enine (sphingosine, **30**) and its saturated derivative sphinganine, several unsaturated derivatives that were identified as sphinga-4,14-dienine **(31)** and the lower homologs hexadecasphing-4-enine **(32)** and heptadecasphing-4-enine. The locations of the double bonds in the diene **31** were

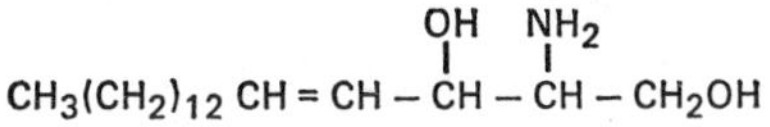

$$CH_3(CH_2)_{12}\,CH=CH-\underset{\underset{OH}{|}}{CH}-\underset{\underset{NH_2}{|}}{CH}-CH_2OH$$

(30)

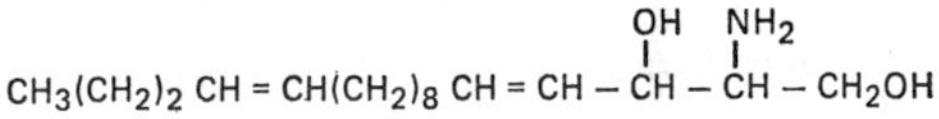

$$CH_3(CH_2)_2\,CH=CH(CH_2)_8\,CH=CH-\underset{\underset{OH}{|}}{CH}-\underset{\underset{NH_2}{|}}{CH}-CH_2OH$$

(31)

$$CH_3(CH_2)_{10}\,CH=CH-\underset{\underset{OH}{|}}{CH}-\underset{\underset{NH_2}{|}}{CH}-CH_2OH$$

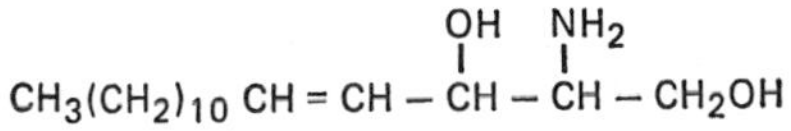

(32)

established (152) by mass spectrometry before and after OsO_4 oxidation. Additional confirmation for the position of the double bonds follows from the isolation of sebacic acid from the $KMnO_4$ oxidation of **31** (152). Mass spectrometry of sphingosines requires trimethylsilylation of both the hydroxyl groups with (153) or without (154) acetylation of the amino group. Similar techniques have been used to identify tetradecasphing-4-enine and hexadecasphing-4-enine as the principal sphingosines from the larvae and adults of *Musca domestica* (155). A sphingosine derivative, tetraacetylphytosphingosine **(33)**, is produced in relatively large quantities by yeast, but it was clearly demonstrated (156) by incubation with either $^{18}O_2$ or $H_2^{18}O$ and subsequent mass-spectrometric analysis of the product, that **33** is not derived from sphing-4-enine by oxidation of the double bond.

Acylation of the amino group of (for example,) sphing-4-enine with a long-chain fatty acid (such as stearic acid) leads to a member of the ceramide family (**34**). The Stockholm group (157) has synthesized a series of ceramides

$$
\begin{array}{c}
\overset{\displaystyle OAc\ \ \ OAc\ \ NHAc}{CH_3(CH_2)_{13}CH - CH - CH - CH_2OAc}
\end{array}
\qquad
\begin{array}{c}
\overset{\displaystyle OH\ \ \ \ NH - CO(CH_2)_{16}CH_3}{CH_3(CH_2)_{12}CH = CH - CH - CH - CH_2OH}
\end{array}
$$

$$
\textbf{(33)} \qquad\qquad\qquad\qquad \textbf{(34)}
$$

by acylation of the appropriate sphingenine or sphinganine, and studied the mass spectra of their derivatives. Both hydroxyl groups could be readily trimethylsilylated, and the resulting derivative is easily gas chromatographed but fails to give a molecular ion upon electron impact. However, the highest mass ion at $m/e \sim 700$ is formed, as is common with trimethylsilyl ethers, by loss of a methyl group from the molecular ion. The whole mass spectrum is quite satisfactory from the point of view of identification of the two fatty-acid residues. The method has therefore been applied (158) to the problem of identifying the ceramides in sphingomyelins from human plasma. The sphingomyelin is hydrolyzed enzymatically and some fractionation of the mixture of ceramides is achieved by TLC of their diacetates on silica gel containing silver nitrate. Each fraction is methanolyzed and trimethyl-silylated and further analyzed by gas chromatography, each peak being identified from its mass spectrum, but with varying degrees of certainty. The complete gas-chromatographic separation of the trimethylsilyl derivatives of the various homologous ceramides is rather difficult, and this appears to be currently the weak point in this promising method.

An attempt to trimethylsilylate the phosphate hydroxyl group in phosphatidyl serines (**35**) is reported (159) to have led to trimethylsilyl

$$
\begin{array}{c}
\overset{\displaystyle OCOR'}{} \qquad\qquad \overset{\displaystyle O}{\overset{\displaystyle \|}{}} \qquad\qquad \overset{\displaystyle NH_3^+}{\nearrow}\\
R - COO - CH_2 - CH - CH_2 - O - P - O - CH_2 - CH\\
\underset{\displaystyle O^-}{\overset{\displaystyle |}{}} \qquad\qquad\qquad \searrow COOH
\end{array}
$$

$$
\textbf{(35)}
$$

derivatives of 1,2 and 1,3 diglycerides, which can be purified by gas chroma-tography and identified by mass spectrometry. In this way, 12 diglyceride derivatives from bovine phosphatidyl serines were identified.

More troublesome than the foregoing phospholipids are the glyco-sphingolipids such as the gangliosides, which are ceramides in which a sugar or a series of sugars is attached to the primary hydroxyl group via a glycoside linkage as in **36**. Sweeley and Dawson (160), in an impressive paper, have shown that following trimethylsilylation, even those glycosphingolipids containing several sugars in an oligosaccharide chain, when admitted on a

$$\underset{\text{(36)}}{CH_3(CH_2)_{12}\ CH=CH-\overset{\overset{\displaystyle OH}{|}}{CH}-\overset{\overset{\displaystyle NHCO(CH_2)_{16}CH_3}{|}}{CH}-CH_2-O-Gly}$$

probe, gave characteristic mass spectra from which their identification was possible. Here, however, the situation arises that molecular weight rather than volatility of the sample is a limiting factor. The completely derivatized gangliosides have molecular weights as high as 3000, and all the available precedents would lead one to expect an ion at m/e (M-15), that is, beyond the range of most mass spectrometers. The Michigan group has developed a method of structural determination using an LKB-9000 with an accelerating voltage of 3.5 kV—that is, a mass range of 1,000. In this case, the attempt to identify the compound from its fragment ions of low mass was successful, but it cannot be expected always to be so. A solution to this dilemma lies in the development of larger magnetic sectors that will permit very heavy ions to be focused without lowering accelerating voltage and so losing sensitivity. A better but more elusive solution, however, would be the replacement of trimethylsilyl by a much lighter blocking group. In a typical trimethylsilylated ganglioside of molecular weight 2500, about 40% of the mass is accounted for by trimethylsilyl groups, and while there is no denying the product's volatility, it is painfully clear to all who use such blocking groups that their benefits are costly.

Finally, the problems of handling phosphates should be considered. In only one case has the mass spectrum of a phosphorus-containing phospholipid been reported (161), and while considerable work has been done on the subject, the development of a technique for converting phosphates to volatile derivatives remains an outstanding problem. In this connection, some work has been reported with nucleotide derivatives. McCloskey et al. (162) demonstrated that esters of phosphoric acid, such as AMP, could be converted to volatile, phosphate-containing derivatives by trimethylsilylation. Biemann's group reported that dinucleoside phosphates could be converted to phosphate-containing volatile derivatives by pertrimethylsilylation (163). Such derivatives give molecular ions, but unfortunately, derivatives of the isomeric dinucleoside phosphates ApU and UpA give identical fragments, and sequence determination is not a simple procedure. Recently, it has been shown by Dolhun and Wiebers (164) that if the terminal 1,2-diol system in ribonucleotides is first protected as the phenylboronate, trimethylsilylation of the remaining hydroxyl groups in the dinucloside phosphate leads to a derivative whose mass spectrum is characteristic of the particular nucleoside sequence present.

VIII. STUDIES IN METABOLISM

In present-day terms, a study of the metabolism of a compound in an organism often does not proceed beyond the first stage, which is identification of some or all of the metabolic products. Successful approaches to this phase of the problem can best be made using an analytical technique that combines high sensitivity with high compound specificity. Mass spectrometry is of course just such a technique and for these reasons, it is being used with great success in the identification of metabolites. The American pharmaceutical companies, which are nothing if not pragmatic, use mass spectrometry as a central technique for the identification of metabolites.

Typical of such work is the study (165) of the metabolites in the rat of the tranquilizer Valium[R]* (diazepam, 37). The mixture of metabolite conjugates obtained from the small intestine four hours after ingestion of the ^{3}H-labeled drug was enzymatically deconjugated and the free metabolites

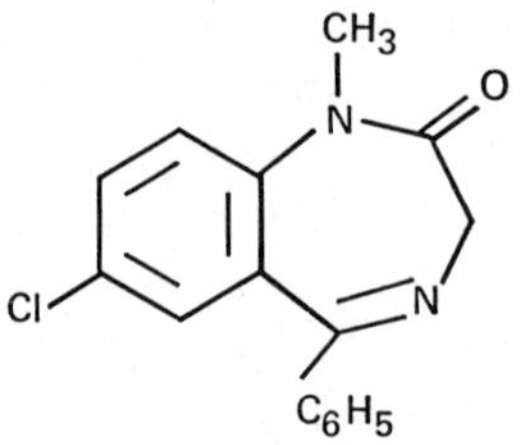

(37) 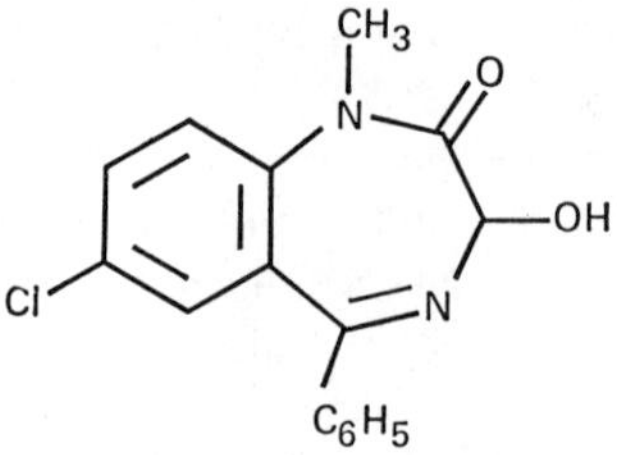   (38)

separated by TLC using ultraviolet light or radioactivity to detect the spots. Each spot was eluted and its high-resolution mass spectrum was measured on a CEC 21-110B double-focusing mass spectrometer. In this way, four metabolites (38-41) of diazepam were identified. The position of the new aliphatic hydroxyl group is deduced from the mass spectra, but nmr is necessary to locate the new aromatic hydroxyl, and even if time-averaged nmr spectra are run, the quantity of metabolite required for this is two orders of magnitude more than for mass spectrometry. A similar study has been carried out by the same group (166) on Librium[R]* (chlordiazepoxide, 42). The same types of conversions are observed here as in the case of diazepam, and in addition, complete removal of the methylamino group and of the N-oxide function is observed.

* Hoffmann LaRoche, Inc.

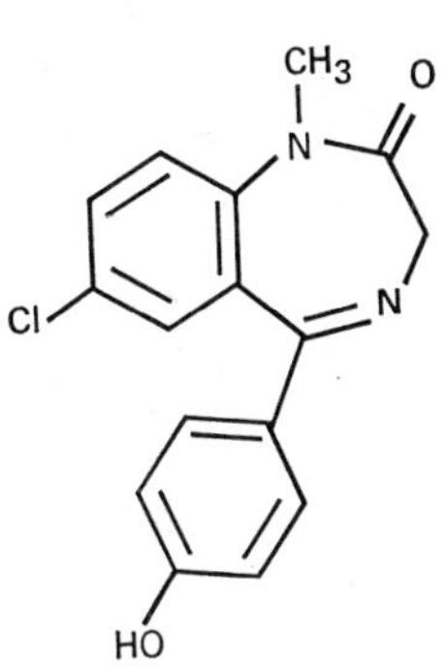

(39)

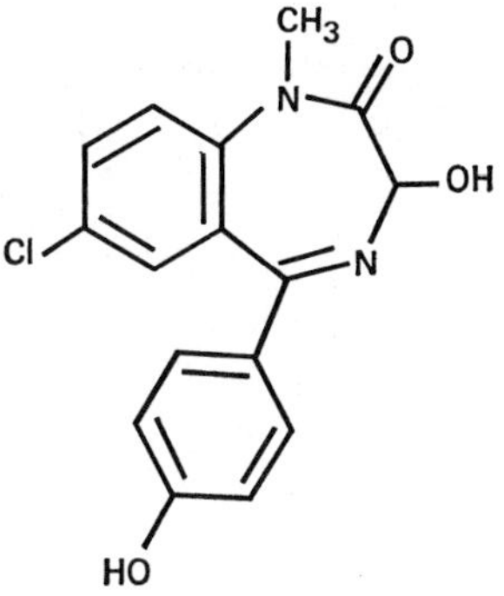

(40)

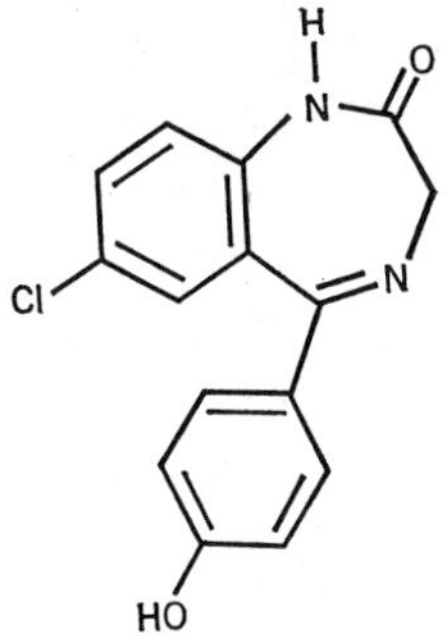

(41)

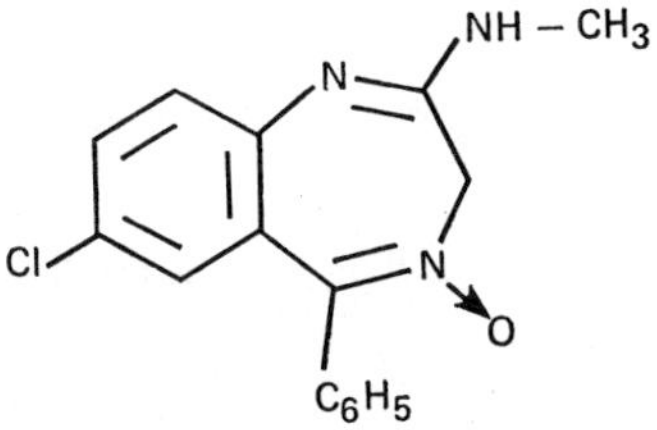

(42)

Using gas chromatography-mass spectroscopy, Horning et al. (167) have painstakingly identified a very large number of the normal constituents of urine and plasma of newborn infants. In this way, they have arrived at a "profile" for such samples—that is, a general quantitative and qualitative knowledge of an average sample. Changes in these profiles can be seen rather readily following ingestion of certain drugs either by the newborn or by the mother prior to delivery. Various types of information are made available by this interesting technique. The metabolism of the drugs in either the mother or the child could presumably be followed, and, more interestingly, the placental transfer of drugs may be studied.

Four metabolites of the progestogen norgestrel (43) were isolated from the urine of patients by TLC, and of these, the reduction products 44 and 45 were identified (70) by comparison of their mass spectra with those of authentic materials. The other two metabolites are monohydroxylated derivatives of 43, but the position of the new hydroxyl group in each case was not established. A major metabolite of norethindrone (46) in $10^4 \times g$ supernatant from rabbit liver was identified (168) by comparison with

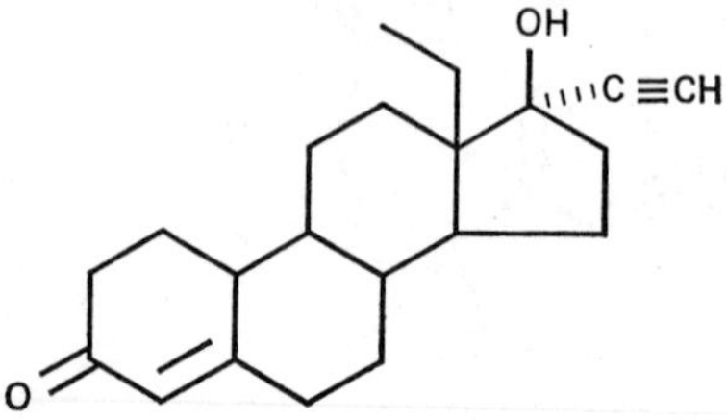

(43)

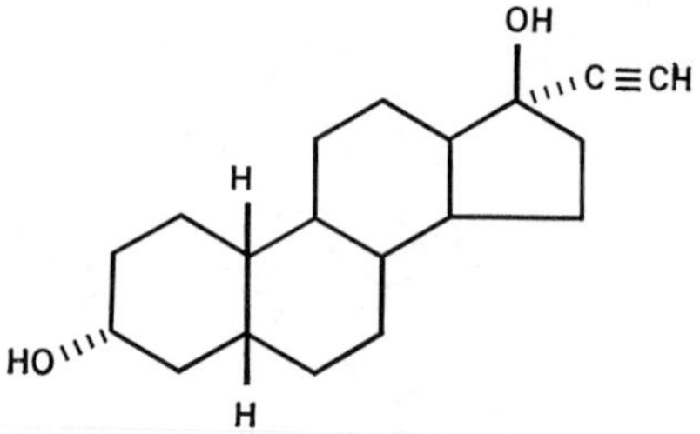

(44)

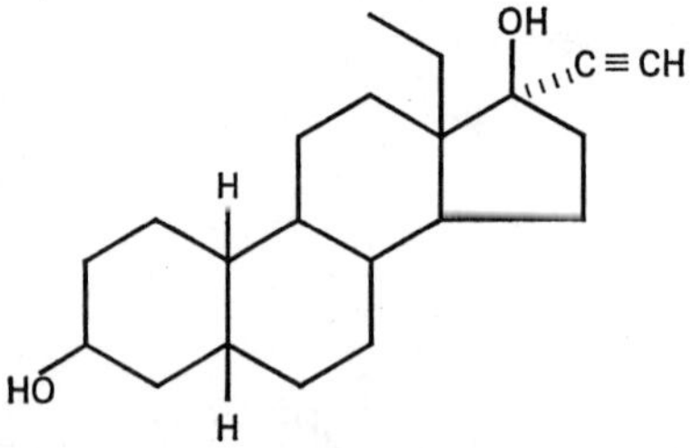

(45)

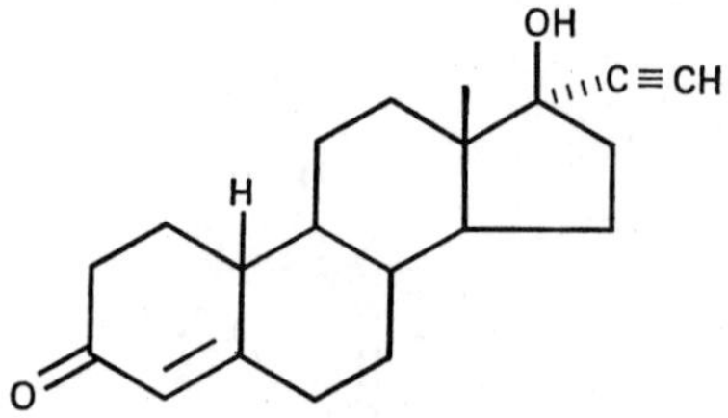

(46)

authentic material as estr-4-en-3,17-dione (47). The similar identification of 48 and 49 as metabolites of norethynodrel (50) has been reported by the same group. (169).

Four metabolites have been isolated (170) from the urine of patients

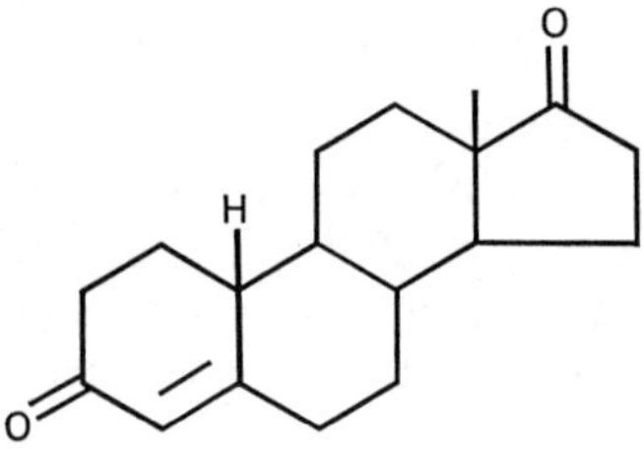

(47)

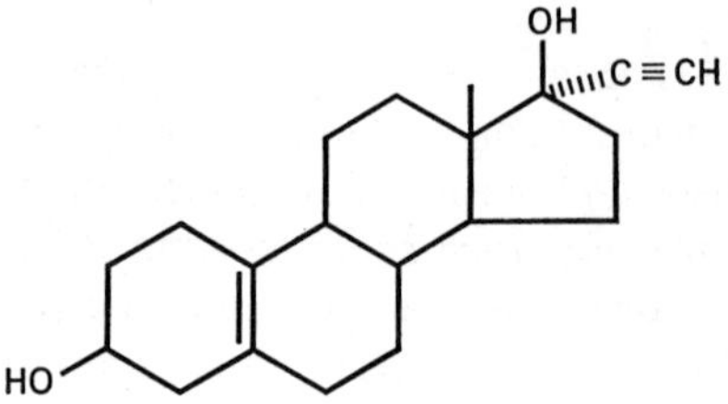

(48)

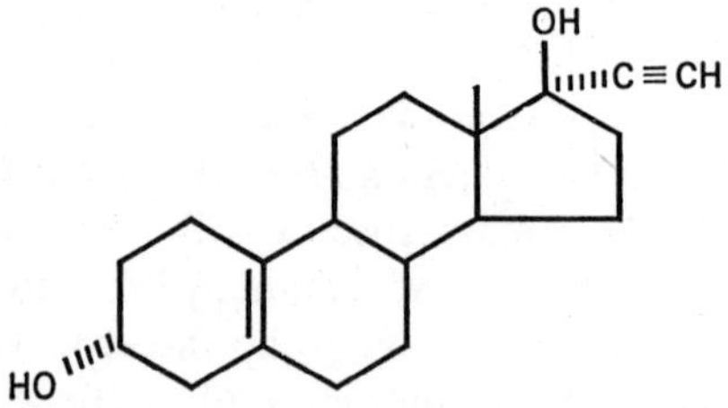

(49)

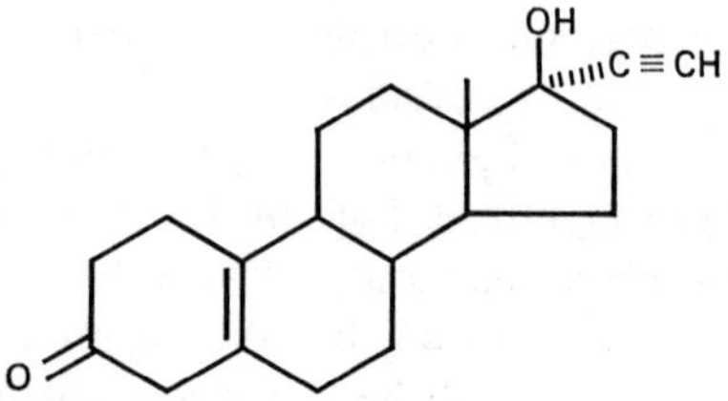

(50)

and dogs dosed with the diuretic, methyl N(o-aminophenyl)-N-(3-dimethyl-aminopropyl) anthranilate. Satisfactory identification of each of these follows from the low-resolution mass spectra. The fluorine-containing metabolites of 2-fluorobenzoic acid in a *Pseudomonas* species were identified (171) by mass spectrometry as 3-fluorocatechol and 2-fluoromuconic acid.

From this representative collection of reports dealing with the application of mass spectrometry to the study of metabolism, a number of interesting facts emerge. In most cases, considerable effort is still being expended on the isolation and purification of each metabolite prior to its mass-spectrometric analysis. Such difficult work is often superfluous in the light of the fact that the mass spectrometer is a vastly superior analytical device than is, say, a thin-layer chromatogram. A low-resolution mass spectrometer used to study a crude metabolite mixture will not, of course, be able to detect isomerizations and may lead to confusion between metabolic degradation and electron-impact induced fragmentation. In general, however, the common metabolic transformations such as are dealt with above will be laid out in some clarity in a low-resolution mass spectrum. Thus, an increase in mass over that of the substrate of 16 amu almost certainly must imply the acquisition by the substrate of an oxygen atom. Other possibilities, such as replacement of oxygen by sulfur, are without precedent and in any event can usually be eliminated from consideration by other means.

In the case of the metabolism of 2-fluorobenzoic acid (171), subsequent work showed that each metabolite could be at least tentatively identified as to its molecular formula from a low-resolution spectrum of an ether extract of the culture medium, and all these tentative formulas could be confirmed by high-resolution mass spectrometry. Furthermore, analysis of this same extract by gas chromatography (172) revealed the presence of a previously undetected metabolite at levels of less than 1 % of those of the other products. Errors of omission of this sort can happen when metabolite identification follows isolation, and although the use of radioactive substrates offers some insurance against this, an additional, very effective safeguard is gas chromatography, particularly when coupled with mass spectrometry.

A second point that should be considered is whether the widespread use of high-resolution mass spectrometers in metabolite identification is justifiable. There can be no doubt that such techniques constitute extremely elegant weapons with which to attack structural problems, and in natural-products chemistry, as is discussed later, they have had a significant impact. Drug-metabolite identification however, is a unique type of problem because the structure of the substrate, by definition, must be known, and one is seeking variations, as it were, on this structure. The question is simply whether a high-resolution mass spectrometer, costing at least three times as much as a low-resolution mass spectrometer, is three times more capable of dealing with these problems. To judge from the examples quoted above, the answer would seem to be no, but one cannot speak for future problems. It is perhaps sufficient at this point merely to pose the question, which must ultimately be answered by each individual investigator with reference to his own circumstances.

Finally, there are some biochemical aspects of this work that should be considered. Identification of the metabolites of a drug is clearly a worthy objective, and the work discussed above shows that this can be done fairly easily with the help of mass spectrometry, among other techniques. The relationship between metabolite structure and the mechanism of the biological activity in question, however, is very unclear and could be quite limited. The preponderance of reactions so far delineated as leading to polar, or more polar, products, such as hydroxylated derivatives, is not inconsistent with the possibility that such derivatives merely define an excretory mechanism, unrelated to the mode of action of the drug. This, coupled with the knowledge that metabolites of a drug frequently fail to possess the full activity of the drug, casts even more doubt on the relevance of metabolite structure with respect to drug action.

IX. STUDIES INVOLVING THE USE OF ISOTOPES

The use of stable isotopes in chemistry and biochemistry is the subject of Chapter 8, "Mass-Spectral Studies Employing Stable Isotopes in Chemistry and Biochemistry" and will therefore be treated only briefly here. The case for stable isotopes in biochemistry has been well made (173) and does not need to be repeated. It is sufficient to note that in terms of biochemistry publications, the stable isotopes most commonly used, in order of popularity, are ^2H, ^{18}O, ^{15}N, and ^{13}C.

If a biological system is in any way involved in an experiment, the resulting dilution of the substrate usually precludes the use of ^{13}C because a hundredfold dilution by ^{12}C of 100% enriched ^{13}C results in a ^{13}C level lower than the natural abundance. The radioactive isotope ^{14}C, on the other hand, does not occur naturally to any significant extent, and the dynamic range for its detection is of the order of 10^{12}. If dilution is not a problem, however, ^{13}C is the more useful isotope because it can often be detected *and* located by the combined techniques of mass (174) and nmr spectroscopy (175). The same results for ^{14}C can usually only be achieved by chemical degradation.

Mass spectrometry can be used to detect ^{14}C, and this has been done both by autoradiographic mass spectroscopy (176) and electron multiplier detection (177). At levels of radioactivity of mc/mM, such detection of

[14]C-containing material ($\sim$5%) is relatively easy, and Occolowitz has demonstrated (177) the use of the technique in studies of fragmentation mechanisms.

An elegant example of the use of ^{13}C, on the other hand, was reported by Waller *et al.* (178) who showed that ^{13}C is incorporated from H[13]COOH into the two methyl groups of ricinine **(51)**—a result previously established with ^{14}C labeling.

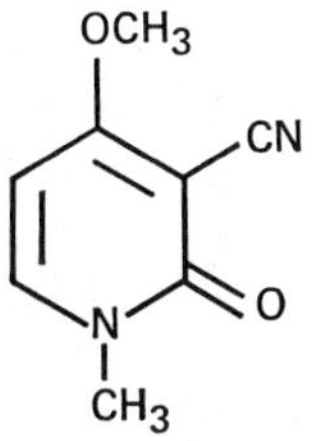

(51)

Nitrogen has no convenient radioisotope, but the stable isotope ^{15}N is readily available, and its use in a variety of biochemical problems has been reported. The techniques devised by Rittenberg (179) for ^{15}N determination by combustion and subsequent mass-spectrometric determination of the resulting [15]N$_2$ gas are now used infrequently, since direct examination of the intact molecule is usually possible. Tovarova *et al.* (180) have shown, by the use of arginine labeled with ^{15}N in the amidine unit, that the amidine nitrogens of streptomycin are derived from arginine, but that the nitrogens of the streptamine moiety have different origins. The incorporation into both nitrogens of ricinine of ^{15}N from HCO[15]NH$_2$ has been observed (178), and it has similarly been demonstrated (181) that the nitrogen of β-nitro-propionic acid produced by *Penicillium atroveretum* is derived from NH$_4^+$ rather than from NO$_2^-$.

Like nitrogen, oxygen also has no convenient radioisotope, but two stable isotopes, [18]O and [17]O, are available. The latter is more difficult to isolate and thus more expensive. It has in common with ^{15}N and ^{13}C, however, the property of giving nmr signals. At present, essentially all the published biochemical work with oxygen isotopes has involved [18]O.

This isotope is especially useful as a means of distinguishing between O$_2$ and H$_2$O as a source of oxygen in biochemical reactions. For example, the enzymatic conversion of monofluoroacetate to glycolate, when carried out in H$_2$[18]O, gives glycolate with [18]O in the new hydroxyl group at the same level of enrichment as the medium (182), showing that atmospheric oxygen is not involved in this reaction. The enzymatic conversion of anthranilic acid to catechol, on the other hand, was shown (183) by use of [18]O$_2$ to proceed with incorporation of atmospheric oxygen.

Similarly, it was shown (184) that in the conversion of cysteine to cysteinesulfinic acid, both oxygens in the product are derived from atmospheric oxygen. A number of oxygenations of steroid hormones catalyzed by adrenal or testicular enzymes have been shown (185) to involve incorporation of oxygen from $^{18}O_2$. Among these are conversion of progesterone to 17α-hydroxyprogesterone and testosterone; 20α,22R-dihydroxycholesterol to pregnenolone acetate; progesterone to deoxycorticosterone; and 17α-hydroxyprogesterone to 11-deoxycortisol. The conversion of progesterone to testosterone acetate in microorganisms also proceeds (186) with incorporation of ^{18}O from $^{18}O_2$, and a Baeyer-Villiger mechanism (52) has been advanced for this reaction.

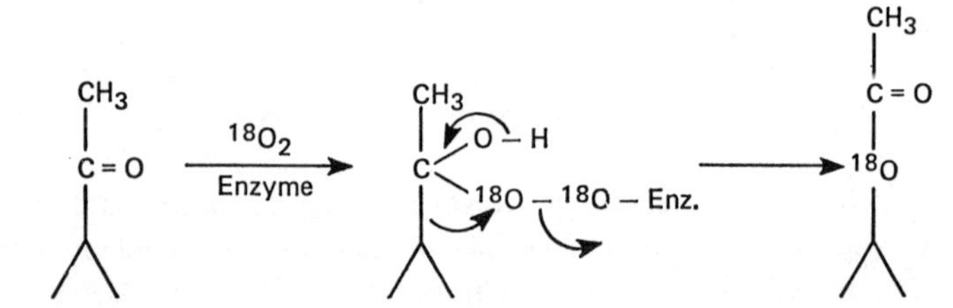

Progesterone   (52)   Testosterone
                                     Acetate

In 1965, Samuelsson (187) demonstrated that considerably more subtle information could be obtained from ^{18}O labeling experiments. Using a mixture of $^{18}O = ^{18}O$ and $^{16}O = ^{16}O$ devoid of $^{18}O = ^{16}O$, he demonstrated that in the conversion of 8,11,14 eicosatrienoic acid (53) to prostaglandin E_1 (27), catalyzed by vesicular gland preparations, an intact molecule of oxygen is incorporated. The enzyme involved must therefore be a dioxygenase, and a conrotatory mechanism for the oxygenation can be postulated:

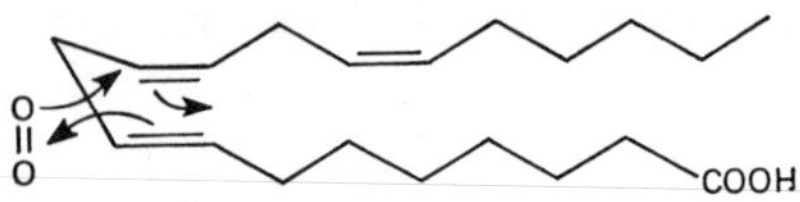

(53)

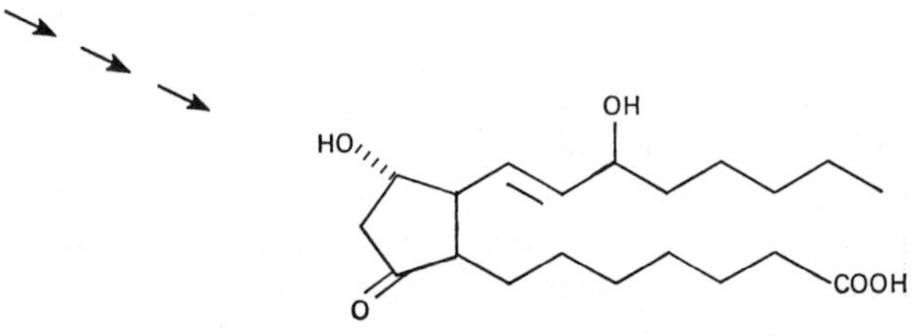

(27)

In just the same way, the conversion of o-fluorobenzoic acid to catechol or 3-fluorocatechol was shown (172) to involve the incorporation of an intact oxygen molecule and the intervention of a dioxygenase.

The oxidative metabolism of naphthalene by rat-liver microsomes to 1,2-dihydro-1,2-dihydroxy naphthalene, on the other hand, has been shown (188) to involve the incorporation of only one atom of oxygen from O_2; the other is presumably derived from water. Thus a monooxygenase is involved in this case, and the intermediacy of a naphthalene-1,2-epoxide was strongly suggested. This was subsequently proved to be correct by Witkop et al. (189), who isolated the epoxide from the reaction mixture.

Deuterium is very widely used as a tracer in chemical and biochemical systems because, being relatively easily enriched, it is cheap, and it can, in addition, often be introduced into a molecule by exchange reactions. Examples of the use of deuterium are legion, and a more comprehensive survey is given in Chapter 8, "Mass-Spectral Studies Employing Stable Isotopes in Chemistry and Biochemistry" and Chapter 11 "Mechanism Studies of Fragmentation Pathways." This section will consider two examples of the use of this isotope in biochemical problems.

The conversion by phenylalanine hydroxylase of p-deuterophenyl-alanine to tyrosine was found by Guroff et al. (190) to proceed with substantial retention of deuterium, which was subsequently found (191) to be in the positions ortho to the new hydroxyl group. The probable intermediary of an arene epoxide in this so-called "NIH shift" was strengthened by the observation (192) that the oxide (54) of p-deuterotoluene was converted to

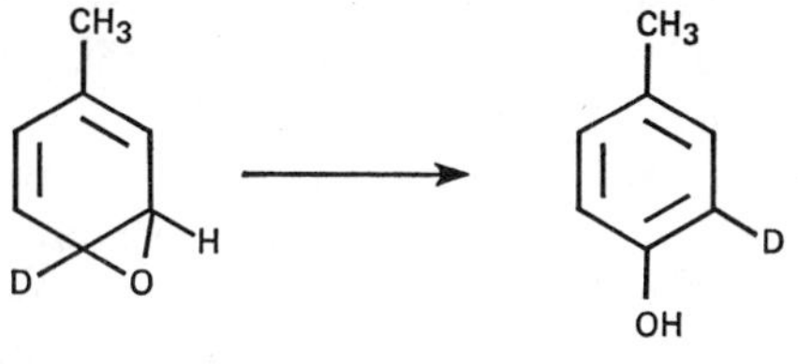

(54)

p-cresol by liver microsomes with 70% retention of the deuterium, as shown, but the mechanism of this rearrangement remains unclear.

An elegant application of deuterium in a biosynthetic problem was reported by Jackman et al. (193), who grew E. coli in a medium containing CD_3-methionine. Analysis by mass spectrometry and nmr of the vitamin K_2 and the ubiquinone produced, revealed that the methyl groups of the methoxyl groups and those of the aromatic rings are methionine-derived, but that the methyl groups of the side chain are not.

Mass spectrometry has opened new areas to biochemists, and the availability of simple methods of ^{18}O and ^{15}N determination has led to results that would otherwise have been very difficult to obtain. At the same time, however, much work of potential interest remains to be done. Almost no work has been reported on the biochemical applications of stable isotopes other than the four discussed here, and the occasional paper, such as that (194) describing the variation of $^{39}K/^{41}K$ ratios during active transport, suggests the interesting future that mass spectrometry could enjoy in this area.

X. THE USE OF MASS SPECTROSCOPY IN STUDYING THE FATE OF PESTICIDES AND INSECTICIDES

The catabolism of artificial pesticides and insecticides in soils is similar in some ways to that of compounds in mammalian systems. The study of either of these often involves the same difficulties, most notably that of sensitivity, and the general metabolic pathways involved have some similarities. Agricultural chemicals are, however, subject to enzymatic breakdown by a host of nonspecific soil microorganisms, and this, together with their inevitable degradation in aerobic, aqueous, irradiated, variable-temperature conditions, tends to complicate the picture. In view of the recent urgency surrounding the study of the fate of pesticides, much effort is being directed to the problem, and the approach used by a number of groups has involved mass spectrometry.

Possibly the first report of the use of mass spectrometry to study such problems was the identification (195) of various metabolites in the soil fungus *Aspergillus fumigatus* of the herbicide simazin (**55**). The herbicide, uniformly labeled with the β emitters ^{14}C or ^{36}Cl, was incubated with the organism for eight days, whereupon the radioactive products were isolated from the medium by chromatography. A major metabolite, which contained both ^{14}C and ^{36}Cl, was identified by mass and nmr spectroscopy as 2-chloro-4-amino-6-ethylamino-s-triazine (**56**), and a second ^{36}Cl-containing metabolite was isolated but not fully characterized.

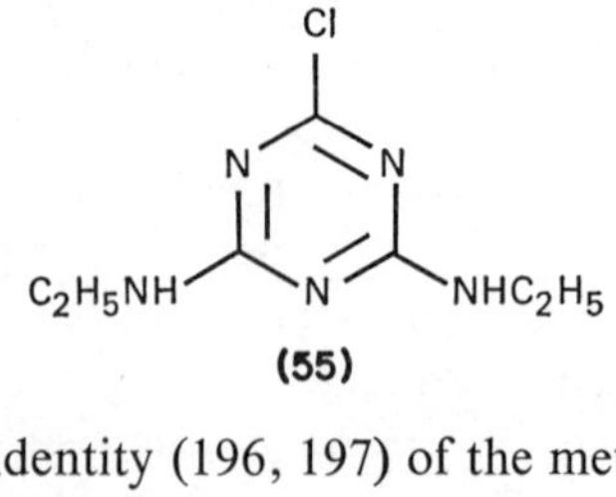

(55) 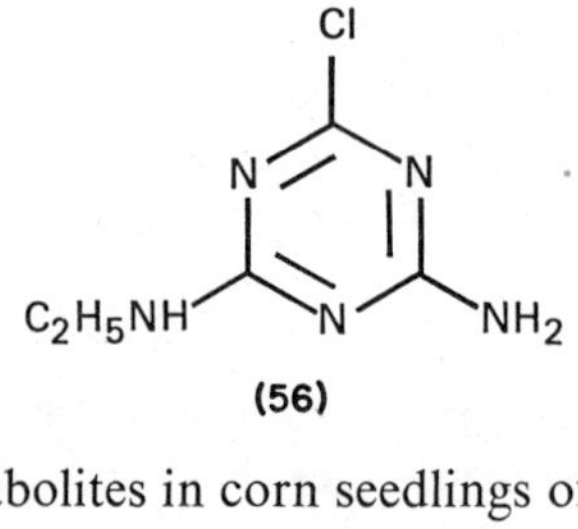   (56)

The identity (196, 197) of the metabolites in corn seedlings of the herbicide diuron (**57**) as 3,4-dichloronitrobenzene, 3-(3,4-dichlorophenyl)-1-methyl urea, and 3,4-dichloroaniline has been confirmed by mass spectrometry during a quantitative study (198) of the problem. The mechanism

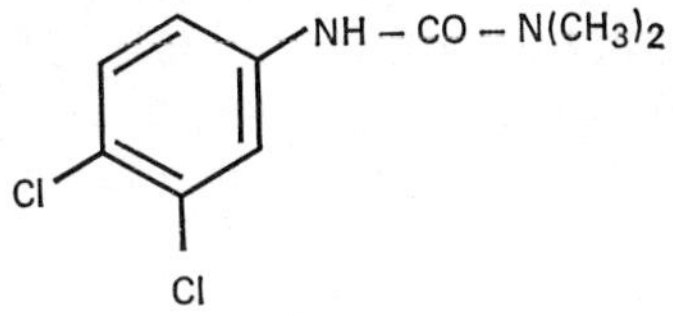

(57)

by which the deuterium-labeled DDT [1,1,1-trichloro-2,2-bis(p-chlorophenyl)-2-deutero-ethane, **58**] is converted by *Aerobacter aerogenes* to the deuterium-containing DDE [1,1-dichloro-2,2-bis(p-chlorophenyl)-2-deuteroethane, **59**] was established by mass spectrometry (199), and the olefin DDE (**60**) was precluded as an intermediate in this transformation.

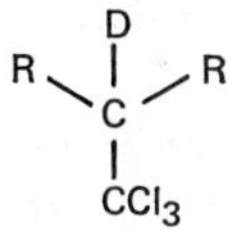

$$R = p{-}ClC_6H_4$$

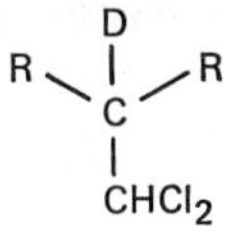

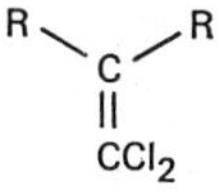

(58)   **(59)**   **(60)**

A major metabolite of the soil fungicide chloroneb (1,4-dichloro-2,5-dimethoxy benzene) has been identified by mass spectrometry (200) as the demethylated product 1,4-dichloro-2-methoxy phenol.

Indirect methods of pesticide disposal have received some attention. For example, DDE, when irradiated in air, is now known (201) to cyclize to the fluorenone (**61**), which was identified by mass spectrometry, and similarly,

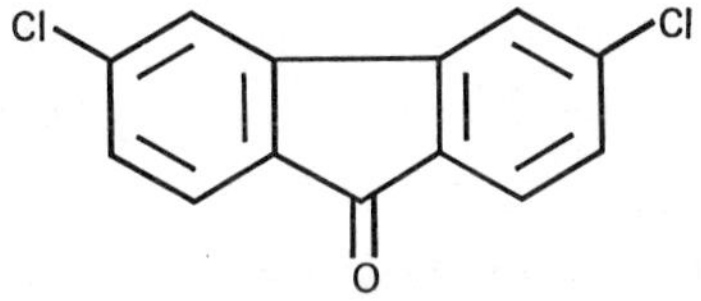

(6l)   **(62)**

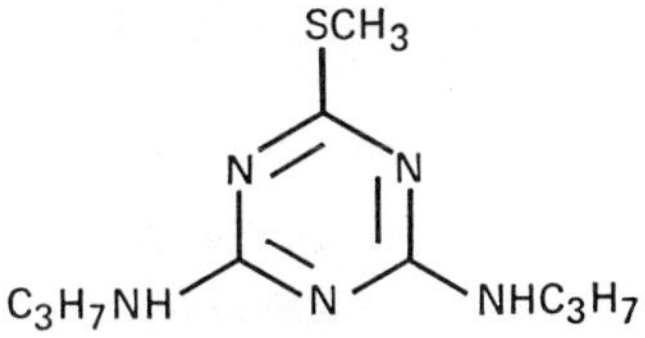

irradiation of the herbicide prometryn (**62**) leads to the formation (202) of the des-thio product **63**.

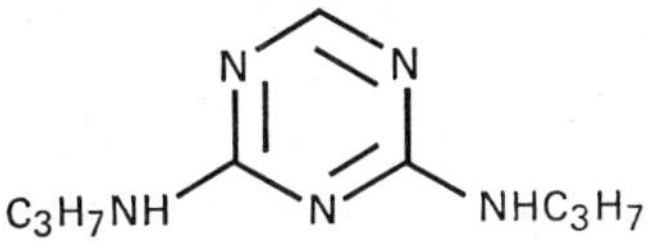

(63)

The mammalian toxicity of pesticides is of obvious importance, and some work has appeared on the metabolism of these compounds in animals. Thus the isolation of a metabolite of 1-(butylcarbamoyl)-2-benzimidazolecarbamate (64) from the urine of rats dosed with the fungicide was followed by its identification (203), using mass spectrometry, as the desbutylcarbamoyl compound (65). Similarly, two of the metabolites of the insecticide lindane

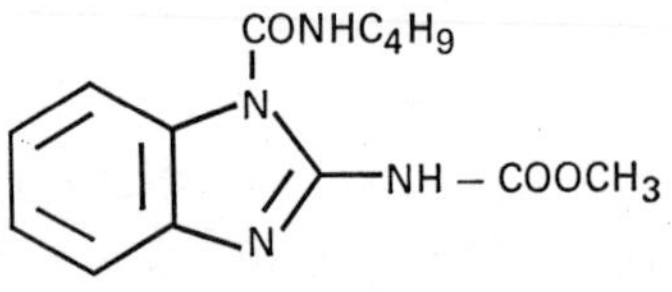

(64) 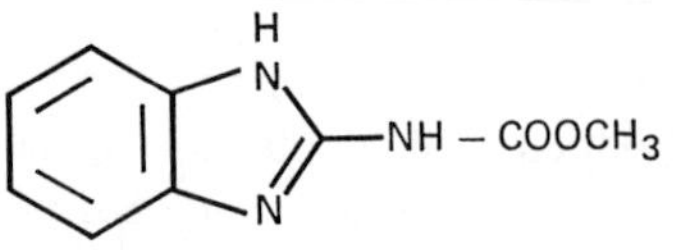   **(65)**

(1,2,3,4,5,6-hexachlorocyclohexane) in susceptible and resistant houseflies have been tentatively identified (204) by gas chromatography-mass spectroscopy as isomers of pentachlorocyclohexene.

Metabolism of the organophosphorus pesticides has been studied (205) with some success by gas-chromatographic techniques. Mass spectra of some 23 phosphorus-containing pesticides have been recorded (206), and the method has been suggested as a complement to the gas-chromatographic technique.

Work on pesticide metabolism is of somewhat recent origin, and the next few years may see considerable effort invested in this area. As in drug metabolism, the potential of mass spectrometry, particularly when coupled to gas chromatography, is enormous and is now recognized by workers in the area. It seems, therefore, that a great deal of published work may soon begin to appear from laboratories using these techniques.

XI. IDENTIFICATION OF BIOLOGICALLY SIGNIFICANT NATURAL MATERIALS

As is usual with new physical methods of chemical analysis, the initial most vigorous exploitation has been by organic chemists working with natural products. The annual number of papers in which mass spectrometry is used as a routine analytical technique has, in five years, grown far beyond the breadth of any review of the subject.

The use of mass spectrometry in organic chemistry tends to fall into two categories, those in which it constitutes a means of determining (or confirming) the molecular weight and/or the molecular formula and those in which it is used—often alone—as a method of structure determination. In the second case, the reward for a careful study of mass spectra is a high degree of sensitivity, permitting structural analysis at the submilligram level. This is more difficult to do if other techniques, such as nmr, are used, and almost impossible if chemical degradation is involved. It is felt by many practitioners that the mass spectrum alone, if correctly interpreted, should lead to a unique structure, but instances of this having been achieved are still relatively rare. This section will deal with some natural product structure determinations that have rested heavily, if not entirely, on mass spectrometry.

A very elegant demonstration of the power of mass spectrometry in *de novo* structure determination was reported by Penders and Arens (207), who isolated 3–6 mg of an unknown peptide from pig neurohypophysis (the posterior lobe of the pituitary). Hydrolysis gave tyrosine and valine, and from the high-resolution mass spectrum of the dinitrophenyl derivative, a molecular formula of $C_{23}H_{26}N_4O_9$ was established. Similar identification of the major fragment ions led to the definitive formulation of the original peptide as O-acetyltyrosylvaline. The use of mass spectrometry in peptide sequencing is now feasible, as is described in Chapter 9, "Mass Spectrometry in Peptide Chemistry," but as has been pointed out (208), it is not as simple a proposition as some mass spectrometrists suggest, and it may be no more sensitive than some of the highly sensitive "wet" techniques that are available.

Much of the most satisfactory application of mass spectrometry to the structure determination of very small quantities of material has been in the area of insect secretions. Typical of the work is the characterization of the arthropod defensive secretion, glomerin (209), as the quinazolone (**66**).

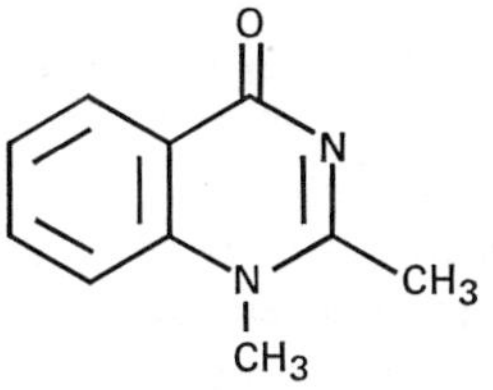

(66)

This structure follows directly from the low-resolution mass spectrum, which shows it to have a molecular weight of 174 and an $^{12}C/^{13}C$ ratio consistent with either $C_{10}H_{10}N_2O$ or $C_{10}H_8NO_2$, the latter being ruled out by the even mass of the molecular ion. The losses, supported by metastable ions, of units such as CH_3CN, HCN, CO, and so on permit only structure **66**, which was confirmed by IR and nmr spectra, and finally, total synthesis. The amount of natural product used here was not reported, but a few milligrams was clearly sufficient. Perhaps most impressive here is the amount of information that was derived from a low-resolution spectrum. This is often a classroom exercise, and it is reassuring to see it successfully applied in practice.

A structure determination in which mass spectrometry played an important role was that of the juvenile hormone (210), one of the factors necessary for the postembryonic development of the giant silkworm moth. The hormone gave a molecular ion at m/e 294 ($C_{18}H_{30}O_3$) and a mass spectrum compatible with a β-methyl-α,β-unsaturated methyl ester. Hydrogenation gave $C_{18}O_{36}O_2$, but if a poisoned catalyst was used, a mixture of compounds of molecular weight 296 and 298 was obtained. Both of these lost 87 amu ($C_5H_{11}O$) readily. In this way, the third oxygen was located at C_{10}. The positions of the double bonds were determined by OsO_4 oxidation and time averaged nmr studies, and the final structure (67) was ultimately confirmed by synthesis (211).

(67)

The ease with which materials can be isolated and identified by gas chromatography-mass spectroscopy is difficult to exaggerate. There are now many published reports of the successful use of this technique by workers in various disciplines such as botany, entomology, and so on. For example, a defensive secretion of the willow-feeding larva (*Coleoptera: chrysomelidae*) was, without any preparation, admitted via a gas chromatograph to a mass spectrometer, whereupon the only significant component was identified at once as salicylaldehyde (212). Similarly, gas chromatography-mass spectroscopy of the secretion from the scent gland of the larvae (*Hemiptera: Pyrrhocoridae*) permitted the identification (213) of *n*-dodecane, *n*-tridecane, *n*-pentadecane, *n*-hexanal, hex-2-en-1-al, hex-2-en-4-on-1-al, oct-2-en-1-al, and oct-2-en-4-on-1-al. Noncoupled gas chromatography-mass spectroscopy is perhaps more difficult, but has been used with success in problems of this sort. Thus nitrogen bases eliminated by the desert locust were analyzed and collected by gas chromatography and identified (214) by high-resolution mass spectrometry of their dansyl (5-dimethylaminonaphthalenesulfonyl) (215) derivatives as ammonia, dimethylamine, Δ^1-pyrroline, 1,4-diaminobutane, and spermine. The same group used coupled gas chromatography-mass spectroscopy to identify (216) the neutrals, acetone, methyl acetate, and ethyl acetate, from the same species. Gas chromatography was used to isolate a population aggregating pheronome from the Southern Pine Beetle (217), and mass spectrometry served to establish its molecular weight as 142, but no structure was proposed. The traces of C_2–C_{12} monocarboxylic acids produced in methane-water mixtures submitted to a corona discharge (possible prebiotic conditions) were analyzed by gas chromatography and their structures confirmed separately by mass spectrometry (218). The same techniques have been used to identify volatile components of tomatoes (219), soybeans (220), cranberries (221), and last, but certainly not least, whiskey (222) (Bourbon and rye).

An intriguing glimpse of a chemically unexplored area, the sea, is afforded by the work of Ashworth and Cormier (223), who isolated a blue-luminescent odiferous compound from the marine acorn worm and identified it as a phenol by ultraviolet spectroscopy, a dibromophenol by mass spectrometry, and 2,6-dibromophenol by nmr.

Paper chromatography of urine from patients with Parkinson's disease and from some schizophrenics gives, after spraying with Ehrlich's reagent, a pink spot not found in controls. This now notorious pink spot was considered by its discoverers (224) to be given by 3,4-dimethoxy-β-phenethyl-amine, but a subsequent report (225) included mass spectra which, it was suggested, were more compatible with p-hydroxyphenethylamine than with the dimethoxy compound. A good deal of IR and ultraviolet spectroscopic evidence was adduced in support of this conclusion, but it was reported in a still later paper (226) that a study of the mass spectrum of its dansyl derivative supported the original dimethoxyphenethylamine structure. It appears that both reports may be correct and that this confusion is symptomatic of the difficulties of working with extremely small amounts of unstable material. Finally, the detection (227) of very small amounts of the 3,4-dimethoxy-β-phenethylamine in tea, may reveal the pink spot to have been artifactual in any case.

Mass spectrometry has been shown (228) to be a useful technique for the analysis and identification of N-γ-glutamyl-2-deoxy-2-acetamido-β-D-glucosylamine derivatives that contain the hexosamine-aminoacid linkage thought to be important in many glycoproteins. The various mycobactins from *Mycobacterium phlei* have been shown to have the same nucleus but different side chains. Mass spectra have been obtained (229) from the aluminum complex of *Mycobacterium phlei*. The α-kansamycolones, isolated from *Mycobacterium kansasii*, were shown (230) by mass spectrometry to be ketones of the type R_1-CO-R_2, where R_1 is n-$C_{23}H_{47}$ and R_2 is a C_{53}, C_{55}, or C_{57} chain containing two cyclopropane rings.

In view of the vast amount of work of this sort that has been completed, it seems to be fair to say that mass spectrometry has found a place in organic chemical analysis. Whether it can fulfill its promise completely and supplant other, less sensitive, techniques, is doubtful, and the current trend appears to be to the increasing of the sensitivity of the other spectroscopic techniques. A serious weakness in present mass-spectrometric techniques is in the method of ion formation. Many of the molecules that are biologically interesting are not sufficiently volatile as they stand, and while the small highly polar molecules such as nucleotides can be dealt with by chemical modification (160–163), the higher-molecular-weight materials will probably require some radically different physical handling such as Dole's electrospraying technique (231) or Milne's molecular-beam approach (232). Deflection of ions of such a high mass (e.g., $\geqslant 10,000$) cannot now be done by any commercial instrument, and much development in this area will be necessary, involving perhaps superconducting solenoids or very large radius systems.

XII. CONCLUSION

In this chapter I have attempted a review of mass spectrometry with respect to its applicability and applications to medical and biochemical problems. Its applicability to both is high, but at the time of writing, applications, particularly to the former, are still being developed. The combination of technical and financial difficulties posed by a mass spectrometer in a clinical environment must, understandably, work against the acceptance of the technique. On balance however, just to judge from the examples quoted in this chapter, it would seem that mass-spectrometry systems rarely if ever fail to justify their expense in a research laboratory. A clinical laboratory with a large throughput is a different matter, but even here, the smaller, less costly spectrometers discussed with respect to gas analysis have established themselves, and it would seem that a simple, cheap, general-purpose mass spectrometer with limited capabilities might be of great value in the clinical environment. In other areas, expense is less of a problem, and one anticipates various instrumental developments. The great resolving power race is not yet over, unfortunately, although more and more attention is being paid to inlet systems, gas chromatographic coupling, and ionization techniques. Progress of this sort, coupled with research on the problems of very-high-molecular-weight materials, referred to above, promises an interesting future for mass spectrometry in the areas of medicine and biochemistry.

REFERENCES

1. F. A. Miller, A. Hemingway, A. O. Nier, R. T. Knight, E. B. Brown, and R. L. Varco, *J. Thoracic Surgery*, **20,** 714 (1950).
2. K. Biemann, *Life Sci. Space Res.*, **3,** 77 (1965).
3. A. B. Foster, *Lab. Practice*, **18,** 743 (1969).
4. F. A. J. M. Leemans and J. A. McCloskey, *J. Amer. Oil Chem. Soc.*, **44,** 11 (1967).
5. F. W. McLafferty, *Science*, **151,** 641 (1966).
6. V. G. Manusadzhyan, *Izv. Akad. Nauk. Arm. SSR Biol. Nauk.*, **18,** 69 (1965). B.A., **47,** 105847.
7. E. Stenhagen, *Chimia* (*Aarau*), **20,** 346 (1966).
8. G. H. Kydd and F. A. Hitchcock, *Fed. Proc.*, **8,** 89 (1949).

9. K. T. Fowler and P. Hugh-Jones, *Brit. Med. J.*, 1205 (1957).

10. J. S. Robertson, W. E. Siri, and H. B. Jones, *J. Clin. Invest.*, **29**, 577 (1950).

11. F. Siehoff, K. Muysers, and G. Worth, *Klinisches Wochenschrift*, **41**, 662 (1963).

12. F. Meade, N. Pearl, and M. J. Saunders, *Scand. J. Resp. Dis.*, **48**, 354 (1967).

13. K. Muysers, L. Delgmann, and U. Smidt, *Pfluger's Arch. Gesamte Physiol. Menschen Tiere*, **299**, 185 (1968).

14. C. E. Brion and W. B. Stewart, *Nature*, **217**, 946 (1968).

15. J. J. Osborn, J. O. Beaumont, J. C. A. Raison, J. Russell, and F. Gerbode, *Surgery*, **64**, 1057 (1968).

16. J. W. Brantigan, V. L. Gott, M. L. Vestal, G. L. Fergusson, and W. H. Johnston, *J. Appl. Physiol.*, **28**, 375 (1970).

17. S. Woldring, G. Owens, and D. C. Woolford, *Science*, **153**, 885 (1966).

18. G. Owens, L. Belmusto, and S. Woldring, *J. Neurosurg.*, **30**, 110 (1969).

19. R. G. Bickel, C. F. Diener, and H. L. Brammell, *Aerospace Med.*, **41**, 203 (1970).

20. J. Roman and W. H. Brigden, *Aerospace Med.*, **37**, 1213 (1969).

21. A. Wolstenholme, *Nature*, **203**, 1284 (1964).

22. C. A. Evans and G. H. Morrison, *Anal. Chem.*, **40**, 869 (1968).
    C. A. Evans, *Diss. Ab.*, *B*, **29**, 1950 (1968).

23. J. L. Hardwick and C. J. Martin, *Helv. Odont. Acta*, **11**, 62 (1967).

24. J. P. Yurachek, G. G. Clemena, and W. W. Harrison, *Anal. Chem.*, **41**, 1666 (1969).

25. I. A. Skul'skii and I. V. Burovina, *Dok. Akad. Nauk.*, **170**, 1185 (1966). B.A., **48** 120539.

26. S. Kluge and H.-J. Dietze, *Z. Naturforsch.*, **23b**, 1393 (1968).

27. O. H. Howard, *Am. Ind. Hyg. Assn. J.*, **29**, 355 (1968).

28. E. Weinig and P. Zink, *Arch. Toxikol.*, **22**, 255 (1967).

29. A. Costopanagiotis and H. Budzikiewicz, *Monatsh. Chem.*, **96**, 1800 (1965).

30. H.-F. Grützmacher and W. Arnold, *Tetrahedron Lett.*, 1365 (1966).

31. R. T. Coutts and R. A. Locock, *J. Pharm. Sci.*, **57**, 2096 (1968).

32. D. M. Baer, *Amer. J. Clin. Path.*, **44**, 114 (1965).

33. H. M. Fales and N. Law. Unpublished work. See H. M. Fales, G. W. A. Milne, and M. L. Vestal, *J. Amer. Chem. Soc.*, **91**, 3682 (1969).

34. G. Bohn and G. Rücker, *Arch. Toxikol.*, **23**, 221 (1968).

35. E. Weinig, G. Machbert, and P. Zink, *Arch. Toxikol.*, **22**, 115 (1966).

36. C. J. W. Brooks, *Biochem. J.*, **107**, 13p (1968).

37. H. Adlercreutz, *Abbh. Deut. Akad. Wiss. Berlin*, 121 (1967). Akad. Verlag (Berlin) 1969.

38. K. Sjövall, J. Sjövall, K. Maddock, and E. C. Horning, *Anal. Biochem.*, **14**, 337 (1966).

39. J. Å. Gustafsson and J. Sjövall, *Eur. J. Biochem.*, **8**, 467 (1969).

40. J. Sjövall and R. Vihko, *Acta Endocrin.*, **57**, 247 (1968).

41. O. Jänne and R. Vihko, *Ann. Med. Exp. Fenn.*, **46**, 301 (1968).

42. A. Aakvaag, *Acta Endocrin.*, Suppl., 119, 97 (1967); J. P. Rapp and K. B. Eik-Nes, *Anal. Biochem.*, **15**, 386 (1966).

43. A. Salokangas and H. Adlercreutz, *Acta Endocrin.*, Suppl. 119, 97 (1967); *Ann. Med. Exp. Fenn.*, **46**, 158 (1968).

44. H. Adlercreutz and T. Luukkainen, in Lipsett (Ed.), *Gas Chromatography of Steroids in Biological Fluids*, Plenum Press, N.Y., 1965, 215; H. Adlercreutz and T. Luukkainen in Eik-Nes and Horning (Eds.), *Gas Phase Chromatography of Steroids*, Springer-Verlag, Berlin.

45. H. Adlercreutz, A. Salokangas, and T. Luukkainen, *Mem. Soc. Endocrinol.*, **16**, 89 (1967).

46. T. Luukkainen and H. Adlercreutz, *Ann. Med. Exp. Fenn.*, **45**, 264 (1967).

47. H. Adlercreutz, C.-J. Johansson and T. Luukkainen, *Ann. Med. Exp. Fenn.*, **45**, 269 (1967).

48. H. Adlercreutz and T. Luukkainen, *Adv. Biosciences*, **3**, 53 (1969).

49. H. Adlercreutz, O. Jänne, T. Laatikainen, B. Lindström, T. Luukkainen, and R. Vihko, *Bull. Schweiz. Akad. Med. Wiss.*, in press (1969).

50. T. Luukkainen, C. A. Gemzell, and H. Adlercreutz, *Acta Endocrin.*, Suppl., 119, 75 (1967).

51. M. G. Horning, D. Rybiski, E. C. Horning, G. L. Jordan, Jr., and C. P. Schaffner, in *Drugs Affecting Lipid Metabolism*, Plenum Press (1969), p. 531.

52. C. C. Sweeley, L. H. Elliott, I. Fries, and R. Ryhage, *Anal. Chem.*, **38**, 1549 (1966).

53. C.-G. Hammar and B. Holmstedt, *Agressologie*, **9**, 109 (1968).

54. C. H. L. Shackleton, R. W. Kelly, P. M. Adhikary, C. J. W. Brooks, R. A. Harkness, P. J. Sykes, and F. L. Mitchell, *Steroids*, **12**, 705 (1968).

55. M. G. Horning, E. M. Chambaz, C. J. W. Brooks, A. M. Moss, E. A. Boucher, E. C. Horning, and R. M. Hill, *Anal. Biochem.*, **31**, 512 (1969).

56. J. Sjövall and K. Sjövall, *Steroids*, **12**, 359 (1968).

57. P. Eneroth, B. Gordon, R. Ryhage, and J. Sjövall, *J. Lipid Res.*, **7**, 511 (1966).

58. T. Luukkainen and H. Adlercreutz, *Biochim. Biophys. Acta*, **107**, 579 (1965).

59 A. L. Siegel, H. Adlercreutz, and T. Luukkainen, *Ann. Med. Exp. Fenn.*, **47**, 22 (1969).

60. H. Adlercreutz, M. Ikonen, and T. Luukkainen, *Scan. J. Clin. Lab. Invest.*, Suppl. 110, 128 (1969).

61. H. Adlercreutz, M. Ikonen, and T. Luukkainen, *Scan. J. Clin. Lab. Invest.*, Suppl. 108, 39 (1969).

62. H. Adlercreutz, M. Ikonen, and T. Luukkainen, *Proc. VIIth Int. Cong. Clin. Chem.* (1969).

63. E. Kuss, *Hoppe-Seyler's Z. Physiol. Chem.*, **349**, 1234 (1968).

64. H. Adlercreutz, O. Jänne, T. Laatikainen, B. Lindström, T. Luukkainen, and R. Vihko, *Bull. Schweiz. Akad. Med. Wiss.*, *1969*, in press.

65. H. Adlercreutz, T. Luukkainen, and W. Taylor, *Eur. J. Steroids*, **1**, 117 (1966).

66. E. H. Venning and J. S. L. Browne, *Proc. Soc. Exp. Biol.*, **34**, 792 (1936).

67. J. Sjövall, *Mem. Soc. Endocrinol.*, **16**, 243 (1967).

68. T. A. Miettinen, *Ann. Med. Exp. Fenn.*, **46**, 172 (1968).

69. J. F. Weiss, G. Galli, and E. G. Paoletti, *J. Neurochem.*, **15**, 563 (1968).

70. D. C. DeJongh, J. D. Hribar, P. Littleton, K. Fotherby, R. W. A. Rees, S. Shrader, T. J. Foell, and H. Smith, *Steroids*, **11**, 649 (1968); P. Littleton, K. Fotherby, and K. J. Dennis, *J. Endocrin.*, **42**, 591 (1968).

71. H. Funasaki and J. R. Gilbertson, *J. Lipid Res.*, **9**, 766 (1968).
72. See, for example, M. N. Galbraith, D. H. S. Horn, Q. N. Porter, and R. J. Hackney, *Chem. Comm.*, 971 (1968).
73. B. E. Gustafsson, J.-Å. Gustafsson, and J. Sjövall, *Acta Chem. Scand.*, **20**, 1827 (1966).
74. T. J. Scallen, W. J. Dean, E. D. Loughran, and B. V. Vora, *J. Lipid Res.*, **10**, 121 (1969).
75. P. D. G. Dean and R. T. Aplin, *Steroids*, **8** 565 (1966).
76. L. J. Schoenfield and J. Sjövall, *Acta Chem. Scand.*, **20**, 1297 (1966).
77. T. Suda, H. F. DeLuca, H. Schnoes, and J. W. Blunt, *Biochem. Biophys. Res. Commun.*, **35**, 182 (1969).
78. J. W. Blunt, H. F. DeLuca, and H. K. Schnoes, *Biochem.*, **7**, 3317 (1968).
79. A. Salaque, M. Barbier, and E. Lederer, *Comp. Biochem. Physiol.*, **19**, 45 (1966).
80. D. R. Idler and U. H. M. Fagerlund, *J. Amer. Chem. Soc.*, **77**, 4142 (1955).
81. B. A. Knights and W. Laurie, *Phytochem.*, **6**, 407 (1967).
82. D. R. Idler, A. Saito, and P. Wiseman, *Steroids*, **11**, 465 (1968).
83. L. L. Jackson and D. S. Frear, *Phytochem.*, **7**, 651 (1968).
84. L. N. Standifer, M. Devys, and M. Barbier, *Phytochem.*, **7**, 1361 (1968).
85. B. A. Knights, *Phytochem.*, **7**, 1707 (1968).
86. B. A. Knights, *J. Gas Chrom.*, **5**, 273 (1967).
87. K. Schubert, G. Rose, and C. Hörhold, *Biochim. Biophys. Acta*, **137**, 168 (1967).
88. M. Lenfant, R. Ellouz, B C. Das, E. Zissmann, and E. Lederer, *Eur. J. Biochem.* **7**, 159 (1969).
89. C. J. W. Brooks, E. C. Horning, and J. S. Young, *Lipids*, **3**, 391 (1968).
90. E. M. Chambaz, G. Maume, B. Maume, and E. C. Horning, *Anal. Lett.*, **1**, 749 (1968).
91. B. Maume, W. E. Wilson, and E. C. Horning, *Anal. Lett.*, **1**, 401 (1968).
92. J. G. Allen, G. H. Thomas, C. J. W. Brooks, and B. A. Knights, *Steroids*, **13**, 133 (1969)
93. H. M. Fales and T. Luukkainen, *Anal. Chem.*, **37**, 955 (1965).
94. R. Ryhage, *Anal. Chem.*, **36**, 759 (1964).
95. J. T. Watson and K. Biemann, *Anal. Chem.*, **36**, 1135 (1964).
96. J. Oro, D. W. Nooner, and S. A. Wikstrom, *J. Gas Chrom.*, **3**, 105 (1965).
97. H. Dannenberg and R. Richter, *Hoppe-Seyler's Z. Physiol. Chem.*, **349**, 565 (1968).
98. J. Avigan, G. W. A. Milne, and R. J. Highet, *Biochim. Biophys. Acta*, **144**, 127 (1967).
99. H. J. O'Neill, L. L. Gershbein, and R. Y. Scholz, *Biochem. Biophys. Res. Commun.*, **35**, 946 (1969).
100. P. P. Nair and Z. Luna, *Arch. Biochem. Biophys.*, **127**, 413 (1968).
101. R. F. Vesonder, L. J. Wickersham, and W. K. Rohwedder, *Can. J. Chem.*, **46**, 2628 (1968).
102. E. Gelpi and J. Oro, *J. Am. Oil. Chem. Soc.*, **45**, 144 (1968).
103. J. Skorepa, P. Hrabak, P. Mares, and A. Linnarson, *Biochem. J.*, **107**, 318 (1968).
104. J. A. McCloskey and M. J. McClelland, *J. Amer. Chem. Soc.*, **87**, 5090 (1965).
105. C. J. Argoudelis and E. G. Perkins, *Lipids*, **3**, 379 (1968).
106. C. B. Johnson, *N. Z. J. Sci.*, **12**, 27 (1969).

107. J. A. McCloskey, R. N. Stillwell, and A. M. Lawson, *Anal. Chem.*, **40**, 233 (1968).

108. P. Capella and C. M. Zorzut, *Anal. Chem.*, **40**, 1458 (1968).

109. G. Odham, *Arkiv. Kemi*, **25**, 543 (1966).

110. G. Odham, *Fette Seifen Anstrichm.*, **69**, 164 (1967).

111. E. O. A. Haahti and H. M. Fales, *J. Lipid Res.* **8**, 131 (1967).

112. A. Dolev, W. K. Rohwedder, T. L. Mounts, and H. J. Dutton, *Lipids*, **2**, 33 (1967).

113. J. H. Baxter and G. W. A. Milne, *Biochim. Biophys. Acta*, **176**, 265 (1969).

114. C. E. Mize, J. Avigan, D. Steinberg, R. C. Pittman, H. M. Fales, and G. W. A. Milne, *Biochim. Biophys. Acta*, **176**, 720 (1969).

115. C. E. Mize, D. Steinberg, J. Avigan, and H. M. Fales, *Biochim. Biophys. Res. Commun.*, **25**, 359 (1966).

116. D. Steinberg, J. Avigan, C. E. Mize, J. H. Herndon, Jr., H. M. Fales, and G. W. A. Milne, *Path. Europ.*, **3**, 450 (1968).

117. E. G. Trams, H. M. Fales, and A. E. Gal, *Biochem. Biophys. Res. Commun.*, **31**, 973 (1968).

118. E. G. Trams, W. L. Stahl, and J. Robinson, *Biochim. Biophys. Acta*, **163**, 472 (1968).

119. M. W. Goldblatt, *Chem. Ind.*, **52**, 1056 (1933); U.S. v. Euler, *Arch. Exp. Path. Pharmakol.*, **175**, 78 (1934).

120. S. Bergström, R. Ryhage, B. Samuelsson, and J. Sjövall, *Acta Chem. Scand.*, **16**, 501 (1962); *J. Biol. Chem.*, **238**, 3555 (1963).

121. S. Bergström and B. Samuelsson, *Ann. Rev. Biochem.*, **34**, 101 (1965).

122. M. Hamberg and B. Samuelsson, *Biochem. Biophys. Res. Commun.*, **34**, 22 (1969).

123. E. G. Daniels, J. W. Hinman, B. E. Leach, and E. E. Muirhead, *Nature*, **215**, 1298 (1967).

124. B. Samuelsson, *Prog. Biochem. Pharmacol.*, **3**, 59 (1967).

125. F. Vane and M. G. Horning, *Anal. Lett.*, **2**, 357 (1969).

126. R. F. Muraca, J. S. Whittick, G. D. Davies, Jr., P. Friis, and K. Folkers, *J. Amer. Chem. Soc.*, **89**, 1505 (1967).

127. H. Morimoto, T. Shima, I. Imada, M. Sasaki, and A. Ouchida, *Justus Liebig's Ann. Chem.*, **702**, 137 (1967).

128. D. Misiti, H. W. Moore, and K. Folkers, *J. Amer. Chem. Soc.*, **87**, 1402 (1965).

129. B. C. Das, M. Lounasmaa, C. Tendille, and E. Lederer, *Biochem. Biophys. Res. Commun.*, **21**, 318 (1965).

130. W. T. Griffiths, *Biochem. Biophys. Res. Commun.*, **25**, 596 (1966).

131. P. J. Rietz, F. S. Skelton, and K. Folkers, *Int. Z. Vitaminforsch.*, **37**, 405 (1967).

132. B. S. Strauch, H. M. Fales, R. C. Pittman, and J. Avigan, *J. Nutr.*, **97**, 194 (1969).

133. H. A. Lloyd, E. A. Sokoloski, B. S. Strauch, and H. M. Fales, *Chem. Commun.*, 299 (1969).

134. D. E. Minnikin and N. Polgar, *J. Chem. Soc.*, (C), 2107 (1966).

135. H. Demarteau-Ginsberg, E. Lederer, R. Ryhage, S. Ställberg-Stenhagen, and E. Stenhagen, *Nature*, **183**, 1117 (1959).

136. E. Gelpi, J. Oro, H. J. Schneider, and E. O. Bennett, *Science*, **161**, 700 (1968).

137. R. F. Struck, J. L. Frye, and Y. F. Shealy, *J. Agr. Food Chem.*, **16**, 1028 (1968).

138. A. G. Douglas, K. Douraghi-Zadeh, and G. Eglinton, *Phytochem.*, **8**, 285 (1969).

139. G. Eglinton and D. H. Hunneman, *Phytochem.*, **7**, 313 (1968)

140. J. L. Laseter, D. J. Weber, and J. Oro, *Phytochem.*, **7**, 1005 (1968).

141. E. Bursell and A. N. Clements, *J. Insect. Physiol.*, **13**, 1671 (1967).

142. A. Greichus and Y. A. Greichus, *Exp. Parasitol.*, **21**, 47 (1967).

143. W. Grosch, *Z. Lebensmittel Unterforsch.*, **137**, 216 (1968).

144. G. Eglinton, A. G. Douglas, J. R. Maxwell, J. N. Ramsay, and S. Ställberg-Stenhagen, *Science*, **153**, 1133 (1966).

145. P. Haug, H. K. Schnoes, and A. L. Burlingame, *Science*, **158**, 772 (1967).

146. P. Haug, H. K. Schnoes, and A. L. Burlingame, *Chem. Commun.*, 1130 (1967).

147. P. Haug, H. K. Schnoes, and A. L. Burlingame, *Geochim. Cosmochim. Acta*, **32**, 358 (1968).

148. A. L. Burlingame and B. R. Simoneit, *Science*, **160**, 531 (1968).

149. G. Eglinton, D. H. Hunneman, and K. Douraghi-Zadeh, *Tetrahedron*, **24**, 5929 (1968).

150. A. L. Burlingame, M. Calvin, J. Han, W. Henderson, W. Reed, and B. R. Simoneit, *Science*, **167**, 751 (1970); W. G. Meinschein, E. Cordes, and V. J. Shiner, Jr., *Science*, **167**, 753 (1970); R. C. Murphy, G. Preti, M. M. Nafissi-V, and K. Biemann, *Science*, **167**, 755 (1970); P. I. Abell, G. H. Draffan, G. Eglinton, J. M. Hayes, J. R. Maxwell, and C. T. Pillinger, *Science*, **167**, 757 (1970); C. Ponnamperuma, K. Kvenvolden S. Chang, R. Johnson, G. Pollock, D. Philpott, I. Kaplan, J. Smith, J. W. Schopf, C. Gehrke, G. Hodgson, I. A. Breger, B. Halpern, A. Duffield, K. Krauskopf, E Barghoorn, H. Holland, and K. Keil, *Science*, **167**, 760 (1970); J. Oro, W. S. Updegrove, J. Gilbert, J. McReynolds, E. Gil-av, J. Ibanez, A. Zlatkis, D. A. Flory, R. L. Levy and C. Wolf, *Science*, **167**, 765 (1970); B. Nagy, C. M. Drew, P. B. Hamilton, V. E. Modzeleski, M. E. Murphy, M. M. Scott, H. C. Urey, and M. Young, *Science* **167**, 770 (1970); and S. R. Lipsky, R.J. Cushley, C. G. Horvath, and W. J. McMurray, *Science*, **167**, 778 (1970).

151. R. Ryhage, *J. Dairy Res.*, **34**, 115 (1967).

152. A. J. Polito, T. Akita, and C. C. Sweeley, *Biochem.*, **7**, 2609 (1968).

153. A. J. Polito, J. Naworal, and C. C. Sweeley, *Biochem.*, **8**, 1811 (1969).

154. K.-A. Karlsson, *Acta Chem. Scand.*, **19**, 2425 (1965).

155. L. L. Bieber, J. D. O'Connor, and C. C. Sweeley, *Biochim. Biophys. Acta*, **187**, 157 (1969).

156. S. R. Thorpe and C. C. Sweeley, *Biochem.*, **6**, 887 (1967).

157. B. Samuelsson and K. Samuelsson, *J. Lipid Res.*, **10**, 41 (1969).

158. B. Samuelsson and K. Samuelsson, *J. Lipid Res.*, **10**, 47 (1969).

159. G. Casparrini, M. G. Horning, and E. C. Horning, *Anal. Lett.*, **1**, 481 (1968).

160. C. C. Sweeley and G. Dawson, *Biochem. Biophys. Res. Commun.*, **37**, 6 (1969).

161. J. H. Duncan, W. J. Lennarz, and C. C. Fenselau, *Biochemistry*, **10**, 927 (1971).

162. J. A. McCloskey, A. M. Lawson, K. Tsuboyama, P. M. Krueger, and R. N. Stillwell, *J. Amer. Chem. Soc.*, **90**, 4182 (1968).

163. D. F. Hunt, C. E. Hignite, and K. Biemann, *Biochem. Biophys. Res. Commun.*, **33**, 378 (1968).

164. J. J. Dolhun and J. L. Wiebers, *J. Amer. Chem. Soc.*, **91**, 7755 (1969)

165. M. A. Schwartz, P. Bommer, and F. M. Vane, *Arch. Biochem. Biophys.*, **121**, 508 (1967).

166. M. A. Schwartz, F. M. Vane, and E. Postma, *Biochem. Pharmacol.*, **17**, 965 (1968).

167. M. G. Horning, L. D. Waterbury, E. C. Horning, and R. M. Hill, *The Foeto-Placental Unit*, Proc. Int. Symp., Milan, 1968, p. 305.

168. K. H. Palmer, J. F. Feierabend, B. Baggett, and M. E. Wall, *J. Pharmacol. and Exp. Therap.*, **167**, 217 (1967).

169. K. H. Palmer, F. T. Ross, L. S. Rhodes, B. Baggett, and M. E. Wall, *J. Pharmacol. and Exp. Therap.*, **167**, 207 (1969).

170. D. L. Smith and M. F. Grostic, *J. Med. Chem.*, **10**, 375 (1967).

171. P. Goldman, G. W. A. Milne, and M. T. Pignataro, *Arch. Biochem. Biochem. Biophys.*, **118**, 178 (1967).

172. G. W. A. Milne, P. Goldman, and J. L. Holtzman, *J. Biol. Chem.*, **243**, 5374 (1968).

173. F. A. White, *Mass Spectrometry in Science and Technology*, Wiley, New York, 1968, p. 317 ff.

174. S. Meyerson and E. K. Fields, *Science*, **166**, 325 (1969).

175. M. Tanabe and G. Detre, *J. Amer. Chem. Soc.*, **88**, 4515 (1966).

176. H. Knöppel and W. Beyrich, *Tetrahedron Lett.*, 291 (1968).

177. J. L. Occolowitz, *Chem. Commun.*, 1226 (1968).

178. G. R. Waller, R. Ryhage, and S. Meyerson, *Anal. Biochem.*, **16**, 277 (1966).

179. D. Rittenberg, A. S. Keston, F. Rosebury, and R. Schoenheimer, *J. Biol. Chem.*, **127**, 291 (1939).

180. I. I. Tovarova, E. Ya. Kornitskaya, S. A. Pliner, V. A. Puchkov, N. S. Wul'fson, and A. S. Khokhlov, *Izv. Akad. Nauk. USSR.*, Ser. Biol., **6**, 911 (1966). *B.A.* **48**, 97631.

181. P. D. Shaw and J. A. McCloskey, *Biochem.* **6**, 2247 (1967).

182. P. Goldman and G. W. A. Milne, *J. Biol. Chem.*, **241**, 5557 (1966).

183. S. Kobayashi, S. Kuno, N. Itada, O. Hayaishi, S. Kozuka, and S. Oae, *Biochem. Biophys. Res. Commun.*, **16**, 556 (1964).

184. J. B. Lombardini, T. P. Singer, and P. D. Boyer, *J. Biol. Chem.*, **244**, 1172 (1969).

185. H. Nakano, C. Takemoto, H. Sato, and B.-I. Tamaoki, *Biochim. Biophys. Acta*, **152**, 186 (1968).

186. H. Nakano, H. Sato, and B.-I. Tamaoki, *Biochim. Biophys. Acta*, **164**, 585 (1968).

187. B. Samuelsson, *J. Amer. Chem. Soc.*, **87**, 3011 (1965).

188. J. L. Holtzman, J. R. Gillette, and G. W. A. Milne, *J. Biol. Chem.*, **242**, 4386 (1967); *J. Amer. Chem. Soc.*, **89**, 6341 (1967).

189. D. M. Jerina, J. W. Daly, B. Witkop, P. Zaltzman-Nirenberg, and S. Udenfriend, *J. Amer. Chem. Soc.*, **90**, 6525 (1968).

190. G. Guroff, C. A. Reifsnyder, and J. Daly, *Biochem. Biophys. Res. Commun.*, **24**, 720 (1966).

191. J. Daly, D. Jerina, and B. Witkop, *Arch. Biochem. Biophys.*, **128**, 517 (1968).

192. D. M. Jerina, J. W. Daly, and B. Witkop, *J. Amer. Chem. Soc.*, **90**, 6523 (1968).

193. L. M. Jackman, I. G. O'Brien, G. B. Cox, and F. Gibson, *Biochim. Biophys. Acta*, **141**, 1 (1967).

194. U. Zimmerman and K. Wagener, *Z. Naturforsch*, **22b**, 707 (1967).

195. P. C. Kearney, D. D. Kaufman, and T. J. Sheets, *J. Agr. Food Chem.*, **13**, 369 (1965).

196. R. L. Dalton, A. W. Evans, and R. C. Rhodes, *Proc. 18th Meeting SWC.*, Dallas, Tex., 1965, pp. 72–8.

197. H. Geissbuhler, C. Haselbach, H. Aebi, and L. Ebner, *Weed Res.*, **3**, 277 (1963).

198. J. H. Onley, G. Yip, and M. H. Aldridge, *J. Agr. Food Chem.*, **16**, 426 (1968).

199. J. R. Plimmer, P. C. Kearney, and D. W. Von Endt, *J. Agr. Food Chem.*, **16**, 594 (1968).

200. W. K. Hock and H. D. Sisler, *J. Agr. Food Chem.*, **17**, 123 (1969).

201. J. R. Plimmer and U. I. Klingebiel, *Chem. Commun.*, 648 (1969).

202. J. R. Plimmer, P. C. Kearney, and U. I. Klingebiel, *Tetrahedron Lett.*, 3891 (1969)

203. J. A. Gardiner, R. K. Brantley, and H. Sherman, *J. Agr. Food Chem.*, **16**, 1050 (1968).

204. W. T. Reed and A. J. Forgash, *Science*, **160**, 1232 (1968).

205. L. Giuffrida, N. F. Ives, and D. C. Bostwick, *J. Assn. Offic. Anal. Chem.*, **49**, 8 (1966).

206. J. Damico, *J. Assn. Offic. Anal. Chem.*, **49**, 1027 (1966).

207. Th. J. Penders and J. F. Arens, *Exp.*, **22**, 722 (1966).

208. K. L. Agarwal, G. W. Kenner, and R. C. Sheppard, *J. Amer. Chem. Soc.*, **91**, 3096 (1969).

209. H. Schildknecht and W. F. Wenneis, *Z. Naturforsch.*, **21b**, 552 (1966).

210. H. Röller, K. H. Dahm, C. C. Sweeley, and B. M. Trost, *Angew. Chem.*, **79**, 190 (1967). *Intern. Ed. Engl.*, **6**, 179 (1967).

211. K. H. Dahm, B. M. Trost, and H. Röller, *J. Amer. Chem. Soc.*, **89**, 5293 (1967)

212. J. B. Wallace and M. S. Blum, *Ann. Ent. Soc. Amer.*, **62**, 503 (1968).

213. D. H. Calam and A. Youdeowei, *J. Insect Physiol.*, **14**, 1147 (1968).

214. M. M. Blight, *J. Insect Physiol.*, **15**, 259 (1969).

215. C. R. Creveling, K. Kondo, and J. W. Daly, *Clin. Chem.*, **14**, 302 (1968).

216. M. M. Blight, J. F. Grove, and A. McCormick, *J. Insect. Physiol.*, **15**, 11 (1969).

217. J. A. A. Renwick and J. P. Vite, *Contribn. Boyce Thompson Inst.*, **24**, 65 (1968).

218. W. V. Allen and C. V. Ponnamperuma, *Curr. Mod. Biol.*, **1**, 24 (1967).

219. R. G. Buttery and R. M. Seifert, *J. Agr. Food Chem.*, **16**, 1053 (1968).

220. L. R. Mattick and D. B. Hand, *J. Agr. Food Chem.*, **17**, 15 (1969).

221. R. J. Croteau and I. S. Fagerson, *J. Food Sci.*, **33**, 386 (1968).

222. J. H. Kahn, E. G. LaRoe, and H. A. Conner, *J. Food Sci.*, **33**, 395 (1968).

223. R. B. Ashworth and M. J. Cormier, *Science*, **155**, 1558 (1967).

224. A. J. Friedhoff and E. van Winkle, *Nature*, **194**, 897 (1962).

225. A. A. Boulton, R. J. Pollitt, and J. R. Majer, *Nature*, **215**, 132 (1967).

226. C. R. Creveling and J. W. Daly, *Nature*, **216**, 190 (1967).

227. J. R. Stabenau, C. R. Creveling, and J. Daly, *Amer. J. Psychiatry*, **127**, 71 (1970).

228. L. Mester, A. Schimpl, and M. Senn, *Tetrahedron Lett.*, 1697 (1967).

229. A. J. White and G. A. Snow, *Biochem. J.*, **108**, 593 (1968).

230. A.-M. Miquel, B. C. Das, and A.-H. Etemadi, *Bull. Soc. Chim. Fr.*, 2342 (1966).

231. M. Dole, L. L. Mack, R. L. Hines, E. C. Mobley, L. D. Ferguson, and M. B. Alice, *J. Chem. Phys.*, **49**, 2240 (1968).

232. T. A. Milne, *Int. J. Mass Spec. Ion Physics*, **3**, 153 (1969).

Mechanism Studies of Fragmentation Pathways

MAURICE M. BURSEY AND MICHAEL K. HOFFMAN

*Venable Chemical Laboratory, The University of North Carolina,
Chapel Hill, North Carolina*

Le plus grand dérèglement de l'esprit, c'est de croire les choses parce qu'on veut qu'elles soient, et non parce qu'on a vu qu'elles sont en effet. BOSSUET

I. Introduction 373
II. Identification of Composition 375
III. Identification of Participating Structural Components 377
IV. Correlation with Empirical Models 387
V. Correlations with Theoretical Models 397
VI. Reactivities of Ions 407
References 411

I. INTRODUCTION

It is the purpose of this review to examine the methods by which mechanisms of reactions in the mass spectrometer may be explored.

Following the very early demonstrations of the utility of mass spectrometry for structure determination, popularizers of the technique began to adapt the electron-pushing methods of organic chemists to rationalize mechanisms. These first studies often had only the backing of similarity in the reactivities of members of a homologous series, and the parallels between characterization of this functional-group reactivity and the first classification of organic compounds by functional groups in the nineteenth century are strong. The recognition of functional-group patterns in rearrangement processes emphasized the importance of being able to predict such events, and so the value of mechanistic interpretations of mass spectra grew, not only in the eyes of the avid students of mass spectrometry, but also in the eyes of physical-organic chemists who followed events in the field from time to time.

Many of the interested generalists, caught up in the avalanche of descriptive material published in the late 1950s and early 1960s, which built on itself—as any rationalization must rightly do if there is to be progress—became less and less aware of the essentially empirical nature of the correlations and the *ex post facto* nature of the rationalizations proposed for them. Emphasis on this sort of explanation has led oftentimes to structures in the literature for which only the most meager evidence exists. An accumulation of such structures, and the tolerant acceptance of them for analytical purposes of structure determination, led to a belief by much of the world that there was really firm evidence for them.

There is, in fact, only slight evidence for most of the structures postulated, and one could anticipate that a person could build a career on disproving most of the straw men in the literature.

Consider then the difficulty with mechanistic mass spectrometry. If we do not know the structure of the reactant ion, nor the structures of the products, how can we evaluate the pathway between them?

The answer lies in considering the fact that parts of the mechanism may be evaluated with the knowledge of parts of the structural information to be learned in principle about the system. We draw an analogy to the information about, say, the hydrolysis of ethyl iodide, for which an organic chemist finds himself satisfied with a well-developed body of information. Information from kinetics suggests bimolecularity, that from optical studies in appropriately analogous systems suggests inversion at carbon, and the enthalpy and entropy of activation at different temperatures can be correlated with other model reactions. A satisfactory understanding is available. But not for everyone, for the establishment of atomic locations in the activated complex, the contributions of pathways with distortions, the involvement of specific vibrational and rotational states of the ethyl iodide, water, and products, and how modifications in solvent change all of these things are all unknown features of the mechanism, and not trivial ones either. The point is that actually few mechanisms, those under study by the molecular kineticists, are well understood, although the solution man thinks he understands them well enough.

The gaseous ion is by virtue of its nature of generation more difficult of access, more likely to decompose in a surfeit of ways to gaseous ionic tars, and thus less attractive to those enamored of ground-glass flasks. But the problems faced by mechanicians in mass-spectral studies are not much different in principle from those of the organic chemist who works with solutions. True, even less is known about these chemical systems, and less can probably be known, at least with the analytical instrumentation on hand today. Yet at least some structural aspects of the decomposition of an ion can be worked out, and these constitute mechanistic information.

We must look at the kinds of experiments we can do, either with a mass spectrometer or the equivalent of pencil and paper, to evaluate this information. But first we note that there are problems in defining the structures of starting materials and products. For example, in some experiments we observe products of a decomposition and ascribe some deduced property to the ion reacting. We must keep in mind that the property might belong to the ion as it reacts, the transition state, not necessarily the ion as it originally is formed. Further, reacting ions may be different in structure from those that do not fragment; indeed, the different structure might be the cause of their lack of reactivity. We must choose terms carefully, and further, must be satisfied sometimes to know more about the activated complex than about what preceded it.

II. IDENTIFICATION OF COMPOSITION

The first step of any classical structural determination has been the determination of the empirical formulas of isolable species associated with a reaction. Surely the same should apply in mass-spectrometric studies as well, to the extent that one can determine the formulas of ions associated with a process. Assumptions of composition are never entirely safe: consider the recent observation that the [M − 30] peak in the spectrum of nitrobenzene is not due solely to the loss of NO from the molecular ion [1]

$$C_6H_5NO_2^{+\cdot} \rightarrow C_6H_5O^+ + NO\cdot \qquad [1]$$

but that a fraction of the ion current at this m/e ratio is apparently $C_6H_7N^{+\cdot}$ and may be explained by the production of aniline in the source of the instrument from nitrobenzene by an unknown reductant, possibly water [2] (1).

$$C_6H_5NO_2 \rightarrow C_6H_5NH_2 \rightarrow C_6H_5NH_2^{+\cdot} \qquad [2]$$

Neglecting to determine this composition could have caused confusion in the interpretation of appearance potentials derived by some methods.

There are two important methods for the determination of molecular formulas of ions, both of which are discussed in satisfactory detail in introductory textbooks (2–4). They are briefly reviewed below.

With a low-resolution instrument, empirical compositions can be determined (or approximated as a function of reproducibility and mass discrimination by the instrument) by the analysis of isotopic composition. The statistical distribution of isotopes found on the planet permits the application of the binomial coefficient formula to the isotopic distribution of a set of peaks corresponding to one chemical formula. Following McLafferty (3), one finds for a typical organic compound containing C, H, N, and O, that the composition of an ion is determined by [3] and [4]:

$$\frac{[A + 1]}{[A]} = 0.011N_C + 0.0036N_N \qquad [3]$$

$$\frac{[A + 2]}{[A]} = 0.000060N_C^2 + 0.0020N_O \qquad [4]$$

where A is the nominal mass of the ion composed of the most abundant isotopes, $A + 1$ and $A + 2$ are the masses of those greater by one and two respectively, $[B]$ is the relative intensity of the peak at mass B (where B may be A, $A + 1$, or $A + 2$ in this case), and N_C, N_N, and N_O are respectively the number of carbon, nitrogen, and oxygen atoms contained in one ion. The distribution of peak intensities for chlorine- and bromine-containing ions is illustrated in several texts for most substitution patterns expected in common organic compounds (2, 3).

In practice, there are difficulties with this approach: the typical low-resolution mass spectrometer, designed to perform several kinds of experiments fairly well, may give peak heights reproducible to 1% for reasonably intense ions. If several peaks related by the loss or gain of hydrogen atoms cluster around the peak of interest, then corrections must be made for their contributions in terms of heavy isotopes to it. This in itself creates a problem, for if the peak of interest is of unknown composition, then the other peaks may be too. If the fragment peak to be analyzed by this method is actually a mixture of nominally similar peaks,* then the method fails, because the isotopic analysis will not make sense in terms of a single composition.

This method of analysis lends itself easily to computer analysis, and programs have been written for the calculation of isotope distribution patterns for species containing several different polyisotopic elements (5), and for the analysis of patterns in terms of a limited number of elements.

High-resolution data are obviously the key to a more general solution to the problem of elemental composition, for they may be interpreted directly from a knowledge of mass defects of isotopes constituting the ion (6). All of the practical problems associated with low-resolution spectral identification of composition can be answered by the high-resolution approach, provided that the resolution of the instrument is sufficiently great.

There is no question that computer handling of high-resolution data is a working answer to compiling it in a usable form; high-resolution data

---

* The nominal mass is the sum of the integers best approximating the atomic masses of the constituent isotopes of the ion.

have been handled with more publicity than low-resolution data as well. The printout has been arranged simply as a table of peak identities, or in some more digestible form. Methods of arranging data to simplify interpretation have included the element map (7) in its original form, in which ion compositions are arranged two-dimensionally with a rough indication of intensity; heteroatomic plots (8), a series of graphs illustrating intensities of peaks with a defined set and number of heteroatoms in sequence; and topographic element maps (9), in which a computer-drawn three-dimensional representation of the sequence of heteroatomic plots combines all of their information into a single illustration.

Errors can occur in the computer handling of high-resolution data, and it is of value to insure against improper assignment of peak identities by combination of the advantages of the high-resolution and low-resolution approaches—that is, to use both isotopic distributions and mass defects to identify composition of ions. The combination has been reported (10), along with the application of anticipated fragmentation patterns as a check on the structure of the proposed ion. Rather than the use of anticipated fragmentation patterns, it would appear that the application of the learning-machine approach to spectral interpretation (11, 12) would be useful as an ancillary method here. Since the learning-machine approach has been used with roughly 90% success in structural identification of unknown organics by computer, the possibility of uncovering new routes of structural identification not based on present recognition of spectral correlations with composition exists.

III. IDENTIFICATION OF PARTICIPATING STRUCTURAL COMPONENTS

The most widely applied method for identifying which atoms are involved in a reaction in the mass spectrometer draws on a technique widely applied in solution chemistry. Just as isotopic tracer experiments have been successfully applied to the identification of atoms participating in solution mechanisms and to the elucidation of intermediates and activated complexes there, and just as deuterium and other heavy-atom isotope effects are today being used to pinpoint the more subtle aspects of solvolysis mechanisms, so labeling experiments are used to study mechanisms of mass-spectral decompositions. The most frequently used technique is that of deuterium labeling. It has been used to great advantage in defining the general geometry involved in elimination processes as well as rearrangements.

One of the most widely quoted early examples of deuterium labeling was the application to the identification of the migrating hydrogen in the typical β cleavage with hydrogen rearrangement in the spectra of carbonyl compounds—the McLafferty rearrangement. The spectra of sec-butyl-2-d and -3-d acetate show that the rearrangement of one hydrogen can be accounted for by a six-membered cyclic transition state, and that when a second hydrogen is transferred at least the β-hydrogen must be implicated [5, 6] (13). These data do not, of course, apply to the

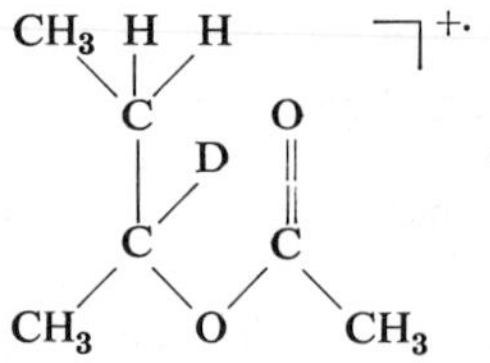

m/e 60, 0.88%

$\longrightarrow$ m/e 61, 2.20%; m/e 62, 1.62%   [5]

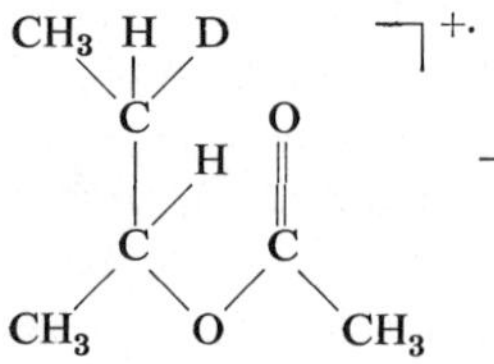

m/e 60, 0.85%

$\longrightarrow$ m/e 61, 2.94% ; m/e 62, 1.28%   [6]

postulated enolic form for the analogous reactions of ketones, just as they are ambiguous here with respect to a four- or six-membered transition state.

Another widely studied example, one in which deuterium labeling has been used to study rearrangements not specifically involving a hydrogen atom and therefore to be distinguished from hydrogen transfer like the previous case, is the $C_7H_7^+$ ion found in the spectra of most compounds containing a benzyl or a methylphenyl group. This ion conveniently decomposes by loss of acetylene to $C_5H_5^+$ [7], and the loss of labeled acetylene from labeled $C_7H_7^+$ allows some insight into the amount of rearrangement before fragmentation. Early studies on labeled ethylbenzenes and toluenes that produced a singly deuterated $C_7H_7^+$ ion showed a 2:5 loss of $C_2HD:C_2H_2$, which could be accommodated by a tropylium, not a benzyl, structure for the precursor $C_7H_7^+$ (14). More recent labeling work on the structure of doubly labeled $C_7H_7^+$ formed from benzyl-α-d_2 chloride and alcohol and ethyl-benzene-α-d_2 has been interpreted to mean that the positional identity of the two deuterons and five protons are lost before the $C_7H_7^+$ ion fragments further (15). The isotopic distributions of $C_5H_5^+$ formed from the decompositions of $C_7H_7^+$ produced from the mass-spectral decomposition of bibenzyl-$\alpha,\alpha,\beta,\beta$-$d_4$ support complete scrambling of hydrogens before fragmentation of $C_7H_7^+$, but the scrambling of the same ion from diphenyl-methane-α,α-d_2 supports tropylium-1,2-d_2 as the precursor of the $C_5H_5^+$ ion

(15). In the spectrum of bis(vinylcyclopentadienyl-α-d)iron, the $C_7H_7^+$, $C_7H_7Fe^+$, and $(C_7H_7)_2Fe^+$ ions each lose C_2H_2. The scrambling in each case is different; this observation suggests that rearrangement proceeds at a rate competitive with fragmentation in these cases and is influenced by the presence of the iron atom (16).

Of course, all of the results above for $C_7H_7^+$ could be explained equally well by hydrogen-atom migration and exchange on an intact (or slightly modified) carbon-atom skeleton. This objection can be met by using heavy-atom labeling to study the migrations of heavy atoms.

The mass spectrum of toluene-α-^{13}C still supports the postulate that the $C_7H_7^+$ ion from toluene has the symmetrical tropylium structure, for a nearly random loss of $C_2H_2/^{12}C^{13}CH_2$ was observed in the $C_7H_7^+ \rightarrow C_5H_5^+$ transition (17). Again, the singly ^{13}C-labeled compound does not differentiate between a tropylium ion arising from simple 1,2-insertion of the α-carbon [7] and a tropylium formed by random insertion of the α carbon between any two carbons of the benzene ring [8]. Rinehart and his students have presented

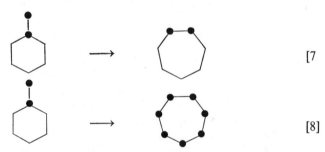

[7

[8]

preliminary data for the fragmentation of toluene-α,1-$^{13}C_2$; all of the carbon atoms appear to have lost their positional integrity before the $C_7H_7^+$ ion fragments to $C_5H_5^+$, an observation consistent with random insertion of the side chain into the ring (18).

We emphasize that carbon and deuterium labeling results do not necessarily lead to the same patterns; this point will be discussed later.

We return to a discussion of hydrogen migrations to illustrate another problem of interpretation; our example is that of cyclic transition states in rearrangement and elimination reactions of ions. The nature of the cyclic transition state in hydrogen rearrangements of aliphatic esters and amines has also been investigated by deuterium labeling techniques (19). Besides α-fission, the most characteristic reaction of ethers and amines is the hydrogen rearrangement pictured in [9]. The hydrogen transferred to the heteroatom was shown to originate from at least the α, β, γ, and δ positions of the original compounds; in fact, at least 75% of hydrogen transfer is that of a

$$\overset{\displaystyle H}{\overset{\displaystyle \frown}{R-C}}-\overset{+}{X}=C \rightarrow H-\overset{+}{X}=C \qquad [9]$$

hydrogen originally on the β position in some cases. Likewise, double hydrogen-transfer reactions of the esters of butyl alcohol were shown to involve transfers of hydrogen originally attached to every carbon atom of the butyl group; those attached to the β and γ positions of the original group were more frequently transferred [10] (20). This need not imply that the deuterium originally at a given position is still at that same position just before the transfer with elimination occurs.

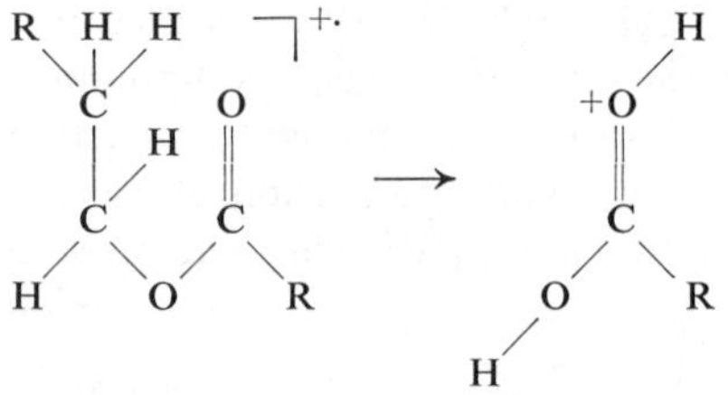

$$[10]$$

Similarly, the elimination of HX (X = OH, halogen) has been observed in mass-spectral decompositions. Dehydration of alcohols is a general process, occurring in primary, secondary, and tertiary aliphatic alcohols; in primary alcohols it is frequently important, and in secondary and tertiary alcohols it competes with α-cleavage of the C–C bond (21). The elimination of water was first assumed to parallel the process in solution, a 1,2-process; but it was shown that at high voltage the hydrogen lost was for the most part one that had been originally attached to the 4-position. The loss from 1-butanol-2,2-d_2 is <1% 1,2; from 1-pentanol-3,3-d_2, 6 ± 2% 1,3; from 1-hexanol-4,4-d_2, 90 ± 1% 1,4; from 1-heptanol-5,5-d_2, 5 ± 2% 1,5 elimination (22). The principal route is consonant with a six-membered transition state. On the other hand, dehydrohalogenation, a frequently observed reaction of alkyl halides, appears to proceed through a five-membered transition state: 1-chlorobutane-3,3-d_2 eliminates 93% DCl; 1-chloropentane-3,3-d_2, 72% DCl; and 1-chloropentane-4,4-d_2, 18% DCl (23). Our problem with these data is to interpret the significance of the apparent multiple sites of hydrogen transfer.

We may consider either that there are multiple sites for hydrogen abstraction, involving a series of three-membered, four-membered, five-membered, or six-membered transition states; or that there is one site and that hydrogen-atom migration up and down the alkyl chain occurs before the atom is actually transferred, with elimination of the neutral molecule. If the rate of hydrogen interchange on the alkyl chain is slow compared to the rate of hydrogen transfer, and if the hydrogen is transferred exclusively from one site on the chain, the deuterium labeling results are specific and easily interpreted; if the rate of hydrogen interchange on the alkyl chain is competitive with that of hydrogen transfer, then even though the transfer is site-specific, contributions from hydrogens originally attached to other positions will appear to a greater or lesser extent. Of course these are extreme pictures, and both may occur simultaneously.

The second hypothesis, that of specific transfer preceded by hydrogen migration along the chain, has experimental support (24). The rationale for the experiment is based on the understanding that if scrambling proceeds at a rate near that of fragmentation, then looking at isotopic distributions in products of fragmentation formed after two different periods of time will show decreasing scrambling with time, for if the ion decomposes slowly, it also has less energy, so that hydrogen scrambling on the ring is similarly less efficient. In fact, this experiment can be done with a high-resolution instrument, in which decompositions in the source may be compared with those in the first and second field-free regions (see Chapter 12, "Some Aspects of Metastable Transitions"). Results for various labeled isopropyl butyl ethers indicate that this is the case for hydrogen migration from the original α and δ positions; and a mechanism of hydrogen transfer involving at most transfer from the β and γ positions was proposed for the rearrangement in these ethers. This experiment in essence was an examination of the relative amounts of scrambling in ions containing more (normal ions) or less (metastable ions) internal energy. One could also carry out an experiment in which ionizing voltage was varied, comparing the amount of scrambling when high-energy ions decompose to that found for the decomposition of low-energy ions. Parallel results were in fact found here, although the reaction is almost independent of ionizing energy at high voltage for the isopropyl butyl ether example (24).

Hydrogen rearrangements between positions seem prominent at high ionizing voltage, and have been observed to increase at low voltage (25), probably as a general consequence of quasi-equilibrium theory (see below). Low-voltage studies of the McLafferty rearrangement of 2,2,4,4-d_4-heptanone-3 can be rationalized in a similar fashion (26, 27). Hence it is important not to extrapolate high-voltage results to low-voltage, since there are considerations that can cause substantial changes in the amount of rearrangement occurring.

The problem of hydrogen rearrangement in aliphatic systems would appear to complicate studies proposed to elucidate more subtle mechanistic details of abstraction of hydrogen or elimination of hydrogen-containing molecules from aliphatic molecules. Yet this is seemingly not the case, for data have been presented that support the retention of at least some stereochemical specificity in dehydration or dehalogenation reactions of cyclic and acyclic systems! In the decomposition of cyclohexanol and cyclohexyl chloride, the site selectivity for loss of H_2O and HCl was fairly similar, 1,4/1,3 elimination being in the ratio of about 70:30 after statistical correction for both compounds [11], [12] (28). Again, the apparent multiplicity of origins for the hydrogen transferred may be a disguised randomization of hydrogens before transfer (29). In spite of this problem, the 1,4 elimination appears to be nearly specific for the hydrogen *cis* to the hydroxyl group in the

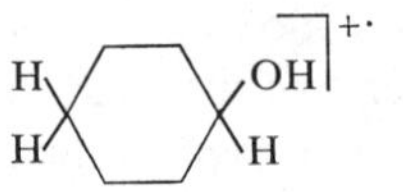

$$\text{[structure]} \longrightarrow C_6H_{10}^{+\cdot} + H_2O \qquad\qquad [11]$$

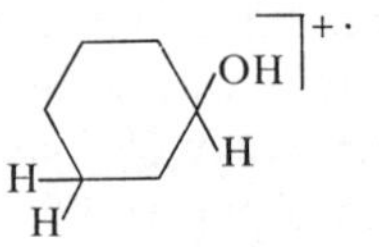

$$\text{[structure]} \longrightarrow C_6H_{10}^{+\cdot} + H_2O \qquad\qquad [12]$$

original molecule, the *cis/trans* ratio being about 80:20 (30). The apparent 1,3 dehydration seems to proceed either without stereospecificity through a common intermediate or else after scrambling; the 1,3 dehydration of *cis*-3,5-d_2-*cis*-cyclohexanol and of a mixture of this and *cis*-3,5-d_2-*trans*-cyclohexanol produces the same [M—H_2O]/[M—HOD] ratio, 4 at a nominal 9 eV, and 2.5 at 75 eV. On the other hand, the 1,3 dehydrohalogenation of *cis*-3,5-d_2-*cis*-cyclohexyl chloride and an approximately equimolar mixture of this and *cis*-3,5-d_2-*trans*-cyclohexyl chloride leads to different [M—HCl]/[M—DCl] ratios for the two cases, 1.85 and 3.70, respectively. It thus appears that here again the favored process, in this case 1,3 elimination, is fairly stereospecific, the *cis* elimination being favored (30).

Stereospecificity even in acyclic systems has been suggested, in the loss of DCl or HCl from the molecular ion of (S,R)-2-deuterio-4-chloropentane [13] and (S,S)-2-deuterio-4-chloropentane [14], which differ only in the interchange of a proton and a deuteron available for 1,3 elimination. The [M—HCl]/[M—DCl] ratio for the first compound was 2.69 and for the

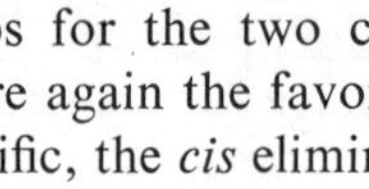

$$\text{[structure]} \longrightarrow C_5H_{10}^{+\cdot} + C_5H_9D^{+\cdot} \qquad\qquad [13]$$

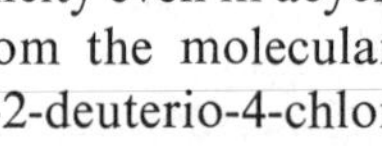

$$\text{[structure]} \longrightarrow C_5H_{10}^{+\cdot} + C_5H_9D^{+\cdot} \qquad\qquad [14]$$

second it was 2.86 at a nominal 9 eV; at 70 eV the difference between the two compounds was less (31). The fact that the results cluster around a value other than unity points to the existence of an isotope effect; the fact that the results differ for the two compounds was taken to mean that the two transition states **1** and **2** are of different energy, so that different amounts of the two

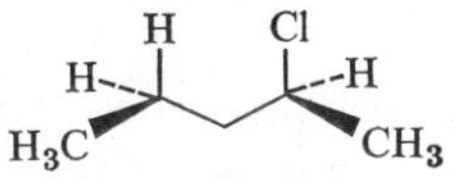

(1) 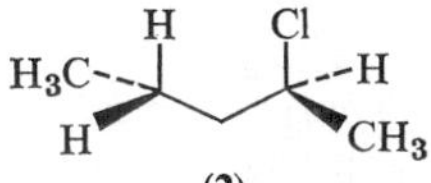 (2)

compounds proceed through them. Alternately, the data might be taken as slightly different isotope effects for the two species, reflecting slightly different extents of hydrogen transfer in the activated complexes. Either argument demands a stereochemical difference in the ions at the time of hydrogen transfer with elimination, of course.

This problem of interpreting results by two pictures requires that more information on mass-spectral isotope effects be collected. The definition of an isotope effect on the surface requires sorting out of different contributions, commonly called the π and Γ effects; these have been reviewed (32). A usable definition implying the experiments required is Djerassi's (33): atoms of deuterium per atom of protium transferred for the hypothetical case in which equal numbers of deuterium and protium are available. By this definition the isotope effect in the McLafferty rearrangement of γ-mono-, di-, and trideuterated methyl butyrates is 0.88 (33). In other words, k_H/k_D may be considered the inverse of this value, or 1.14, subject to the difficulty of defining rates in mass spectral reactions (see below). Further values for isotope effects in the McLafferty rearrangements of aliphatic ketones, aliphatic esters, aromatic compounds, and thiol esters vary between 0.5 and 1.0 atom D/atom H (34). An especially attractive study of mass-spectral isotope effects involved the mass-spectral aromatization of 4-deuterio-N-alkyl-3-cyano-1,4-dihydropyridines [15] (35). The value of k_H/k_D, taken as [M—1]/[M—2] at voltages so low that further decomposition was negligible, was found to be inversely related to the ionizing voltage; it was 1.25 at 20 eV and rose smoothly to 7.85 at about 11.2 eV,

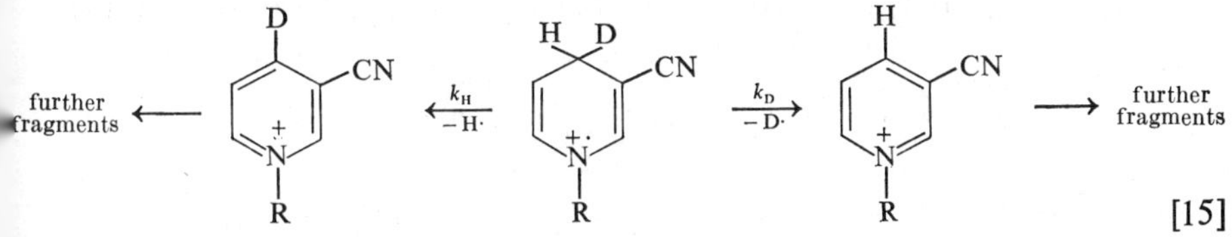

[15]

very near the appearance potential of the ion; the large value represents a quantal, rather than purely classical, isotope effect. Very large quantal effects then can operate at low ionizing energy when the difference between C–H and C–D zero point energies in the respective activated complexes will determine the isotope effect, particularly as the average internal energy of molecular ions approaches the activation energy for loss of D, the higher of the two.

Besides isotopic labeling with stable isotopes [and interesting examples of unstable isotopes in which their activity was detected by autoradiography of the photographic plate! (36)], one may also take advantage of the fact that a p-fluoro substituent in an aromatic ring alters the intensities of ions produced by fragmentation of bonds in other groups substituted on the ring very little; and so if one studies phenylated compounds, the p-fluoro substituent can produce a label, p-fluorophenyl, that functions in the same way as, for example, deuteriophenyl (37). The mass difference of 18 units is convenient for some kinds of studies, but of course the substitution of a different atom for another causes more complications than interchanging isotopes of the same element. One cannot compare the cracking pattern of hexaphenylbenzene, or a labeled hexaphenylbenzene, with that of benzene or a labeled benzene with the idea of making definitive statements about the cracking of benzene; but interesting comparisons can be made that in some cases seem to speak to the problem of interpreting deuterium scrambling in aromatic systems in terms of ring rearrangements. Scrambling observed in the fragmentation of partially deuterated benzene (38), pyridine (39), furan, and thiophene (40) has been suggested to proceed through the valence isomers postulated, and in some cases isolated, from photochemical studies of the neutral molecules. This is an attractive hypothesis, but the data could also be explained by hydrogen exchange on a carbon skeleton where positional integrity was maintained. Similar fragmentation routes for the phenylated compounds would tend to support valence-bond isomerization of the ring atoms themselves, since the migration of heavy-atom groups (41, 42) generally seems to follow different rules from that of hydrogen. Hydrogen/ deuterium labeling results for specifically labeled thiophenes suggested that the hydrogen atoms have equilibrated on the ring before the formation of HCS^+ (40). Before specifically labeled tetraphenylthiophene ions produce $C_6H_5CS^+$ or $FC_6H_4CS^+$, there is partial equilibration [16]; even further

$$\begin{array}{c} C_6H_5 \diagdown\diagup C_6H_5 \\ FC_6H_4 \diagup\hspace{-0.5em}\diagdown C_6H_4F \\ S \end{array}$$

[16]

$$[M—C_6H_6S]^{+\cdot} + [M—FCH_{65}S]^{+\cdot} \qquad\qquad [C_6H_5CS]^+ + [FC_6H_4CS]^+$$

$$1 \qquad : \qquad 1.1 \qquad\qquad\qquad 1 \qquad : \qquad 3$$

equilibration is found before the formation of other fragment ions. On the other hand, the formation of HCO$^+$ from specifically labeled furans is not preceded by equilibration, practically all of the hydrogen in the HCO$^+$ ion being that originally attached to positions 2 and 5 of the furan ring. Again there is a parallel in the phenylated system, for the labeled tetraphenylfuran produces the benzoyl ion and the benzoyl radical before much equilibration can occur [17] (43). The similarity of scrambling results in these two cases

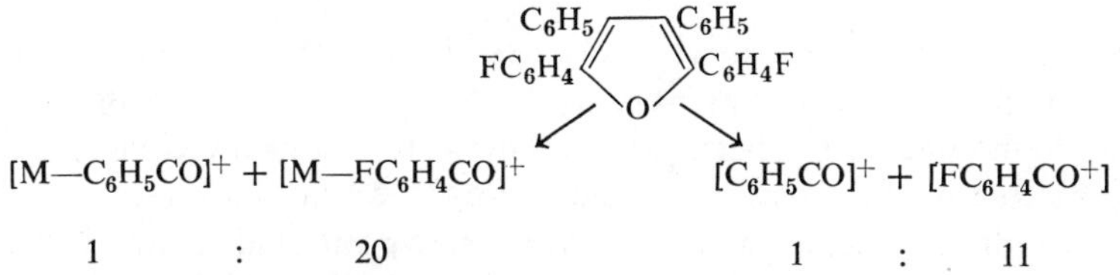

$$C_6H_5 \quad\quad C_6H_5$$
$$FC_6H_4 \diagup\!\!\!\!\diagdown\; C_6H_4F$$
$$O$$

$$[M{-}C_6H_5CO]^+ + [M{-}FC_6H_4CO]^+ \qquad\qquad [C_6H_5CO]^+ + [FC_6H_4CO^+] \qquad [17]$$

$$1 \quad : \quad 20 \qquad\qquad\qquad\qquad 1 \quad : \quad 11$$

would tend to support the suggestion of isomerization of the thiophene system through valence-isomer formation, rather than hydrogen scrambling on a carbon skeleton where atoms retain their positional identity.

Results for the six-membered rings, on the other hand, differ for the parent compound labeled with deuterium and the fully phenylated compound labeled with the *p*-fluoro group. Although there is equilibration in specifically labeled pyridine, the labeled pentaphenylpyridine shows results in accord with only a slight approach to equilibration, loss of C_6H_6 and FC_6H_5 from the [M—1] ion being heavily weighted in favor of the latter when there is little reason to expect any particular influence of the fluorine atom on intensity [18]. The pyrazine fragmentation even more strongly indicates a

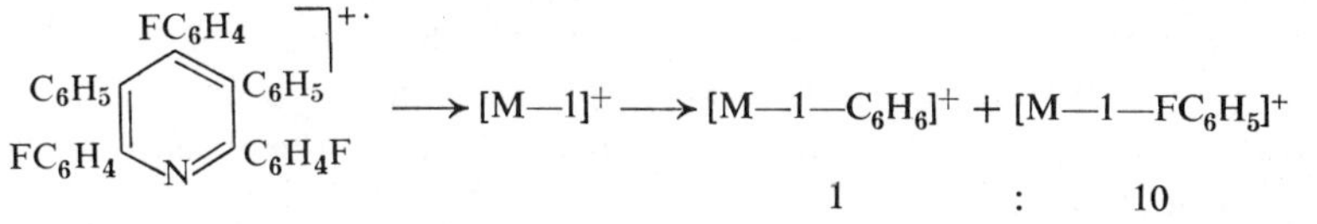

$$\begin{array}{c} FC_6H_4 \\ C_6H_5 \diagup\!\!\!\diagup\; C_6H_5 \\ FC_6H_4 \diagdown_N\diagdown C_6H_4F \end{array}^{+\cdot} \longrightarrow [M{-}1]^+ \longrightarrow [M{-}1{-}C_6H_6]^+ + [M{-}1{-}FC_6H_5]^+ \quad [18]$$

$$1 \quad : \quad 10$$

preference for fragmentation without substituent scrambling, since the monosubstituted product ion is formed in overwhelming preference [19].

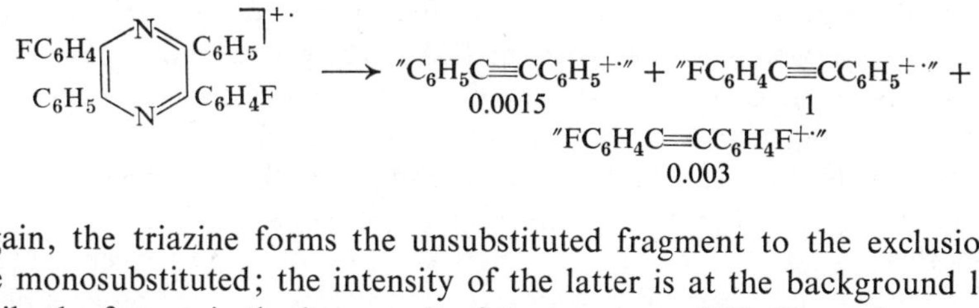

$$\begin{array}{c} FC_6H_4 \diagdown_N\diagdown C_6H_5 \\ C_6H_5 \diagdown_N\diagup C_6H_4F \end{array}^{+\cdot} \longrightarrow \;''C_6H_5C{\equiv}CC_6H_5^{+\cdot}{''} + {''}FC_6H_4C{\equiv}CC_6H_5^{+\cdot}{''} + \quad [19]$$
$$0.0015 \qquad\qquad\qquad 1$$
$$''FC_6H_4C{\equiv}CC_6H_4F^{+\cdot}{''}$$
$$0.003$$

Again, the triazine forms the unsubstituted fragment to the exclusion of the monosubstituted; the intensity of the latter is at the background level, while the former is the base peak of the spectrum [20]. These data suggest that almost no randomization of any sort occurs in the phenylated compounds. This means that either the phenyl substituent influences reactivity of the system to increase the rate of fragmentation relative to that of positional scrambling, or else that deuterium labeling in the simple compounds does

$$C_6H_5 \underset{C_6H_5}{\overset{N}{\bigcirc}} C_6H_4F^{+\cdot} \longrightarrow \text{only } ''C_6H_5C \equiv CC_6H_5^{+\cdot}'' \qquad [20]$$

not tell the whole story. The relative rates of scrambling and fragmentation in the thiophenes and furans were not strongly affected by going from the parent compound to the perphenylated derivative, and the same statement could be true of these systems; if that were the case, then these data would point to hydrogen exchange on a skeleton where positional identity is retained as the mechanism operating in pyridine molecular ions (44). But the results here raise more questions than they answer, and a series of ^{13}C labeling experiments would seem to be the answer with the most authority.

When phenylated compounds are significantly more stable than the parent compounds from which they are derived, the p-fluoro label is the simplest way to study scrambling in fragmentation. It has been applied to several systems where deuterium labeling would be difficult in the parent compound for a number of reasons. In these studies, $C_4Ar_4^{+\cdot}$ is produced by several reactions; it decomposes to $C_2Ar_2^{+\cdot}$, and the scrambling found in the product is dependent on the substitution pattern of the original molecule but essentially independent of the fragment eliminated to get to $C_4Ar_4^{+\cdot}$ from the molecular ion (45). Scrambling is nearly, but not quite, complete, each carbon of the C_4 group becoming attached to each of the others; in this the scrambling resembles that found for ^{13}C labeling to study scrambling before fragmentation of 1-butene (46). But the interpretation of the data illustrates the problems that must be considered even in the absence of those listed for deuterium labeling.

Is the scrambling occurring in $C_4Ar_4^{+\cdot}$ or in its molecular-ion precursor (47)? If it is occurring in the molecular ions, then quite different kinds of molecular ions are rearranging to the same extent, an extent determined only by the substitution pattern, not the class of compound or the size of the group lost or the energetics of formation of $C_4Ar_4^{+\cdot}$ or even the number of steps required to form $C_4Ar_4^{+\cdot}$. Molecular ions decomposing by another route, direct formation of $C_2Ar_2^{+\cdot}$, show scrambling ranging in extent from complete scrambling to less than 1% scrambling. Are the two processes completely uncoupled? Are the two sets of molecular ions unable to interact with each other, or does scrambling in one set say something in very general terms about the rate of scrambling to be expected before another process occurs at the same time scale in the other set?

Do scrambling data for molecular ions speak to the structure of all the molecular ions in the source? No; they give information only about the process undergone by those molecular ions with sufficient energy to decompose by a given route. If one uses metastable ions for scrambling studies, then he must speak only of the scrambling of a set of ions with only a limited range of energies, and the actual case on either side of this energy band could be different.

Do scrambling data for molecular ions of a certain energy speak to the structures of those ions? They speak only to the extent that they indicate the maximum scrambling through which an ion has passed before it fragments, and not necessarily to the actual symmetry of stable or metastable structures that the ion may possess before fragmentation. The scrambling data can then be reconciled with the symmetry of a transition state rather than the symmetry of a stable ionic form.

Do scrambling data for molecular ions of a certain energy speak to the structures of specific activated complexes for fragmentation? They speak only to the extent that randomization and fragmentation have similar transition states. If randomization is a multistep process prior to fragmentation, then randomization is described inaccurately by a single transition-state structure.

A single scrambling experiment is therefore liable to several interpretations, and needs to be supported by other kinds of evidence before much can be said about its interpretation. The identification of participating structural components is simple only when labeling results are clean and associated directly with the atom to which a bond is broken (or formed, in the case of rearrangements). Otherwise, several possibilities are open for the interpretation of the results, and further information is needed.

IV. CORRELATION WITH EMPIRICAL MODELS

Correlations of ionization potentials with empirical substituent constants such as the sigma constants for aromatic substituents (48) or the δ_K values for additive alkyl substituent effects (49) were observed somewhat in parallel with correlations of peak intensities at normal ionizing voltages (50). The first explanation of the peak-intensity correlations was based on the assumption that a steady-state approximation may be applied to the kinetics of ions in the source (51). This approach has been summarized (52). It leads to the result that the intensity ratio Z/Z_0, where Z is $[A^+]/[M^{\cdot +}]$, the ratio of peak heights for a series of compounds in which the molecular ion $M^{\cdot +}$ produces the same A^+, corresponds to the ratio of average rate constants k/k_0, which then were related by a Hammett correlation [21]. The equivalence of Z/Z_0

and k/k_0 required that A^+ decompose

$$\log \frac{Z}{Z_0} = \log \frac{k}{k_0} = \rho\sigma \qquad [21]$$

at the same rate; it would be more reasonable that A^+ decompose at a rate determined by the variable energetics of its formation, in which case $\log Z/Z_0$ would not be equivalent to $\log k/k_0$ but would still be related to it by a $\rho\sigma$ relationship, since the major decomposition rates of A^+ would be correlated by a $\rho\sigma$ relationship.

This approach was successful in explaining the shape of the correlation line when the ion is formed by two pathways (53), and the lack of any sort of correlation when the ion retains the substituent (54), and so it was applied to other systems with some measure of success. It was extremely limited in its utility, since only the least fraction of mass-spectral reactions could be correlated by it when the defined limitations were imposed. Some attempts at extension were made; some narrowly defined reactions of molecules with aliphatic substituents were seen to give correlations, but even in these reactions, complications resulting from previously unconsidered subtleties were noted (55.)

Application of this approach to low-voltage spectra began to point up the difficulties with the simple assumptions of the original hypothesis. Perhaps the most objectionable point in the original derivation is that the steady-state reaction hypothesis, while obviously valid on an experimental level, is not really pertinent at the molecular level, where non-interacting ions have essentially reacted to completion (or in better terms, almost to the extent that they ever will) by the time the sample is analyzed. This point, more than any other, weakens the original argument; another important consideration is that $[M^{\cdot+}]/[\text{total ion intensity}]$ correlations with σ have been found, and that $[A^+]/[M^{\cdot+}]$ correlations may therefore be represented as $[A^+]/[\text{total ion intensity}]$ correlations, in keeping with the correlation anticipated from quasi-equilibrium theoretical considerations (56).

One may generalize all of the possible substituent effects on ion intensities, as McLafferty has done in terms used by the organic chemist (57). These include the distribution of the internal energy values of the precursor ions; the stabilization of the product ion, which empirically seems to be more important than stabilization of the product neutral; bond strength of all bonds cleaved or formed in the reaction; competitive reactions introduced by the substituent either by involvement of the substituent itself in a decomposition or by lowering the energy for another competitive process of less importance in the unsubstituted compound; and secondary decompositions of the product ion. This large number of factors makes it difficult to interpret substituent effects, for a single substituent will exert an effect in each of these directions, and they are difficult to disentangle. Some of them have been sorted out—for example, the product-ion-stability argument. The reaction of bibenzyls to give substituted $C_7H_7^+$ ions or unsubstituted ions [22] offers an apparently clean example of the second effect, stabilization of

$$C_7H_7^+ + YC_7H_6^\cdot \leftarrow C_6H_5CH_2CH_2C_6H_4Y^{+\cdot} \rightarrow C_7H_7^\cdot + YC_7H_6^+ \qquad [22]$$

the product ion (58). Others are observed, for example, as the energy distribution effect known as the degree-of-freedom effect (59); this causes deviations from the linearity of a Hammett correlation for metastable ion intensities for very complex substituents containing many atoms, like the dimethylamino and phenyl groups that produce metastable intensities for the decarbonylation of benzoyl ion from substituted benzophenones considerably lower than the values anticipated from the correlation line (60).

There is merit in attempting to outline the general effects of substituents in terms of rate theory (61). The theory we employ is an early approximation that is quite crude, and we will miss some effects that more nearly exact forms of the theory include (for example, there is no provision for effects involving rotation, and internal rotation can be important in determining rates). This is most easily drawn from the early equations correlating rate with energy (62, 63) in excess of the minimum required to effect a reaction, [23], where k is the rate of decomposition of ions containing energy E, E_0 is

$$k(E) = A\left(\frac{E - E^0}{E}\right)^{n-1} \qquad [23]$$

the minimum energy required to effect the reaction, n is the effective number of harmonic oscillators in the ion, and A is a type of frequency factor. We may decompose each of these energy, size, and entropy terms into its components and also consider how the range of E values will vary for an ion. At a minimum, then, a substituent would be expected to influence each of the following effects, and thus influence the reaction rate of its formation:

1. The ionization potential of the molecule.
2. The appearance potential of each fragment.
3. The ionization efficiency dependence on voltage.
4. The back-reaction potential.
5. The density of occupied energy states.
6. The fraction of excited states with sufficient energy to decompose that actually do decompose.
7. The frequency factor.
8. The effective number of degrees of freedom.

Substituent effects on the ionization potential are well documented. The class investigated by the largest number of workers is the effect of substituents in aromatic rings on ionization potentials. There is frequently in these cases a correlation with σ, or better σ^+, values, and many examples have been noted (64–68). This of course implies that the usual description of solution processes in terms of electronic effects of substituents is borne out by the simplest process in which solvent effects have been removed, and is satisfying.

Substituent effects on the ionization potentials of small molecules successively substituted by alkyl groups have been fit to a simple predictive scheme based on a set of constants, the δ_K constants (49), and it has been shown that these values may be used to predict the ionization potentials of other sets of compounds with a high degree of accuracy (69).

Other types of substituent effects on factors influencing rates have to be considered insofar as they influence the rate or rates of formation of an ion and also the rate or rates of its decomposition; this point doubly complicates the interpretation of peak intensities. There have been several examples of substituent effects on appearance potentials of fragment ions (56, 57, 70). Among the earliest were σ correlations with potentials at which the $[M—CH_3]^+$ ion from substituted anisoles appeared (70), and the determination of structural characteristics of $YC_7H_6^+$ ions: if $YC_7H_6^+$ from meta- and para-substituted benzyl compounds has the same appearance potential, then positional identity is lost and the ion may have a tropylium structure; but if the appearance potential is different for the two isomers, then the structures of the ions at the appearance potential differ (70, 71). One sometimes finds that appearance potentials fit a σ correlation (70), and sometimes the fit is very poor (72). It is disconcerting when the preceding sentence holds for the same ion in the same series of compounds investigated by different groups (56, 67). This sort of result points up the need of understanding the flaws of different methods of determining appearance and ionization potentials.

The techniques to be used easily with an instrument producing ions by electron impact include a number of older methods that are capable only of giving rough values even with a calibrated instrument, like the linear-extrapolation technique (73) and the initial-break method (74). The extrapolated-voltage-difference method is based on measuring differences between the ionization efficiency curve of a standard and the curve of an unknown, and extrapolating the differences to zero ion current (75). The energy-compensation technique gives quick results with a slightly larger error; it consists of finding the voltage at which the ion intensity is some small fraction, such as 1 %, of the intensity at 50 V for the unknown and a standard. The difference between these allows one to calculate the ionization potential of the unknown, provided that the ionization efficiency curves are fairly comparable (76). The critical-slope method assumes that the probability of ionization is proportional to the square of the difference between the electron energy and the ionization potential, and evaluates the ionization potential as that potential at which a semilogarithmic plot of the ionization efficiency has the slope $\frac{2}{3}kT$; the method is claimed to be self-consistent to 0.02 eV (77). The semilogarithmic plot method calls for plotting the logarithm of the relative ion current against electron energy; this plot is linear from about 1 % (depending on the compound and instrument, it seems to be higher or lower) to perhaps 0.05 %. If a standard and an unknown are treated in this fashion, the average difference in voltage over this region is the difference in their ionization potentials (78). Derivative methods have been studied with an eye to removing the effect of the electron-energy spread (79). The second derivative of the ionization efficiency curve should have a maximum corresponding to the ionization potential (80).

These methods in general become less reliable for the determination of appearance potentials, because the efficiency of daughter ion production may have a voltage dependence rather different from the ionization-efficiency curve of the standard used. The techniques that do not make use of the whole curve are suspect in this case; and even those that employ the whole curve are often difficult to interpret. An instructive dialogue of opinions on the reliability of these methods, and on the general question of the applicability of standard methods for determination of molecular ionization potentials to the measurement of fragment appearance potentials, appeared recently (81, 82).

The effect of a substituent on the appearance potential of a fragment may be conveniently decomposed into the additive effects on the molecular ionization potential and the additional energy necessary to produce the fragment ion from the molecular ion. This difference between the appearance potential of the fragment and the ionization potential of the molecule may be taken as a sort of activation energy for the process. It is in fact the energy required to produce molecular ions that decompose in less than the residence time of ions in the source, such as 10^{-6} sec, not the energy required to produce molecular ions that will eventually decompose to the daughter ion studied. This means that extra energy has to be put into the ion above the true activation energy for the process before the daughter ion actually appears in the spectrum; the extra energy is termed the kinetic shift (83). This shift is often assumed to be small, but large kinetic shifts have been reported (84). It would appear that if such an effect is crucial to an argument, it should be studied in detail; the mechanician really cannot afford to neglect it in any case (85).

The correlation of relative ion intensities with the difference between the molecular ionization potential and the appearance potential of the fragment of interest probably can say with some security that the kinetic shift effect is unimportant for interpretation of the data; here one considers essentially the same reaction and only slight perturbations on its rate, so that the kinetic shift should be subject to a similar perturbation in the same direction, and the correlation should not be destroyed. Such a correlation of [A.P.–I.P.] and relative ion intensity has been found in a system where neither the [A.P.–I.P.] nor the ion intensity ratio correlates with σ constants (86). It is instructive to note that the correlation holds for a case in which the same daughter ion is formed from several parents. One would think that this might be the most important reason, then, for the correlation of ion intensities with Hammett σ constants observed earlier, for both the ionization potentials of, say, acetophenones, and the appearance potentials of CH_3CO^+ can be correlated at least roughly with σ constants (56, 57). The activation energy has been cited by several groups as the most important factor influencing a correlation of intensities (61, 87), particularly when the correlation line for relative ion intensities and σ values has a positive slope. A simple correlation between relative ion intensities and the [A.P.–I.P.] difference implies an energy distribution of molecular ions (88), which is not too different in shape from the elaborately worked out model for the energy distribution of propane molecular ions (89). Much more work remains to be done in the derivation of models of energy distributions from mass spectral data, but this crude prototype model suggests that no great violation of intuitive pictures of energy distribution occurs when mathematical approaches to substituent effects are used to produce numbers. If [A.P.–I.P.] values are in fact the most important factor in determining relative ion intensities, then one does not need to search for a structural interpretation of correlations of σ values with reactions like the loss of CHO from substituted phenols [24] (87) and the loss of the substituent from substituted biphenyls [25] (90).

$$YC_6H_4OH^{+\cdot} \rightarrow YC_5H_4^+ + CHO^\cdot \qquad [24]$$

$$YC_6H_4C_6H_5^{+\cdot} \rightarrow Y^\cdot + C_{12}H_9^+ \qquad [25]$$

The ionization efficiency of at least one set of aromatic compounds has been correlated with σ constants: above the onset potential for the molecular ion, the number of molecules ionized per number in the source rises with voltage, and the rate of rise is a function of σ constants (91). This correlation has been observed only for voltages slightly above the ionization potential, where no fragmentation occurs. It has not been extended to higher voltages, or even to the voltage region just above the first fragmentation of the molecular ion.

Substituents can also influence the back-reaction potential, that is, the potential through which the ground states of the product ion and product neutral species must be raised to return them to the energy of the activated complex for the reaction. Consideration of the point becomes important when thermochemical information is drawn from the appearance potential of an ion; generally the formation of product species in some rotational, vibrational, or even electronically excited state has to be considered, and the formation of products with translational energy has been dealt with only infrequently as discussed in Chapter 12. The thermochemical information about an ion must be considered to set only an upper limit to the heat of formation of the ion. One has no way of knowing whether this excess energy in the ion and neutral is divided according to any rule. For purposes of calculations of various sorts, all of the excess kinetic energy has been assumed to be retained by the ion (92–94). For calculations of other sorts, the energy has been assumed to be equipartitioned between ionic and neutral products (95). The latter is certainly closer to the truth but is probably still badly in error, since in general the statistical theory of dissociations cannot be applied to ion-molecule reactions involving skeletal rearrangement through an intimate reaction complex (96), a process that seems to be a fair model for dissociation of a reaction complex from a unimolecular reaction.*

A further, very important way of influencing the reaction rate is the alteration of the density of energy states in the ion. Consider that the density of states will be determined by the spacing of rotational, vibrational, and electronic levels of the ion; then a substituent will in general alter these spacings in a complex fashion—for example, the microwave and fingerprint infrared regions of the spectrum will be changed greatly as one goes from p-bromobenzophenone to benzophenone. The energy manifold of the molecular ion will therefore be different in these two systems. It could be that a substituent could significantly change the density of states from that of a model compound in this fashion, although in general the application of statistical considerations alone seems enough to generate a density-of-states function, and a crucial test of a substituent dependence here has not been demonstrated.

Perhaps more important in this sense, the density of states may be affected by a substituent that introduces states into the energy manifold that correspond to ionization at a different site in the molecule. Under these considerations the entire reactivity of the ion may be altered. At low enough ionizing voltages, some reactions may be suppressed. At higher voltages, ions may be considerably reduced in intensity. The reduction in intensity of rearrangement peaks due to the McLafferty rearrangement (97, 98) and to the loss of NO from nitrobenzenes (99) when a substituent with an ionization potential below that of the reacting system is introduced into the molecule is such an example. If, on the other hand, the substituent at the far-removed site is of a higher ionization potential, then the reactivity seems not to be

---

* See Chapter 12 for a discussion of energy distribution in the decompositions of doubly charged ions.

greatly altered. The system has been pictured (98) as if charge were localized (if charge is the driving force instead of an accumulation of vibrational energy) at some point for several vibrations, then transferred to another site. Presumably the apparent residence time at a site is a function of the difference between the local ionization potentials, or in more tangible quantities, the ionization potentials of the two model ions representing the two parts of the molecule in competition for the charge. For in general, it appears to be true that removal of the electron from the highest filled molecular orbital produces a picture of the ion that can be brought into concord with the ionic reactivity, to a first approximation. The influence of substituents on the energy of the highest filled molecular orbital depends on their ability to interact with it; in cases in which they do not actually determine which portion of the molecule carries the charge, the magnitude of effects fits qualitatively predictable models: amino acids resemble simple amines in their ionization potentials when the carboxyl group is well removed (100), and thioureas methylated at different sites show a substituent effect proportionate to the proximity of the methyl group to the site of ionization (101). On the other hand, recent studies of metal chelates suggest that such a model—ionization from the highest filled molecular orbital, Koopmans' theorem (102)—is in serious disagreement with trends in ionization potentials (103, 104). It has been suggested that the data point to ionization from another than the highest filled molecular orbital, perhaps an orbital on the chelate ring. But it has also been suggested that the ordering of levels is so strong a function of MO parameters and the type of calculation used that the original calculations are incorrect in their predictions of which orbital is in fact the highest filled molecular orbital (105). Data with mixed ligands indicate that each chelate ring influences the ionization potential to the same extent, a result inconsistent with the earlier observation on amino acids (100) that the site of the substituent with lowest ionization potential determines the ionization potential of the whole molecule—if ionization comes from a chelate ring (106, 107). One dislikes to think that molecules containing metal atoms are somehow more mysterious than those containing carbon, and since this reactivity can be brought into alignment with all the other available data on ionization potentials and reactivity if the accuracy of the calculations is challenged, that seems the acceptable route for the moment. There is, then, reason to accede to intuition in predicting that reactivity can be governed to a startling degree by substituents that lower the energy of the ground state of the ion by affecting an orbital distant from the reactive site, and that the substituent effect can be estimated beforehand by consideration of suitable model compounds.

Additionally, substituent effects on energy distributions may involve the creation of isolated states in the system, states that for some reason are incapable of efficient interaction with other states by energy transfer to various modes. This would correspond to the creation of two kinds of molecular ions incapable of being transformed into each other, at least on the mass-spectrometric time scale. There may possibly be two kinds of toluene molecular ions that decompose at different rates (108). Actual examples of the formation of two distinct kinds of molecular ions as a result of a substituent effect on a normal model compound are not well documented; one example has been the use of isolated states to explain large deviations between calculated and observed ion abundances in the spectra of phenyl benzyl ethers (72). It is conceivable that other explanations could be introduced here later, as our understanding of reactivity improves.

Thus the substituent may alter the density of states in the molecular ion by spacing of levels, by adding in low-energy processes, and by creating states that are incapable of interaction with those states that produce the desired reaction. As a result, we may have substituted molecules that may have sufficient energy to decompose but that do not decompose by the desired route in the time scale of the mass spectrometer. This is because we have siphoned off molecules by letting them react in some other fashion, for reactions of ions are usually competitive reactions, or by bottling up the energy in some fashion so that reaction is impossible. It would appear, then, that correlations with σ constants for ion intensities cannot be expected if one looks at a reaction in which there are some processes of lower activation energy that proceed rapidly; to be safe, it would seem advisable to choose a reaction that is the lowest-energy process by a very great margin.

We have also to consider the fact that the frequency factor A may itself be a function of substituent; this seems appropriate enough if we consider that the frequency factor for a simple cleavage corresponds to the conversion of a stretching mode into an activated complex. Since rate is directly proportional to the frequency factor irrespective of the energy of the species above the critical energy for reaction, this could be in principle an important factor. The effect that it will play will vary in magnitude depending on the transmission of the substituent effect to the bond broken—that is, how great a variation in the stretching frequency substituents can introduce relative to the mean value of the stretching frequency for the series. The carbonyl-stretching frequency of aromatic ketones is only slightly affected by substituents, 20 or 30 cms^{-1} out of 1700 cms^{-1} for a range of σ of about 1 (109). If this were a model for typical reactions, the effect would be miniscule and would probably be ignored. On the other hand, the cleavage of aromatic ketones to acyl ions, which can be correlated with σ constants (51), involves a bond closer to the substituted ring, and the effect of substituent on stretching frequency is more important. Increasing its importance by a factor of 10 over the influence on carbonyl stretching would make it responsible for a 10 to 15% variation in the intensity of product ions, less than those actually observed in general in the ketones by at least an order of magnitude.

Finally, we consider that substituents influence the exponential term, the effective number of degrees of freedom, in the rate equation, if the substituent has a different number of atoms than the group it replaces in the model compound. For a large molecule, a change from a one-atom to a two-atom substituent does not make a large relative change in this factor. That is not the case if the molecule is rather small, and so substituent-effect studies in small molecules should be conducted, if one wants to avoid this complication, without changing the number of atoms in the substituent. Even so, one may still change the *effective* number of oscillators; how one generally relates the effective number of oscillators to the total number of oscillators is unclear. Some effects of increasing the complexity of the substituent on intensities have been noted in metastable ion intensity correlations (59, 60), as we noted above. These are more susceptible to the degree-of-freedom effect than normal ion intensities, where the effect cannot be sorted out in benzophenone spectra. But in a small molecule, any change is a big change, and the effect might appear more prominently.

In general, then, it appears premature to say that substituent effects can be used for evaluating details of reaction mechanisms, except in the most general sense. More likely, substituent effects should be the subject of research themselves. Given the large range of effects that a substituent can have on the magnitude of an ion current, it is not surprising that so few correlations of, say, aromatic reactivity with σ constants have appeared in the literature. The study of ion-intensity results not correlated by substituent constants, in order to analyze what sort of interfering effect has caused the anticipated correlation to disappear, is beginning to become popular (72, 110–112); in fact, there is renewed interest in what the real significance of a good correlation is (87, 88). As we noted earlier, most workers feel that the basis of a good correlation is the so-far-unexplained correlation of the activation energy with relative ion intensity under conditions in which other substituent effects are minimized. The principal work of the next few years will likely be the improvement of models for these reactions to get a fuller understanding of their operation. Information for these models will come not only from [A.P.–I.P.] data, but also from the simpler comparative studies of *meta* and *para* substituent effects on ion intensities over a wide range of voltages (wide-range electron-impact kinetics) (113–115). Interesting parallels with solution chemistry in selected systems point up the importance of the [A.P.–I.P.] difference in qualitative and semi-quantitative predictions of reactivity; thus, for example, the formation of benzoyl ion from substituted benzophenones can be correlated with a new set of *ortho* substituent constants drawn from gas-phase reactivities (116), and the energies of fragment ions appear to be predictable in some systems to the extent that steric inhibition of resonance fits the qualitative interpretation of ion intensities (117, 118): the intensity of the product ion is 20 times greater in [26] than it is in [27]. The

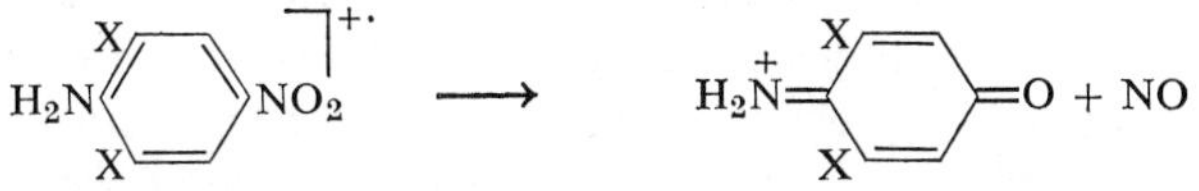

real understanding will come, of course, from theoretical considerations of such results.

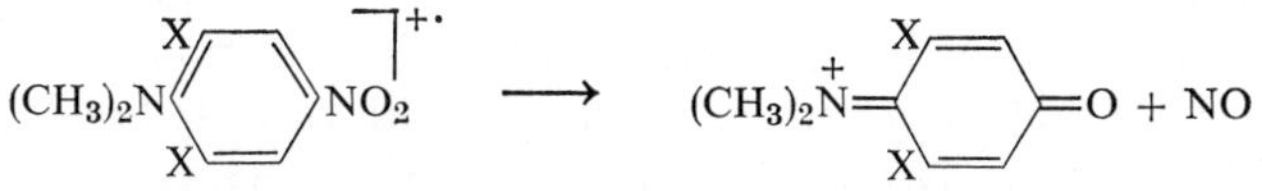

V. CORRELATIONS WITH THEORETICAL MODELS

The energetics of ion formation and decomposition can, at least in the simplest cases, be calculated with sufficient accuracy to make certain types of calculations useful for the production of numbers that can be compared with experimental observations in order to draw conclusions about ionic structure. The most appropriate calculations seem to be the semiempirical type, where data from one member of a series of compounds (the series can be very generally defined) can be extended to others with a high degree of accuracy. Such methods as the equivalent orbital method (119) have been used to calculate accurately the ionization potentials of alkanes (120, 121), and simplified approaches have been used to calculate the ionization potentials of many compounds with functional groups (122–124). The simplified method of group equivalents (123) permits the calculation of energies of conjugated systems containing multiple bonds, and so the method can be used to calculate the ionization potentials of aromatic compounds to see whether the observed ionization potentials agree with the cyclic structure or linear structures of the same formula (125, 126). Agreement between observed and calculated heats of formation of fragment ions, if the fit is acceptable for one proposed structure but not for another, can be taken as support for the structure giving the best fit; in this fashion $C_6H_4^{+\cdot}$ from benzene is best pictured as a linear ion, not the benzyne ion, and $C_8H_6^{+\cdot}$ from naphthalene is likewise linear (125). These particular results, of course, support other empirical studies that arrived at the same conclusion (127).

Though the combination of the group-equivalent method with equivalent-orbital calculations provides a very strong and general method for getting rough theoretical values of ions assumed to have a given structure, the problems that we mentioned earlier in connection with the interpretation of appearance potentials have to be taken into account in comparing data from an instrument with results from pencil-and-paper calculations. To repeat, there is a kinetic shift in appearance potentials because one must produce molecular ions that decompose to the desired product *in* 10^{-6} *sec* (83), and in some cases it has been shown to be large (84); if it is large, then the observed appearance potential seriously overestimates the heat of formation of the daughter ion. To repeat a second point, there is no good way of estimating the back-reaction energy, and so one is unsure whether his observed value for the heat of formation of the daughter ion corresponds to a relatively cold structure with n bonds or a very hot structure with $(n + 1)$ bonds. One can only hope that the calculation of energies for model ions will give results so different in energy that the observed heat of formation will fall close enough to the lowest value calculated to warrant an assignment of structure represented by the lowest value. This is not often the case, of course; and then the comparison of calculated and observed values becomes more or less unsatisfactory as rationalization in terms of cold structures becomes more or less tenuous.

The question of calculation of reaction pathways, or the prediction of the energies of transition states, is a further problem in the correlation of theoretical models. Several approaches have been used in this area. One of these is the use of HMO calculations for the study of the McLafferty reaction (128). Here the input into the calculations was nonempirical; charge density, overlap populations, and energies were calculated for the molecular ion and then were compared with similar figures for the neutral molecule and the photochemical excited (n, π^*) neutral molecule. These calculations did not provide support for a mechanism of concerted redistribution of electrons in a six-membered transition state, and so a two-step process was suggested as a reasonable alternate*. Again, there are problems associated with simplified calculations; ideally one might want to use SCF-LCAO calculations, but the difficulty of SCF calculations in a molecule as large as 2-pentanone, the molecule used for these studies, is too great to permit serious consideration of such more elegant approaches at the moment.

Advantage has been taken of the inherent ability of the perturbation molecular orbital (129) method to distinguish between the relative energies of two systems even when the approximations used would make a calculation of total energy very bad; and so somewhat more trust can be placed in these calculations than might be expected on simple consideration of their roughness. They have been used to establish a set of selection rules for mass spectra that parallel Dewar's statements (130) on thermolytic and photolytic processes (131). Where the Woodward-Hoffmann rules (132, 134) rely on

---

* A planar configuration of the transition state was assumed for calculations; this has not been demonstrated to be the path of minimum energy, though it seems reasonable.

symmetry properties to predict reactivity, Dewar's approach allows the energetics of the processes to show through as the underlying consideration in establishing reactivity rules.

Reactions may be readily classified (131) as occurring from vibrationally excited ground states or from electronically excited states as follows.

Class I reactions are those that occur from low-lying doublet or singlet states. For purposes of identification, all reactions of molecular ions (and other radical cations) that give rise to significant metastable ion peaks, and decompositions of even-electron ions that were initially formed in low-lying singlet states, are included here.

Class II reactions are those that occur from excited electronic states. They are expected to include most of the decay processes of doubly ionized molecules, and also reactions of molecular ions for which metastable peaks either cannot be observed or are extremely weak.

Thus, as a rough guide, fast processes imply excited electronic states; slow processes imply low-lying or ground electronic states of ions in this classification. This is qualitatively reasonable, though one can think of exceptions to the statement, and the abundance of metastable peaks has been correlated with structural features, not electronic features, in another study (135). Perturbation molecular orbital (PMO) theory (131) and the Woodward-Hoffmann rules (132) predict that Class I reactions should follow selection rules for thermal reactions. Those of Class II should follow the predictions made for excited-state reactivity. In PMO terminology, thermal electrocyclic reactions take place through aromatic states, and electrocyclic reactions of excited states of closed-shell molecules take place through antiaromatic transition states; an easy extension to mass-spectral processes can be made. In the symmetry arguments of Woodward and Hoffmann, electrocyclic transformations within odd-electron systems should follow the same stereochemical course as the even-electron system containing one further electron, and even-electron charged systems should behave in the same way as neutral systems containing the same number of electrons.

Several applications of the rules have been worked out. Consider the predictions of both approaches to electrocyclic reactions of the butadiene system in the mass spectrometer, which we summarize in the Woodward-Hoffmann style:

π e$^-$ in polyene ion	process	stereochemistry
3	Δ	conrotatory
3	hν	disrotatory

Thus the butadiene ion, with 3π electrons, resembles the neutral molecule with 4π electrons.

The cyclobutenedicarboxylic-acid/muconic-acid system is predicted, therefore, to have the following interrelations in the mass spectrometer, whether ionization occurs from the π system or from the nonbonding electron on oxygen:

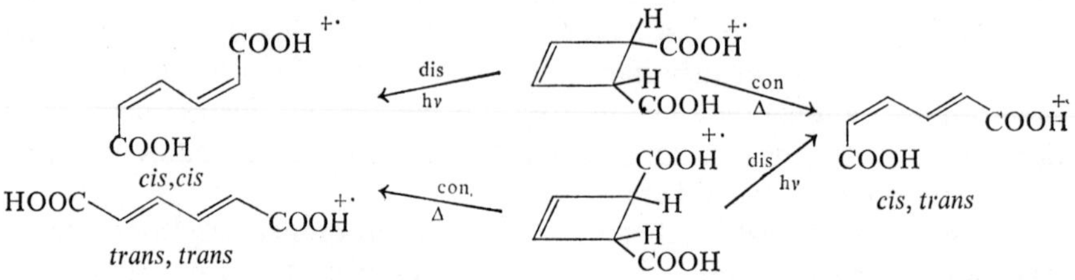

Spectra obtained over the whole range of voltages from 11 to 70 eV indicate that equilibration of structures in this system occurs by a disrotatory process (133). The Woodward-Hoffmann rules classify this equilibration as that type proceeding through electronically excited states; the PMO interpretation of the disrotatory process is a formal similarity between the transition state and cyclobutadiene radical cation, an antiaromatic species (corresponding to electronic excitation). All this is in accord with the observation that metastable-ion intensities for processes closely related to the molecular ion are all extremely weak, a point that characterizes the reaction as Class II, or occurring from some electronically excited state; so the selection rules predict the stereochemical course of this reaction correctly.

PMO calculations also indicate that the benzene radical cation is at least an order of magnitude less aromatic than benzene, and in fact less stable than the open-chain hexatriene radical cation. Hence the theory predicts that there should be ring opening when benzene is ionized, a statement in accord with the thermochemical (125) and spectral (127) evidence for the course of the reaction.

The mass spectrum of hexahelicene (136) has an intense peak at m/e 300; this corresponds to the loss of C_2H_4 from the molecular ion. An intense metastable peak at m/e 274.5 is observed for the transition m/e 328 → 300, and so this reaction may be placed in Class I, the category corresponding to low-lying or ground electronic state reactivity. The PMO method may be applied to the reaction. Hexahelicene may be converted through different transition states to two structures, 3 and 4, that have m/e 300, with the loss of small molecules. The PMO method indicates that 3 is aromatic with respect to 4, and if the reaction produces 3, the reaction would be in agreement with the lack of negative residuals when the mass spectrum of coronene is subtracted from that of hexahelicene. There are two routes that may be pictured for the transformation of hexahelicene to coronene [28], [29]. The first

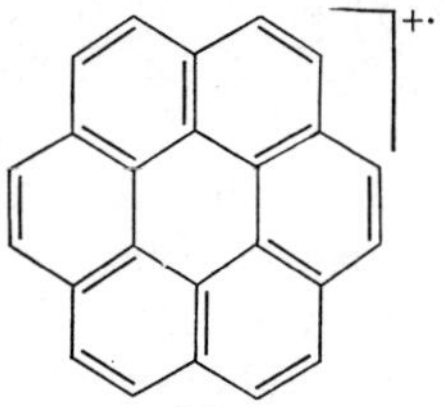

(3) 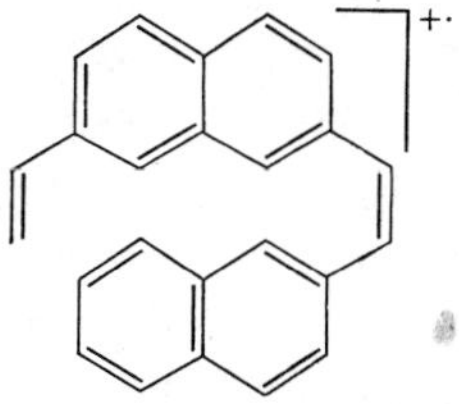   (4)

route [28], a twist through a symmetrical conformation, introduces a great amount of strain, but such a process is known to occur in the racemization of resolved hexahelicene about its melting point (137). The second route

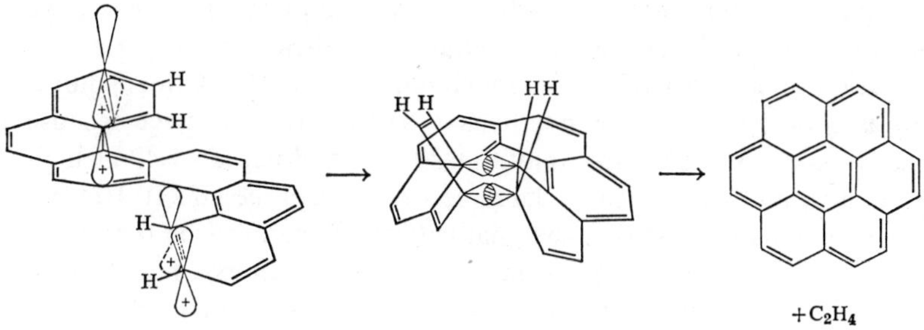

[28]

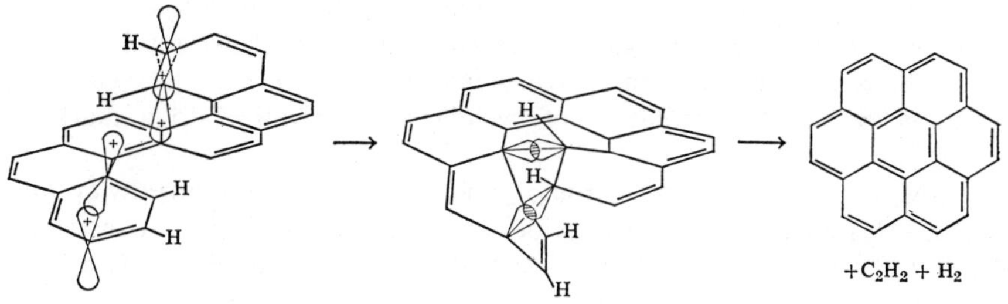

$+C_2H_4$

$+C_2H_2 + H_2$

[29]

produces one set of carbon orbitals with inverted phase from the ground state, making the transition state antiaromatic. Now the Class I reaction should parallel thermal reactivity. In [28], the molecule may readily lose ethylene as a molecule by a simple bond shift; but in [29], the m/e 300 species may be obtained only after the loss of acetylene and hydrogen as discrete units. There is no evidence for the loss of acetylene in the mass spectrum, even when a sensitive technique like metastable defocusing (138–140) is applied. This evidence suggests that the mass-spectral decomposition of hexahelicene takes place through the aromatic transition state; and the selection rule correlates this with the ground-state character suggested by the presence of the metastable. The thermal analogy fully confirms the prediction, for only ethylene was detected when a sample of hexahelicene was pyrolyzed to coronene and other products.

In addition to the HMO method mentioned earlier (128), the PMO method has also been applied to the McLafferty rearrangement (136). The transition state is considered as a heteronuclear five-electron five-orbital cyclic conjugated system, by allowing homoconjugation between the α and γ carbons. A path is then suggested on the basis of the rule for thermal electrocyclic rearrangements. Extension of the arguments to models for alkyl transfer demonstrates why these should not occur.

The retro-Diels-Alder reaction of deuterium-labeled 4-vinylcyclohexene radical cation to produce butadiene radical cation and presumably butadiene has been found to produce selectively butadiene radical cation containing the vinyl group by a factor of, at the maximum, 1.85 (141). If this is not a deuterium-isotope effect (it does not behave like an isotope effect because the selectivity goes through a maximum at 20.6 V, and is even 1.49 at high electron voltage, an extraordinarily large secondary isotope effect), frontier electron calculations suggest that the charged butadiene must be formed in its first excited state, and thus explain the selective charge retention in conformity with consideration of only bonding orbitals in vinylcyclohexene radical cation, since a strong metastable ion for the process is observed (136). The possibility that stereochemical differences may account for the selectivity in charge distribution was also considered; the energy difference between *s-cis* and *s-trans* conformations of butadiene is slight (141), but on the other hand, the conformation of an olefinic side chain can influence the course of retro-Diels-Alder fragmentation of cyclohexene derivatives to a marked extent (142).

The loss of Br from 1-bromo- and 1,1-dibromo-2,3-dimethylcyclopropanes parallels pyrolytic behavior, and can be described as concerted disrotatory processes. The Woodward-Hoffmann rules would classify a disrotatory ring opening for this ion as a thermal process; one needs to confirm that the disrotatory process expected for the neutral molecules, also a thermal process, has not occurred prior to ionization (143).

The hydrogen-loss patterns from hydrocarbon ions have been taken as evidence of structure (144). Recently a correlation of hydrogen losses was established with the direction of rotatory processes; a two-step loss of two hydrogen atoms can be associated with a *trans*-configuration in bond formation prior to H loss, but loss of H_2 in a single step is associated with a *cis* configuration formed in such cyclizations (145). Cyclization of diphenylmethyl cation at high voltage thus appears to be *cis*; cyclization of stilbene and diphenylamine radical cations appears to be *trans*. These processes can be correlated by the Woodward-Hoffmann rules with excited-state processes and have parallels in photochemistry of solutions. The cyclization of 1,4-diphenylbutadiene also appears to follow the Woodward-Hoffmann rules for an excited-state process at high voltage, and losses from diarylethylenes of various types also can be rationalized in terms of a parallel with photochemical behavior (146). Evidence for similar behavior in various decompositions in the 1,2-diphenylcyclobutane series has been observed; here the analogy with excited-state chemistry holds in spite of the fact that a metastable ion appears for the transition (147).

The previous discussion has dealt primarily with models for the energetics of reactions; we turn now to the more generally and widely explored area of kinetics, particularly the statistical interpretation of reaction rates starting from defined energetics. The most generally applied statistical argument assumes that the relation between ground state and transition state may be treated by equilibrium methods, though of course the kinetics are not equilibrium kinetics, since ions once formed have no opportunity to transfer their energy to other species. The consideration that reaction occurs when sufficient energy has accumulated to convert a particular vibration into a dissociation, then, explains the name, quasi-equilibrium theory (148), which has been applied to the approach.

The application of the simplest form of the theory is illustrated in a readable fashion by the patiently worded explanation of Kiser (149); the reader is referred to his book for a step-by-step numerical illustration. The collision of electrons with molecules produces an ensemble of molecular ions with a distribution of internal energies. Some of these energies exceed the appearance potentials of daughter ions. Depending on the amount by which the ion exceeds the activation energy for dissociation, the rate constant at which ions of that particular energy will decompose by the route to give that daughter ion is given by [23] (p. 389). Of course, if the energy of the ion exceeds the activation energies for the formation of two different decompositions of the molecular ion, the two processes will be in competition with each other. If the ion has more than enough energy to fragment twice, then one must consider consecutive reactions. Now if the energy distribution of the molecular ions is known, and the rate constant dependence of all the possible reactions of the molecular ions and the daughter ions on energy can be calculated from [23]*, then one can do the following;

1. Tabulate rate constants for competing and/or consecutive reactions as a function of the energy in the molecular ion.

2. Set up the usual simple expressions for calculating the amount of product formed or the amount of starting material consumed, by integration

* Again, the dependence predicted by this equation is only a rough (and frequently poor) approximation; more complex dependences give better fits.

of the rate-dependence law for consecutive and/or competing unimolecular reactions.

3. Assume a time on the order of microseconds at which the reaction will be stopped to examine the amounts of products and starting material.

4. Calculate the amounts of products and starting material at this time, by using the k's from 1 and the time from 3 in the exponential expressions in 2 for each energy.

5. Multiply the amounts of products and starting material for each energy times the probability that the molecular ion has this energy.

The result should be the mass spectrum. The metastable-ion spectrum can be calculated in principle in the same way by calculating the amount of further reaction in the time it takes to traverse the field-free region where the metastable reaction takes place.

As outlined, the calculation of a spectrum is a very straightforward but mathematically tedious process, exactly the problem for which a computer is suited. In practice some awkward complications arise.

The theory as usually applied has no provision for electronically excited-state contributions to reactivity. If enough were known about excited states of ions, appropriate values could be inserted in the various equations for these contributions. They are almost universally neglected, since in addition to the lack of knowledge of molecular parameters of excited states of ions, there is little information on the relative transition probabilities involved. But excited-state reactivities do seem to be important, particularly since some seem to serve as isolated states in defiance of the accessibility of all states built into the quasi-equilibrium model (150). Not all cases appear to be isolated, however (151).

The energy distribution of molecular ions is an unknown function. In the beginning, some very crude models were used to approximate it, even simple step functions giving fairly acceptable results when inserted into calculations (152). More accurately considered functions have lately been derived (89, 153), and the derivation of rough approximations to fit theory better to experiment has been the objective of a number of recent studies included in papers on applications to large organic molecules (72, 85, 110, 154). These draw on the results of older applications to simpler systems like hydrocarbons and monofunctional compounds (83, 155–162), but combine an intuitive approach to the shape of the distribution with the agreement found or not found when the distribution proposed is actually tested against experimental results of peak intensities. As we noted earlier, some attempt has been made to fit substituent effects to a distribution derived from the Hammett equation dependences found, but there are many approximations, probably serious ones, involved in it (88). A more accurate calculation requires consideration of transition probabilities; when a molecule is ionized, the transition probability to any particular energy of the molecular ion will be governed not only by the energy spread of the bombarding electrons but also by the availability of energy states of a given energy to which the molecular ion may be promoted. In quasi-equilibrium theory, if electronically excited states of the ion are formed, it crosses to highly excited vibrational and rotational states of the ion before it fragments, so that the reactivity resembles ground-state reactivity. What one needs to determine the distribution of ions with respect to energy, however, is the electronic spectrum of the gaseous molecular ion (with due corrections for the fact that formation is by collision of an electron with the neutral molecule, not by collision of a photon with the ground-state ion). Thus if there is an energy band corresponding to an excited state of the ion centered at 2 eV above the ground state of the ion, but no band at 3 eV, we would expect to have many molecular ions with 2 eV internal energy, but few with 3 eV. There is no efficient way, after the initial collision, that energy can be transferred from or to them (except by emission of radiation). Thus a substituent, for example, can alter the density of states in a fashion resembling its influence on the shape of an ultraviolet spectrum.

Additional uncertainty arises in the selection of numbers for the constants besides the activation energy used in [23]. In principle the frequency factor A can be assigned by considering the stretching frequency that corresponds to the breaking of a bond if the fragmentation is considered an overstretching of the bond. In the absence of the infrared spectrum of the molecular ion, the infrared spectrum of the neutral from which it is derived is considered a good guide for these frequencies. On the other hand, if the reaction involves rearrangement, then entropy factors must be considered; the reaction does not proceed as rapidly because of the high degree of organization needed to reach the transition state. The more tightly organized the transition state, the less likely it is, and the lower the frequency factor (163). Earlier studies allowed a fairly arbitrary reduction in the value used for the frequency factor, but it has recently been found that one can make the reduction less arbitrary by studying a large molecule with two competing processes, a simple cleavage and a rearrangement, occurring from the molecular ion. The only adjustable parameter for getting a good fit to peak intensities observed over a voltage range (if the energy distribution is defined by an appropriate function) is the frequency factor of the rearrangement, and so one arrives at a value that gives a "best fit" (85, 164). Thus typical values for frequency factors for simple cleavages are of the order of $5 \times 10^{13}/$ sec, but a six-membered transition state (the loss of methanol from methyl o-toluate) is set up less frequently so that the value is only of the order of $3 \times 10^{10}/$sec; a four-membered transition state, the one for the loss of ethane from the molecular ion of diethyl ketone, is even more difficult to reach because of its "tightness," and so the very greatly reduced value of $3 \times 10^6/$ sec seems quite reasonable (164). This seems like a very useful way to learn something about the size of transition states in mechanisms, and should be applied to such problems. Before it becomes an acceptable tool, of course, we will need to have a better feeling of what disagreement of theory with experiment can mean in general.

There is also the problem of the exponential term, which unfortunately is not simply the number of oscillators in the molecule, but rather some fraction of these. There is very little justification for the fraction used, whether one applies the theory to gaseous unimolecular decompositions or to mass spectra. One has a fraction of the total number of oscillators that may be considered an effective number of oscillators, the number inserted that seems to work best. There has been an approach used to study whether the number that works has any relation to structure or reactivity, that is, whether the effective number of oscillators is the number of effective oscillators (165). It has not provided any unifying concepts in kinetics on which we can draw here. Indeed, to get the best fit, one finds that the idea of a uniform number of oscillators provides a very poor approximation to the number of states that seem to be available just above threshold for the reaction. Since this will be important in low-voltage spectra, there are several routes out of the problem (166, 167). One can try another approach to calculating the number of states at low voltage, or one can arbitrarily modify the number of oscillators for values just above the appearance potentials of ions in order to retain the applicability of [23] in some form. Though there is no theoretical justification for the latter, it is much easier, and has been applied to improve correlations, particularly for metastable ions (85).

The quasi-equilibrium approach, then, appears to be a more generally applicable tool for interpretation of reactivity in the mass spectrometer than organic chemists have realized. The simple approach emphasized above is not the best, and one is obliged to recall that the lensmakers of Jena sold the world most of its telescopes because they took greater pains to make the best telescopes in earlier days. But anything that lets us see the moon more clearly is worthy of close examination in our present state of knowledge. Even qualitative considerations that can be drawn from it will be of value; some of these have been suggested in the discussion on substituent effects in consideration there of [23]. Another important consideration from [23] involves the relative effectiveness of simple cleavages and rearrangements at low and high voltages (25). The dependence of the rate constant of a reaction on energy is scaled by its dependence on A, the frequency factor. If two processes of equal importance at high ionizing voltage have different frequency factors, then it is possible at low voltage that the rate constant for the process with the lower frequency factor will be greater if the activation energy for the process is lower. This results from the fact that at low ionizing voltages the contribution of the energy term, $([E - E_0]/E)^{n-1}$, is overriding: the ratio of the terms for two different E_0 values can be large, and the process of lower activation energy can be faster even though it has a low frequency factor. On the other hand, at much higher average energies the ratio tends toward 1, and the process with a low frequency factor will be considerably slower than the one with a high frequency factor. As a consequence, we can expect that in low-voltage spectra, processes with low frequency factors and low activation energies will compete more effectively with other processes than they do at high voltages. In other words, rearrangements compete more effectively with simple cleavages at low voltages. As expected, peaks due to rearrangement ions are larger relative to ion intensities produced by simple cleavages at low voltages (25). Not all rearrangements produce ions, of course; and deuterium-labeling results indicate that rearrangement of hydrogens before cleavage is more favored in some cases at low voltage as well.

It should be noted that the quasi-equilibrium theory's applicability to complex organic molecules that contain heteroatoms where the charge may be considered more or less to reside has been questioned, and an alternate approach bearing some support in the temperature dependence of spectra of certain kinds of compounds, where certain intermediate ions in decomposition schemes are more prominent at low temperatures and low ionizing energies, has been proposed (168).

VI. REACTIVITIES OF IONS

The reactivities of ions in the mass spectrometer may be divided into reactions of the ion alone and reactions with other species. The reactions of individual ions most applicable to structure and reaction pathways are those that can be most easily ascribed to that particular ion, *viz.*, metastable ions. It is impossible to consider the general significance of metastable peaks outside of the quasi-equilibrium theory, and the pictures we have drawn here from theory must be borne in mind in considering what metastable peaks mean. A general discussion of their significance is given in Chapter 12, and we call attention here only to the proposals for their use in determining structures and mechanisms, not to studies of metastable ions themselves.

Ions of the same empirical formula but of different structure give different amounts of metastable products in two competing reactions in general, and if the differences are great enough the ions may be sorted out into groups that correspond to different structures (169). For ions of a similar structure formed from members of a homologous series, the intensity of the metastable ion decreases exponentially with the reciprocal of the number of degrees of freedom of the precursor (59), and is related to electronic effects in the precursor as well (60). In this fashion, information about the structures of ions may be gained (170, 171), but if the two competing reactions compared originate from different states of the ion, or from different structures, then the metastable abundances relative to each other will be very dependent on the energy content of the molecular ions, so that false differences between structures could be ascribed (172). Often, as in the case of alkanes, when normal spectra are quite similar, reactivities at the metastable time scale are considerably different, and distinction between isomers can be made from a consideration of these differences (173). On the other hand, the abundance of collision-induced metastable ions appears to be independent of the energy distribution of the precursor ion (174). This observation suggests that collision imparts much more energy than the activation energy for a metastable decomposition; it may be applied to the characterization of ion structures, since the energy distribution of the precursor ion is unimportant in influencing intensities, as is the case with ordinary metastables. Collision-induced metastable decompositions apparently involve so great an energy transfer that rearrangements are effectively suppressed, as [23] and previous discussion of quasi-equilibrium theory require for higher energy processes (175).

In typical spectra with normal metastables, the strength of a metastable intensity has been correlated with rearrangement of the precursor ion. A process that produces an abundant ion does not involve rearrangement if the corresponding metastable ion is not abundant (135). Often the most abundant ions arise from reaction pathways unrecognized in normal spectra, the loss of methyl from 2-hexanone being mostly a loss from the end of the alkyl group [30] (176). All this repeats that the energetically most favored process is not necessarily the one that produces the most abundant fragment ion in the metastable mass spectrum.

$$R-\overset{O^{+\cdot}}{\underset{}{\|}}-R \longrightarrow R^{\cdot} + \overset{O^{+}}{\underset{R}{\|}} \qquad [30]$$

The reactivities of ions in ion-molecule reactions provide an attractive route for studying their structure and reactivity. We view two recent developments in this field, chemical ionization mass spectrometry (177) and ion cyclotron resonance spectrometry (178), as being of the greatest significance for those interested in mechanisms of fairly complex molecules. Chemical ionization mass spectrometry has been reviewed recently (179). It has been used primarily with an eye to its analytical utility, the production of different kinds of mass spectra from typical molecules and the less easily handled natural products (180) through ionization with CH_5^+ and $C_2H_5^+$ ions from methane at high pressure. Interest in the reactivities of other ions as reactants has begun to spread; the alkyl ions become progressively less reactive in hydride abstraction as one goes from $C_2H_5^+$ to $C_3H_7^+$ to t-$C_4H_9^+$. Hydroxylic solvents form solvated proton species in which the polymerization is a function of pressure. They have the interesting property, apparently, of effecting selective ionization of functional groups (179). The temperature of the ion source has a great effect on the appearance of chemical ionization mass spectra (181), and in studies of the ionization of t-amyl acetate and benzyl acetate with isobutane as the reagent ion, activation energies and preexponential factors for the formation of benzyl ion and t-amyl ion from the protonated benzyl acetate and t-amyl acetate precursors (formed by proton transfer from the t-butyl ion, which comprises 95% of the total ionization) can actually be extracted. Equilibria between the protonated dimers of the two esters could be found, as well as equilibria between the protonated esters and residual water and between benzyl acetate and $C_3H_3^+$. Rate constants for the total reactions of ions produced from isobutane and methane with the two esters were found to have extraordinarily large values. When the benzyl group was substituted in the *para* position, activation energies for the formation of substituted benzyl ions from the protonated ester ions were found to decrease as the electron-donating ability of the substituent increases; a correlation with σ^+ values was obtained (182). The activation energies are not exactly equal to the differences in ion energies, but they are obviously related to the relative energies of the ions. For the equilibria to form protonated dimers and to form the ester-$C_3H_3^+$ complex, the entropy changes are positive, an unexpected result.

Early examples of the applications of ion cyclotron resonance to reactivity and structure have been summarized in a review (183). Of particular interest is the possibility of allowing ions of unknown structure to react with various neutral species to determine reactivity. If one has neutrals of different reactivity, then ions of similar composition but different structure will react differently with some of them, and again one has the possibility of sorting out unknown ions of the same formula from different sources and assigning structural differences to them (184, 185). So, for example, with the information that the proton affinities of ammonia, isobutylene, and propene decrease in that order, one expects to find chemical differences between the $C_2H_5O^+$ ions 5 and 6 and that produced by protonation of ethylene oxide (7). In the

$$CH_3C\overset{\overset{+}{O}H}{\diagdown H} \qquad CH_2\!\!=\!\!\overset{+}{O}\!-\!CH_3 \qquad \overset{\displaystyle H}{\underset{CH_2\!-\!CH_2}{\overset{O}{\diagup\!\!\!\!\overset{+}{\diagdown}}}}$$

$$(5) \qquad\qquad (6) \qquad\qquad (7)$$

ion cyclotron resonance experiment (184) **5** and **7** transferred a proton to ammonia and isobutylene but not propene; **6** transferred a proton to ammonia but not isobutylene or propene. The chemical distinction between **6** on the one hand and **5** and **7** on the other is demonstrated, and since the result was that **6** transfers a proton less readily than **5**, the direction of results can be taken in support of the structures actually proposed.

An interesting application of this concept has been made to a study of the McLafferty rearrangement and the double McLafferty rearrangement (186). Seven reactions were designed to distinguish between the keto and enol forms of acetone, the keto form presumably being formed by ionization of acetone directly and the enol form by fragmentation of 2-hexanone and 1-methylcyclobutanol [31–37]. The enol forms from these two compounds

$$CH_3\overset{+\cdot}{C}OCH_3 + CH_3COCH_3 \rightarrow (CH_3)_2C\overset{+}{=}OCOCH_3 + CH_3^{\cdot} \qquad [31]$$

$$CH_3\overset{+\cdot}{C}OCH_3 + CH_3CO(CH_2)_3CH_3 \rightarrow CH_3\overset{+\cdot}{C}O(CH_2)_3CH_3 + CH_3COCH_3 \qquad [32]$$

$$CH_3\overset{+\cdot}{C}(OH)CH_2 + CH_3CO(CH_2)_3CH_3 \rightarrow CH_3C(\overset{+}{O}H)(CH_2)_3CH_3 + CH_3COCH_2 \cdot \qquad [33]$$

$$CH_3\overset{+\cdot}{C}OHCH_2 + \square^{\bullet} OH \longrightarrow \underset{CH_2}{\overset{H_3C}{\diagdown}}C\overset{+}{-}O\overset{CH_3}{-}C\underset{CH_2 \cdot}{\overset{\diagup}{\diagdown}} + H_2O + CH_2{=}CH_2 \qquad [34]$$

$$CH_3\overset{+\cdot}{C}OHCH_2 + \square^{\bullet} OH \longrightarrow \diamondsuit{=}\overset{+}{O}{-}C\underset{CH_3}{\overset{CH_2}{\diagup}} + H_2O + CH_3 \cdot \qquad [35]$$

$$CH_3\overset{+\cdot}{C}OCH_3 + CH_3(CH_2)_3CO(CH_2)_3CH_3 \longrightarrow (CH_3)_2C\overset{+}{=}O\underset{CH_2 \cdot}{\overset{OH}{\overset{|}{C}}}CH_3 + 2C_3H_6 \qquad [36]$$

$$CH_3\overset{+\cdot}{C}OHCH_2 + CH_3(CH_2)_3CO(CH_2)_3CH_3 \rightarrow$$
$$CH_3(CH_2)_3C(\overset{+}{O}H)(CH_2)_3CH_3 + CH_3COCH_2 \cdot \qquad [37]$$

were indistinguishable on the basis of these seven reactions. It was also impossible to distinguish between the enol form and the species generated by the double McLafferty rearrangement, although metastable characteristics suggest different structures for the products of single and double McLafferty rearrangements (170). Deuterium labeling (187) seems to indicate that the double McLafferty peak has the enol form, since the formation of (M + 2) in the spectrum of 4-nonanone-1,1,1-d_3 is not brought about by m/e 59, the double McLafferty rearrangement ion, while (M + 1) *is* produced by m/e 59 [38], [39]. Other experiments showed that discrimination against deuterium transfer cannot be the explanation. Presumably deuterium never appears on

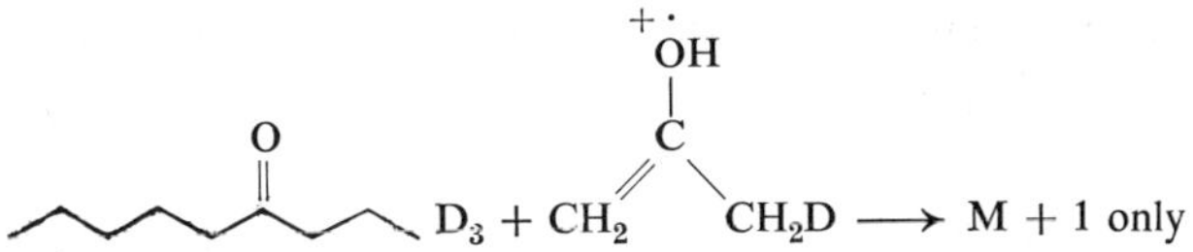

$$\underset{\text{O}}{\overset{}{\parallel}} \quad \text{D}_3 + \text{CH}_2 \underset{}{\overset{\overset{+\cdot}{\text{OH}}}{\underset{\parallel}{\text{C}}}} \text{CH}_2\text{D} \longrightarrow \text{M} + 1 \text{ only} \qquad [38]$$

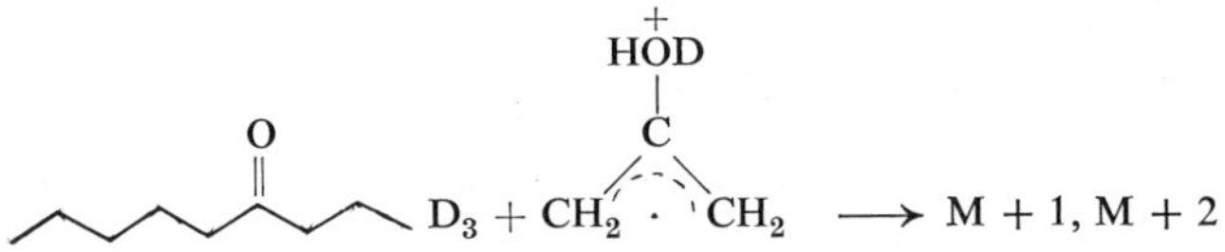

$$\underset{\text{O}}{\overset{}{\parallel}} \quad \text{D}_3 + \text{CH}_2 \underset{}{\overset{\overset{+}{\text{HOD}}}{\underset{}{\text{C}}}} \text{CH}_2 \longrightarrow \text{M} + 1, \text{M} + 2 \qquad [39]$$

oxygen; the reason is associated with the preference for hydrogen abstraction from a secondary, as opposed to a primary, position by oxygen in the first step of the double rearrangement.

Finally, ion cyclotron resonance appears to be of use in the study of collision-induced reactions (188). The reactions of p-chloroethylbenzene in a mixture of N_2 and this compound (10:1) at a pressure of 10^{-5} torr have been worked out, and the technique appears to be of promise in further study of reactions at higher pressures.

The study of structure and the study of mechanism in the mass spectrometer are then closely tied to each other. One cannot study reaction mechanisms without knowing something about structure, and the tools for studying structure are dependent on an understanding of mechanism. It takes careful thought not to be caught in circular reasoning.

Indeed, the techniques we have for studying mechanisms and structures of ions should be more the subject of study, rather than their applications. We are still discovering what the limitations of different techniques are, as this review has tried to point out. Some interpretations of results have changed in the last few years, and more will undoubtedly change. It will be years before a text equivalent to the introductions to physical organic chemistry on our shelves can be written.

But there never was a real hunter who did not feel that the elusiveness of his prey and the discomforts of his stalking were not a part of the joy of the chase.

ACKNOWLEDGMENTS

We are indebted to Dr. Ralph C. Dougherty, Dr. Alex G. Harrison, and Dr. Dudley H. Williams, for making available manuscripts of papers prior to their publication.

REFERENCES

1. J. H. Beynon, J. A. Hopkinson, and G. R. Lester, *J. Mass Spectrom. Ion Phys.*, **2**, 291 (1969).

2. K. Biemann, *Mass Spectrometry: Organic Chemical Applications*, McGraw-Hill, New York, 1962.

3. F. W. McLafferty, *Interpretation of Mass Spectra*, Benjamin, New York, 1966.

4. G. Spiteller, *Massenspektrometrische Strukturanalyse Organischer Verbindungen*, Verlag Chemie, Weinheim, 1966.

5. D. Rosenthal, personal communication.

6. J. H. Beynon, *Mass Spectrometry and Its Applications to Organic Chemistry*, Elsevier, Amsterdam, 1960; J. H. Beynon, R. A. Saunders, and A. E. Williams, *The Mass Spectra of Organic Molecules*, Elsevier, Amsterdam, 1968.

7. K. Biemann, P. Bommer, and D. M. Desiderio, *Tetrahedron Lett.*, 1725 (1964).

8. A. L. Burlingame, *13th Annual Conference on Mass Spectrometry and Allied Topics, ASTM, Committee E-14*, 1965, St. Louis, Missouri.

9. R. Venkataraghavan and F. W. McLafferty, *Anal. Chem.*, **39**, 278 (1967).

10. R. Venkataraghavan, R. D. Board, R. Klimowski, J. W. Amy, and F. W. McLafferty, *15th Annual Conference on Mass Spectrometry and Allied Topics, ASTM, Committee E-14*, 1967, Denver, Colorado, p. 93.

11. P. C. Jurs, B. R. Kowalski, T. L. Isenhour, and C. N. Reilley, *Anal. Chem.*, **41**, 690, 695 (1969).

12. See also A. M. Duffield, A. V. Robertson, C. Djerassi, B. G. Buchanan, G. L. Sutherland, E. A. Feigenbaum, and J. Lederberg, *J. Amer. Chem. Soc.*, **91**, 2977 (1969); R. Venkataraghavan and F. W. McLafferty, *15th Annual Conference on Mass Spectrometry and Allied Topics, ASTM, Committee E-14*, 1967, Denver, Colorado, p. 98; and A. Mandelbaum, P. Fennessey, and K. Biemann, *15th Annual Conference on Mass Spectrometry and Allied Topics, ASTM, Committee E-14*, 1967, Denver, Colorado, p. 111.

13. F. W. McLafferty and M. C. Hamming, *Chem. Ind.* (London), 1366 (1958).

14. P. N. Rylander, S. Meyerson, and H. M. Grubb, *J. Amer. Chem. Soc.*, **79**, 842 (1957).

15. S. Meyerson, H. Hart, and L. C. Leitch, *J. Amer. Chem. Soc.*, **90**, 3419 (1968).

16. D. T. Roberts, Jr., W. F. Little, and M. M. Bursey, *J. Amer. Chem. Soc.*, **90**, 973 (1968).

17. S. Meyerson and P. N. Rylander, *J. Chem. Phys.*, **27**, 901 (1957).

18. K. L. Rinehart, Jr., A. C. Buchholz, G. E. Van Lear, and H. L. Cantrill, *J. Amer. Chem. Soc.*, **90**, 2983 (1968).

19. C. Djerassi and C. Fenselau, *J. Amer. Chem. Soc.*, **87**, 5747, 5752 (1965).

20. C. Djerassi and C. Fenselau, *J. Amer. Chem. Soc.*, **87**, 5756 (1965).

21. R. A. Friedel, J. L. Schultz, and A. G. Sharkey, *Anal. Chem.*, **28**, 940 (1956).

22. W. Benz and K. Biemann, *J. Amer. Chem. Soc.*, **86**, 2375 (1964).

23. A. M. Duffield, S. D. Sample, and C. Djerassi, *Chem. Commun.*, 193 (1966).

24. G. A. Smith and D. H. Williams, *J. Amer. Chem. Soc.*, **91**, 5254 (1969).

25. D. H. Williams and R. G. Cooks, *Chem. Commun.*, 663 (1968).

26. A. N. H. Yeo, R. G. Cooks, and D. H. Williams, *Chem. Commun.*, 1269 (1968).

27. W. Carpenter, A. M. Duffield, and C. Djerassi, *J. Amer. Chem. Soc.*, **90**, 160 (1968).

28. M. M. Green and J. Schwab, *Tetrahedron Lett.*, 2955 (1968).

29. R. S. Ward and D. H. Williams, *J. Org. Chem.*, **34**, 3373 (1969).

30. M. M. Green and R. J. Cook, *J. Amer. Chem. Soc.*, **91**, 2129 (1969).

31. M. M. Green, *J. Amer. Chem. Soc.*, **90**, 3872 (1968).

32. P. Natalis, *Bull. Soc. Chim. Belges*, **73**, 389 (1964).

33. D. H. Williams, H. Budzikiewicz, and C. Djerassi, *J. Amer. Chem. Soc.*, **86**, 284 (1964).

34. J. K. MacLeod and C. Djerassi, *J. Amer. Chem. Soc.*, **89**, 5182 (1967).

35. B. J.-S. Wang and E. R. Thornton, *J. Amer. Chem. Soc.*, **90**, 1216 (1968).

36. H. Knöppel and W. Beyrich, *Tetrahedron Lett.*, 291 (1968).

37. M. M. Bursey, R. D. Rieke, T. A. Elwood, and L. R. Dusold, *J. Amer. Chem. Soc.*, **90**, 1557 (1968).

38. K. R. Jennings, *Z. Naturforsch.*, **22a**, 454 (1967).

39. D. H. Williams and J. Ronayne, *Chem. Commun.*, 1129 (1967).

40. D. H. Williams, R. G. Cooks, J. Ronayne, and S. W. Tam, *Tetrahedron Lett.*, 1777 (1968).

41. P. Brown and C. Djerassi, *Angew. Chem. Intern. Ed. Engl.*, **6**, 477 (1967).

42. R. G. Cooks, *Org. Mass Spectrom.*, **2**, 481 (1969).

43. T. A. Elwood, P. F. Rogerson, and M. M. Bursey, *J. Org. Chem.*, **34**, 1138 (1969).

44. T. A. Elwood and M. M. Bursey, *J. Org. Chem.*, **35**, 793 (1970).

45. M. M. Bursey and T. A. Elwood, *J. Amer. Chem. Soc.*, **91**, 3812 (1969).

46. G. G. Meisels, J. Y. Park, and B. G. Giesser, *J. Amer. Chem. Soc.*, **91**, 1555 (1969).

47. P. Bommer and K. Biemann, *Ann. Rev. Phys. Chem.*, **16**, 481 (1965).

48. H. Baba, I. Omura, and K. Higasi, *Bull. Chem. Soc. Japan*, **29**, 521 (1956).

49. J. J. Kaufman and W. S. Koski, *J. Amer. Chem. Soc.*, **82**, 3262 (1960).

50. F. W. McLafferty, *Anal. Chem.*, **31**, 477 (1959).

51. M. M. Bursey and F. W. McLafferty, *J. Amer. Chem. Soc.*, **88**, 529 (1966).

52. M. M. Bursey, *Org. Mass Spectrom.*, **1**, 31 (1968).

53. M. M. Bursey and F. W. McLafferty, *J. Amer. Chem. Soc.*, **88**, 4484 (1966).

54. M. M. Bursey and F. W. McLafferty, *J. Amer. Chem. Soc.*, **89**, 1 (1967).

55. M. M. Bursey and E. S. Wolfe, *Org. Mass Spectrom.*, **1**, 543 (1968).

56. M. S. Chin and A. G. Harrison, *Org. Mass Spectrom.*, **2**, 1073 (1969).

57. F. W. McLafferty, *Chem. Commun.*, 956 (1968).

58. F. W. McLafferty and M. M. Bursey, *J. Amer. Chem. Soc.*, **90**, 5299 (1968).

59. F. W. McLafferty and W. T. Pike, *J. Amer. Chem. Soc.*, **89**, 5951 (1967).

60. M. L. Gross and F. W. McLafferty, *Chem. Commun.*, 254 (1968).

61. R. G. Cooks, I. Howe, and D. H. Williams, *Org. Mass Spectrom.*, **2**, 137 (1969).

62. O. K. Rice and H. C. Ramsperger, *J. Amer. Chem. Soc.*, **49**, 1617 (1927).

63. L. S. Kassel, *J. Phys. Chem.*, **32**, 225 (1928).

64. A. Streitwieser, Jr., *Progr. Phys. Org. Chem.*, **1**, 1 (1963).

65. A. G. Harrison, P. Kebarle, and F. P. Lossing, *J. Amer. Chem. Soc.*, **83**, 777 (1961).

66. G. F. Crable and G. L. Kearns, *J. Phys. Chem.*, **66**, 436 (1962).

67. A. Buchs, G. P. Rossetti, and B. P. Susz, *Helv. Chim. Acta*, **47**, 1563 (1964).

68. A. Foffani, S. Pignataro, B. Cantone, and F. Grasso, *Z. Physik. Chem.* (Frankfurt), **42,** 221 (1964).

69. J. J. Kaufman, *J. Phys. Chem.*, **66,** 2269 (1962).

70. J. M. S. Tait, T. W. Shannon, and A. G. Harrison, *J. Amer. Chem. Soc.*, **84,** 4 (1962).

71. F. Meyer and A. G. Harrison, *Can. J. Chem.*, **42,** 1762 (1964).

72. R. S. Ward, R. G. Cooks, and D. H. Williams, *J. Amer. Chem. Soc.*, **91,** 2727 (1969).

73. R. H. Vought, *Phys. Rev.*, **71,** 93 (1947).

74. H. D. Smyth, *Proc. Roy. Soc. London*, **102,** 283 (1922).

75. J. W. Warren, *Nature*, **165,** 810 (1950).

76. R. W. Kiser and E. J. Gallegos, *J. Phys. Chem.*, **66,** 947 (1962).

77. R. E. Honig, *J. Chem. Phys.*, **16,** 105 (1948).

78. F. P. Lossing, A. W. Tickner, and W. A. Bryce, *J. Chem. Phys.*, **19,** 1254 (1951).

79. R. F. Winters, I. H. Collins, and W. L. Courchene, *J. Chem. Phys.*, **45,** 1931 (1966).

80. J. D. Morrison, *J. Chem. Phys.*, **21,** 1767 (1953); **22,** 1219 (1954).

81. F. H. Dorman, *J. Chem. Phys.*, **50,** 1042 (1969).

82. A. G. Harrison, *J. Chem. Phys.*, **50,** 1043 (1969).

83. W. A. Chupka, *J. Chem. Phys.*, **30,** 191 (1959).

84. I. Hertel and Ch. Ottinger, *Z. Naturforsch.*, **22a,** 40 (1967).

85. A. N. H. Yeo and D. H. Williams, *J. Amer. Chem. Soc.*, **92,** 3984 (1970).

86. M. A. Baldwin and A. G. Loudon, *Org. Mass Spectrom.*, **2,** 549 (1969).

87. T. W. Bentley, R. A. W. Johnstone, and D. W. Payling, *J. Amer. Chem. Soc.*, **91,** 3978 (1969).

88. R. P. Buck and M. M. Bursey, *Org. Mass Spectrom.*, **3,** 387 (1970).

89. M. L. Vestal, *J. Chem. Phys.*, **43,** 1356 (1965).

90. M. M. Bursey and P. T. Kissinger, *Org. Mass Spectrom.*, **3,** 395 (1970).

91. F. T. Deverse and A. B. King, *J. Chem. Phys.*, **41,** 3833 (1964).

92. J. H. Beynon, A. E. Fontaine, and G. R. Lester, *J. Mass Spectrom. Ion Phys.*, **1,** 1 (1968).

93. J. L. Occolowitz and G. L. White, *Aust. J. Chem.*, **21,** 997 (1968).

94. R. G. Gillis, G. J. Long, A. G. Moritz, and J. L. Occolowitz, *Org. Mass Spectrom.*, **1,** 527 (1968).

95. D. H. Williams, R. G. Cooks, and I. Howe, *J. Amer. Chem. Soc.*, **90,** 6759 (1968).

96. J. L. Franklin and M. A. Haney, *J. Phys. Chem.*, **73,** 2857 (1969).

97. T. Wachs and F. W. McLafferty, *J. Amer. Chem. Soc.*, **89,** 5044 (1967).

98. A. Mandelbaum and K. Biemann, *J. Amer. Chem. Soc.*, **90,** 2975 (1968).

99. T. H. Kinstle and W. R. Oliver, *J. Amer. Chem. Soc.*, **91,** 1864 (1969).

100. H. J. Svec and G. A. Junk, *J. Amer. Chem. Soc.*, **89,** 790 (1967).

101. M. Baldwin, A. Kirkien-Konasiewicz, A. G. Loudon, A. Maccoll, and D. Smith, *Chem. Commun.*, 574 (1966).

102. T. Koopmans, *Physica*, **1,** 104 (1933).

103. S. M. Schildcrout, R. G. Pearson, and F. E. Stafford, *J. Amer. Chem. Soc.*, **90,** 4006 (1968).

104. C. Reichert and J. B. Westmore, *Inorg. Chem.*, **8**, 1012 (1969); G. M. Bancroft, C. Reichert, J. B. Westmore, and H. D. Gesser, *Inorg. Chem.*, **8**, 474 (1969).

105. E. A. Magnusson, K. A. Thomson, and A. G. Wedd, *Chem. Commun.*, 842 (1969).

106. H. F. Holtzclaw, Jr., R. L. Lintvedt, H. E. Baumgarten, R. G. Parker, M. M. Bursey, and P. F. Rogerson, *J. Amer. Chem. Soc.*, **91**, 3774 (1969).

107. M. M. Bursey and P. F. Rogerson, *Inorg. Chem.*, **9**, 676 (1970).

108. F. Meyer and A. G. Harrison, *J. Chem. Phys.*, **43**, 1778 (1965).

109. H. H. Freedman, *J. Amer. Chem. Soc.*, **82**, 2454 (1960).

110. I. Howe and D. H. Williams, *J. Amer. Chem. Soc.*, **90**, 5461 (1968).

111. F. W. McLafferty and T. Wachs, *J. Amer. Chem. Soc.*, **89**, 5043 (1967).

112. I. Howe, D. H. Williams, D. G. I. Kingston, and H. P. Tannenbaum, *J. Chem. Soc. (B)*, 439 (1969).

113. P. Brown, *J. Amer. Chem. Soc.*, **90**, 4459 (1968).

114. P. Brown, *Org. Mass Spectrom.*, **2**, 1085 (1969).

115. P. Brown, *Org. Mass Spectrom.*, **2**, 1317 (1969).

116. K. K. Lum and G. G. Smith, *J. Org. Chem.*, **34**, 2095 (1969).

117. M. M. Bursey, *J. Amer. Chem. Soc.*, **91**, 1861 (1969).

118. M. M. Bursey and M. K. Hoffman, *J. Amer. Chem. Soc.*, **91**, 5023 (1969).

119. J. Lennard-Jones and G. G. Hall, *Disc. Faraday Soc.*, **10**, 18 (1951).

120. G. G. Hall, *Proc. Roy. Soc. London*, **A205**, 541 (1951).

121. J. Lennard-Jones and G. G. Hall, *Trans. Faraday Soc.*, **48**, 581 (1952).

122. G. G. Hall, *Trans. Faraday Soc.*, **49**, 113 (1953); **50**, 319 (1954).

123. J. L. Franklin, *J. Chem. Phys.*, **22**, 1304 (1954).

124. J. Lennard-Jones and G. G. Hall, *Proc. Roy. Soc. London*, **A213**, 102 (1952).

125. P. Natalis and J. L. Franklin, *J. Phys. Chem.*, **69**, 2935 (1965).

126. P. Natalis and J. L. Franklin, *J. Phys. Chem.*, **69**, 2943 (1965).

127. J. Momigny, L. Brakier, and L. D'Or, *Bull. Classe Sci. Acad. Roy. Belg.*, **48**, 1002 (1962).

128. F. P. Boer, T. W. Shannon, and F. W. McLafferty, *J. Amer. Chem. Soc.*, **90**, 7239 (1968).

129. M. J. S. Dewar, *The Molecular Orbital Theory of Organic Chemistry*, McGraw-Hill, New York, 1969, Chapter 8.

130. M. J. S. Dewar, *Tetrahedron, Suppl.* **8**, 75 (1966).

131. R. C. Dougherty, *J. Amer. Chem. Soc.*, **90**, 5780 (1968).

132. R. B. Woodward and R. Hoffmann, *J. Amer. Chem. Soc.*, **87**, 395 (1965).

133. M. K. Hoffman, M. M. Bursey, and R. E. K. Winter, *J. Amer. Chem. Soc.*, **92**, 727 (1970).

134. R. Hoffmann and R. B. Woodward, *Accounts Chem. Res.*, **1**, 17 (1968).

135. F. W. McLafferty and R. B. Fairweather, *J. Amer. Chem. Soc.*, **90**, 5915 (1968).

136. R. C. Dougherty, *J. Amer. Chem. Soc.*, **90**, 5788 (1968).

137. M. S. Newman, R. S. Darlak, and L. Tsai, *J. Amer. Chem. Soc.*, **89**, 6191 (1967).

138. K. R. Jennings, in R. Bonnett and J. G. Davis (Eds.), *Some Newer Physical Methods in Structural Chemistry*, United Trade Press, Ltd., London, 1967, p. 105.

139. M. Barber, K. R. Jennings, and R. Rhodes, *Z. Naturforsch.*, **22a**, 15 (1967).

140. T. W. Shannon, T. E. Mead, C. G. Warner, and F. W. McLafferty, *Anal. Chem.*, **39**, 1748 (1967).

141. E. P. Smith and E. R. Thornton, *J. Amer. Chem. Soc.*, **89**, 5079 (1967).

142. S. W. Staley and D. W. Reichard, *J. Amer. Chem. Soc.*, **90**, 816 (1968).

143. M. S. Baird and C. B. Reese, *Tetrahedron Lett.*, 2117 (1969).

144. S. W. Staley, J. P. Erdman, and T. J. Henry, *16th Annual Conference on Mass Spectrometry and Allied Topics*, ASTM, Committee E-14, 1968, Pittsburgh, Penna., p. 147.

145. R. A. W. Johnstone and S. D. Ward, *J. Chem. Soc. (C)*, 1805 (1968).

146. R. A. W. Johnstone and S. D. Ward, *J. Chem. Soc. (C)*, 2540 (1968).

147. M. L. Gross and C. L. Wilkins, *Tetrahedron Lett.*, 3875 (1969).

148. H. M. Rosenstock, M. B. Wallenstein, A. L. Wahrhaftig, and H. Eyring, *Proc. Natl. Acad. Sci. U.S.*, **38**, 667 (1952).

149. R. W. Kiser, *Introduction to Mass Spectrometry and Its Applications*, Prentice-Hall, Englewood Cliffs, N.J., 1965, Chapter 7.

150. N. Garcia and R. C. Dougherty, Seventeenth Annual Conference on Mass Spectrometry and Allied Topics, Dallas, Texas, May 18–23, 1969; R. C. Dougherty, personal communication.

151. A. N. H. Yeo, R. G. Cooks, and D. H. Williams, *J. Chem. Soc. (B)*, 149 (1969).

152. M. Vestal, A. L. Wahrhaftig, and W. H. Johnston, "Theoretical Studies in Basic Radiation Chemistry," ARL-62-426, September, 1962.

153. See also W. A. Chupka and M. Kaminsky, *J. Chem. Phys.*, **35**, 1991 (1961).

154. I. Howe and D. H. Williams, *J. Amer. Chem. Soc.*, **91**, 7137 (1969).

155. A. Kropf, E. M. Eyring, A. L. Wahrhaftig, and H. Eyring, *J. Chem. Phys.*, **32**, 149 (1960).

156. E. M. Eyring and A. L. Wahrhaftig, *J. Chem. Phys.*, **34**, 23 (1961).

157. M. Vestal, A. L. Wahrhaftig, and W. H. Johnston, *J. Chem. Phys.*, **37**, 1276 (1962).

158. J. Collin, *Bull. Soc. Roy. Sci. Liège*, **7**, 520 (1956).

159. L. Friedman, F. A. Long, and M. Wolfsberg, *J. Chem. Phys.*, **27**, 613 (1957).

160. A. B. King and F. A. Long, *J. Chem. Phys.*, **29**, 374 (1958).

161. W. A. Chupka and J. Berkowitz, *J. Chem. Phys.*, **32**, 1546 (1960).

162. B. Steiner, C. F. Giese, and M. G. Inghram, *J. Chem. Phys.*, **34**, 189 (1961).

163. B. S. Rabinovitch and D. W. Setser, *Adv. Photochem.*, **3**, 1 (1964).

164. A. N. H. Yeo and D. H. Williams, *Chem. Commun.*, 956 (1969).

165. N. B. Slater, *Phil. Trans. Roy. Soc. (London)*, **A246**, 57 (1953).

166. H. M. Rosenstock and M. Krauss, in F. W. McLafferty (Ed.), *Mass Spectrometry of Organic Ions*, Academic Press, New York, 1963, p. 1.

167. H. M. Rosenstock, *Adv. Mass Spectrom.*, **4**, 523 (1968).

168. G. Spiteller and M. Spiteller-Friedmann, *Ann. Chem.*, **690**, 1 (1965).

169. T. W. Shannon and F. W. McLafferty, *J. Amer. Chem. Soc.*, **88**, 5021 (1966).

170. F. W. McLafferty and W. T. Pike, *J. Amer. Chem. Soc.*, **89**, 5953 (1967).

171. W. T. Pike and F. W. McLafferty, *J. Amer. Chem. Soc.*, **89**, 5954 (1967).

172. J. L. Occolowitz, *J. Amer. Chem. Soc.*, **91**, 5202 (1969).

173. F. W. McLafferty and T. A. Bryce, *Chem. Commun.*, 1215 (1967).

174. W. F. Haddon and F. W. McLafferty, *J. Amer. Chem. Soc.*, **90**, 4745 (1968).

175. F. W. McLafferty and H. D. R. Schuddemage, *J. Amer. Chem. Soc.*, **91**, 1866 (1969).

176. F. W. McLafferty, D. J. McAdoo, and J. S. Smith, *J. Amer. Chem. Soc.*, **91**, 5400 (1969).

177. M. S. B. Munson and F. H. Field, *J. Amer. Chem. Soc.*, **88**, 2621 (1966).

178. J. D. Baldeschwieler, *Science*, **159**, 263 (1968).

179. F. H. Field, *Accounts Chem. Res.*, **1**, 42 (1968).

180. H. M. Fales, G. W. A. Milne, and M. L. Vestal, *J. Amer. Chem. Soc.*, **91**, 3683 (1969).

181. F. H. Field, *J. Amer. Chem. Soc.*, **91**, 2827 (1969).

182. F. H. Field, *J. Amer. Chem. Soc.*, **91**, 6334 (1969).

183. J. D. Baldeschwieler, in G. A. Olah and P. von R. Schleyer (Eds.), *Carbonium Ions*, Wiley-Interscience, New York, 1968, p. 413.

184. R. C. Dunbar, unpublished results cited in Ref. 178.

185. J. L. Beauchamp, *17th Annual Conference on Mass Spectrometry and Allied Topics, ASTM, Committee E-14*, 1969, Dallas, Texas.

186. J. Diekman, J. K. MacLeod, C. Djerassi, and J. D. Baldeschwieler, *J. Amer. Chem. Soc.*, **91**, 2069 (1969).

187. G. Eadon, J. Diekman, and C. Djerassi, *J. Amer. Chem. Soc.*, **91**, 3986 (1969).

188. F. Kaplan, *J. Amer. Chem. Soc.*, **90**, 4483 (1968).

Some Aspects of Metastable Transitions

K. R. JENNINGS

Department of Chemistry, The University,
Sheffield S3 7HF, England

I.	Introduction	420
II.	The Observation of Metastable Transitions	420
	A. In the Field-Free Region Before the Magnetic Sector	420
	B. During Acceleration (Single-Focusing Instrument)	422
	C. Within the Magnetic Sector	423
	D. In the Final Field-Free Region	423
	E. In the Field-Free Region before the Electric Sector	425
	F. Ion Kinetic Energy Spectroscopy	427
	G. During Acceleration (Double-Focusing Instrument)	427
	H. Within the Electric Sector	428
	I. Consecutive Metastable Transitions	428
	J. Collision-Induced Metastable Transitions	429
	K. In 180° Magnetic Deflexion Instruments	430
	L. In Time-of-Flight Instruments	430
III.	Kinetic Aspects of Metastable Transitions	431
	A. Quasi-Equilibrium Theory	431
	B. Competing Fragmentation Processes	433
	C. Kinetic Shifts and Heats of Formation of Ions	435
	D. Applications of Intensity Measurements	436
	E. Metastable Transitions of Partly Deuterated Compounds	439
IV.	Fragmentation Maps	441
	A. Experimental Procedure	441
	B. Structural Applications	442
V.	The Shapes of Metastable Peaks	443
	A. General Considerations	443
	B. In the Field-Free Region Before the Magnetic Sector: Peak Widths	445
	C. Peak Shapes	447
	D. In the Field-Free Region Before the Electric Sector	450
	E. Effects of β-Slit Width on Peak Shapes	452
	F. Applications to the Energetics of Fragmentation Processes	453
	G. Substituent Effects on Peak Shapes	455
VI.	Conclusion	456
	References	456

419

I. INTRODUCTION

The recorded mass spectrum of a substance may be looked on as the result of a series of competing, consecutive unimolecular decompositions of excited molecular and fragment ions. For each reaction, the rate constant increases in a complicated manner with increase in internal energy of the decomposing ion so that, in general, reactions will be characterized by a range of rate constants. Typically, ions spend about 10^{-6} sec in the ion source and rather more than 10^{-5} sec after acceleration and before collection. Therefore, reactions for which k exceeds 10^6/sec occur in the source, and daughter ions give rise to a peak at the appropriate mass-to-charge ratio; if k is appreciably less than 10^5/sec, the parent ion will be collected before reaction occurs. Metastable transitions are those decompositions for which k lies within the range 10^5–10^6/sec and which therefore occur between the source and the collector. By no means all reactions occur such that a significant number of ions decompose with rate constants within this range, and in some cases, the minimum rate constant exceeds 10^6/sec so that no metastable transitions occur.

In this chapter, the various conditions under which metastable transitions can be observed and characterized are described. Applications in organic, analytical, and physical chemistry are then discussed. Metastable transitions observed in cycloidal instruments and those produced by field ionization are not considered, but these topics are discussed in Chapter 7, "Newer Ionization Techniques," to which the reader is referred.

II. THE OBSERVATION OF METASTABLE TRANSITIONS

A. In the Field-Free Region before the Magnetic Sector

In a mass spectrum obtained under normal operating conditions in a magnetic deflexion instrument, metastable transitions give rise to diffuse peaks of low intensity, usually occurring at nonintegral mass-to-charge ratios and referred to as "metastable peaks" (1). The factors affecting peak shapes are discussed in detail in a later section, but most metastable peaks are approximately gaussian in shape, ranging from sharp and narrow to broad and rounded. These peaks arise from the collection of daughter ions formed in metastable transitions that occur in the field-free region immediately before the magnetic analyzer in a single or double focusing instrument (regions 2 and 4, respectively—see Figure 1). Suppose m_2^+ ions are formed in the reaction

$$m_1^+ \rightarrow m_2^+ + (m_1 - m_2) \qquad [1]$$

where $(m_1 - m_2)$ may be one or more neutral fragments. Ions of mass m_2 have the velocity v normally associated with ions of mass m_1, that is,

$$v = (2V_0e/m_1)^{\frac{1}{2}} \qquad [2]$$

where V_0 is the accelerating potential and e is the electronic charge. The

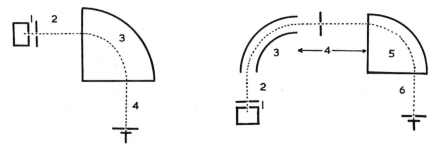

Figure 1. Regions in which metastable transitions occur in single- and double-focusing instruments.

radius R of trajectories of ions of mass-to-charge ratio m/e in a magnetic field H is given by

$$R = mv/He \qquad [3]$$

so that for m_2^+ ions moving with velocity v

$$R = \left(\frac{m_2^2}{m_1} \cdot \frac{2V_0}{H^2e}\right)^{\frac{1}{2}} \qquad [4]$$

This may be compared with the normal expression for m_1^+ ions collected without decomposition:

$$R = \left(m_1 \cdot \frac{2V_0}{H^2e}\right)^{\frac{1}{2}} \qquad [5]$$

That is, by comparing [4] and [5], it is seen that the m_2^+ ions formed in the metastable transitions behave as normal ions of apparent mass m_2^2/m_1. This assumes that no internal energy is converted into translational energy of the fragments during the decomposition, but even if the peak is broadened due to energy release (2), the center of the peak lies very close to m_2^2/m_1. In the more general case in which $m_1^{p+} \rightarrow m_2^{q+}$, the daughter ions behave as ions of apparent mass pm_2^2/qm_1 and usually give rise to broad peaks.

Because of the relatively low intensities and diffuse nature of metastable peaks, they are not always readily observable in a normal mass spectrum. Under typical operating conditions of high resolution and sensitivity low enough to be able to record all normal peaks on scale, and using 70 eV electrons, only a few of the most intense metastable peaks will be identifiable. The metastable transitions responsible for these peaks are usually readily assigned, since both parent and daughter ions are usually fairly prominent in the normal mass spectrum. For example, a metastable peak at $m/e = 35.4$ in the mass spectrum of benzene is ascribed to the process $C_6H_6^{+\cdot} \rightarrow C_4H_4^{+\cdot} + C_2H_2$ ($m/e = 52^2/78 = 35.4$), since both the molecular ion and the $m/e = 52$ ion are prominent in the mass spectrum. This peak is clearly observable in mass spectra obtained in relatively low performance instruments, since it occurs in a region of the mass scale where normal peaks are absent. On the other hand, the peak arising from the process $C_6H_6^{+\cdot} \rightarrow C_6H_5^+ + H\cdot$ occurs at $m/e = 77^2/78 = 76.01$, and would not be observed in a low performance instrument owing to the presence of a normal peak at $m/e = 76$ arising from $C_6H_4^{+\cdot}$. If a high-performance, single-focusing instrument is employed, the resolution is then sufficiently high to enable one to detect such peaks at high sensitivity and slow scanning speeds. While such a procedure is adequate for recording metastable transitions in the mass spectra of fairly simple molecules, it becomes increasingly difficult to assign metastable peaks with certainty in the mass spectra of larger molecules. Low-intensity peaks often overlap with each other and with peaks due to normal fragment ions, making accurate mass measurement and an unambiguous assignment very difficult if not impossible.

B. During Acceleration (Single-Focusing Instrument)

Under normal operating conditions, metastable transitions that occur in regions 1, 3, and 4 of a single-focusing magnetic deflexion instrument are very difficult to detect. If the decomposition occurs in region 1, the final velocity of the m_2^+ ions will depend on the potential difference V' through which m_1^+ has fallen before the decomposition (1), since the final energy of the m_2^+ ion is given by

$$\frac{m_2}{m_1} V'e + (V_0 - V')e = \tfrac{1}{2}m_2v^2 \qquad [6]$$

This leads to the following expression for R:

$$R = \left[\frac{2((m_2^2/m_1)V' + m_2(V_0 - V'))}{H^2e} \right]^{\frac{1}{2}} \qquad [7]$$

which reduces to [4] when $V' = V_0$, and to [5] when $V' = 0$. In general, V' ranges from zero to V_0, so that the resulting daughter ions behave as if their masses ranged from m_2^+ to $(m_2^2/m_1)^+$. As a result of this, they are not brought to focus at the collector slit at a single value of V or H and so contribute to the background continuum between m_2^+ and $(m_2^2/m_1)^+$. In a double-focusing instrument, metastable transitions occurring in region 1 are not detected under normal operating conditions, since the daughter ions will not be transmitted by the electric sector.

C. Within the Magnetic Sector

Very similar arguments apply to the collection of daughter ions resulting from metastable transitions that occur in the magnetic analyzer. If an m_2^+ ion is formed just as an m_1^+ ion leaves the magnetic field, the m_2^+ ion will be collected as an m_1^+ ion, since no further mass separation occurs. On the other hand, if the decomposition occurs just as the m_1^+ ion enters the field, the m_2^+ ion will be collected at an apparent mass of m_2^2/m_1. If the decomposition occurs within the magnetic field (3), the m_2^+ ions are not brought to focus at a single value of V or H, and contribute to the background continuum between m_1^+ and $(m_2^2/m_1)^+$. In practice, this is only observable if the neutral species lost in the decomposition is very light—such as H or H_2—since otherwise, m_2^+ ions will tend to be lost to the walls of the analyzer tube. The first observation of metastable transitions of this type was reported for the reaction

$$CH_3OH^{+\cdot} \rightarrow CH_2OH^+ + H\cdot$$

in the mass spectrum of methanol, in which the continuum stretched from the $m/e = 32$ peak to the $m/e = 30.1$ metastable peak.

D. In the Final Field-Free Region

If a metastable transition occurs in the field-free region between the magnetic sector and the collector (regions 4 and 6, respectively, for single- and double-focusing instruments), the daughter ions are collected at the same mass-to-charge ratio as the parent ions and the occurrence of such transitions cannot usually be detected under normal operating conditions. However, if the fragmentation is accompanied by the release of internal energy as translational energy of the fragments (3), as in the reaction

$$C_6H_5NO_3^{+\cdot}(m/e\ 139) \rightarrow C_6H_5O_2^+(m/e\ 109) + NO \qquad [8]$$

in the mass spectrum of o-nitrophenol, the width of the parent ion peak ($m/e = 139$) is measurably greater than the widths of the $m/e = 134$ and 136 peaks of xenon. This was attributed to the collection of $m/e = 109$ ions, formed in region 6 of a double-focusing instrument, with components of translational energy perpendicular to the plane of the ion beam.

A sensitive ion detector that can record normal ions and daughter ions of metastable transitions occurring in the field-free regions before and after the magnetic sector has recently been described by Daly (4), and is shown schematically in Figure 2. The retarding grid B allows ions of energy below a

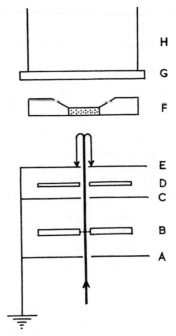

Figure 2. Metastable ion detector. (*a*) Resolving slit. (*b*) Retarding grid. (*c, e*) Slits at ground potential. (*d*) Secondary electron suppressor. (*f*) Glass scintillator. (*g*) Glass window. (*h*) Photomultiplier.

certain value to be rejected. A glass scintillator is held in a metal electrode F, the potential V_F of which can be varied. All ions of less energy than eV_F are turned back and strike plate E so that secondary electrons are emitted; these are accelerated and pass through the aluminium coating, so causing emission of light from the glass scintillator. This is detected and recorded by the photomultiplier. When $V_F \geq V_0$, the accelerating potential, a normal mass spectrum is recorded. If $V_F < V_0$, normal ions are not recorded, but daughter ions formed in metastable transitions occurring anywhere before the final earthed slit can be detected, providing $V_F > V_0 \times m_2/m_1$. The m_2^+ ions formed from m_1^+ ions in the field-free region before the magnetic sector are detected at an apparent mass of m_2^2/m_1, and those formed after leaving the magnetic field are detected at an apparent mass of m_1^+. The different decompositions of a given parent ion may be found by making use of the retarding grid B. The magnetic field is adjusted so that the parent ion is collected; V_F is then made a little less than V_0 so that normal ions are not recorded. The voltage V_B on the retarding grid is then increased slowly and the ion current is monitored, this current being due entirely to the different daughter ions originating from the single parent ion. Daughter ions of mass m_2 will be removed by the grid when $V_B > V_0 \times m_2/m_1$, so that steps in the plot of ion current against retarding grid voltage allow one to identify the masses of the daughter ions. Since the normal spectrum can be suppressed using this detector, the sensitivity for recording products of metastable transitions is very high. It has been used to distinguish between *cis-* and *trans*-butene-2, which give significantly different "metastable spectra" but produce almost identical normal spectra (5).

E. In the Field-Free Region before the Electric Sector

In a double-focusing instrument, the introduction of an electric sector between the source and the magnetic sector gives rise to two further regions in which metastable transitions may occur (regions 2 and 3 in Figure 1). In region 2, if an ion m_1^+ fragments to give an ion m_2^+ without the release of internal energy, the resulting translational energy of the m_2^+ ions is $V_0e \times m_2/m_1$. In normal operation, the potential difference E_0 across the plates of the sector is such that these ions are lost and only ions of translational energy very close to V_0e are transmitted by the sector. The range of energies transmitted depends on the instrument geometry and slit widths; in a standard MS902 instrument (A.E.I. Scientific Apparatus, Ltd.), ions with translational energies within the range $(V_0 \pm 0.66\% \, V_0)e$ are transmitted when $V_0/E_0 \simeq 14.8$. It is possible to arrange for the m_2^+ ions formed between the source and electric sector to be transmitted by increasing the ratio V/E to the value m_1V_0/m_2E_0 either by (1) increasing the accelerating voltage to $V_0 \times m_1/m_2$ at constant sector voltage E_0, so that after fragmentation, the energy of m_2^+ ions is V_0e, or (2) decreasing the sector voltage to $E_0 \times m_2/m_1$ at constant accelerating voltage V_0, so that ions of translational energy $V_0e \times m_2/m_1$ are transmitted. In either case, normal ions strike the outer plate of the sector and are lost.

In the first method (6–9), the magnetic field is adjusted so that m_2^+ ions formed in the source are collected. The accelerating voltage is then scanned and the normal beam is lost, but since the electric sector voltage is unchanged, the mass scale is constant. When $V = V_0 \times m_1/m_2$, m_2^+ ions formed in region 2 will be transmitted by the sector and a signal will be observed. If several precursors of m_2^+ ions exist, signals will be observed whenever $V = V_0 \times m_P/m_2$, where m_P is the mass of the precursor ion. An example of this showing the precursors of the C_3H^+ ion in the mass spectrum of toluene is shown in Figure 3. The advantage of this method is that precursors can be identified directly using

$$\frac{V}{V_0} = \frac{m_1}{m_2} \qquad\qquad [9]$$

and by using a linear ramp voltage generator, this information can be obtained very rapidly. The resolution obtainable in defining the precursor,

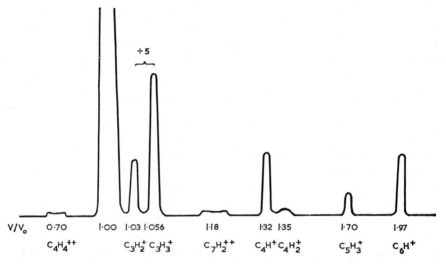

Figure 3. Precursors of the C_3H^+ ion in the mass spectrum of toluene; a plot of signal intensity as the accelerating voltage is scanned. For singly charged precursors, $(V/V_0) = (M/37)$ where M is the mass of the precursor ion. For doubly charged precursors, which give the broad peaks, $(V/V_0) = (M/74)$.

that is, the accuracy with which V/V_0 can be measured, varies with the instrument and the operating conditions, but in many cases is limited by the inherent width of the metastable peaks. In favorable cases, the resolution (10% valley definition) of peaks given by m_2^+ ions formed from different precursors rarely exceeds 50, so that if m_1/m_2 is large, there is sometimes an uncertainty of ± 1 amu in m_1^+ at high molecular weights.

The above method suffers from two disadvantages. In many instances, the tuning of an ion source depends quite critically on the applied accelerating voltage, owing to field penetration effects. These effects will clearly vary very considerably during a scan of the accelerating voltage, and the greater the value of V/V_0, the greater this variation becomes. As a result, weak signals can often be enhanced considerably by retuning the source, although care must be taken to allow for any slight change in V_0 brought about by the retuning. Consequently, relative intensities of metastable peaks obtained in this way are of little quantitative significance. A more serious disadvantage of the method is that it cannot be used to study metastable transitions in which the mass of the parent ion is considerably greater than that of the daughter ion; for most instruments, the practical limit for m_1/m_2 is 5–10, since the source will not operate satisfactorily over a wide range of accelerating voltages.

In the second method, the accelerating voltage is fixed at V_0 and the electric sector voltage E is scanned (10). This has the advantage that source tuning conditions are constant so that extraction efficiencies for ions of different mass will not vary. Its major disadvantage is that the mass scale changes as E is scanned so that it is no longer possible to tune to a particular daughter ion and obtain precursors from a simple voltage scan. It is therefore less well-suited than the first method for exploratory work, but is preferable for work concerned with metastable peak intensities.

F. Ion Kinetic Energy Spectroscopy

An interesting development of the second method is ion kinetic energy spectroscopy (11), in which an electron multiplier is inserted into the ion beam immediately after the intermediate slit. The ion beam can therefore be subjected to energy analysis by the electric sector, but no mass analysis is performed. The electric sector voltage E is scanned at fixed V_0 and the voltage ratio E_0/E is put equal to m_1/m_2. If this can be measured sufficiently accurately, and there is no interference from overlapping peaks, m_1^+ and m_2^+ can often be identified from the normal mass spectrum.

G. During Acceleration (Double-Focusing Instrument)

If an ion m_1 decomposes in the accelerating region after falling through a potential difference V, the resulting m_2^+ ion is then subjected to a further accelerating potential of $(V_0 - V)$ so that the final translational energy of m_2^+ ions formed in this region is given by

$$Ve \cdot \frac{m_2}{m_1} + (V_0 - V)e \cdot m_2 \qquad [10]$$

Consequently m_2^+ ions formed in this region will be transmitted by the electric sector when the accelerating voltage is between V_0 and $V_0 \times m_1/m_2$

so that decompositions within the accelerating field give rise to a low-intensity background continuum when metastable transitions in the first field-free region of a double-focusing instrument are under observation.

H. Within the Electric Sector

Ions that decompose within the electric sector are usually lost to the negative plate of the sector or are not transmitted by the intermediate slit between the electric and magnetic sectors. If the mass of the neutral species is only a very small fraction of the mass of the fragmenting ion, however, the path followed by the daughter ion differs very little from that of normal ions, and under these conditions, the daughter ion can pass through the intermediate slit and reach the collector. This was demonstrated (12) for the metastable transition

$$C_7H_8^{+\cdot} \rightarrow C_7H_7^+ + H\cdot \qquad [11]$$

by changing the V/V_0 ratio. When this ratio is 92/91, $C_7H_7^+$ ions formed in the first field-free region are collected at $m/e = 91$ on the mass scale. If it is slowly reduced, $C_7H_7^+$ ions formed within the electric sector at increasingly later points in the flight path are collected until when $V = V_0$, $C_7H_7^+$ ions formed in the second field-free region are collected at $m/e = 90.01$ on the mass scale. For values of V/V_0 between 92/91 and 1.00, a peak is observed at a point on the mass scale between $m/e = 91.0$ and 90.01, the exact position moving to lower values of m/e as the ratio V/V_0 is reduced. In an instrument of Mattauch-Herzog geometry, the intermediate slit is much wider than in instruments of Nier-Johnson geometry, so that daughter ions formed within the electric sector are much more readily collected (13), giving rise to a higher background continuum and broader metastable peaks due to metastable transitions occurring in the second field-free region. A more detailed account (13a) of metastable transitions occurring within the electric sector of an instrument of Mattauch-Herzog geometry has recently been given.

I. Consecutive Metastable Transitions

In the above discussion, it has been assumed that m_1^+ ions give m_2^+ ions in a single step with the loss of a single neutral species. While this is almost certainly true in the vast majority of metastable transitions, weak metastable peaks can sometimes be observed that appear to require the loss of two neutral species, that is, $m_1^+ \rightarrow m_2^+ \rightarrow m_3^+$. Since the observation of a metastable transition in a single field-free region gives information only on the ions entering and leaving that particular field-free region, it is necessary to use two field-free regions to demonstrate the occurrence of consecutive metastable transitions (14). In the mass spectrum of toluene, for example, the sequence $C_7H_7^+ \rightarrow C_5H_5^+ \rightarrow C_3H_3^+$ was shown to occur by increasing the accelerating voltage to $V_0 \times 91/65$ so that $C_5H_5^+$ ions formed from $C_7H_7^+$ ions in the first field-free region of a double-focusing instrument were transmitted by the electric sector; these ions alone therefore reached the second field-free region where decomposition to $C_3H_3^+$ ions gave a low-intensity peak at $m/e = 39^2/65 = 22.3$. The occurrence of this sequence was suspected from the observation of a weak peak at $m/e = 16.8$, corresponding to $C_7H_7^+ \rightarrow C_3H_3^+$, in the normal mass spectrum of toluene. A number of other observations of consecutive metastable transitions have now been reported (15–19).

J. Collision-Induced Metastable Transitions

A further type of metastable transition that may also be observed in a magnetic deflexion instrument is a collision-induced fragmentation (20). If the pressure in the analyzer tube is appreciably above 1×10^{-6} torr, ions with insufficient energy to fragment may collide with molecules of the background gas and so convert a small fraction of their translational energy into internal energy, thereby causing the ion to fragment. For example, if the translational energy of an ion is 8000 eV, a glancing collision may be sufficient to convert a few eV into internal energy so that the ion may undergo one or more unimolecular decompositions subsequent to the collision. Since in such a collision, an ion will change its flight-path very little and will suffer a very small fractional reduction in its translational energy prior to decomposition, daughter ions can be observed using exactly the same methods as those employed in studying normal metastable transitions. A metastable peak may be shown to arise from a collision-induced process by studying the pressure dependence of the intensity relative to the intensities of normal peaks; the relative abundance of a metastable peak arising from a uni-molecular process is independent of pressure of sample, whereas that of a metastable peak arising from a collision process rises. In this way, the peak at $m/e = 16.9$ in the mass spectrum of n-butane was shown (20) to be due to the collision-induced process $C_3H_7^+ \rightarrow C_2H_3^+ + CH_4$.

Since it is unlikely that the conversion of the translational energy of a high-velocity ion to internal energy will depend critically on the nature of the neutral collision partner, collision-induced metastable transitions may also be studied by introducing a trace of a second gas into the analyzer tube or more simply by gently baking the tube so as to raise the background pressure. This method was used (21) to show that the products arising from the colli-sion-induced decomposition of molecular ions of several aromatic compounds were very similar to those formed by electron impact, as is to be expected, since in each case, the products arise from relatively highly excited ions. This contrasts with the much lower energy processes that give rise to normal metastable peaks (*vide infra*). A preliminary study of the use of collision-induced metastable transitions in the determination of structures of organic ions has been reported, and the results suggest that it may well prove to be a useful additional technique (22).

K. In 180° Magnetic Deflexion Instruments

In a 180° magnetic deflexion instrument, there is no field-free region between the source and the magnetic analyzer, but it is nevertheless possible to observe metastable peaks due to the decomposition of ions in the early part of the magnetic sector (23, 24). If m_1^+ ions give rise to m_2^+ ions, the peak is observed at an apparent mass of m_2^2/m_1, so that the behavior is very similar to that found for metastable transitions that occur in the field-free region before the magnetic analyzer in a 90° sector instrument.

L. In Time-of-Flight Instruments

Metastable transitions are not observed in a standard time-of-flight mass spectrometer, since if no translational energy is released during the fragmentation process, undissociated m_1^+ ions, newly formed m_2^+ ions, and the neutral species $(m_1 - m_2)$ will all reach the detector simultaneously. They can be distinguished, however, by placing a retarding grid a short distance in front of the detector to which a small retarding potential V' is applied (25, 26). If V_0 is the applied accelerating voltage, the translational energies of the three types of species, after passing through the grid, are

$$m_1^+: \quad (V_0 - V')e$$

$$m_2^+: \quad \frac{m_2}{m_1}(V_0 - V')e \qquad\qquad [12]$$

$$(m_1 - m_2): \quad V_0 e$$

Since the neutral fragment is not retarded, its flight time is equal to that of an m_1^+ ion in the absence of a retarding field. Undecomposed m_1^+ ions are retarded somewhat and their flight time therefore rises, but m_2^+ ions will be retarded still more so that their flight time is even greater. As the retarding potential is increased, therefore, a single peak at m_1^+ on the mass scale splits into three components due to the different flight times of the three species, as illustrated in Figure 4. A simplified equation due to McLafferty et al. (26)

allows rapid calculation of the masses of the fragment ion and neutral species from a measurement of the flight times:

$$\frac{t_1 - t_n}{t_2 - t_n} = \frac{[1 - \{1 - (m_1/m_2)(1 - V'/V_0)\}^{1/2}](V'/V_0)^{1/2}}{[1 - \{1 - (m_1/m_2)(1 - V'/V_0)\}^{1/2}]} \qquad [13]$$

where t_1, t_2, and t_n are the flight-times of m_1^+, m_2^+, and the neutral species, respectively. Since the effective values of V_0 and V' are usually not quite

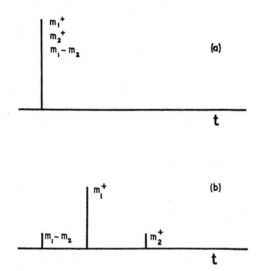

Figure 4. Signals from a time-of-flight instrument (a) with no retarding potential and (b) with a retarding potential. The neutral species are not retarded and so appear at the same point on the time axis in each case; because of its lower energy, the daughter ion is retarded to a greater extent than the parent ion.

equal to the applied values, a calibration procedure is necessary to eliminate the effects of fringing fields. More general expressions for use with instruments containing several grids have been developed by Hunt and McGee (27).

III. KINETIC ASPECTS OF METASTABLE TRANSITIONS

A. Quasi-Equilibrium Theory

In its semiclassical form, the quasi-equilibrium theory (QET) (28–30) of mass spectra predicts that the specific rate constant $k(E)$ for a given fragmentation reaction varies with the internal energy E according to the equation

$$k(E) = \frac{S}{h} \frac{W^{\ddagger}(E - E_0)}{\rho(E)}$$ [14]

in which h is Planck's constant, $W^{\ddagger}(E - E_0)$ is the number of states of the activated complex configuration with energy $\leqslant (E - E_0)$, $\rho E \, dE$ is the number of states of the reactant ion having energy between E and $E + dE$, and S is the symmetry factor, that is, the number of identical reactions (same activated complex and same activation energy E_0) yielding identical products. The rate constant is therefore a function only of the internal energy of the ion, and the assumption is made that whichever internal degrees of freedom are excited on ionization, rapid randomization of the excitation energy among all degrees of freedom occurs. When $E < E_0$, $k(E) = 0$, and when $E = E_0$, the minimum value of $k(E)$ is given by

$$k(E)_{\text{min}} = \frac{S}{h\rho(E_0)}$$ [15]

A quantum-mechanical treatment of the reaction coordinate removes this sharp discontinuity at $E = E_0$, but $k(E)$ rises very rapidly with increasing internal energy in the region of $E = E_0$. In principle, therefore, it should be possible to calculate $k(E)$ for a given fragmentation as a function of internal energy, but except in a few relatively simple cases, insufficient information is available to make use of [14]. Nevertheless, the theory predicts that for each fragmentation reaction, there can be a full range of rate constants from approximately zero to very high values, provided that the quasi-equilibrium hypothesis holds. It is not yet certain whether the hypothesis is valid for very small ions (three or four atoms), but in its most refined form, it can account for the observation of metastable transitions in the mass spectrum of methane (30) and is thought to be valid for larger ions. A case in which the hypothesis is clearly invalid is that in which the ion is formed on a repulsive potential-energy surface, as appears to be the case for a number of halogenated compounds; in such cases, the minimum rate constant is likely to be in the region of 10^{12}–10^{13}/sec.

For the qualitative discussion of the fragmentation of large ions, a very-much-simplified approximate form of the QET is used. The ion is looked on as a set of N degenerate harmonic oscillators, and in its simplest form, this model leads to the following relationship between $k(E)$ and E:

$$k(E) = \nu \left(\frac{E - E_0}{E} \right)^{N-1}$$ [16]

in which v is identified as a frequency factor. Although quantitative predictions made using [16] are quite incorrect, especially near threshold, the equation is useful in qualitative discussions of fragmentation reactions in terms of the three parameters v, E_0, and N. Because of the approximate nature of the equation, much better agreement between calculations and experiment is obtained if N is replaced by $N/2$ or $2N/3$, so that the "effective number" of oscillators (*not* the number of "effective oscillators") is always much less than N (28). For large ions formed with a range of internal energies, each fragment should occur with a continuous range of rate constants from less than 10^4/sec to greater than 10^6/sec, so that peaks arising from parent ions, daughter ions, and metastable transitions should be observable in principle. This is in agreement with Ottinger's observations on a number of metastable transitions, and the bulk of the experimental evidence currently available supports a continuous distribution of rate constants rather than a set of discrete rate constants (31). In practice, relatively few of the many fragmentation reactions that occur in the source give rise to intense metastable peaks, for the reasons discussed below.

B. Competing Fragmentation Processes

In many cases, parent ions fragment in several different ways, giving rise to daughter ions m_1^+, m_2^+, m_3^+ ... , and one can characterize each reaction in terms of the parameters (v_1, E_1), (v_2, E_2), (v_3, E_3) ···. If it is assumed that the fragmentations all take place from the same electronic state of the parent ion, they will be in effective competition with each other. The simple case of two competing fragmentations can be considered with the aid of Figure 5. If the frequency factors are very similar, the shapes of the log k versus E curves will be very similar as shown in Figure 5a. When the internal energy of the parent ion equals E_1, reaction 1 will occur sufficiently rapidly for it to give rise to a metastable peak. As the internal energy of the parent ion increases, $k(E)_1$ rises very rapidly and is always greater than $k(E)_2$, so that for an internal energy of E_2, sufficient for reaction 2 to give rise to a metastable peak, $k(E)_1 \gg k(E)_2$ and the parent ions will undergo reaction 1 in the source rather than reaction 2 several microseconds later in a field-free region. If fragmentation reactions are in true competition with each other, therefore, it is always the lowest energy process that gives rise to the most intense metastable peak. In this particular case, reaction 2 would never compete favorably with reaction 1, and the intensity of the peak due to m_1^+ ions would be much greater than that due to m_2^+ ions.

A situation that frequently arises is illustrated in Figure 5b: in this case, reaction 3 has the lower energy of activation, but now the frequency factor for reaction 3 is appreciably less than that for reaction 4. The lowest energy process is frequently a rearrangement reaction or a molecular-elimination reaction for which one might expect a low frequency factor on steric grounds. The loss of a radical, by what is usually assumed to be a simple bond cleavage, normally requires more internal energy, but is expected to have a higher frequency factor than does a rearrangement reaction. In this case, an appreciable range of energies in the region of E_3 can give rise to metastable transitions so that an intense metastable peak will be observed for reaction 3.

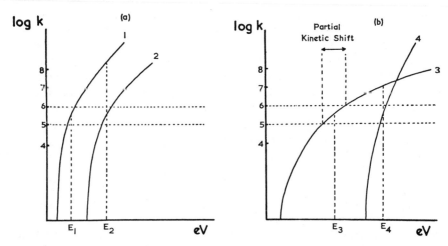

Figure 5. Log k versus E curves for different types of fragmentation reaction. In (a), reactions 1 and 2 have similar frequency factors so that at all energies, reaction 1 is faster than reaction 2. In (b), reaction 3 has a much lower frequency factor than that of reaction 4, so that despite its lower energy of activation, it is slower than reaction 4 at high energies.

When the internal energy of the parent ion is approximately equal to E_4, $k(E)_3 > k(E)_4$, but the difference is not as much as between $k(E)_1$ and $k(E)_2$ in Figure 5a so that a low intensity metastable peak may be observed for reaction 4. At still higher internal energies, however, $k(E)_4 > k(E)_3$, so that if the majority of parent ions are formed with high internal energies, reaction 4 will be much faster than reaction 3 in the source, and the peak arising from $m_4{}^+$ fragment ions will be much more intense than that arising from $m_3{}^+$ fragment ions, that is, just the reverse of the relationship between the metastable-peak intensities. It is clear, therefore, that in a 70 eV spectrum, one cannot necessarily infer from the observation of an intense metastable peak that the reaction responsible for the peak is the major fragmentation pathway of the parent ion in the source; still less can one infer that if a metastable peak for a particular reaction is of very low or zero intensity that this reaction is of little or no importance in the source. Whereas in most cases, the relative activation energies largely control the relative intensities of the metastable peaks, the relative frequency factors are the important parameters at higher internal energies.

A few cases have been reported of the observation of metastable peaks for two fragmentations of the same parent ion that differ significantly in their energies of activation (29). If the processes are in effective competition, only the lower energy process should give rise to a metastable peak, and the observation of two metastable peaks suggests that the fragmentations occur from different sets of electronic states between which interconversion is forbidden. Several fragmentation reactions of the benzene molecular ion, apparently differing in activation energy by as much as 2 eV, have been observed to give metastable peaks of comparable intensity (32, 33). This was originally interpreted as being due to fragmentation from isolated electronic states, but it has been shown recently that fragmentation from a common state is possible if the very large kinetic shifts (*vide infra*) are taken into account (30). More recently, the intensities of metastable peaks in the mass spectra of fluoroethylene and 1,1-difluoroethylene have been interpreted (34) in terms of fragmentation from isolated electronic states. In each case, the loss of HF is the lowest energy process and gives rise to an intense metastable peak, but fairly intense metastable peaks are observed for the loss of a hydrogen atom, which, especially in the CH_2CF_2 case, is a much higher energy process.

C. Kinetic Shifts and Heats of Formation of Ions

One of the problems involved in applying the QET to the fragmentation of larger ions is the evaluation of E_0 for a particular process, that is, the minimum internal energy required to pass through the activated complex configuration. Values based on appearance potential measurements of fragment ions are likely to be high, since the internal energy must be sufficiently above the minimum E_0 for $k(E)$ to be in the region of 10^6/sec, and this excess energy is referred to as the "kinetic shift" (35). Since $k(E)$ for a metastable transition is less than that for a process giving a normal fragment ion, it follows that the kinetic shift for a metastable transition will be less than that for a fragment ion. The appearance potential of a metastable transition when subtracted from the appearance potential of the normal fragment ion can therefore give at least a partial measure of the kinetic shift. The difference in these two measurements clearly depends on the instrument design and on the operating conditions and has no quantitative significance in most cases. For example, the partial kinetic shift determined in this way for the reaction

$$C_6H_5CN^{+\cdot} \rightarrow C_6H_4^{+\cdot} + HCN \qquad [17]$$

has been measured as 1.3 eV (36), 0.6 eV (37), and 0.7 eV (38) in three different laboratories. If a process gives rise to an intense metastable peak, this may suggest that k rises slowly with E so that a large partial kinetic shift should be measurable. This has been shown to be true in a number of cases in which partial kinetic shifts ranging from 0.25—1.0 eV were found for processes giving rise to intense metastable peaks (38). How much of the total kinetic shift is measured by this method clearly depends on the form of the log k versus E curve, but for curves similar to that shown for reaction 3 in Figure 5b, it is likely that the total kinetic shift is appreciably larger than the measurable part. It therefore follows that thermochemical calculations based on the appearance potentials of fragment ions formed in processes giving intense metastable peaks are likely to be significantly in error. Calculations based on the appearance potentials of metastable transitions will be less in error, and this approach has been used (38, 39) to show that the heat of formation of the $C_4H_4^{+\cdot}$ ion in the mass spectrum of benzene is not more than 12.7 eV, in good agreement with the value obtained from the ionization potential of vinyl acetylene; this is to be compared with a value of approximately 13.5 eV based on the appearance potential of the $C_4H_4^{+\cdot}$ fragment ion.

D. Applications of Intensity Measurements

In the simple case in which m_1^+ ions fragment by only one route to give m_2^+ ions, the relative intensities of peaks due to m_1^+ ions, m_2^+ ions, and metastable ions, m^*, depend primarily on the energy distribution in the initially formed m_1^+ ions. In general, as the average internal energy rises, the intensity ratios $I(m_2^+)/I(m_1^+)$, $I(m_2^+)/I(m^*)$, and $I(m^*)/I(m_1^+)$ all increase; in different mass spectra, therefore, if the same m_1^+ ion is formed from different precursors, these ratios provide a guide to the average internal energy distributions of m_1^+ ions in the two cases. This presupposes that there is only one route in each case to the formation of m_1^+ and m_2^+ ions, and that subsequent decomposition of m_2^+ ions is negligible. An interesting correlation of this type has been observed by McLafferty and Pike (40) for a number of fragmentation reactions in the mass spectra of alcohols and ketones. Taking the case of primary alcohols as an example, all yield abundant $C_2H_5O^+$ ions, which fragment by two possible routes (41):

$$C_2H_5O^+ \rightarrow H_3O^+ + C_2H_2 \qquad m^* = 8.0 \qquad\qquad [18]$$

$$\rightarrow HCO^+ + CH_4 \qquad m^* = 18.7 \qquad\qquad [19]$$

It is found that a plot of log $[I(m^*)/I(C_2H_5O^+)]$ against the reciprocal of the number of vibrational degrees of freedom of the original molecular ion gives a good straight line for each reaction. The plots for the two reactions have similar slopes but different intercepts. Similar plots for the two reactions can be obtained for secondary alcohols and terminal diols; for each type of alcohol, the slopes of the plots given by the two reactions are the same, but for a given reaction, the slopes of plots given by different types of alcohol are different. The general trends can be explained by assuming that the initial excitation energy distribution is determined largely by the type of functional group present in the molecule rather than on the actual size of the molecule; in this case, therefore, for a given type of molecule, the initial excitation energy is expected to be partitioned between the fragment ion and the neutral species roughly in proportion to the number of vibrational degrees of freedom of each species. The fraction of the original excitation energy retained by the fragment ions m_1^+ will therefore on average fall as the size of the original molecules rises, thereby leading to a reduction in the ratio $I(m^*)/I(m_1^+)$. While this can account for the general trends observed, it does not readily lead to an explanation of the good linearity of the plots obtained for a number of different fragmentations.

In an earlier study (42), the $C_7H_7^+$ ion was generated from a number of different precursors so that its apparent heat of formation varied from 208–242 kcal/mole. It was found that as the apparent heat of formation increased—that is, as the internal energy of the ion increased, the ratio $I(m^*)/I(m_2^+)$ decreased, where the reaction studied was

$$C_7H_7^+ \rightarrow C_5H_5^+ + C_2H_2 \qquad [20]$$

This trend was interpreted by suggesting that as the average internal energy of the $C_7H_7^+$ ions increases, the average rate constant for [20] increases so that an increasing fraction of the $C_7H_7^+$ ions decompose in the source. In a similar study of the fragmentation of $C_6H_5CO^+$ ions generated from different molecules (43), the reaction

$$C_6H_5CO^+ \rightarrow C_6H_5^+ + CO \qquad [21]$$

was used to demonstrate that the intensity ratios $I(m_1^+)/I(m^*)$ and $I(m_1^+)/I(m_2^+)$ are better explained in terms of a degrees-of-freedom effect than a heat-of-formation effect, and it was pointed out that the observations on [20] could be similarly explained.

If an ion undergoes two or more fragmentation reactions that give rise to reasonably intense metastable peaks, it has been suggested that the relative intensities of the metastable peaks will be constant when the same ion is produced from different precursors, so that the intensity ratio can be used to characterize an ion structure (41). In a study of the $C_2H_5O^+$ ion, [18] and [19] were used to show that different precursors gave rise to ions of different structure. The following table indicates the intensity ratios found for different types of precursor:

TABLE 1

Molecule	$I(m^* = 8.0)/$ $I(45^+)$	$I(m^* = 18.7)/$ $I(45^+)$	$I(m^* = 8.0)/$ $I(m^* = 18.7)$	K.E. for $m^* = 18.7$
A CH_3OCH_2—Y	<0.1	12.3 ± 1.8	<0.01	<0.1 eV
B HO—CH_2CH_2—Y	55 ± 27	30 ± 14	1.8 ± 0.2	0.50 ± 0.04 eV
C $CH_3CH(OH)$—Y	10.1 ± 2.6	5.3 ± 1.2	1.9 ± 0.1	0.44 ± 0.02 eV
D CH_3CH_2O—Y	13.7 ± 5.3	6.9 ± 2.3	2.0 ± 0.1	0.52 ± 0.05

It is clear that $C_2H_5O^+$ ions formed from Type A molecules differ very considerably from the other types, and it is plausible to suggest that these ions have the structure $CH_3O^+ = CH_2$. Similarly, $C_2H_5O^+$ ions formed from Type B compounds have very high relative intensities of metastable peaks, which distinguish them from ions formed from Type C and D compounds, and a protonated oxirane structure has been proposed for these ions. For $C_2H_5O^+$ ions formed from compounds of Type B, C, and D, the metastable peak for [19] is broadened, owing to the release of approximately half an electron volt of internal energy as kinetic energy of the fragments.

A variety of other competing fragmentations have also been studied using this technique in an attempt to distinguish between different possible structures of the fragmenting ions (40, 44), but in most cases, it has been assumed that the intensity ratios of the metastable peaks are independent of the internal energy distributions of the fragmenting ions. While this may be approximately valid for metastable transitions that are truly in competition with each other and that occur from a relatively narrow range of internal energies, it will not necessarily be valid if the fragmentations occur from different isolated electronic states (44a) or if either of the fragmentations has a low frequency factor, which would mean that a fairly broad range of energies could give rise to metastable transitions. Since many rearrangement processes that give rise to metastable peaks are thought to have low frequency factors, it is possible to demonstrate the effect on the intensity ratio of changing the internal energy distribution in a number of ways. Williams and Yeo (45) have shown that for the two processes

$$C_3H_7O^+ \rightarrow C_3H_5^+ + H_2O \qquad [22]$$

$$C_3H_7O^+ \rightarrow CH_3O^+ + C_2H_4 \qquad [23]$$

which give rise to metastable transitions in the mass spectrum of diethyl ether, the ratio of the metastable peak intensities $I(m^*[22])/I(m^*[23])$ increases by approximately a factor of 10 when observations are made in the second field-free region rather than the first field-free region of an MS9 double-focusing instrument. Variation of the electron-beam energy and the accelerating voltage also caused significant changes in the metastable peak intensity ratio; in all cases, the ratio increased when experimental conditions were altered so as to reduce the average internal energy of the $C_3H_7O^+$ ion. Similar effects have been observed (46) for the intensity ratio of the metastable peaks given by the two fragmentations of the $C_7H_6F^+$ ion in the mass spectrum of p-fluorobenzyl chloride:

$$C_7H_6F^+ \rightarrow C_5H_4F^+ + C_2H_2 \qquad [24]$$

$$C_7H_6F^+ \rightarrow C_7H_5^+ + HF \qquad [25]$$

The intensity ratio $I(m^*[24])/I(m^*[25])$ was measured for the first field-free region of an MS9 instrument and was typically ~ 15 under normal operating conditions. This ratio was found to decrease when any one of the following was increased: the accelerating voltage, the ion repeller voltage, the source temperature, and the electron energy—that is, whenever the average internal energy of the $C_7H_6F^+$ ion was increased. These two sets of observations provide clear evidence that the intensity ratios of metastable peaks do vary with the average internal energy of the fragmenting ion so that fairly small changes in the ratio do not necessarily imply different structures of the fragmenting ions.

E. Metastable Transitions of Partly Deuterated Compounds

Observations on the metastable transitions that occur in the mass spectra of partly deuterated compounds strongly suggest that in the absence of any specific isotope effect, the time interval between formation of an ion and its decomposition in the flight tube is sufficiently long for H–D randomization to be complete in many cases. For example, the metastable-peak intensities for the loss of C_2H_2, C_2HD, and C_2D_2 from the molecular ion of 1,4 dideuterobenzene are in the ratio of 6:8:1, as expected for complete H–D scrambling (33). Similar scrambling is found in the mass spectrum of mono-2-deuteropyridine (47), and it is possible that in each case, this occurs by randomization of the carbon atoms via structures such as benzvalenes and prismanes and their aza-analogues. In the case of loss of HCN and DCN from the 2,4,6-trideuterobenzonitrile molecular ion, early work (48) appeared to indicate that the loss of DCN is favored at low electron energy, but a more complete study (48a) showed that this ratio may have been affected by the presence of impurities and that HCN and DCN are lost statistically at all electron energies, probably due to rapid isomerization via a common intermediate. Similar results have been obtained for loss of HCN and DCN in the 70 eV spectra of bicyclic aromatic compounds (49). In a study of meta-stable transitions in the mass spectra of some aromatic carboxylic acids (50), metastable peaks were found corresponding to the loss of both ˙OD and ˙OH from the molecular ion of C_6H_5COOD, the relative intensities of the two peaks being in the ratio 1:2. Rapid exchange of the two hydrogen atoms in the ortho positions with the deuterium atom in the carboxylic group is postulated to rationalize this ratio, and a similar explanation was invoked to explain metastable peak intensities in the mass spectrum of phthalic acid-d_2.

In contrast to the above results, strong isotope effects and little H–D scrambling are found in the metastable transitions given by partly deuterated propanes (51, 52). One of the fragmentation routes is the loss of CH_4 by the molecular ion of propane, but this gives only a low-intensity metastable peak in the mass spectrum of C_3H_8; the corresponding loss of CD_4 by $C_3D_8^{˙+}$ gives a much more intense metastable peak. In the mass spectra of partly deuterated propanes, a very intense metastable peak for the loss of CH_4 by $CH_3CD_2CH_3^+$ ions was found, but no peak corresponding to the loss of CD_4 or CD_3H from $CD_3CH_2CD_3^+$ could be detected. Similarly, no meta-stable peak for the loss of CH_3D from $CH_3CD_2CH_3^+$ was observed. These strong isotope effects are in agreement with the observations of a similar effect on the loss of D and H atoms from $CH_3CD_2CH_3^+$ and $CH_3CHDCH_3^+$ (53) and from a series of partly deuterated methanes (54). Metastable transitions in the mass spectra of propane and partly deuterated propanes have been reinvestigated using the first field-free region of a double-focusing mass spectrometer. The very strong isotope effects were confirmed, and quantitative agreement with quasi-equilibrium theory calculations was obtained (55).

While the above results appear to indicate that H/D scrambling does not occur in the propane molecular ion, observations on the metastable peaks in the mass spectra of a number of partly deuterated aliphatic ketones indicate considerable H/D scrambling in the alkyl chain before fragmentation occurs (56). For example, the loss of propene by the molecular ion of $CH_3CD_2 \cdot CO \cdot CD_2CH_2CH_2CH_3$ in the McLafferty rearrangement is expected to proceed via a specific γ-hydrogen atom rearrangement, leading to the loss of C_3H_6, and retention of all deuterium ions in the resulting fragment ion. Whereas this is very nearly the case in the 70 eV spectrum, significant quantities of C_3H_5D and a little $C_3H_4D_2$ are lost in the 10 eV spectrum. Metastable transitions observed in the first field-free region of an MS9 double-focusing instrument indicate that C_3H_5D is now the main fragment to be lost, together with C_3H_6, but in the second field-free region, metastable peaks of comparable intensities are given for the loss of the species C_3H_6, C_3H_5D, $C_3H_4D_2$, and $C_3H_3D_3$ together with a small peak arising from the loss of $C_3H_2D_4$. These four observations (70 eV, 10 eV, m_1^*, and m_2^*) are of the product distributions given by the fragmentation of ions of decreasing internal energy and consequently increasing average lifetime. In this case, increasing H/D scrambling as the average lifetime of the ion increases appears to be a more plausible rationalization of the observations than an explanation in terms of energy-dependent isotope effects.

IV. FRAGMENTATION MAPS

A. Experimental Procedure

When a routine low-resolution mass spectrum of a compound is obtained, the number of metastable peaks that can be assigned with certainty is relatively limited. They can normally be used to establish the occurrence of a number of rearrangement reactions, which give rise to the more intense metastable peaks, but the routes to many of the abundant fragment ions cannot be inferred from metastable transitions in this way. If metastable transitions are observed in the first field-free region of a double-focusing instrument, it is possible to operate at a much higher electron-multiplier gain owing to the fact that the normal ions are not collected (8). Since the mass of the daughter ion is always known, a very large number of metastable transitions with intensities covering a range of 10^4 can be observed and readily assigned. As the number of atoms in the fragmenting ion rises, the rate of change of k with E falls for given values of v and E_0, so that a more or less broad range of energies will give rise to rate constants in the range 10^5–10^6/sec for almost all fragmentation processes. In view of the increased sensitivity available when making observations in the first field-free region, it is therefore reasonable to assume that, for ions containing a large number of atoms, all important fragmentations will give rise to a metastable peak. Precursors for each fragment ion can be established by tuning in to the particular ion of interest and then scanning the accelerating voltage as previously described. In this way, a complete fragmentation map can be built up indicating both the precursor ions of a given daughter ion and all daughter ions of a given precursor (33). If this is done manually for each ion, it is very time-consuming and also necessitates maintaining a reasonably constant sample pressure over a long period. A semiautomatic method employing a linear ramp generator to scan the accelerating voltage has been used to speed up the acquisition of data (57, 58).

For fairly small molecules, a simple flow chart is adequate to represent the different fragmentation paths established by this technique, and this has been used to illustrate the fragmentation paths in the mass spectra of benzene

(33) and the fluoroethylenes (34), for example. A flow chart becomes increasingly difficult to construct as the size of the molecule increases; in the mass spectrum of n-butane, for example, 61 metastable transitions were identified (57), many of which arise from the loss of H· or H_2. For substances of much higher molecular weight, several hundred metastable transitions may be detected, so that the collection, processing, and presentation of data become of considerable importance, and are best carried out using computer techniques.

At present, the most satisfactory method of obtaining data on metastable transitions is the semiautomatic voltage scan coupled with the manual tuning of the magnet for each daughter ion. The linearity of the voltage scan is used to define the mass scale for precursor ions, and the ion current during the voltage scan is either recorded in analog form on the ultraviolet recorder chart, or in digital form by a slow digitizer that is then used to calculate the voltage centroid for each metastable peak. The data read into the computer consists of the nominal mass of each fragment ion, the accelerating voltage ratio, the accurate mass of each ion studied, and the maximum number of heteroatoms lost in the neutral species in each step. All possible fragmentation pathways are then identified in terms of the low-resolution data, and a redundancy check is made to remove routes that contain no additional information that is not found in the other routes that are retained. For the remaining routes, the accurate masses of the daughter ions are subtracted from that of the molecular ion, thereby establishing the probable atomic compositions of the neutral species lost.

B. Structural Applications

This approach was used (57) to study the metastable transitions in a peptide derivative, molecular weight 638, when 81 different possible fragmentation pathways were reduced to 33 by the redundancy check. For example, several began with $638 \rightarrow 494$ but were almost all duplicated by others that began with $638 \rightarrow 623 \rightarrow 494$. In such cases, the pathway with more steps is the better one to retain for structural studies, and for this particular molecule, one of the pathways retained was

$$638 \rightarrow 623 \rightarrow 494 \rightarrow 437 \rightarrow 338 \rightarrow 310 \rightarrow 267$$

Reference to the high resolution spectrum shows that the units lost in this fragmentation sequence are

$$-CH_3-C_6H_{11}NO_2-C_2H_3NO-C_5H_9NO-CO-C_2H_5N$$

leaving the residue $C_{18}H_{35}O$. It is reasonable to suppose that the molecule is built up of some arrangement of these units, and in this case, the six units could give rise to 180 different possible structures. Since any arrangement of the units must also satisfy other fragmentation pathways, many of these possible structures can be ruled out. In this way, consideration of only six fragmentation sequences has resulted in only two possible arrangements of the six units in the molecule. In a similar study on cortisone (58), the 86 original pathways were reduced to 56 after the redundancy check, and when used in conjunction with the high-resolution data, the number of possible structures was less than 20. It seems likely that this approach, coupled with the use of high-resolution data, should be particularly useful for further computer interpretation of mass spectra.

Having obtained and processed the considerable amount of data on metastable transitions, one is faced with the problem of presenting this in a compact and readily assimilated form. One example of how this may be done is shown in the fragmentation map for Tomatidine in Figure 6 (58). A line drawing of the normal mass spectrum is shown at the bottom of the map, and the units lost by each precursor ion to give each daughter ion are shown in the map. The precursors for a given ion are found by following the vertical line above the fragment peak in the mass spectrum until it is above any horizontal line on the left of it. Each horizontal line to the right then gives a precursor ion, the end of the horizontal line terminating in a dot, thereby allowing ready identification of all precursors. Similarly, the way in which a particular ion fragments is found by following the vertical line above the fragment ion peak in the mass spectrum and then looking at the neutral species given at the left hand of any horizontal line which begins in a dot on the chosen vertical line. It should clearly be possible to program the computer to print out a map of this type and so provide detailed mechanistic and structural information very rapidly.

V. THE SHAPES OF METASTABLE PEAKS

A. General Considerations

It has already been mentioned that metastable peaks observed in both the first and second field-free regions of a double-focusing mass spectrometer exhibit a considerable range of widths and shapes. The narrowest metastable peaks are little wider than the normal peaks in a mass spectrum, but the most commonly observed metastable peaks are considerably wider and are approximately gaussian in shape. Even these, when examined under slow-scanning conditions, exhibit a range of shapes, some possessing smoothly-rounded maxima, others being more triangular, with a sharp maximum. Rather less common are the somewhat wider peaks, which may extend over

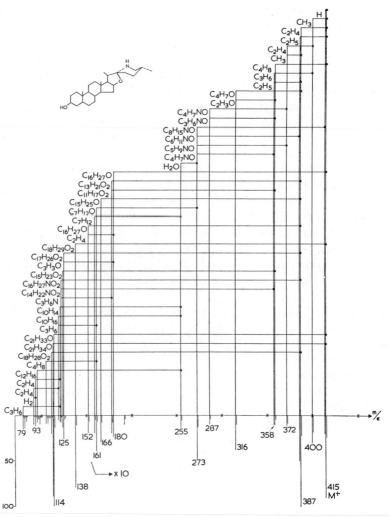

Figure 6. Fragmentation map of Tomatidine built up from observed metastable transitions.

several mass numbers, that have a more or less flat top to the peak, sometimes with a slight maximum at each end of the flat-topped region. Other peaks, especially those given by doubly charged ions fragmenting to give two singly charged ions, consist of two maxima with a central concave section or "dish" top, and recently several examples of metastable peaks exhibiting three maxima have been found (67). The shape of a peak resulting from a given metastable transition varies with the operating conditions in a particular instrument and can differ very considerably when observed in different magnetic deflexion instruments. Some typical peak shapes are shown schematically in Figure 7.

There are two major reasons for the wide variety of widths and shapes of metastable peaks. In almost every fragmentation reaction, a small amount of internal energy will be released as translational energy of the products and will be partitioned between the charged and neutral species in the inverse ratio of their masses. If the fragmentation occurs in a field-free region, the

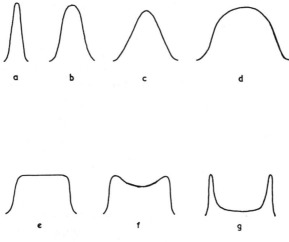

Figure 7. Schematic representation of shapes of metastable peaks broadened owing to the release of internal energy during fragmentation. The discrimination of the instrument against ions possessing components of translational energy perpendicular to the direction of the ion beam increases along the series $(a) \rightarrow (g)$.

resultant energy spread introduced into the ion beam causes the ions to be brought to focus at slightly different points on the mass scale, thereby increasing the *width* of the peak. The recorded *shape* of the peak depends largely on the discrimination effects within the instrument, which clearly vary with operating conditions and from instrument to instrument. In the following discussion, factors affecting metastable peak widths and shapes in (1) the field-free region before the magnetic analyzer and (2) the field-free region before the electric sector of a double-focusing instrument, are considered.

B. In the Field-Free Region Before the Magnetic Sector: Peak Widths

In deriving the simple relationship $m^* = m_2^2/m_1$, it was assumed that no internal energy was released as translational energy of the fragments, but if T eV is released and partitioned between m_2^+ and the $(m_1 - m_2)$ neutral species in accordance with the laws of conservation of energy and momentum, the m_2^+ ions are given a velocity increment additional to v_1, the velocity of the fragmenting ion, m_1. The two limiting values of v_2, the velocity of the m_2^+ ions, are related to v_1 by the equation

$$\frac{v_2}{v_1} = 1 \pm \left(\frac{\mu T}{Ve}\right)^{\frac{1}{2}} \qquad [26]$$

where $\mu = (m_1 - m_2)/m_2$, V is the accelerating potential, and e is the electronic charge (2). These limiting values will be reached only if all the translational energy is released in the direction of the ion beam (x direction). In general, there will be components in the y direction (perpendicular both to the direction of the ion beam and the direction of the magnetic field) and the z direction (parallel to the magnetic field), so that the velocity in the x direction, v_2, will lie between these two values. Equation [26] can be used to calculate the limiting positions on the mass scale at which m_2^+ ions will be collected, using the equation

$$m^* = \left(\frac{m_2^2}{m_1}\right)\left(\frac{v_2}{v_1}\right)^2 = \left(\frac{m_2^2}{m_1}\right)\left[1 \pm \left(\frac{\mu T}{Ve}\right)^{\frac{1}{2}}\right]^2 \qquad [27]$$

The limiting values of m^* have been equated to the ends of the flat-topped region of a metastable peak, and this leads to the following expressions for the width d of the flat-topped region and the translational energy release T:

$$d = \left(\frac{4m_2^2}{m_1}\right)\left(\frac{\mu T}{Ve}\right)^{\frac{1}{2}} \qquad [28]$$

$$T = \frac{Ve\,d^2 m_1^2}{16\mu m_2^4} \qquad [29]$$

These equations were tested using the flat-topped metastable peak given by the loss of NO by the molecular ion of o-nitrophenol, that is, $139^+ \rightarrow 109^+$, for which the predicted value of $m^* = 85.5$. The observed peak was symmetrical about this value, and observations of the width of the flat-topped region of the peak at accelerating voltages of 8 kV, 6 kV, 4 kV, and 2 kV confirmed that $d \geqslant V^{-\frac{1}{2}}$ and led to a value of $T = 0.76$ eV. This was in excellent agreement with a value of 0.74 eV obtained using a retarding-grid method (2).

The same argument, suitably modified for the case in which

$$m_1^{++} \rightarrow m_2^+ + (m_1 - m_2)^+ \qquad [30]$$

has been used to determine the amount of translational energy released in the decomposition of a number of doubly charged ions given by benzene (59) and other aromatic hydrocarbons (60) from the equations

$$d = \left(\frac{m_2^2}{m_1}\right)\left(\frac{\mu T}{2Ve}\right)^{\frac{1}{2}}$$ [31]

$$T = \frac{Ve\, d^2 m_1^2}{32\, \mu m_2^4}$$ [32]

Since both particles formed in the fragmentation are charged, both may be detected in principle, although the peak given by the lighter ion is often masked by peaks due to normal ions. Both are observable for the reaction

$$C_6H_6^{++} \rightarrow C_5H_3^+ + CH_3^+$$ [33]

and lead to a value of about 2.7 eV for T. Since the coulombic repulsion of the two charges in a cyclic ion of approximately the same size as the benzene molecule would lead to an estimated translational energy release of over 5 eV, it is clear that the $C_6H_6^{++}$ ion has an open-chain structure (61). Of the 2.7 eV of translational energy released, about 2.2 eV will be carried by the CH_3^+ ion, in good agreement with the value obtained from satellite peak* measurements (62).

C. Peak Shapes

While the above treatment can account for the widths of metastable peaks, it does not account for the shapes. Calculations by both Elliott (63) and Flowers (64) attempted to allow for the release of energy in directions perpendicular to the ion path at different points in the field-free region, and in each case, the predicted peak shape was one of two fairly sharp maxima between which was a relatively deep minimum, that is, a very highly concave central region of the peak. While this shape is approached by some of the peaks observed for the decomposition of doubly charged ions, the majority of broad metastable peaks are relatively flat-topped—that is, the intensity in the center of the peak is greater than the calculations suggest. Flowers suggested that peaks with a concave central region could be attributed to metastable transitions in which the translational energy released is sharply peaked about a particular value. The flat-topped peaks have been attributed to metastable transitions in which a range of translational energies is released; the different values of T would lead to a range of values of d ([28] and [29]).

---

* Fragment ions formed with excess translational energy within the source give rise to satellite peaks on the high mass side of normal peaks if a mass spectrum is obtained, using a single-focusing instrument, by scanning the accelerating voltage at fixed magnetic field strength. The difference in accelerating voltages required to bring to focus the thermal ions and those carrying excess translational energy directly measures this excess translational energy.

Although this may be a factor in determining the shapes of some meta-stable peaks, and has recently been suggested as an explanation of the shape of the peak given by the metastable transition

$$CH_2OH^+ \rightarrow CHO^+ + H_2 \qquad\qquad [34]$$

in the mass spectrum of methanol (65), an alternative explanation in terms of discrimination effects is probably of more general applicability (3, 13). A detailed quantitative account is given in Ref. 3, together with the supporting experimental evidence, but the basis of the treatment can be summarized qualitatively in the following terms (see Figure 8).

The ion beam travels in the xz plane, and after leaving the source slit in a single-focusing instrument or the intermediate β-slit of a double-focusing instrument, it travels through only one further slit, the collector slit, before collection. Discrimination effects will take place primarily at this slit. If translational energy is released only in the x direction, no ions are lost and the daughter ions are collected at an apparent mass equal to either the high or low mass end of the metastable peak. If the energy is released only in the xy plane, no ions will be lost, but the daughter ions will be brought to focus at an apparent mass lying between the two extreme values, the exact position depending on the magnitudes of the x and y components and on the point in the flight path at which fragmentation occurs. In the general case in which the energy released has a component in the z direction, some ions will be lost, since there is no focusing in the z direction and the collector slit is of finite length. Ion losses will increase in a given case if either the accelerating voltage or the collector-slit length is decreased. This is borne out by observing broad peaks at different accelerating voltages in the MS9 (collector slit 0.080 in. long) and in the MS12 (collector slit 0.200 in. long) (3). The broad peak given by the ion of NO from the molecular ion of o-nitrophenol is essentially flat-topped in the MS9 at accelerating voltages between 2 kV and 8 kV, whereas in the MS12, the peak is appreciably convex in the central region, owing to the loss of fewer ions with components of energy in the z direction. A convex top is also observed for the metastable peak given by the same metastable transition in the mass spectrum of p-nitrophenol and recorded on a Hitachi RMU-6A instrument at an accelerating voltage of 2.5 kV (13). On all three instruments, however, despite the variations in shape of the peak, the energy release calculated from the width of the peak is in the region of 0.72–0.76 eV. In a double-focusing instrument of Mattauch-Herzog geometry, such as the CEC 21-110B, a relatively wide intermediate β-slit is used between the electric and magnetic analyzers; this allows daughter ions from fragmentations occurring in the final regions of the electric sector to pass through the β-slit and so reach the collector. This produces a much broader image and greatly increases the width of the metastable peak, leading to spuriously high values for T. For the loss of NO by the molecular

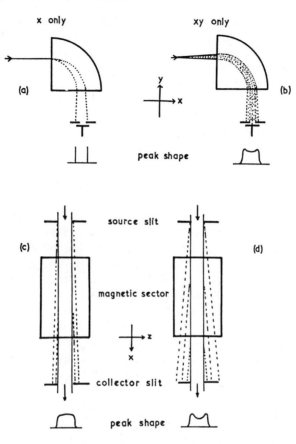

Figure 8. Schematic representation of discrimination effects. In (*a*), energy is released in the *x*-direction only so that the peak splits into two narrow components. In (*b*), energy release occurs only in the *xy* plane; the peak becomes broad and its exact shape depends on the masses of the fragments and the energy released. In (*c*) and (*d*), the ion beam is viewed looking along the *y*-axis so that a ribbon of ions is seen. The dotted lines represent the loss of ions formed with components of energy in the *z*-direction (*c*) at high accelerating voltage and (*d*) at low accelerating voltage; the greater losses in (*d*) lead to the concave top of the peak as opposed to the convex top found in (*c*). It is clear that a reduction in the length of the collector slit has the same effect on discrimination as a reduction in the accelerating voltage.

ion of *p*-nitrophenol, an apparent value of 1.77 eV was obtained (13). For a given instrument operating under fixed conditions, ion losses and hence discrimination effects will decrease as T decreases and as m_2 increases. Consequently, if a relatively small amount of energy is released in a fragmentation in which a light neutral fragment is lost from a heavy ion, the discrimination effects at the collector slit will be too small to give rise to a flat-topped peak, and the peak-shape will be approximately gaussian. In practice, it is difficult to obtain reliable values of T under about 0.2 eV.

The detailed calculations on metastable peak-shapes predict that the slope of the sides of the peak should be very steep, whereas observed peaks invariably have sides of a lower slope. This suggests that there is always a small range of translational energies released, even in fragmentations giving rise to narrow gaussian peaks, since explanations in terms of fringing magnetic fields and decompositions occurring within the magnetic field do not appear to account for the observed slopes (3).

D. In the Field-Free Region Before the Electric Sector

For metastable transitions occurring in the first field-free region of a double-focusing instrument, daughter ions are transmitted only when $V/V_0 = m_1/m_2$, providing that no translational energy is released during the fragmentation. In this type of observation, the magnetic field is fixed and the release of energy causes m_2^+ ions to be transmitted over a small range of values of V; the peak is therefore broadened, not along the normal mass scale obtained by sweeping the magnetic field, but along the "precursor mass scale," defined by the ratio of V/V_0. Using arguments very similar to those used above (66), one can derive an equation analogous to [27], but in terms of the voltage ratio required to transmit ions formed in fragmentations in which the energy is released entirely in the x direction:

$$\frac{V}{V_0} = \frac{(m_1/m_2)}{[1 \pm (\mu T/Ve)^{1/2}]^2} \qquad [35]$$

If the length of the flat-topped region or the separation of the two maxima is D volts,

$$D = 4(\mu TVe)^{1/2} = 4\left(\frac{\mu T m_1 V_0 e}{m_2}\right)^{1/2} \qquad [36]$$

and

$$T = \frac{D^2}{16\mu V} = \frac{m_2 D^2}{16\mu m_1 V_0} \qquad [37]$$

The corresponding expressions for the formation of a singly charged daughter ion from a doubly charged precursor are

$$D = 2(2\mu TVe)^{1/2} = 2\left(\frac{\mu T m_1 V_0 e}{m_2}\right)^{1/2} \qquad [38]$$

and

$$T = \frac{D^2}{8\mu V} = \frac{m_2 D^2}{4\mu m_1 V_0} \qquad [39]$$

These expressions were tested using the reaction

$$C_3H_4F^+ \rightarrow C_3H_3^+ + HF \qquad [40]$$

in the mass spectrum of 2-fluoropropene, and [33] in the mass spectrum of benzene. Observations in the second field-free region and use of [29] and [32] led to values of T of 0.57 and 2.67 eV respectively for these reactions. It was found that the values of T calculated from [37] and [39] using data obtained from metastable transitions that occurred in the first field-free region were appreciably less than these values and were a function of V_0. A plot of the

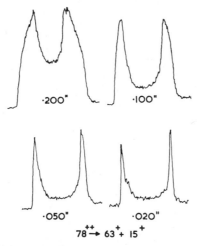

Figure 9. The effect of the width of the β-slit on the shape of the peak is given by the metastable transition $C_6H_6^{++} \rightarrow C_6H_3^+ (+ CH_3^+)$ in the mass spectrum of benzene.

calculated value of T against V_0 gives a straight line that, when extrapolated to $V_0 = 0$, gives a value of T in excellent agreement with values obtained using [29] and [32]. A plot for [33] is shown in Figure 9.

The explanation for the variation of the calculated value of T with V_e is again in terms of discrimination effects. Equations [35]–[39] are strictly valid only for the case in which all ions formed with release of energy in the y and z directions are lost, but owing to the finite widths and lengths of both the intermediate β-slit and the final collector slit, this condition is never realized in practice. It is most nearly fulfilled for light ions carrying considerable excess translational energy when working with low accelerating voltage. In the limit, as the accelerating voltage tends to zero, this condition will become true for all ions, and values of D and T calculated from these equations will be accurate. Under normal operating conditions, while ions formed with appreciable components of energy in the z direction will be lost, many ions with components of energy in the y direction will pass through the β-slit (0.20 in. width in an MS9 instrument), and the effect of collecting these is to cause the intensity maxima or "wings" of the peak to move towards the center of the peak. At very high accelerating voltages, the peaks become approximately gaussian in shape as the number of ions lost decreases.

E. Effects of β-Slit Width on Peak Shapes

An alternative method of increasing the discrimination at the intermediate β-slit is to reduce the slit width, either by replacing the standard slit

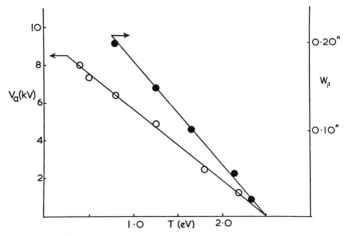

Figure 10. Variation of the apparent value of the energy release (T eV) in the reaction $C_6H_6^{++} \rightarrow C_5H_3^+ + CH_3^+$ in the mass spectrum of benzene with variation of accelerating voltage V_a and width of the β-slit, w_β.

with a fixed slit of narrower width or by fitting a continuously variable slit (67, 68). The effect of reducing the slit width is very similar to that of reducing the accelerating voltage: there is a marked reduction in the intensity of the central portion of the peak, and the separation of the intensity maxima increases (see Figure 10). A plot of the apparent value of T against the β-slit width at a fixed accelerating voltage is linear, and the extrapolated value for zero slit width is in excellent agreement with the value obtained by extrapolating to zero accelerating voltage (67) (see Figure 9). This suggests that the effects of collecting ions with appreciable components of energy in the z direction are negligible, presumably because the long path-length between the entrance to the electric sector and the collector slit ensures that most of these ions are lost.

THE SHAPES OF METASTABLE PEAKS 453

The increased discrimination obtainable by working with a relatively narrow β-slit (e.g., 0.02 in.) at low accelerating voltage leads to peak shapes for certain metastable transitions that are very different from those observed in a normal mass spectrum. For example, the flat-topped metastable peaks in the mass spectra of o- and p-nitrophenol ($139^+ \rightarrow 109^+ + NO$) become peaks with two very sharp intensity maxima, the intensity of the central region being approximately 50% of these maxima. A peak of similar shape is obtained for the metastable transition arising from [34]. In each case, the sharpness of the intensity maxima suggests that the energy release is sharply peaked about a given value and that the flat-topped peaks commonly observed are due to incomplete discrimination rather than to the release of a range of translational energies (67). While flat-topped peaks become concave-topped, other peaks that normally appear to be very approximately gaussian-shaped become flat-topped as the discrimination increases, and it is reasonable to suppose that they would become concave-topped if the discrimination could be further increased without losing too much sensitivity. An example of such a peak is that given by the reaction

$$C_3H_7^+ \rightarrow C_3H_5^+ + H_2 \qquad [41]$$

in the mass spectra of many hydrocarbons (67, 68). In other cases, peaks that are approximately flat-topped under normal operating conditions, such as those given by the loss of HF from the molecular ion of fluorobenzene and from the $C_2H_4F^+$ ion in the mass spectrum of ethyl fluoride, are resolved to give three intensity maxima, suggesting that a narrow gaussian peak is superimposed on a broader concave-topped peak. It is possible that the same neutral species is lost, but the structures or energy levels of the precursor or daughter ions may differ (67). The improved discrimination at the β-slit is also useful in improving the resolution of peaks given by precursors of very similar masses, although the variation of both the sensitivity and the ability to resolve precursor ions with the width of the β-slit depends on the intrinsic width of the metastable peak (69).

F. Applications to the Energetics of Fragmentation Processes

Broad metastable peaks have been used in a number of studies of the energetics of fragmentation reactions. In an investigation of the fragmentation of doubly charged ions given by aliphatic and alicyclic hydrocarbons (70), it was found impossible to interpret the energies released in terms of simple coulombic charge repulsion (71). The energies carried by CH_3^+ ions formed in these reactions were in good agreement with satellite peak measurements for C_6 hydrocarbons, but the satellite-peak method gives higher values for

C_3–C_5 hydrocarbons, suggesting that they may be formed by other routes. The fragmentation

$$HNCO^+ \rightarrow HCO^+ + N \qquad [42]$$

has been observed in the mass spectra of several compounds (13, 72) to give a broad metastable peak indicating the release of about 0.54 eV, and it has been suggested that this is due to the fragmentation occurring via a spin-forbidden predissociation (72). A similar mechanism was suggested earlier (73) in a detailed study of the possible states involved in the metastable transition

$$H_2S^+ \rightarrow S^+ + H_2 \qquad [43]$$

These have been discussed further in a recent review (74). A similar mechanism may also explain the observation of broad metastable peaks in the mass spectra of nitrogen and oxides of nitrogen (75, 76, 77).

A detailed study has been made of the metastable peaks observed in the mass spectra of CH_3OH, CD_3OH, CH_3OD, and CD_3OD (71, 78). All give broad metastable peaks centered at masses 27.12, 27.27, 26.28, and 26.46 respectively, which arise from the following four fragmentation processes:

$$CH_2OH^+ \rightarrow CHO^+ + H_2 \quad (\text{or } 2H^{\cdot}) \text{ in } CH_3OH \qquad [44]$$

$$CH_2OD^+ \rightarrow CHO^+ + HD \,(\text{or } H^{\cdot} + D^{\cdot}) \text{ in } CH_3OD \qquad [45]$$

$$CD_2OH^+ \rightarrow CDO^+ + HD \,(\text{or } H^{\cdot} + D^{\cdot}) \text{ in } CD_3OH \qquad [46]$$

$$CD_2OD^+ \rightarrow CDO^+ + D_2 \quad (\text{or } 2D^{\cdot}) \text{ in } CD_3OH \qquad [47]$$

The energies released in these processes are respectively 1.42 eV, 1.64 eV, 1.27 eV, and 1.48 eV. If one assumes that the fragmenting ions and products are in comparable excited states, then using the standard enthalpies of formation of $H^{\cdot}$ and $D^{\cdot}$, subtraction of [46] from [44], and [47] from [45] gives

$$D(C–D) - D(C–H) = 0.15 \quad \text{or} \quad 0.19 \text{ eV}$$

$$D(C–D) - D(C–H) = 0.16 \quad \text{or} \quad 0.20 \text{ eV}$$

where D is the bond dissociation energy. The lower values are based on the assumption that two atoms are eliminated; the higher values are obtained by assuming that a molecule is eliminated. Observations on other metastable peaks led to $D(O^+–H) - D(O^+–D) = 0.18$ eV or 0.22 eV, depending on whether a molecule or two atoms were eliminated. In this work, the translational energy release derived from the width of a metastable peak is identified with the energy of activation of the reverse reaction. The fragmenting ion is looked on as being in a modified electronic state rather than in a highly vibrationally excited ground electronic state, so that there are essentially no vibrational quanta to distribute among products on fragmentation. The energy release is therefore precisely determined and appears entirely as translational energy in this model. It is possible, however, that this assumption may not be generally applicable. A gaussian-shaped metastable peak is observed for the process

$$C_6H_5C_2H_5^{+\cdot} \rightarrow C_6H_6^{+\cdot} + C_2H_4 \qquad [50]$$

for which a calculated energy of activation for the reverse reaction of about 0.75 eV is obtained using standard heats of formation (37). It could be argued that the enthalpy of formation of $C_6H_6^{+\cdot}$ is incorrect in this case, since the structure is unknown, but otherwise, the excess energy must be retained as internal energy of the fragments. A further case in which the assumption does not appear to hold is in the reaction

$$C_2H_3F^{+\cdot} \rightarrow C_2H_2^{+\cdot} + HF \qquad [51]$$

observed in the mass spectrum of fluoroethylene (34). A broad metastable peak is given by this fragmentation, indicating the release of 0.68 eV of translational energy, but from appearance potential measurements, the enthalpy of formation of the $C_2H_2^+$ ion is calculated to be 15.1 eV, compared with a value of 13.75 eV calculated from the ionization potential of acetylene and the appearance potential of the ion in the mass spectrum of ethylene. This appears to require an energy of activation for the reverse reaction of $(15.1 - 13.75) = 1.35$ eV, of which only half is obtained from the metastable peak width. In the fragmentation, the remaining excess energy is assumed to be retained as internal energy of the two fragments.

G. Substituent Effects on Peak Shapes

Relatively few studies have been made of substituent effects on the shapes and widths of broad metastable peaks. While the loss of NO by the molecular ions of o- and p-nitrophenol molecular ions gives a flat-topped metastable under normal operating conditions, the same process in the mass spectrum of the meta-isomer gives a gaussian peak (2). In an investigation of the process $M^{\cdot+} \rightarrow (M-NO)^+$ in the mass spectra of a series of para-substituted nitrobenzenes, the width of the metastable peak increases as the electron-donating ability of the substituent rises (79). This implies a correlation between the product ion stabilization and the amount of translational energy released. A similar correlation was not found in an investigation of substituent effects in the metastable peaks found in the mass spectra of substituted 2-pyrones and coumarins (80). This could be explained by assuming that the translational energy release is not simply related to the energy of activation of the reverse reaction and so does not necessarily reflect the stabilities of the product ion.

VI. CONCLUSION

This review has attempted to show that metastable transitions are of interest in a wide variety of fields; analytical, structural, kinetic, mechanistic, and energetic information may be derived from them in appropriate circumstances. Since they have been studied extensively only during the last five years, many applications still await further study. Only a little over a decade ago, metastable peaks were regarded as a nuisance and could be removed with the aid of a "metastable suppressor." It is interesting that in many of the recent studies, "pure metastable spectra" are obtained by suppressing the normal peaks! During the next few years, it seems very probable that metastable transitions will continue to find an ever-widening range of applications.

REFERENCES

1. J. A. Hipple, R. E. Fox, and E. U. Condon, *Phys. Rev.*, **69**, 347 (1946).
2. J. H. Beynon, R. A. Saunders, and A. E. Williams, *Z. Naturforsch.*, **20a**, 180 (1965).
3. J. H. Beynon and A. E. Fontaine, *Z. Naturforsch.*, **22a**, 334 (1967).
4. N. R. Daly, A. McCormick, and R. E. Powell, *Rev. Sci. Inst.*, **39**, 1163 (1968).
5. N. R. Daly, A. McCormick, and R. E. Powell, *Org. Mass Spectrom.*, **1**, 167 (1968).
6. M. Barber and R. M. Elliott, *12th Annual Conference on Mass Spectrometry and Allied Topics*, ASTM, Committee E-14, 1964, Montreal, Canada, p. 150.
7. K. R. Jennings, *J. Chem. Phys.*, **43**, 4176 (1965).
8. K. R. Jennings, in R. Bonnett and J. G. Davies (Eds.), *Some Newer Physical Methods in Structure Determination*, United Trade Press, London, 1967, p. 105.
9. K. R. Ryan, L. W. Sieck, and J. H. Futrell, *J. Chem. Phys.*, **43**, 1832 (1965).
10. J. H. Beynon, J. W. Amy, and W. E. Baitinger, *Chem. Commun.*, 723 (1969).
11. J. H. Beynon, R. M. Caprioli, W. E. Baitinger, and J. W. Amy, *Internatl. J. Mass Spec. Ion Phys.*, **3**, 313 (1969).
12. J. H. Beynon, R. A. Saunders, and A. E. Williams, *Nature*, **204**, 67 (1964).
13. T. W. Shannon, F. W. McLafferty, and C. R. McKinney, *Chem. Commun.*, 478 (1966).
13a. P. Schulze and A. L. Burlingame, *J. Chem. Phys.*, **49**, 4870 (1968).
14. K. R. Jennings, *Chem. Commun.*, 283 (1966).
15. J. Seibl, *Helv. Chim. Acta*, **50**, 236 (1967).
16. H. Budzikiewicz, F. van der Haar, and H. H. Inhoffen, *Justus Liebig's Ann. Chem.*, **701**, 23 (1967).
17. R. M. Letcher and S. H. Eggers, *Tetrahedron Lett.*, 3541 (1967).
18. E. Capsi, J. Wicha, and A. Mandelbaum, *Chem. Commun.*, 1161 (1967).
19. E. Tajima and J. Seibl, *Int. J. Mass Spec. Ion Phys.*, **3**, 245 (1969).
20. H. M. Rosenstock and C. E. Melton, *J. Chem. Phys.*, **26**, 314 (1957).

21. K. R. Jennings, *Intl. J. Mass Spec. Ion Phys.*, **1**, 227 (1968).

22. W. F. Haddon and F. W. McLafferty, *J. Amer. Chem. Soc.*, **90**, 475 (1968).

23. N. D. Coggeshall, *J. Chem. Phys.*, **37**, 2167 (1962).

24. A. S. Newton, *J. Chem. Phys.*, **44**, 4015 (1966).

25. R. E. Ferguson, K. E. McCullogh, and H. M. Rosenstock, *J. Chem. Phys.*, **42**, 100 (1965).

26. F. W. McLafferty, R. S. Gohlke and R. C. Golesworthy, *12th Annual Conference on Mass Spectrometry and Allied Topics, ASTM, Committee E-14*, 1964, Montreal, Canada, p. 331.

27. W. W. Hunt and K. E. McGee, *J. Chem. Phys.*, **41**, 2709 (1964).

28. H. M. Rosenstock and M. Krauss, in F. W. McLafferty (Ed.), *Mass Spectrometry of Organic Ions*, Academic Press, New York, 1963, Chap. 1.

29. H. M. Rosenstock and M. Krauss, in R. M. Elliott (Ed.), *Advances in Mass Spectrometry*, Vol. II, Institute of Petroleum, London, 1963, p. 251.

30. M. L. Vestal, in P. Ausloos (Ed.), *Fundamental Processes in Radiation Chemistry*, Interscience, New York, 1968, Chap. 2.

31. I. Hertel and Ch. Ottinger, *Z. Naturforsch.*, **22a**, 1141 (1967).

32. Ch. Ottinger, *Z. Naturforsch.*, **20a**, 1229 (1965).

33. K. R. Jennings, *Z. Naturforsch.*, **22a**, 454 (1967).

34. K. R. Jennings, *Org. Mass. Spectrom.*, **3**, 85 (1970).

35. W. A. Chupka, *J. Chem. Phys.*, **30**, 191 (1959).

36. I. Hertel and Ch. Ottinger, *Z. Naturforsch.*, **22a**, 40 (1967).

37. R. G. Cooks, I. Howe, and D. H. Williams, *Org. Mass Spectrom.*, **2**, 137 (1969).

38. R. D. Hickling and K. R. Jennings, *Org. Mass Spectrom.* 3, 1499 (1970).

39. J. H. Beynon, J. A. Hopkinson, and G. R. Lester, *Intl. J. Mass Spec. Ion Phys.*, **2**, 291 (1969).

40. F. W. McLafferty and W. T. Pike, *J. Amer. Chem. Soc.*, **89**, 5951, 5953, 5954 (1967).

41. T. W. Shannon and F. W. McLafferty, *J. Amer. Chem. Soc.*, **88**, 5021 (1966).

42. K. R. Jennings and J. H. Futrell, *J. Chem. Phys.*, **44**, 4315 (1966).

43. R. G. Cooks and D. H. Williams, *Chem. Commun.*, 627 (1968).

44. F. W. McLafferty and T. A. Bryce, *Chem. Commun.*, 1215 (1967).

44a. J. L. Occolowitz, *J. Amer. Chem. Soc.* **91**, 5202 (1969).

45. A. N. H. Yeo and D. H. Williams (submitted for publication).

46. A. F. Whiting and K. R. Jennings (unpublished results).

47. D. H. Williams and J. Ronayne, *Chem. Commun.*, 1129 (1967).

48. R. G. Cooks, R. S. Ward, and D. H. Williams, *Chem. Commun.*, 850 (1967).

48a. A. N. H. Yeo, R. G. Cooks, and D. H. Williams, *J. Chem. Soc. B*, 149 (1969).

49. R. G. Cooks, I. Howe, S. W. Tam, and D. H. Williams, *J. Amer. Chem. Soc.*, **90**, 4064 (1968).

50. J. H. Beynon, B. E. Job, and A. E. Williams, *Z. Naturforsch.*, **20a**, 883 (1965).

51. C. Lifshitz and M. Shapiro, *J. Chem. Phys.*, **45**, 4242 (1966).

52. C. Lifshitz and M. Shapiro, *J. Chem. Phys.*, **46**, 4912 (1967).

53. Ch. Ottinger, *J. Chem. Phys.*, **47**, 1452 (1967).

54. Ch. Ottinger, *Z. Naturforsch.*, **20a,** 1232 (1965).

55. M. Vestal and J. H. Futrell, *J. Chem. Phys.*, **52,** 978 (1970).

56. A. N. H. Yeo, R. G. Cooks, and D. H. Williams, *Chem. Commun.*, 1269 (1968).

57. M. Barber, W. A. Wolstenholme, and K. R. Jennings, *Nature*, **214,** 664 (1967).

58. M. Barber, B. N. Green, W. A. Wolstenholme, and K. R. Jennings, in E. Kendrick (Ed.), *Advances in Mass Spectrometry*, Vol. IV, p. 89, Institute of Petroleum, London, 1968.

59. W. Higgins and K. R. Jennings, *Chem. Commun.*, 99 (1965).

60. W. Higgins and K. R. Jennings, *Trans. Faraday Soc.*, **62,** 97 (1966).

61. J. H. Beynon and A. E. Fontaine, *Chem. Commun.*, 717 (1966).

62. J. Olmsted III, K. Street, Jr., and A. S. Newton, *J. Chem. Phys.*, **40,** 2114 (1964).

63. R. M. Elliott, private communication.

64. M. C. Flowers, *Chem. Commun.*, 235 (1965).

65. K. C. Smyth and T. W. Shannon, *17th Annual Conference on Mass Spectrometry and Allied Topics*, *ASTM*, *Committee E-14*, 1969, Dallas, Texas, p. 89.

66. M. Barber, K. R. Jennings, and R. Rhodes, *Z. Naturforsch.*, **22a,** 15 (1967).

67. S. Jones and K. R. Jennings, *17th Annual Conference on Mass Spectrometry and Allied Topics*, *ASTM*, *Committee E-14*, 1969, Dallas, Texas, p. 95.

68. J. H. Beynon, A. E. Fontaine, J. A. Hopkinson, and A. E. Williams, *Intl. J. Mass Spec. Ion Phys.*, **3,** 143 (1969).

69. S. Jones, K. R. Jennings, and R. M. Elliott, unpublished work.

70. M. Barber and K. R. Jennings, *Z. Naturforsch.*, **24a,** 134 (1969).

71. J. H. Beynon, in E. Kendrick (Ed.), *Advances in Mass Spectrometry*, Vol. IV, Institute of Petroleum, London, 1968, p. 123.

72. C. G. Rowland, J. H. D. Eland, and C. J. Danby, *Chem. Commun.*, 1535 (1968).

73. F. Figuet-Fayard and P. M. Guyon, *Mol. Phys.*, **11,** 17 (1966).

74. H. M. Rosenstock, in E. Kendrick (Ed.), *Advances in Mass Spectrometry*, Vol. IV, Inst. of Petroleum, London, 1968, p. 523.

75. A. S. Newton and A. F. Sciamanna, *J. Chem. Phys.*, **44,** 4327 (1966).

76. A. S. Newton and A. F. Sciamanna, *J. Chem. Phys.*, **50,** 4868 (1969); **52,** 327 (1970).

77. R. J. Coleman, J. S. Delderfield, and B. G. Reuben, *Intl. J. of Mass Spec. and Ion Phys.*, **2,** 25 (1969).

78. J. H. Beynon, A. E. Fontaine, and G. R. Lester, *Intl. J. of Mass Spec. and Ion Phys.*, **1,** 1 (1968).

79. M. M. Bursey and F. W. McLafferty, *J. Amer. Chem. Soc.*, **88,** 5023 (1966).

80. M. M. Bursey and R. L. Dusold, *Chem. Commun.*, 712 (1967).

Author Index

Each entry consists of two numbers. The number in parentheses is that of the appropriate reference and is followed by the number(s) of the page(s) on which the reference is found.

Aakvaag, A., (42)337
Abell, P. I., (150)346
Abrahams, P. W., (18)144
Abrahamsson, S., (19)82,83
Aczel, T., (44)33
Adhikary, P. M., (54)338
Adlercreutz, H., (30)98; (37)334,338;
  (43)337; (44)337; (45)337; (46)337;
  (47)337; (48)337; (49)337; (50)337;
  (58)339; (59)339; (60)339;(61)339;
  (62)339; (64)339; (65)339
Aebi, H., (197)358
Agadzhanyan, Ts. E., (48)302,303
Agarwal, K. L., (122)316; (126)316;
  (208)361
Ahearn, A. J., (21)19
Akita, T., (152)347
Akopyan, M. N., (103)315
Alakhov, Yu. B., (18)292,295,297,300,301,
  302,304,315; (40)300,301; (45)301,313,
  314; (51)304,305,306,307,313; (52)304,
  305,306,307,313,314; (54)304,305,313;
  (84)311,312,313,316,318,319,320;
  (85)311,313; (138)320
Aldanova, N. A., (51)304,305,306,307,
  313; (54)304,305,313; (65)309;
  (66)309; (84)311,312,313,316,318,319,
  320; (85)311,313; (129)318,319
Aldridge, M. H., (198)359
Alice, M. B., (77)210; (231)364
Allen, J., (19)144
Allen, J. G., (92)342
Allen, W. V., (218)363
Amy, J. W., (10)377; (10)427; (11)427;
  (14)8; (15)16; (38)28
Anders, L. R., (42b)191
Andersson, B. A., (29)294, 315
Andersson, C.–O., (11)291
Annis, J. L., (4)92
Anthony, K. V., (19)95
Aplin, R. T., (75)341; (93)312,313,315;
  (109)315; (110)315; (111)315;
  (112)315
Aquilanti, V., (66)205
Arens, J. F., (28)294; (56)304,319;
  (113)315; (114)315; (207)361
Argamone, F., (83)311
Argoudelis, A. D., (45)248; (105)343
Arima, K., (83b)311
Arnold, W., (30)333
Arsenault, G. P., (26)84; (49)303,312,314
Ashworth, R. B., (223)363

Aspen, A. J., (34)237
Asselineau, J., (97)313
Audier, H. E., (106)270
Aust, S. D., (34)237
Avigan, J., (98)342; (114)344; (115)344;
  (116)344; (132)345
Axenrod, T., (57)201; (59)202; (61a)202;
  (94)274

Baba, H., (48)387
Bacon, V., (115)315
Baer, D. M., (32)333
Baggett, B., (168)351; (169)352
Bailey, E., (58)110,116
Baird, M. S., (143)402
Baitinger, W. E., (10)427; (11)427
Baker, K. M., (116)315
Baldeschwieler, J. D., (42b)191; (178)408;
  (183)409; (186)410
Baldwin, M. A., (86)392; (101)394
Bancroft, G. M., (104)394
Banner, A. E., (7)46,48; (8)46,48
Barber, M., (6)425; (21)70; (25)292,313,
  315; (26)292,314; (27)84; (57)441,442;
  (58)110,116; (58)441,443; (66)450;
  (70)453; (74)310; (90)312,314; (101)314;
  (139)401
Barbier, M., (79)341; (84)341
Barghoorn, E., (150)346
Barnes, W. T., (135)319
Barrett, G. C., (137)319
Barron, R. P., (35)189
Baty, J. D., (58)202
Baumgarten, H. E., (106)394
Baxter, J. H., (113)344
Bayer, E., (32)98; (50)303; (95)313,319;
  (96)313,319
Bazinet, M. L., (4)44,61; (16)8; (18)61;
  (99)314
Beak, P., (57)256
Beauchamp, J. L., (42b)191; (185)409
Beaumont, J. O., (15)329
Becker, D. A., (40b)190
Becker, E. W., (23)95
Beckey, H. D., (1)180; (10)219; (13)184;
  (14)16; (16)185; (17)185; (18)186;
  (20)186; (21)186; (22)186; (23)186; (24)186;
  (25)187; (26)188; (27)188; (31)188;
  (32)188; (34)189; (36)189; (37)189;
  (40a),189
Belmusto, L., (18)330
Bennett, E. O., (136)346

Bentley, T. W., (87)392,396
Benz, W., (22)380
Bergstrom, S., (120)345; (121)345
Berkowitz, J., (161)404
Berlin, K. D., (79)264
Bernal, G. E., (70)206
Bethe, H., (64)204,205
Bevan, K., (83a)311
Beynon, J. H., (1)375; (2)421,446,455;
    (3)78; (3)423,448,449,450; (4)218,255;
    (6)377; (9)7; (10)427; (11)78; (11)427;
    (12)50; (12)219; (12)428; (13)219;
    (28)188; (39)436; (50)440; (61)447;
    (68)452,453; (71)453,454; (78)454;
    (92)393
Beyrich, W., (15)220; (36)384; (176)354
Bickel, R. G., (19)330
Bieber, L. L, (155)347
Bieber, M., (14)94
Biemann, K., (1)43; (1)290,295,297; (2)44;
    (2)328; (2)376; (4)78; (5)14; (7)290;
    (7)377; (8)290; (9)290; (10)7; (10)93,96,
    109; (11)7,8; (11)116; (12)8; (12)78;
    (12)377; (13)79; (18)81; (22)380; (23)82;
    (24)84; (26)84; (28)85; (32)86,296;
    (33)96,98,109; (36)98,114; (42)31; (43)33,
    35; (46)33; (47)386; (49)35; (49)303,312,
    (46)33; (47)386; (49)35; (49)303,312,
    314; (50)35; (53)40; (59)305; (60)202;
    (64)115,116; (88)312,314; (95)342;(98)314;
    (98)393,394; (150)346; (163)349,364
Bier, K., (23)95
Binum, E., (58)305
Black, D. R., (5)92; (43)96,103,105,106
Blair, C. M., (15)131,143; (16)131
Blight, M. M., (214)362; (216)362
Blum, M. S., (212)362
Blumenfeld, A., (110)272
Blumer, M., (40)96,100,105,106
Blunt, J. W., (77)341; (78)341
Board, R. D., (10)377
Bochkarev, V. N., (18)292,295,297,300,
    301,302,304,315; (24)292,293,309;
    (36)298; (65)260; (65)309; (66)309;
    (91)312,315
Bodanszky, M., (123)316
Boer, F. P., (128)398,402
Boettger, H. G., (19)61
Bogdanova, I. A., (77)310
Bohn, G., (34)333
Bolker, H. J., (126)279
Bolton, H. C., (11)48
Bommer, P., (1)43; (7)377; (10)7; (47)386;
    (49)35; (50)35; (98)314; (165)350
Bonham, J., (57)256
Bose, A, K., (35a)237; (35b)237; (64)260;
    (88)267
Bostwick, D. C., (205)360
Boucher, E. A., (29)98; (55)338
Boulton, A. A., (225)363
Bowen, H. C., (25)71,73,74
Boyer, P. D., (184)356
Brakier, L., (127)397,400

Brammell, H. L., (19)330
Brandt, W. W., (11)94
Brantigan, J. W., (16)329
Brantley, R. K., (203)360
Breger, I. A., (150)346
Breslow, R., (93)268
Bricas, E., (26)292,314; (31)294,298,299,
    314,315
Brickstock, A., (20)82
Bridges, J. C., (65)115
Brigden, W. H., (20)330
Briggs, L. H., (69)309,310
Brion, C. E., (14)329; (75)209
Brix, P., (34)23
Brooks, C. J. W., (13)94; (36)334; (54)338;
    (55)338; (89)342; (92)342
Broquist, H. P., (34)237
Brown, E. B., (1)327,328
Brown, P., (33)189; (41)384; (59)256;
    (106a)315; (113)396; (114)396; (115)396
Brown, R. F. C., (124)278
Browne, J. S. L., (66)339
Brunée, C., (46)96,104
Bryce, T. A., (44)438; (173)408
Bryce, W. A., (78)391
Buchanan, B. G., (7)125,126,144,168,175,
    176; (12)377; (17)143,147,175; (20)144;
    (21)144; (22)144,175; (23)146,147,175;
    (31)154,155,162,175; (34)155,160,161,
    162,163,167,175; (35)162,163,167,175;
    (36)175; (38)176
Buchholz, A. C., (18)379; (69)261,262;
    (70)261; (73)262
Buchs, A., (34)155,160,161,162,163, 167,
    175; (35)162,163,167,175; (38)176;
    (67)390
Buck, R. P., (88)392,396,404
Budzikiewicz, H., (1)218,255; (2)90,218,
    220,255; (3)218,255; (6)78; (7)78; (8)78·
    (10)78; (16)429; (24)146; (26)148;
    (28)75; (29)153; (29)333; (30)153; 33,
    155,159; (33)383
Bultemann, H. J., (46)96,104
Buonocore, V., (136)319
Burghoff, H. Z., (23)95
Burgus, R., (8)15; (9)15; (10)15
Burikov, V. M., (18)292,295,297,300,301,
    302,304,315; (61)308; (76)310; (77)310
Burlingame, A. L., (8)377; (9)47,56,72,73,
    74; (10)47,56,73,75; (13)8; (13a)428;
    (16)16; (16)220; (22)10,70,75; (24)71;
    (51)39; (145)346; (146)346; (147)346;
    (148)346; (150)346
Burovina, I. V., (25)331
Burrus, F. J., (54)106
Bursell, E., (141)346
Bursey, M. M., (16)379; (17)220; (37)384;
    (43)385; (44)386; (45)386; (51)387, 395;
    (52)387; (53)388; (54)388; (55)388;
    (58)388,389; (79)455; (80)455; (88)392,
    396,404; (90)392; (106)394; (107)394;
    (117)397; (118)397; (133)400
Burwell, R. L., Jr., (119)277

Butcher, M., (124)278
Buttery, R. G., (44)103; (219)363

Caglioti, L., (83)311
Calam, D. H., (213)362
Calvert, J. G., (118)277
Calvin, M., (150)346
Campbell, A. J., (14)52,70
Cantone, B., (68)390
Cantrill, H. L., (18)379; (70)261
Capella, P., (16)94; (108)343; (127)258
Caprioli, R. M., (11)427; (31)231
Capsi, E., (18)429
Carles, J., (108)271
Carpenter, W., (27)149; (27)381
Carrick, A., (20)62,74
Casparrini, G., (159)348
Cassinelli, G., (83)311
Cayley, A., (2)125
Cermăk, V., (53b)255; (73)207,208; (74)207, 208
Chait, E. M., (15)16; (106)315
Chambaz, E. M., (55)338; (90)342
Chang, S., (150)346
Chapman, J. R., (58)110,115,116; (61)257; (137)319
Chin, M. S., (56)388,390,392
Chupka, W. A., (35)435; (83)391,398,404; (153)404; (161)404
Churchman, C. W., (36)175
Chuvilin, A. V., (16)291
Clayton, E., (24)224; (24)331; (25)71,73, 74; (25)224
Clemena, G. G., (24)331
Clements, A. N., (141)346
Coburn, T. B., (65)115
Coffen, D. L., (95)269
Coggeshall, N. D., (23)430
Coleman, R. J., (77)454
Collin, J., (158)404
Collins, J. H., (79)391
Condon, E. U., (1)420,422
Cone, C., (26)84; (42)31; (49)303,312,314; (88)312,314
Conner, H. A., (222)363
Cook, R. J., (30)382
Cooks, R. G., (25)381,406,407; (26)381; (37)435,455; (40)384; (42)384; (43)437; (48)439; (48a)439; (49)440; (56)440; (61)389,392; (72)390,395,396,404; (95)393; (151)404
Cope, A. C., (128)281
Copier, H., (56)304,319; (114)315
Corbato, F. J., (30)86; (31)86
Cordes, E., (150)346
Corey, E. J., (1)124
Cormier, M. J., (223)363
Cornforth, J. W., (37)241; (38)242
Cornforth, R. H., (38)242
Corval, M., (6)218
Costopanagiotis, A., (29)333
Cotter, J. L., (56)255

Courchene, W. L., (79)391
Coutant, J. E., (20)8
Coutts, R. T., (31)333
Cox, G. B., (44)246; (193)357
Crable, G. F., (66)390
Crawford, L. R., (21)82,83
Cree, R., (39)96,99
Creech, B. G., (18)95
Crespi, H. L., (42a)245; (42b)245
Creveling, C. R., (215)362; (226)363; (227)363
Croteau, R. J., (221)363
Curtis, R. W., (62)309
Curtius, H.-Ch., (34)98
Cushley, R. J., (150)346

D'Or, L., (127)397,400
Daggett, M. M., (30)86
Dahm, K. H., (210)362; (211)362
Daley, R. C., (30)86
Dalgarro, A., (53b)255
Dalgliesh, C. E., (15)94
Dalton, R. L., (196)358
Daly, J. W., (29)229; (189)357; (190)357; (191)357; (192)357; (215)362; (226)363; (227)363
Daly, N. R., (4)424; (5)425
Damico, J. N., (35)189; (206)360
Danby, C. J., (72)454
Daniels, E. G., (123)345
Dannenberg, H., (97)342
Darlak, R. S., (137)401
Das, B. C., (19)292,301,314; (26)292,314; (31)294,298,299,314,315; (74)310; (79)310; (80)310; (88)342; (92)312,316,317; (117)316; (118)316; (119)316; (129)345; (230)363
Das, K. G., (35a)237; (64)260; (88)267
Davies, G. D., Jr. (126)345
Davies, J. S., (83a)311
Dawson, G., (14)94; (160)349,364
Day, E. A., (3)44
Day. J. C., (5)92
de Haas, G. H., (87)311,316,320; (131)282
de Vries, J. X., (63)309,319
Dean, F. M., (85)266; (86)266
Dean, H. M., (63)204
Dean, P. D. G., (75)341
Dean, W. J., (74)341
DeGraff, B. A., (118)277
DeJongh, D. C., (14)94; (70)340,351
Delany, E. B., (13)50,71,73
Delderfield, J. S., (77)454
Delfino, A. B., (34)155,160,161,162,163, 167,175; (35)162,163,167,175; (38)176
Delgmann, L., (13)329
Deluca, H. F., (77)341; (78)341
Demarteau-Ginsburg, H., (135)345
DeMore, W. B., (109)272
Dempster, A. J., (2)78
Denisov, Yu. V., (24)292,293,309
Dennis, K. J., (70)340,351
Desiderio, D. M., (1)43; (7)15,19; (7)377;

Desiderio, D. M., (Cont'd.), (8)15; (9)15; (10)15; (11)7,8; (25)19; (37)28,34,35; (39)28; (40)28; (43)33,35; (49)35; (50)35; (98)314
Detre, G., (175)354
Deverse, F. T., (91)392
Devys, M., (84)341
Dewar, M. J. S., (129)398; (130)398
Diekman, J., (186)410; (187)410
Diener, C. F., (19)330
Dietze, H.-J., (26)331
Dijkstra, G., (28)294; (56)304,319; (114)315
Dine-Hart, R. A., (56)255
Djerassi, C., (1)218,255; (2)218,220,255; (2)290; (3)218,255; (6)78; (7)78; (8)78; (10)78; (12)377; (17)143,147,175; (19)380; (20)380; (23)146,147,175; (23)380; (24)146; (26)148; (27)149; (27)381; (28)75; (29)153; (30)153; (31)154,155,162,175; (33)155,159; (33)383; (34)155,160,161,162,163,167, 175; (34)383; (35)162,163,167,175; (38)176; (39)176; (41)384; (59)256; (61b)204; (72)262; (74)262,263; (75)263; (76)263; (99)269; (186)410; (187)410
Doane, W. M., (111)273
Dole, M., (77)210; (231)364
Dolev, A., (112)343
Dolhun, J. J., (164)349
Dorman, F. H., (81)391
Dorsey, J. A., (6)92
Dougherty, R. C., (131)398,399; (136)400, 402; (150)404
Douglas, A. G., (138)346; (144)346
Douraghi-Zadeh, K., (138)346; (149)346
Draffan, G. H., (76)210; (150)346
Drew, C. M., (150)346
Ducluda, A., (83b)311
Dudenbostel, B. F., Jr., (2)4,5; (3)4,5,6
Duffield, A. M., (12)377; (17)143,147,175; (23)146,147,175; (23)380; (27)149; (27)381; (31)154,155,162,175; (34)155, 160,161,162,163,167,175; (35)162,163, 167,175; (38)176; (150)346
Dunbar, R. C., (184)409,410
Duncan, J. H., (161)349,364
Dunn, T. F., (8)15; (9)15; (10)15
Dupir, J. F., (106)270
Dusold, L. R., (37)384; (80)455
Dutton, H. J., (112)343

Eadon, G., (187)410
Earle, N. R., (7)15,19
Ebert, A. A., (7)92
Eble, T. E., (45)248
Ebner, L., (197)358
Edwards, D. J., (18)144
Eggers, S., (17)429; (30)188
Eglinton, G., (138)346; (139)346; (144)346; (149)346; (150)346
Ehmann, W. J., (116)276
Eik-Nes, K. B., (42)337
Eisman, S. H., (12)127

Eland, I., (93)312,313,315
Eland, J. H. D., (72)454
Elliott, L. H., (52)338
Elliott, R. M., (6)425; (63)447; (69)453
Elliott, W. H., (61)114
Ellouz, R., (88)342
Elwood. T. A., (37)384; (43)385; (44)386; (45)386
Eneroth, P., (17)94; (57)339
Enke, C. G., (17)58
Erdman, J. P., (144)402
Eriksson, H., (51)254; (60)112
Etemadi, A.-H., (230)363
Evans, A. W., (196)358
Evans, C. A., (22)331
Eyring, E. M., (155)404; (156)404
Eyring, H., (148)403; (155)404

Faesel, J., (63)309,319
Fagerlund, U. H. M., (80)341
Fagerson, I. S., (221)363
Fairweather, R. B., (135)399,408
Fales, H. M., (33)333; (47)33; (54)195,200; (55)199; (56)200; (57)201; (58)202; (59)202; (61a)202; (80)211; (93)342; (105)315; (111)343; (114)344; (115)344; (116)344; (117)344; (132)345; (133)345; (180)409
Fano, R. M., (31)86
Feierabend, J. F., (168)351
Feigenbaum, E. A., (7)125,126,144,168,175, 176; (10)126,168,175; (12)377; (17)143, 147,175; (20)144; (21)144; (22)144,175; (23)146,147,175; (31)154,155,162,175; (34)155,160,161,162,163,167,175; (35)162,163,167,175; (37)175; (38)176
Feigina, M. Yu., (24)292,293,309; (51)304, 305,306,313; (54)304,305,313; (65)309; (84)311,312,313,316,318,319,320; (85)311, 313
Fennessey, P. V., (12)377; (24)84; (28)85; (32)86; (34)87; (42)31; (53)40; (54)40
Fenselau, C. C., (19)380; (20)380; (26)148; (61b)204; (161)349,364
Fergus, B. J., (69)309,310
Ferguson, C. D., (77)210
Ferguson, E. E., (53b)255
Ferguson, L. D., (231)364
Ferguson, R. E., (25)430
Ferguson, G. L., (16)329
Fessel, H. H., (13)291; (17)292,293,295, 296,297,300,315
Field, F. H., (40b)190; (43)191,192; (44)192, 203; (45c)192; (48)192; (51)193; (52)195; (53)195,200; (56)200; (67)205; (69)205; 177)408; (179)408,409; (181)409; (182)409
Fields, E. K., (7)218; (120)277; (121)277; (122)277; (174)354
Figuet-Fayard, F., (73)454
Flath, R. A., (43)96,103,105,106
Flikweert, J. P., (28)294
Fliszár, S., (108)271
Flory, D. A., (150)346

Flowers, M. C., (64)447
Foell, T. J., (70)340,351
Foffani, A., (68)390
Folkers, K., (126)345; (128)345; (131)345
Follman, H., (32)234
Foltz, R. L., (26)72
Fonina, L. A., (51)304,305,306,307,313
Fontaine, A. E., (3)423,448,449,450;
  (61)447; (68)452,453; (78)454; (92)393
Forgash, A. J., (204)360
Forward, G. C., (132)282
Foster, A. B., (3)328
Foster, N. G., (78)264
Fotherby, K., (70)340,351
Fowler, K. T., (9)328
Fox, J. A., (45)248
Fox, R. E., (1)420,422
Franek, F., (34)297,312,316,320; (87)311,
  316,320; (131)282
Franklin, J. L., (45c)192; (67)205; (96)393;
  (123)397; (125)397,400; (126)397
Frear, D. S., (83)341
Freedman, H. H., (109)395
Friedel, R. A., (21)380
Friedhoff, A. J., (224)363
Friedman, L., (47)192; (53a)255; (53b)255;
  (159)404
Fries, I., (52)338; (61)114
Friis, P., (126)345
Fritts, B. K., (1)3
Fry, A., (102)269
Frye, J. L., (137)346
Fukumoto, K., (39)242
Funasaki, H., (71)341
Funke, P. T., (35a)237; (35b)237; (64)260;
  (88)267
Funke, W. R., (128)281
Futrell, J. H., (9)425; (42)437; (42a)191;
  (49)192; (55)440

Gal, A. E., (117)344
Galbraith, M. N., (72)341
Gallegos, E. J., (76)390
Galli, A., (66)205
Galli, G., (69)339
Gallop, P. M., (54a)304,305,316
Gapp, F., (8)290; (32)296; (60)202
Garcia, M., (150)404
Gardiner, J. A., (203)360
Gardiner, R. A., (34)237
Geddes, A. J., (86)311,314
Gehrke, C., (150)346
Geissbuhler, H., (197)358
Gelpi, E., (102)342; (136)346
Gemzell, C. A., (50)337
Gerbode, F., (15)329
Gero, S. D., (92)312,316,317; (117)316;
  (118)316; (119)316
Gershbein, L. L., (99)342
Gesser, H., (12)94
Gesser, H. D., (104)394
Giardini-Duidoni, A., (66)205
Giardino, N. A., (35)23,24

Gibson, F., (44)246; (193)357
Giese, C. F., (162)404
Giesser, G. G., (46)386
Giessner, B. G., (83)266
Gifford, A. P., (4)5; (5)6
Gil-Av, E., (150)346
Gilbert, E. J., (59)202
Gilbert, J., (150)346
Gilbertson, J. R., (71)341
Gillette, J. R., (48)250; (49)251; (188)357
Gillis, R. G., (94)393
Gilpin, J. A., (53)304
Giuffrida, L., (205)360
Glock, G. A., Jr., (29)188
Godbole, E. W., (62)204,205
Gohlke, R. S., (2)92; (3)92; (32)155,159
Goldblatt, M. W., (119)345
Goldenfeld, I. V., (11)219
Goldman, P., (46)249; (47)250; (171)353;
  (172)353,357; (182)355
Gomer, R., (3)181; (5)181; (6)182; (7)182;
  (9)182; (10)183; (12)184; (15)185;
  (39)189
Good, R. H., (2)181
Goodchild, J., (86)266
Goodman, D. S., (37)241; (38)242
Gordon, B., (17)94; (57)339
Gorlenko, V. A., (43)301,302,303,314,316,
  317; (46)302,303; (48)302,303; (127)317,
  318; (128)317
Gott, V. L., (16)329
Gould, E. S., (90)268
Graham, G. M., (86)311,314
Grassmann, W., (94)313
Grasso, F., (68)390
Grayson, M. A., (45)96,103,106; (51)105
Green, B. N., (4)44,61; (5)44,50,52;
  (6)44,61,62; (16)52; (21)70; (58)441,443;
  (99)314; (100)314
Green, M. M., (28)382; (30)382; **(31)3**83
Greene, B. N., (16)8; (17)8; (57)**109**
Greichus, A., (142)346
Greichus, Y. A., (142)346
Griffiths, W. T., (130)345
Grigor'ev, V. V., (104)269
Grigsby, R. D., (17)81
Grishko, N. I., (104)269
Grob, K., (14)80; (53)106,110; (54)106;
  (55)106,110; (59)110
Grosch, W., (143)346
Gross, M. L., (60)389,396,408; (147)403
Grostic, M. F., (170)352
Grove, J. F., (216)362
Grubb, H. M., (14)378; (66)261
Grützmacher, H.-F., (3)290,295,296; (14)291;
  (15)291,297,308; (30)333; (41)300;
  (55)304
Guadagni, D. G., (44)103
Guillemin, R., (8)15; (9)15; (10)15
Guinand, M., (74)310; (75)310; (79)310;
  (80)310
Gur'yanova, E. N., (104)269
Gurney, R. W., (28)21

Guroff, G., (29)299; (190)357
Gustafsson, B. E., (73)341
Gustafsson, J.-A., (28)98; (39)335,341; (51)254; (60)112; (73)341
Guyon, P. M., (73)454

Haahti, E. O. A., (111)343
Hackney, R. J., (72)341
Haddon, W. F., (22)430; (174)408
Hagenmayer, E., (95)313,319
Haggstrom, G., (19)82,83
Halevi, E. A., (42a)245
Hall, E. S., (39)242
Hall, G. G., (119)397; (120)397; (121)397; (122)397; (124)397
Hall, L. D., (75)209
Hall, M. J., (83a)311
Haller, I., (117)276
Halliday, J. S., (14)52,70; (15)52,59,63
Halpern, B., (115)315; (150)346
Hamberg, M., (122)345
Hamilton, P. B., (150)346
Hammar, C.-G., (53)338; (62)114
Hamming, M. C., (13)378; (17)81
Han, J., (150)346
Hand, D. B., (220)363
Hanessian, S., (14)94
Haney, M. A., (96)393
Hansen, H.-J., (101)269
Hanson, J. R., (39)242
Hardwick, J. L., (23)331
Harkness, L., (42b)245
Harkness, R. A., (54)338
Harrison, A. G., (56)388,390,392; (65)390; (70)390; (71)390; (82)391; (108)395
Harrison, R., (123)278
Harrison, W. W., (24)331
Hart, H., (15)379
Hart, T. P., (18)144
Haselbach, C., (197)358
Hasofurther, M. E., (26)72
Hassall, C. H., (83a)311; (67)309
Haug, P., (48)105; (145)346; (146)346; (147)346
Hayaishi, O., (183)355
Hayes, J. M., (24)19,24,28,29,32; (35)88; (150)346
Heaney, H., (123)278
Hedfjäll, B., (63)114,115,116
Heerma, W., (28)294; (56)304,319; (107)315; (108)315; (114)315
Heinz, E., (30)229
Hemingway, A., (1)327,328
Henderson, W., (150)346
Henry, T. J., (144)402
Henze, H. R., (15)131,143; (16)131
Herman, T. S., (12)94
Herman, Z., (74)207,208
Herndon, J. H., Jr., (116)344
Hertel, I., (31)433; (36)435; (84)391,398
Herzog, R., (1)12
Hey, H., (18)186; (32)188
Heyns, K., (3)290,295,296; (14)291; (15)291,

297,308; (41)300; (55)304
Hickling, R. D., (38)435,436
Higasi, K., (48)387
Higgins, R. W., (78)264
Higgins, W., (59)446; (60)447
Highet, R. J., (98)342
Hignite, C. E., (46)33; (163)349,364
Hill, A. W., (85)266
Hill, R. M., (55)338; (167)351
Hilt, E., (1)180
Hine, J., (91)268
Hines, R. L., (77)210; (231)364
Hinman, J. W., (123)345
Hipple, J. A., (1)420,422
Hitchcock, F. A., (8)328
Hites, R. A., (22)82,83; (23)82; (42)31; (64)115,116
Hiz, H., (12)127
Ho, P. P. K., (50)251
Hock, W. K., (200)359
Hodgson, G., (150)346
Hoffman, M. K., (118)397; (133)400
Hoffmann, R., (132)398,399; (134)398
Hogenkamp, H. P. C., (32)234
Holland, H., (150)346
Holmes, J. C., (1)92
Holmstedt, B., (53)338; (62)114
Holtzclaw, H. F., Jr., (106)394
Holtzman, J. L., (48)250; (49)251; (172)353, 357; (188)357
Honig, R. E., (20)19,24; (77)391
Hopkinson, J. A., (1)375; (39)436; (68)452, 453
Hörhold, C., (87)342
Hörman, H., (94)313
Horn, D. H. S., (72)341
Horning, E. C., (13)94; (15)94; (16)94; (18)95; (19)95; (21)95; (38)335; (51)337; (55)338; (89)342; (90)342; (91)342; (159)348; (167)351
Horning, M. G., (15)94; (29)98; (51)337; (55)338; (125)345; (159)348; (167)351
Horvath, C. G., (38)96,99; (150)346
Hougen, F. W., (12)94
Howard, O. H., (27)331
Howe, I., (37)435,455; (49)440; (61)389, 392; (95)393; (110)396,404; (112)396; (154)404
Hrabak, P., (103)342
Hribar, J. D., (14)94; (70)340,351
Hugh-Jones, P., (9)328
Hunneman, D. H., (139)346; (149)346
Hunt, D. F., (163)349,364
Hunt, M. H., (36)24
Hunt, R. H., (6)92
Hunt, W. W., (27)431

Ibanez, J., (150)346
Idler, D. R., (80)341; (82)341
Ikonen, M., (60)339; (61)339; (62)339
Illuminati, G., (103)269
Imada, I., (127)345
Inano, H., (20)221

Inghram, M. G., (15)185; (162)404
Inhoffen, H. H., (16)429
Irving, P., (102a), 315
Isenhour, T. L., (11)377
Isono, M., (83b)311
Issenberg, P., (4)44,61; (16)8; (18)61;
  (99)314
Itada, N., (27)225; (183)355
Itano, H. A., (58)305
Ito, T., (124)316
Ivanov, V. T., (65)309; (66)309; (70)309
Ives, N. F., (205)360
Izdebski, J., (123)316

Jablonski, J. M., (123)278
Jackman, L. M., (44)246; (193)357
Jackson, L. L., (83)341
James, T. H., (26)20, (27)21
Jänne, O., (41)336; (49)337; (64)339
Jansson, P. A., (63)114,115,116
Jardine, F. H., (100)269
Jellum, E., (115)315
Jennings, K. R., (7)425; (8)425,441; (12)16;
  (14)249; (21)429; (33)435,439,441,442;
  (34)435,439,441,442,455; (38)384;
  (38)435,436; (42)437; (46)439; (57)441,
  442; (58)441,443; (59)446; (60)447;
  (66)450; (67)444,452,453; (69)453;
  (70)453; (81)212; (101)314; (102)314;
  (138)401; (139)401
Jerina, D. M., (189)357; (191)357; (192)357
Job, B. E., (50)440
Johansson, C.-J., (47)337
Johnson, C. B., (106)343
Johnson, E. G., (2)13
Johnson, R., (150)346
Johnston, P. V., (37)98
Johnston, W. H., (16)329; (152)404; (157)404
Johnstone, R. A. W., (55)255; (80)265;
  (81)266; (82)266; (85)266; (86)266;
  (87)392,396; (126)316; (145)402; (146)403
Jolles, P., (24)292,313,315
Jones, F. N., (128)281
Jones, H. B., (10)329
Jones, J. G. L., (133)283
Jones, J. H., (4)290,295; (93)312,313,315;
  (109)315; (110)315; (111)315; (112)315
Jones, S., (67)444,452,453; (69)453
Jordan, C., (14)129
Jordan, G. L., Jr., (51)337
Julien, J., (106)270
Jung, G., (50)303
Junk, G. A., (5)290; (6)290; (100)394
Jurs, P. C., (11)377

Kahn, J. H., (222)363
Kakinuma, A., (83b)311
Kamerling, J. P., (107)315; (108)315
Kaminsky, M., (153)404
Kammereck, R., (41)244
Kaplan, F., (188)411
Kaplan, I., (150)346
Kappus, G., (46)96,104.

Karlsson, K.-A., (154)347
Kassel, L. S., (63)389
Katz, J. J., (42a)245; (42b)245
Kaufman, D. D., (195)358
Kaufman, J. J., (49)387,390; (69)390
Kazaryan, S. A., (84)311,312,313,316,318,
  319,320; (129)318,319
Kearney, P. C., (195)358; (199)359; (202)359
Kearns, G. L., (66)390
Kebarle, P., (62)204,205; (65)390
Keil, B., (34)297,312,316,320; (87)311,316,
  320; (131)282; (150)346
Keister, D. B., (47)250
Kelly, R. W., (54)338
Kelly, W., (20)82
Kendrick, E., (29)75
Kenner, G. W., (122)316; (126)316; (208)361
Kennicott, P. R., (23)19
Keston, A. S., (179)355
Khanchandani, K. S., (35a)237; (35b)237
Khokhlov, A. S., (65)260; (180)355
King, A. B., (91)392; (160)404
King, G. S., (50)192
King, T. M., (57)305
King, W. H., Jr., (4)5
Kingston, D. G. I., (77)264; (78)310; (112)396
Kinstle, T. H., (9)218,281; (61)257;
  (63)259; (87)267; (99)393
Kirkien-Konasiewicz, A., (101)394
Kiryushkin, A. A., (18)292,295,297,300,
  301,302,304,315; (20)292,297,301,302,
  313; (21)292,295,297,300,301,302,313,
  315; (24)292,293,309; (36)298; (37)298;
  (38)298; (39)299; (40)300,301; (45)301,
  313,314; (46)302,303; (47)302,316;
  (48)302,303; (51)304,305,306,307,313;
  (54)304,305,313; (59)202; (61)308;
  (65)309; (66)309; (70)309; (73)310; (76)310;
  (77)310; (82)311; (84)311,312,316,318,
  319,320; (85)311,313; (91)312,315;
  (127)317,318; (128)317; (138)320
Kiser, R. W., (76)390; (149)403
Kissinger, P. T., (90)392
Klaas, P. G., (3)4,5,6
Klaver, R. F., (8)7
Klayman, D. L., (129)281; (130)281
Kleiner, E. M., (65)260
Klimowski, R. J., (10)377; (13)50,71,73;
  (20)8
Klingebiel, U. I., (201)359; (202)359
Kluge, S., (26)331
Knight, R. T., (1)327,328
Knights, B. A., (81)341; (85)341; (86)341;
  (92)342
Knöppel, H., (14)16; (15)220; (34)189;
  (36)384; (176)354
Knox, K. L., (15)94
Kobayashi, S., (183)355
Koenig, W. A., (32)98; (96)313,319
Kondo, K., (215)362
König, W., (13)291; (50)303; (95)313,319
Konz, W., (63)309,319
Koopmans, T., (102)394

Kornitskaya, E. Ya., (180)355
Korostyshevskii, Z., (11)219
Koski, W. S., (49)387,390
Kovarskaya, B., (110)272
Kowalski, B. R., (11)377
Kozuka, S., (183)355
Kramer, D. A., (20)19,24
Krauskopf, K., (150)346
Krauss, M., (28)431,433; (29)431,435;
    (166)406
Krone, H., (26)188
Kropf, A., (155)404
Krueger, P. M., (6)15; (7)15,19; (41)96,101,
    107; (162)349,364
Kugajevsky, I., (35a)237; (64)260; (88)267
Kung, F. L., (126)279
Kuno, S., (183)355
Kuss, E., (63)339
Kutepova, A. I., (104)269
Kvenvolden, K., (150)346
Kydd, G. H., (8)328

Laatikainen, T., (49)337; (64)339
Lampe, F. W., (45c)192
Lande, S., (127a)317
Landesberg, J. M., (113)274
Laneelle, G., (97)313
LaRoe, E. G., (222)363
Laseter, J. L., (140)346
Laurie, W., (81)341
Law, N., (33)333
Lawson, A. M., (7)15,19; (98)269; (107)343;
    (125)279; (162)349,364
Leach, B. E., (123)345
Lederberg, J., (3)125,141; (4)125,170;
    (5)125; (8)125,170; (9)125,170; (10)126,
    168,175; (12)377; (17)143,147,175;
    (21)144; (22)144,175; (23)146,147,175;
    (27)75; (31)154,155,162,175; (34)155,
    160,161,162,163,167,175; (35)162,163,
    167,175; (38)176; (48)34
Lederer, E., (19)292,301,314; (22)292,304,
    318; (25)292,313,315; (26)292,314;
    (31)294,298,299,314,315; (34)297,312,
    316,320; (43)245; (74)310; (79)310;
    (79)341; (80)310; (87)311,316,320;
    (88)342; (92)312,316,317; (97)313;
    (117)316; (118)316; (119)316; (120)316;
    (123)316; (129)345; (131)282, (135)345
Lee, W. H., (41)244
Leemans, F. A. J. M., (3)14; (4)328,342;
    (49)105; (98)269; (125)279
Leferink, J. V. M., (6)15; (25)19
Leitch, L. C., (15)379
Lenard, J., (54a)304,305,316
Lenfant, M., (43)245; (88)342
Lennard-Jones, J., (119)397; (121)397;
    (124)397
Lennarz, W. J., (161)349,364
Leonard, N. J., (112)273
Lester, G. R., (1)375; (39)436; (78)454;
    (92)393
Levin, M. I., (18)144

Levine, L. P., (70)206
Levitt, M., (29)229
Levsen, K., (18)186; (32)188
Levy, R. K., (12)94
Levy, R. L., (150)346
Liberek, B., (109)315; (110)315
Liedtke, R. J., (39)176
Lifshitz, A., (68)205
Lifshitz, C., (51)440; (52)440, (96)269;
    (97)269
Light, J. F., (6)6
Lindeman, L. P., (4)92
Lindstrom, B., (49)337; (64)339
Lindvedt, R. L., (106)394
Ling, L. C., (44)103
Linnarsson, A., (26)97; (103)342
Lipkin, V. M., (18)292,295,297,300,301,
    302,304,315; (30)294,300,301,314,315,
    316; (40)300,301; (45)301,313,314;
    (51)304,305,306,307,313; (54)304,305,
    313; (84)311,312,313,316,318, 319,320;
    (85)311,313; (138)320
Lipsky, S. R., (6)44,61,62; (16)52; (21)70;
    (38)96,99; (57)109; (100)314; (127a)317;
    (150)346
Little, P. F., (72)207
Little, W. F., (16)379
Littlejohn, D., (42)96, 102
Littleton, P., (70)340,351
Llewellyn, P., (42)96,102
Lloyd, H. A., (54)195,200; (133)345
Locke, D. C., (11)94
Locock, R. A., (31)333
Lombardini, J. B., (184)356
Long, F. A., (159)404; (160)404
Long, G. J., (94)393
Lord, K. E., (25)224
Lossing, F. P., (65)390; (78)391
Loudon, A. G., (86)392; (101)394
Loughran, E. D., (74)341
Lounasmaa, M., (129)345
Lovins, R. E., (135)319
Lüben, G., (63)309,319
Lum, K. K., (116)396
Luna, Z., (100)342
Lund, E. D., (58)256,259
Lundin, R. E., (44)103
Lusibrink, T. R., (23)70
Lusser, M., (102)269
Luukkainen, T., (30)98; (44)337; (45)337;
    (46)337; (47)337; (48)337; (49)337;
    (50)337; (58)339; (59)339; (60)339;
    (61)339; (62)339; (64)339; (65)339;
    (93)342
Lynch, M. F., (11)127,128
Lyubimova, A. K., (41)190; (45a)192

Maas, A., (1)180
Maccoll, A., (101)394
Macdonald, C. G., (68)309; (71)309; (72)309,
    310
Machbert, G., (35)334
Mack, L. L., (77)210; (231)364

Macleod, J. K., (34)383; (186)410
Maddock, K., (38)335
Maddock, R. M., (19)95
Magnusson, E. A., (105)394
Majer, J. R., (225)363
Makita, M., (20)95
Malmstadt, H. V., (17)58
Mancuso, N. R., (11)16; (42)31
Mandelbaum, A., (12)377; (18)429; (24)84; (98)393,394
Manhas, M. S., (89)267
Manusadzhyan, V. G., (6)328; (10)290; (16)291
Marde, Y., (63)114,115,116
Mares, P., (103)342
Marino, G., (136)319
Markey, S. P., (33)87; (35)98,105
Marples, B. A., (133)283
Martin, C. J., (23)331
Marx, M., (72)262; (76)263
Mason, D. J., (45)248
Mason, K. G., (123)278
Mattauch, J., (1)12
Mattick, L. R., (220)363
Maul, J. J., (130)281
Maume, B., (90)342; (91)342
Maume, G., (90)342
Maxwell, J. R., (144)346; (150)346
McAdoo, D. J., (176)408
McCapra, F., (39)242
McCarthy, J., (18)144
McClelland, M. J., (104)343
McCloskey, J. A., (3)14; (4)328,342; (6)15; (7)15,19; (25)19; (33)236; (41)96,101, 107; (41)244; (49)105; (76)210; (98)269; (104)343; (107)343; (125)279; (162)349, 364; (181)355
McCormick, A., (4)424; (5)425; (216)362
McCrea, J. M., (18)17; (22)19
McCullogh, K. E., (25)430
McDaniel, E. W., (53b)255
McEntee, T. E., Jr., (95)269
McFadden, W. H., (3)44; (5)92
McGee, K. E., (27)431
McKinney, C. R., (13)428,448,449
McLafferty, F. W., (3)376; (5)78; (5)328; (8)218; (9)78; (9)377; (10)377; (12)377; (13)16, (13)50,71,73; (13)378; (13)428, 448,449; (14)8; (15)16; (17)220; (20)8; (22)430; (25)84; (25)148; (26)430; (28)153; (32)155,159; (38)28; (40)436, 438; (41)436,437; (44)301,312,313,314; (44)438; (50)387; (51)387,395; (52)39; (53)388; (54)388; (57)388,390,392; (58)388,389; (59)389,396,408; (60)389, 396,408; (62)309; (79)455; (89)312,314; (97)393; (102a)315; (106)315; (111)396; (128)398,402; (135)399,408; (140)401; (169)407; (170)408,410; (171)408; (173)408; (174)408; (175)408; (176)408
McLeister, E., (57)256
McLeod, W. D., Jr., (52)106
McMurray, W. J., (6)44,61,62; (10)7; (16)52;

(17)8; (18)81; (21)70; (28)85; (38)96,99; (50)35; (53)40; (57)109; (100)314; (127a)317; (150)346
McReynolds, J., (150)346
McWilliams, I. G., (11)48
Mead, T. E., (13)16; (39)28; (40)28; (140)401
Meade, F., (12)329
Mees, C. E. K., (26)20; (27)21
Meinschein, W. G., (150)346
Meisels, G. G., (46)386; (83)266
Melton, C. E., (20)429; (65)204,205
Merren, T. O., (4)44,61; (5)44,50,52; (22)70, 75
Merritt, C., Jr., (4)44,61; (16)8; (18)61; (99)314
Merron, T. O., (16)8; (99)314
Mester, L., (228)363
Metzger, J., (114)275
Metzinger, G., (14)16
Meyer, F., (71)390; (108)395
Meyerson, S., (7)218; (14)378; (15)379; (17)379; (36)240; (66)261; (67)261; (68)261; (120)277; (121)277; (122)277; (174)354; (178)355
Michel, G., (74)310; (75)310; (79)310; (80)310
Miettinen, T., (68)339
Migahed, M. D., (1)180
Millard, B. J., (55)225; (60)308; (80)265; (82)266; (85)266; (86)266
Miller, F. A., (1)327,328
Millington, D. S., (55)255; (126)316
Milne, G. W. A., (33)333; (45)301,313,314; (46)249; (47)250; (48)250; (49)251; (54)195, 200; (55)199; (56)200; (57)201; (58)202; (59)202; (61a)202; (94)274; (98)342; (105)315; (113)344; (114)344; (116)344; (129)281; (130)281; (138)320; (171)353; (172)353, 357; (180)409; (182)355; (188)357
Milne, T. A., (78)210; (232)364
Minnikin, D. E., (134)345
Miguel, A.-M., (230)363
Miroshnikov, A. I., (18)292,295,297,300,301, 302,304,315; (35)297,298,299,314; (36)298; (37)298; (38)298; (39)299; (84)311,312,313, 316,318,319,320; (85)311,313
Misiti, D., (83)311; (128)345
Mitchell, F. L., (54)338
Mitchell, J. W., (29)21; (30)21; (31)21
Mize, C. E., (114)344; (115)344; (116)344
Mobley, R. C., (77)210; (231)364
Modzeleski, V. E., (150)346
Mohler, F. L., (63)204
Momigny, J., (127)397,400
Monacelli, F., (103)269
Mondelli, R., (83)311
Money, T., (39)242
Moore, H. W., (128)345
Mootoo, B. S., (39)242
Moran, F. F., (47)192
Morell, F. A., (1)92
Morimoto, H., (127)345
Moritz, A. G., (94)393
Morris, H. R., (86)311,314
Morrison, G. H., (22)331

Morrison, J. D., (21)82,83; (80)391
Mortin, R. B., (83a)311
Moscatelli, E. A., (21)95
Moshonas, M. G., (58)256,259
Moss, A. M., (55)338
Mott, N. F., (28)21
Mounts, T. L., (112)343
Muirhead, E. E., (123)345
Muller, E. W., (2)181; (8)182; (11)184
Muller, M., (34)98
Munson, M. S. B., (40b)190; (44)192,203;
   (53)195,200; (67)205; (69)205; (177)408
Muraca, R. F., (126)345
Muramatsu, I., (123)316
Murphy, M. E., (150)346
Murphy, R. C., (150)346
Murray, J. G., (4)44,61; (5)44,50,52; (16)8;
   (99)314
Muysers, K., (11)329; (13)329

Nafissi-V, M. M., (150)346
Nagy, B., (52)106; (150)346
Nair, P. P., (100)342
Nakano, H., (19)221; (20)221; (21)221,
   223; (22)221,223; (23)223; (185)356;
   (186)356
Natalis, P., (32)383; (125)397,400; (126)397
Naworal, J., (153)347
Neher, E., (79)211
Neher, M. B., (26)72
Neiman, M., (110)272
Nencini, G., (115)275
Newman, M. S., (137)401
Newton, A. S., (24)430; (62)447; (75)454;
   (76)454
Nier, A. O., (1)327,328; (2)13
Nooner, D. W., (96)342
Norman, G., (42b)245
Norris, E., (58)305
Novotny, M., (27)97,106
Nyns, E. J., (50)251

O'Brien, I. G., (44)246; (193)357
O'Connor, J. D., (155)347
O'Neal, M. J., (6)92
O'Neill, H. J., (99)342
Oae, S., (183)355
Occolowitz, J. L., (44a)438; (93)393;
   (94)393; (172)408; (177)354
Ochterbeck, E., (1)180
Odham, G., (109)343; (110)343
Ogihara, Y., (83a)311
Oliver, W. R., (99)393
Olmsted, J., (62)447
Olofson, R. A., (113)274
Olsen, R. W., (9)47,56,72,73,74; (13)8;
   (22)70,75; (24)71
Omura, I., (14)80; (48)387; (53)106
Onley, J. H., (198)359
Ore, O., (13)128
Orlow, V. M., (103)315; (104)315
Oro, J., (96)342; (102)342; (136)346;
   (140)346; (150)346

Osborn, J. A., (100)269
Osborn, J. J., (15)329
Ottenbrite, R. M., (107)270
Ottenheym, H., (63)309,319; (64)309,319
Ottinger, Ch., (31)433; (32)435; (36)435;
   (53)440; (54)440; (84)391,398
Ouchida, A., (127)345
Ovchinnikov, Yu. A., (18)292,295,297,300,
   301,302,304,315; (20)292,297,301,302,
   313; (21)292,295,297,300,301,302,313,
   315; (24)292,293,309; (36)298; (37)298;
   (38)298; (39)299; (40)300,301; (45)301,
   313,314; (46)302,303; (48)302,303;
   (51)304,305,306,307,313; (54)304,305,
   313; (65)309; (66)309; (70)309; (73)310;
   (82)311; (84)311,312,313,316,318,319,
   320; (85)311,313; (91)312,315; (127)317,
   318; (128)317
Ovchinnikova, N. S., (65)260
Owens, E. B., (19)19,26; (32)22; (33)22;
   (35)23,24

Palmer, K. H., (168)351; (169)352
Paoletti, E. G., (69)339
Paoloni, L., (115)275
Park, J. W., (83)266
Park, J. Y., (46)386
Parker, R. G., (106)394
Patton, W., (115)315
Pauschmann, H., (95)313,319
Payling, D. W., (87)392,396
Pearl, N., (12)329
Pearson, R. G., (103)394
Peattie, C. G., (1)3
Penders, Th.J., (28)294; (56)304,319;
   (107)315; (113)315; (114)315; (207)361
Pereira, W., (115)315
Perkins, E. G., (37)98; (105)343
Perry, D., (15)131,143
Perry, W. O., (106)315
Pettersson, B., (15)81,82,84; (16)81,82,84
Pettit, G. R., (33)189; (106a)315
Pfaender, P., (42)300,301
Phillips, D. A. S., (83a)311
Phillips, G. T., (39)242
Philpott, D., (150)346
Pignataro, M. T., (171)353
Pignataro, S., (68)390
Pike, W. T., (40)436,438; (59)389,396,408;
   (170)408,410; (171)408
Pillinger, C. T., (150)346
Pirone, A. J., (71)207
Pittman, R. C., (114)344; (132)345
Plimmer, J. R., (199)359; (201)359; (202)359
Pliner, S. A., (180)355
Polan, M. L., (127a)317
Polgar, N., (134)345
Polito, A. J., (152)347; (153)347
Pollitt, R. J., (225)363
Pollock, G., (150)346
Ponnamperuma, C. V., (150)346; (218)363
Popják, G., (37)241, (38)242
Porter, Q. N., (72)341

Postma, E., (166)350
Powell, R. E., (4)424; (5)425
Powers, P., (27)84; (90)312,314
Preller, H., (94)313
Preti, G., (150)346
Priestley, W., Jr., (2)4,5
Pring, B. G., (60)256
Prox, A., (13)291; (17)292,293,295,296, 297,300,315; (27)294,315; (33)296,309, 311,319; (63)309,319; (64)309,319
Puchkov, V. A., (24)292,293,309; (65)309; (66)309; (70)309; (130)319; (131)319; (132)319; (134)319; (180)355

Quattrochi, A., (68)205

Rabinowitch, B. S., (163)405
Radford, T., (14)94
Raison, J. C. A., (15)329
Ramsay, J. N., (144)346
Ramsperger, H. C., (62)389
Randall, R. B., (26)72
Raper, O. F., (109)272
Rapp, J. P., (42)337
Ready, J. F., (70)206
Reed, W., (150)346
Reed, W. T., 204(360)
Rees, D. I., (47)105
Rees, R. W. A., (70)340,351
Reese, C. B., (143)402
Reichard, D. W., (142)402
Reichert, C., (104)394
Reifsnyder, C. A., (190)357
Reilley, C. N., (11)377
Reinhold, V. N., (36)98,114
Renard, J., (108)271
Renfroe, H. B., (54)255
Renwick, J. A. A., (217)362
Reuben, B. G., (77)454
Reynolds, W., (65)115
Rhodes, L. S., (169)352
Rhodes, R., (66)450; (139)401
Rhodes, R. C., (196)358
Rice, O, K., (62)389
Richards, F. F., (135)319
Richey, H., Jr., (105)270
Richey, J. M., (105)270
Richter, R., (97)342
Richter, W., (59)305
Richter, W. J., (16)220
Ridley, R. G., (20)82
Rieke, R. D., (37)384
Rietz, P. J., (131)345
Rinehart, K. L., Jr., (9)218,281; (18)379; (34)237; (54)255; (63)259; (69)261,262; (70)261; (71)261,263; (73)262
Rist, C. E., (111)273
Rittenberg, D., (31)231; (179)355
Robbins, R. K., (33)189
Roberts, D. T., Jr., (16)379
Robertson, A. V., (12)377; (17)143,147, 175; (23)146,147,175; (72)262; (74)262, 263

Robertson, J. S., (10)329
Robinson, J., (118)344
Rogerson, P. F., (43)385; (106)394; (107)394
Rohwedder, W. K., (38)189; (101)342; (112)343
Röller, H., (210)362; (211)362
Roman, J., (20)330
Ronayne, J., (39)384; (40)384; (47)439
Rose, G., (87)342
Rosebury, F., (179)355
Rosenstock, H. M., (20)429; (25)430; (28)431,433; (29)431,435; (74)454; (148)403; (166)406; (167)406
Rosenthal, D., (5)377
Rosetti, G. P., (67)390
Rosinov, B. V., (18)292,295,297,300,301, 302,304,315; (21)292,295,297,300,301, 302,313,315; (24)292,293,309; (36)298; (37)298; (38)298; (39)299; (40)300,301; (46)302,303; (48)302,302; (51)304,305, 306,307,313; (54)304,305,313; (61)308; (65)309; (70)309; (73)310; (76)310; (77)310; (82)311; (84)311,312,313,316, 318,319,320; (85)311,313; (91)312,315; (127)317,318; (128)317; (129)318,319
Ross, F. T., (169)352
Rowland, C. G., (72)454
Rucker, G., (34)333
Rudolph, P. S., (65)204,205
Russell, C. R., (111)273
Russell, D. W., (72)309,310
Russell, J., (15)329
Rutherford, K. G., (107)270
Ryan, K. R., (9)425
Rybiski, D., (51)337
Ryhage, R., (8)93; (9)93,95,96; (15)81,82, 84; (16)81,82,84; (17)94; (24)96,97; (25)97; (26)224; (36)240; (37)241; (38)242; (52)338; (54)255; (57)339; (61)114; (62)114; (63)114,115,116; (94)342; (120)345; (135)345; (151)346; (178)355
Rylander, P. N., (14)378; (17)379; (68)261

Sabatori, T., (115)275
Saito, A., (82)341
Salague, A., (79)341
Salokangas, A., (43)337; (45)337
Salomone, R., (135)319
Sample, S. D., (23)380
Samuelsson, B., (26)224; (31)98; (120)345; (121)345; (122)345; (124)345); (158)348; (187)356
Samuelsson, K., (31)98; (157)348; (158)348
Sasaki, M., (127)345
Sato, H., (19)221; (20)221; (21)221,223; (22)221,223; (23)223; (185)356; (186)356
Saukov, P. I., (104)269
Saunders, M. J., (12)329
Saunders, R. A., (2)421,446,455; (4)218,255; (6)377; (12)428; (14)220
Saur, W. K., (42a)245; (42b)245
Sautter, W., (95)313,319

Scallen, T. J., (74)341
Schaffner, C. P., (51)337
Schildcrout, S. M., (103)394
Schildknecht, H., (209)361
Schimbor, R. F., (62)259; (63)259
Schimpl, A., (228)363
Schissel, P. O., (14)185; (19)186
Schissler, D. O., (45b),192
Schmid, H., (101)269
Schmid, J., (63)309,319; (64)309,319
Schneider, H. J., (136)346
Schnoes, H. K., (77)341; (78)341; (145)346;
  (146)346; (147)346
Schoenfield, L. J., (76)341
Schoenheimer, R., (179)355
Scholz, R. Y., (99)342
Schopf, J. W., (150)346
Schroepfer, G. J., Jr., (40)244; (41)244
Schroll, G., (31)154,155,162,175; (34)155,
  160,161,162,163,167,175; (38)176
Schubert, K., (87)342
Schuddemage, H. D. R., (175)408
Schultz, J. L., (21)380
Schulze, P., (13a)428; (14)16; (16)16;
  (37)189
Schwab, J., (28)382
Schwartz, M. A., (165)350; (166)350
Sciamanna, A. F., (75)454; (76)454
Scott, A. I., (39)242
Scott, M. M., (150)346
Seibl, J., (8)290; (14)80, (15)429; (19)429;
  (32)296; (53)106;(60)202
Seifert, R. M., (44)103; (219)363
Selva, A., (83)311
Senn, M., (16)220; (25)84; (44)301,312,
  313,314; (62)309; (89)312,314; (228)363
Senning, A., (129)281
Setser, B. W., (163)405
Shackleton, C. H. L., (54)338
Shannon, J. S., (68)309, (69)309; (71)309;
  (72)309,310
Shannon, T. W., (13)16; (13)428,448,449;
  (15)16; (41)436,437; (65)448; (70)390;
  (106)315; (128)398,402; (140)401;
  (169)407
Shapiro, M., (51)440; (52)440; (96)269;
  (97)269
Shapiro, R. H., (95)269
Sharkey, A. G., (21)380
Sharpless, K. B., (25)224
Shasha, B. S., (111)273
Shaw, M. A., (116)315
Shaw, P. D., (33)236; (181)355
Shealy, Y. F., (137)346
Sheets, T. J., (195)358
Sheikh, Y. M., (38)176
Shemyakin, M. M., (18)292,295,297,300,
  301,302,304,315; (20)292,297,301,302,
  313; (21)292,295,297,300,301,302,313;
  315; (23)292,294,295,296,297,298,299,
  301,302,304,311,312,313,314,315,316,
  318; (24)292,293,309; (36)298; (37)298;
  (38)298; (39)399; (40)300,301; (48)302,

303; (51)304,305,306,307,313; (54)304,
  305,313; (65)309; (66)309; (70)309;
  (84)311,312,313,316,318,319,320;
  (85)311,313; (91)312,315; (128)317;
  (129)318,319
Sheppard, R. C., (122)316; (126)316;
  (208)361
Sherman, H., (203)360
Shields, D. J., (25)71,73,74
Shikita, M., (20)221
Shilin, V. V., (61)308
Shima, T., (83b)311; (127)345
Shiner, V. J., Jr., (150)346
Shrader, S., (70)340,351
Shrage, K., (119)277
Shukla, O. P., (35a)237
Shupe, R. D., (79)264
Shvetsov, Yu. B., (18)292,295,297,300,301,
  302,304,315
Sieck, L. W., (9)425
Siegel, A., (71)261,263
Siegel, A. L., (59)339
Siehoff, F., (11)329
Simon, W., (14)80; (53)106,110
Simoneit, B. R., (16)16; (148)346; (150)346
Singer, T. P., (184)356
Siri, W. E., (10)329
Sisler, H. D., (200)359
Sjövall, J., (17)94; (28)98; (38)335; (39)335,
  341; (40)336: (51)254; (56)338; (57)339;
  (60)112; (67)339; (73)341; (76)341;
  (120)345
Sjövall, K., (38)355; (56)338
Skelton, F. S., (131)345
Sketchley, J. M., (123)278
Skorepa, J., (103)342
Skul'skii, I. A., (25)331
Smidt, U., (13)329
Smith, D., (101)394
Smith, D. H., (9)47,56,72,73,74; (22)70,75;
  (24)71
Smith, D. L., (170)352
Smith, E. P., (141)402
Smith, G. A., (24)381
Smith, G. G., (116)396
Smith, H., (70)340,351
Smith, I. C., (20)82
Smith, J., (150)346
Smith, J. S., (176)408
Smith, L., (7)15,19
Smyth, H. D., (74)390
Smyth, K. C., (65)448
Snell, E. E., (50)251
Snow, G. A., (229)363
Snyder, J. J., (34)237
So, R. T., (89)267
Sokoloski, E. A., (133)345
Sparrow, L. G., (50)251
Spencer, J. F. T., (30)229
Sphon, J. A., (35)189
Spiteller-Friedman, M., (30)188; (168)407
Spiteller, G., (4)376; (30)188; (52)255;(168)407

Srinivasan, R., (117)276
Stabenau, J. R., (227)363
Stafford, F. E., (103)394
Stahl, W. L., (118)344
Staley, S. W., (142)402; (144)402
Ställberg-Stenhagen, S., (19)82,83; (135)345; (144)346
Stam, J. G., (87)267
Standifer, L. N., (84)341
Stanier, H. M., (25)71,73,74
Steinberg, D., (114)344; (115)344; (116)344
Steiner, B., (162)404
Stenhagen, E., (7)328; (12)291; (19)82,83; (22)95; (135)345
Stepanov, V. M., (130)319; (131)319, (132)319; (133)319, (134)319
Stevenson, D. D., (84)266
Stevenson, D. P., (45b)192; (46)192
Stewart, R., (92)268
Stewart, W. B., (14)329
Stillwell, R. N., (7)15,19; (76)210; (107)343; (162)349,364
Stjernstrom, N. E., (60)256
Strachey, C., (29)86
Strauch, B. S., (132)345; (133)345
Street, K., Jr., (62)447
Streitweiser, A., Jr., (64)390
Strini, J.-C., (114)275
Struck, R. F., (137)346
Suda, T., (77)341
Sugino, H., (83b)311
Suhadolnik, R. J., (35a)237
Sun, K. K.,(17)292,293,295,296,297,300, 315; (27)294,315
Sundaram, P. K., (50)251
Sung, M., (61)257
Susz, B. P., (67)390
Sutherland, G. L., (6)125,168,175,176; (7)125, 126,144,168,175,176; (12)377; (17)143, 147,175; (20)144; (22)144,175; (23)146, 147,175; (34)154,155, 162,175; (34)155, 160,161,162,163,167,175; (38)176
Sutter, B., (101)269
Svec, H. J., (5)290; (6)290; (100)394
Sweeley, C. C., (14)94; (21)95; (28)227; (52)338; (61)114; (152)347;(153)347; (155)347; (160)349,364; (210)362
Sykes, P. J., (54),338

Tait, J. M. S., (70)390
Tajima, E., (19)429
Takemoto, C., (21)221,223; (185)356
Takemoto, H., (22)221,223
Takeuchi, S., (62)309
Tal'roze, V. A., (41)190; (45a)192
Tam, S. W., (40)384; (49)440
Tamaoki, B.-I., (19)221; (20)221; (21)221, 223; (22)221,223; (23)223; (185)356; (186)356
Tamura, G., (83b)311
Tanabe, M., (175)354
Tannenbaum, H. P., (112)396
Tavares, R., (35b),237

Taylor, W., (65)339
Teeter, R. M., (8)7
Tendille, C., (129)345
Tenschert, G., (18)186; (32)188
Teranishi, R., (5)92; (43)96,103,105,106
Thomas, D. W., (34)297,312,316,320; (87)311,316,320; (92)312,316,320; (119)316; (121)316; (123)316; (124)316; (125)316; (131)282
Thomas, G. H., (92)342
Thomas, O., (67)309
Thomas, W. A., (83a)311
Thomason, E. M., (7)6
Thomson, J. J., (1)78
Thomson, K. A., (105)394
Thornton, E. R., (35)383; (141)402
Thorpe, S. R., (28)227; (156)347
Tickner, A. W., (78)391
Tiernan, T. O., (42a)191, (49)192
Todd, Lord, (78)310
Toi, K., (58)305
Tosato, M. L., (115)275
Tovarova, I. I., (180)355
Trams, E. G., (117)344; (118)344
Trost, B. M., (210)362; (211)362
Tsai, L., (137)401
Tsuboyama, K., (7)15,19; (162)349,364
Tsunakawa, S., (11)16
Tucker, R. B., (65)115
Tullock, A. P., (30)229
Tunnicliff, D. D., (15)8; (41)29,33

Udenfriend, S., (29)229; (189)357
Updegrove, W. S., (48)105; (150)346
Urey, H. C., (150)346

Vacheron, M. J., (80)310
Vale, W., (9)15; (10)15
Van Lear, G. E., (8)218; (18)379; (70)261; (106)315
van Tamelen, E. E., (24)224; (25)224
van Winkle, E., (224)363
Vandenheuvel, W. J. A., (18)95; (19)95
van der Haar, F., (16)429
van Heijenoort, J., (26)292,314; (31)294, 298,299,314,315
Vane, F. M., (125)345; (165)350; (166)350
Varco, R. L., (1)327,328
Varshavskii, Ya. M., (10)290; (16)291; (103)315; (104)315
Vastola, F. J., (71)207
Venkataraghavan, R., (9)377; (10)377; (12)377; (13)50,71,73; (14)8; (25)84; (38)28; (44)301,312,313,314; (52)39; (102a)315
Venning, E. H., (66)339
Vesonder, R. F., (101)342
Vestal, M. L., (16)329; (30)431,432,435; (33)333; (55)199; (55)440; (89)392,404; (105)315; (152)404; (157)404; (180)409
Vetter-Diechtl, H., (59)305
Vetter, W., (9)290; (59)305
Viallard, R., (5)218; (6)218

Vihko, R., (40)336; (41)336; (49)337; (64)339
Vilkas, A., (25)292,313,315; (120)316
Vining, L. C., (81)311
Vinogradova, E. I., (18)292,295,297,300,301, 302,304,315; (21)292,295,297,300,301, 302,313,315; (24)292,293,309; (40)300, 301; (51)304,305,306,307,313; (54)304, 305,313; (65)309; (66)309; (84)311,312, 313,316,318,319,320; (85)311,313; (129)318,319
Vite, J. P., (217)362
Vliegenthart, J. F. G., (107)315; (108)315
Voge, H. H., (84)266
Völlmin, J. A., (14)80; (34)98; (53)106, 110; (55)106,110; (56)106
Volpi, G. G., (66)205
Von Endt, D. W., (199)359
von Euler, U. S., (119)345
Vora, B. V., (74)341
Vought, R. H., (73)390
Vouros, P., (25)19

Wachs, T., (97)393; (111)396
Wadsworth, P. A., (15)8; (41)29,33
Wagener, K., (194)358
Wagner, C. D., (84)266
Wagner, G., (36)189
Wahrhaftig, A. L., (148)403; (152)404; (155)404; (156)404; (157)404
Wall, M. E., (168)351; (169)352
Wallace, J. B., (212)362
Wallenstein, M. B., (148)403
Waller, G. R., (25)97; (36)240; (178)355
Wallington, M. J., (27)84; (90)312,314
Wang, B.J.-S., (35)383
Wanless, G. G., (29)188
Ward, D. N., (9)15; (10)15
Ward, R. S., (29)382; (48)439; (72)390, 395,396,404
Ward, S. D., (81)266; (145)402; (146)403
Warner, C. G., (13)16; (140)401
Warren, J. W., (75)390
Waterbury, L. D., (167)351
Waterfield, M., (135)319
Watson, J. T., (2)44; (4)14; (5)14; (10)93,96, 109; (12)78; (33)96,98,109; (50)105; (95)342
Weber, D. J., (140)346
Webster, B. R., (26)84; (49)303,312,314; (88)312,314
Wedd, A. G., (105)394
Weinig, E., (28)332; (35)334
Weiss, J. F., (69)339
Wells, W. W., (29)95
Wenneis, W. F., (209)361
Westmore, J. B., (104)394
Wexler, S., (68)205
Weygand, F., (13)291; (17)292,293,295, 296,297,300,315; (33)296,309,311,319
White, A. J., (229)363
White, F. A., (173)354
White, G. L., (93)393

Whitesides, G. M., (116)276
Whiting, A. F., (46)439
Whiting, D. A., (132)282
Whittick, J. S., (126)345
Wicha, J., (18)429
Wickersham, L. J., (101)342
Wiebers, J. L., (164)349
Wieland, Th., (63)309,319
Wijtvliet, J., (7)15,19
Wikström, S., (25)97; (63)114,115,116; (96)342
Wilkins, C. L., (147)403
Wilkinson, G., (100)269
Willett, J. D., (24)224; (25)224
Williams, A. E., (2)421,446,455; (4)218, 255; (6)377; (12)428; (13)219; (14)220; (50)440; (68)452,453
Williams, D. H., (1)218,225; (2)218,220, 255; (2)290; (3)218,255; (6)78; (7)78; (8)78; (10)78; (24)146; (24)381; (25)381,406, 407; (26)381; (28)75; (29)153, (29)382; (30)153; (33)155,159; (33)383; (37)435, 455; (39)384; (40)384; (43)437; (45)438; (47)439; (48)439; (48a)439; (49)440; (56)440; (61)389,392; (72)390,395,396, 404; (78)310; (85)391,404,405,406; (95)393; (110)396,404; (112)396; (116)315; (151)404; (154)404; (164)405, 406
Williamson, L., (63)204
Wilson, W. E., (91)342
Winter, R. E. K., (133)400
Winters, R. E., (79)391
Wipke, W. T., (1)124
Wiseman, P., (82)341
Witkop, B., (189)357; (191)357; (192)357
Woldring, S., (17)329; (18)330
Wolf, C., (150)346
Wolf, C. J., (45)96,103,106; (51)105
Wolfe, E. S., (55)388
Wolfsberg, M., (159)404
Wolstenholme, W. A., (21)330; (26)292, 314; (27)84; (31)294,298,299,314,315; (57)441,442; (58)110,116; (58)441,443; (74)310; (81)311; (90)312,314; (97)313; (101)314
Woodgate, P. D., (75)263
Woodward, R. B., (132)398,399; (134)398
Woolford, D. C., (17)329
Woolston, J. R., (20)19,24
Worth, G., (11)329
Wul'fson, N. S., (18)292,295,297,300,301, 302,304,315; (24)292,293,309; (36)298; (65)260; (65)309; (66)309; (70)309, (73)310; (82)311; (91)312,315; (130)319; (131)319; (132)319; (133)319; (134)319; (180)355
Wünsch, E., (94)313

Yarger, K., (15)94
Yeo, A. N. H., (26)381; (45)438; (48a)439; (56)440; (85)391,404,405,406; (151)404; (164)405,406

Yip, G., (198)359
Youdeowei, A., (213)362
Young, J. F., (100)269

Young, J. S., (13)94, (89)342
Young, M., (150)346

Subject Index

Abbreviation of mass spectral files, 83
Aberrations, ion optical, 53
Abstraction of counter-ions from solids, 211
Abu¹-peptidolipin-NA, structure determination of by mass spectrometry, 310
Abundance, natural, of isotopes, 376
of metastable ions, 399
Accelerating voltage, alternator, 114, 338
dependence on of relative metastable peak intensities, 439
effect of on relative metastable peak intensities, 439
relationship of to mass-to-charge ratio and magnetic field strength, 45
scanning of, 46, 425
scanning, tuning of source during, 426
variations in field penetration effects during, 426
variation of sensitivity with, 46
Acceleration of ions, metastable transitions during, 427
Accuracy of, analog-to-digital converters, 63
mass measurement, theoretical limits on, 52
variation of in scanning instruments, 51
of quantitation in spark source mass spectrometers, 331
Acetamidoisobutyrophenone, formation of from aziridines, 273
Acetanilide, carbon-13 labelled, 262
mass spectral fragmentation of, 263
Acetic acid, loss of from molecular ion, in chemical ionization mass spectra of acetates, 189
in field ionization mass spectra of acetates, 189
Acetic anhydride in methanol as acylating agent for peptides, 312
Acetone, electron impact mass spectrum of, measured at 10,000 eV energies, 204
formation of from propene-1-carbon-13, 275
gas chromatography-mass spectrometry of, 92

identification of in desert locust, 362
keto and enol forms of in mass spectra, 410
Acetonitrile, addition of to aziridines, 273
Acetophenones, substituents effects in, 392
bis(p-Acetoxyphenyl) cyclohexylidene methane, use as ovulation promoter, 337
Acetylacetone, condensation of with quanidino groups of arginine in peptides, 305
N-Acetylarginine, mass spectrum of, 304
Acetylene and hydrogen, loss of, from the genzyl cation, 378
from the molecular ion of, benzene, metastable peak for, 422
p-fluorobenzyl chloride, 439
hexahelicene, 401
radiation ionization of, 205
Acetylenic carbons, handling of by DENDRAL notation, 137
Acetyl ions, formation of from acetophenones, 392
N-Acetyl peptides, methyl esters of, mass spectra of, 291
O-Acetyltyrosylvaline, identification of in porcine neurohypophysis, 361
Acids, analysis and identification of by mass spectrometry, 339
Acorn worm, marine, 2,6-dibromophenol in, 363
Acquisition of data, on-line, in real time, 5
Acrolein derivatives, isolation of from Valencia orange peel oil, 259
formation of from propene-1-carbon-13, 275
Activated complexes for fragmentation, structures of, 387
Activation energies, correlation with ion intensity, 392
extraction of from chemical ionization mass spectra, 409
in mass spectral reactions, 391
relation of to metastable peak intensities, 434
Acyclic structures, DENDRAL canons for, 140
DENDRAL program for generation of, 125
Acyclic systems, stereospecificity in elimination reactions of, 382

Acyl, groups, mixed, in peptide mass spectra, 314
  guanidino groups in arginine, thermal instability of, 304
  ions, formation of from aromatic ketones, 395
N-Acyl, groups, effect of on volatility of peptides, 315
  protecting group of peptide derivatives, fragmentation of, 294
Acylated peptides, intensity of aldimine ions in mass spectra of, 293
Acylating agents used in derivation of peptides for mass spectrometry, 312
Acylation of, amino groups in free peptides, 312
  side chain amino groups in lysine and ornithine in peptides, 304
Acyl carbonium ions, formation of in fragmentation of ketones, 147
  loss of carbon monoxide from, 147
N-Acylpeptide esters, containing arginine, mass spectra of, 304
  containing lysine and ornithine, mass spectra of, 304
  conversion of to fully N-methylated derivatives, 316
  fragmentation of, 295
  specific patterns of fragmentation of, 313
N-Acylpeptide methyl esters, fragmentation of in mass spectrometry, 292
  molecular ions from, 292
N-Acylpeptides, reduction of with lithium aluminum hydride or deuteride, 291
Additive alkyl substituent effects, 387
Additives, mass spectra of in chemical ionization conditions, 192
Adenosine, monophosphate, mass spectra of derivatives of, 349
  oxygen-18 labelled, high resolution mass spectrum of, 234
  mass spectrum of, 231
  trimethylsilyl derivative of, mass spectrum of, 15
  triphosphate, oxygen-18 labelled, 234
Adiabatic cooling, problems caused by in jet systems, 210
Advantages of data acquisition systems, 2
Afferent links, multiplicity of in DENDRAL notation, 140
Alanine and glycine, confusion between, following peptide methylation, 316
  confusion of with leucine in peptide mass spectra, 313
  in peptides, specific fragmentation reactions of, 295
Alcohols, aliphatic, chemical ionization mass spectra of, 192
  carbon-carbon bond cleavage in, 380
  confusion of ethers with, 153
  dehydration of in mass spectral fragmentations, 380
  field ionization mass spectra of, 189

fragmentation reactions of, 436
identification of from mass spectra by Heuristic DENDRAL programs, 146
separation, from ketones by gas chromatography, 95
  of by gas chromatography, 94
structure determination of from mass spectra by Heuristic DENDRAL programs, 162
  success of Heuristic DENDRAL programs with, 175
unsaturated, confusion of with ketones and unsaturated ethers, 147
Aldehydes, unsaturated, structure determination of, 259
Aldimine, and amino acid fragment ions in peptide spectra, relative intensities of, 293
  fragment in the mass spectra of peptide derivatives, 292
  ions, intensity of in the mass spectra of trifluoroacetylpeptides, 293
  in peptide mass spectra, decomposition of, 293
  in peptides containing glycine at the c-terminus 293
Aldolase, 245
Aldosterone, methods of quantitating, 337
  in urine, 337
Aldrin, detection of by mass spectrometry, 334
Algae, microscopic, lipids in, 346
Algorithms for conversion of mass to elemental composition, 74
Aliphatic acids, low molecular weight, use of to protect the N-terminus of peptides, 315
  use of to acylate the N-terminus of peptides for mass spectrometry, 293
Aliphatic amino acids, effect of on peptide mass spectra, 295
  peptides containing, mass spectra of, 291
Alkali-labile peptide bonds, 313
Alkaloids, biosynthesis of, 242
Alkanes, in bovine, brain, identification of by mass spectrometry, 342
  feces, identification of by mass spectrometry 342
  dihydroxy, mass spectra of, 343
  enumeration of all possible, 131
  enzymatic hydroxylation of, 229
  ionization potentials of, calculation of, 397
  normal C_6-C_8, field ionization mass spectra of, 186
  number of possible isomers of, 125
  possible centroids of, 131
  total number possible for a given carbon number, 131
Alkyl, carbonium ion transfer in chemical ionization, 192
  halides, dehydrohalogenation in the mass spectra of, 380
  ions, formation of by decarbonylation of acyl carbonium ions, 147
  radicals, monovalent, enumeration of all possible, 131
Alkylation, intermolecular, of the imidazole

ring of histidine in the mass spectrometer, 301
in peptide mass spectra, 295
of peptides, 313
Alveoli, gas pressures in, 330
Amide, bond cleavage in the mass spectra of peptides, absence of resulting ions, 313
bonds, alkali-labile, 313
    cleavage of in the mass spectra of peptide derivatives, 292
    expulsion of the elements of in the mass spectra of cyclic peptides and depsipeptides, 307
    in peptides, exhaustive N-methylation of, 316
groups, in acylated lysine and ornithine in peptides, fragmentation of, 304
    dehydration of in the mass spectra of peptides, 294
Amides, primary, in side chains of peptides, loss of ammonia from, 298
Amidoximes, formation of by photochemical rearrangement of nitrosamines, 274
Amine fragments in peptide mass spectra, 294
Amines, aliphatic, field ionization mass spectra of, 189
    hydrogen rearrangements in, 380
    and amino acids, similar ionization potentials of, 394
    α-cleavage of, 159
    high molecular weight, gas chromatography-mass spectrometry of, 106
    identification of, from mass spectra by Heuristic DENDRAL programs, 146
    using the mass spectra of their dansyl derivatives, 362
    rules for interpretation of mass spectra of, 155
    structure determination of from mass spectra by Heuristic DENDRAL programs, 155
    substitution on α-carbon of, 158
    success of Heuristic DENDRAL programs with, 175
Amino acid, and aldimine fragment ions in peptide spectra, relative intensities of, 293
    analysis, quantitative, value of in peptide sequence determination, 311
    value of in distinguishing leucine and isoleucine, 297
    composition, attempts to determine from mass spectra of peptides, 312
    use of in interpretation of peptide mass spectra, 312
    fragmentation, absence of in arginine-containing peptides, 304
    fragment in the mass spectra of peptide derivatives, 292
    residues in peptides, conversion of under electron impact to other residues, 313
    type of fragmentation in the mass spectra of peptide derivatives, 293
Amino acids, and amines, similar ionization

potentials of, 394
chemical ionization mass spectra of, 202
in cyclic depsipeptides, effect of on fragmentation, 309
effect of, on thermal stability of peptides, 315
    on volatility of peptides, 315
in peptides, effect of on fragmentation, 295
    identification of by mass spectrometry, 300
    sequence of in peptides and proteins, use of mass spectrometry to determine, 289
N-β-Aminoethylcysteine-containing peptides, sequence determination of by mass spectrometry, 304
S-β-Aminoethylcysteine, conversion of cysteine to in peptides, 303
Amino groups, in free peptides, acylation of, 312
    in side chains of lysine and ornithine, acylation of in peptides, 304
2-Amino-2-thiazoline, reaction with nitrogen-15 labelled potassium cyanate, 281
Ammonia, conversion of to benzamide, 273
    dansyl, in desert locusts, 362
    elimination of in the spectra of tyrosine-containing peptides, 301
    formation of from aziridines, 273
    loss of from side chain primary amide groups in peptides, 298
    proton affinity of, 409
Amniotic fluid at term, estrone derivatives in, 339
Amobarbital, determination of, 332
Amplifiers, sample-and-hold, 58
Amplitude of peaks, as a measure of relative abundance, 47, 52
t-Amyl acetate, isobutane chemical ionization mass spectra of, 409
Anagyrine, chemical ionization mass spectrum of, 200
    electron impact mass spectrum of, 200
Analog, recording devices, 61
    tape, recorder for the recording of mass spectra, 61
    recording, of mass spectra, dilation of time scale in, 65
    of mass spectra on, 59
    signal-to-noise levels in, 61
    reproduction of, 61
    use of in electrical recording of high resolution mass spectra, 44
    thresholding of mass spectra, 60
Analog-to-digital, conversion, devices, 79
    fully automatic devices for, 3
    of gas chromatography-mass spectrometry data, 7
    of mass spectral data, 3
    precision of, 63
    semiautomatic devices for, 3
    speed of, 7
    techniques for, 57

converters, 4, 5, 44, 57
  bit capacity of, 66
  choice of for scanning of high resolution
    mass spectra, 62
  computer control of, 57
  galvanometers used as, 5
  sampling rates of, 63
Analysis of mixtures of peptides by mass
  spectrometry, 303
Analyzer of mass spectrometers, collision
  processes in, 212
Androstane derivatives, identification of in
  amniotic fluid, 339
Androst-1,4-dien-3,17-dione, identification
  of by mass spectrometry, 342
Androst-4-en-3,17-dione, 221
  identification of by mass spectrometry, 342
Androst-5-en-3,17-diol, trimethylsilyl ether,
  mass spectrum of, 337
Androsterone, trimethylsilyl ether, mass
  spectrum of, 336
Angolide, structure determination of by
  mass spectrometry, 309
Aniline, -1-carbon-13, high resolution mass
  spectrum of, 263
  formation of from nitrobenzene in the
    mass spectrometer, 375
  mass spectral fragmentation of, 262
Anisoles, fragmentation of, 390
Anthranilate, methyl N(O-aminophenyl)-N-(3-
  dimethylaminopropyl), metabolism
  of, 353
Anthranilic acid, enzymatic conversion to
  catechol, 355
Antiaromatic transition states in electrocyclic
  reactions, 399
Antibiotics, mass spectra of, 309
Apical node, composition of in DENDRAL
  notation, 140
  degree of in DENDRAL notation, 139, 140
  in DENDRAL notation, 140
  as starting point in structure mapping, 136
[A.P.-I.P.] values, correlation with ion
  intensities, 392
Apollo 11 mission, 346
Apparent mass of daughter ions, 425
Appearance, and ionization potentials, difficulties in measurement of, 390
  potential of vinyl acetylene as a fragment
    ion, 436
  potentials, calculation of rate constants
    from, errors in, 436
  σ correlations of, 390
  of daughter ions, 403
  effect of substituents on, 389–391
  interpretation of, 376
  of ions, thermochemical information
    derived from, 393
  effect of substituents on, 389–391
  kinetic shift in, 398
  measurement of, 391
    in electron bombardment sources, 179
    in photoionization sources, 209
    in the study of metastable transitions, 435

  of metastable transitions, 436
Apple cuticle, hydroxylated fatty acids in,
  346
Applications of gas chromatography-mass
  spectrometry, 110
Area of peaks, determination of on photoplates, 31
  as a measure of relative abundance, 47, 52
Arene oxide, intermediacy of, in naphthalene
  metabolism, 357
  in phenylalanine-tyrosine conversion, 357
Arginine, acylated guanidino groups in,
  thermal instability of, 304
  -containing peptides, absence of sequence
    information in the mass spectra of,
    304
  difficulties in reduction of, 291
  at C-terminus of tryptic peptides, 319
  derivatization of for mass spectrometry, 304
  free, attempts to obtain mass spectrum of, 304
  -glycine peptide bonds, lability of to base, 313
  localization of charge in, 304
  nitrogen-15 labelled, 355
  in peptides, derivatization of, 313
    specific fragmentation reactions of, 304
  residues, derivatization of for gas-liquid
    chromatography, 305
  in peptides, conversion of to ornithine
    residues, 305
    modification of for mass spectrometry,
      304
  retention of guanidino group of during
    acylation reactions of peptides, 312
Argon ionization detectors in gas chromatography, 205
Aromatic, acids in shale, 346
  amino acid residues, influence of on de-
    hydration processes, 294
  amino acids, confusion of in peptide mass
    spectra with glycine, 313
  in peptides, specific fragmentation reac-
    tions of, 299
  carboxylic acids, exchange of the ortho
    protons of, 440
  metastable transitions in the mass spectra
    of, 440
  compounds, fragmentation of, 384
  ionization potentials of, 397
  McLafferty rearrangements of, 383
  hydrocarbons, chemical ionization mass
    spectra of, 192
  doubly charged ions in the mass spectrum
    of, 446
  laser-induced ionization mass spectrum
    of, 207
  rings, effect of substituents in on ioniza-
    tion potentials, 389
  enzymatic hydroxylation of, 357
  states in thermal electrocyclic reactions,
    399
  substituents, σ constants for, 387
  substitution reactions, rates of, 269
  systems, influence of phenyl substituents

upon mass spectra of, 386
Arrangement of high resolution mass spectral
data, 377
Arthropod, defense secretion of, glomerin
as, 361
Artificial intelligence, application of to interpretation of mass spectra, 121
research on, 122
use of in computer interpretation of mass
spectra, 88
Aryl carbonates, fragmentation of, 256
N-Aryl groups as markers in fragmentation
studies, 256
Ashing of biological samples for spark source
mass spectrometry, 331
Asparagine residues in peptides, loss of, ammonia from, 298
carboxamide groups from, 298
specific fragmentation reactions of, 297
Aspartate transaminase, cytoplasmic, amino
acid sequence of, 320
structure of peptides formed from by
partial hydrolysis, 313
Aspartic acid, nitrogen-15 labelled, 237, 240
in peptides, specific fragmentation reactions
of, 297
a- and β- peptides of, mass spectra of, 299
residues, in peptides, loss of carbalkoxy
groups from, 298
side reactions of silver oxide with, 316
side chain of, reduction of during sequencing of polypeptides, 291
side reaction of sodium hydride with, 316
Assay procedures for steroids based upon
mass spectrometry, 340
Assignment of metastable peaks, 422
Asymmetrical cystine-containing peptides,
mass spectra of, 303
Atlantic digitizer, 5
Atoms, involved in fragmentation processes,
identification of, 377
ionization potential of, 183
Attenuation of mass spectrometer signals, 61
Auger effect, similarity of Penning ionization
to, 207
Auto-chemical ionization, 189
Autoionization of molecules under high
electrical potential, 184
Automatic, data collection and reduction
systems, 7
determination of amino acid sequence in
peptides from mass spectral data, 314
Automation of mass spectrometer output, 3
Autoradiography-mass spectrometry, use of
in detection of carbon-14, 220, 354,
384
Azatropylium ion, structure of, 262
Azide, phenyl, carbon-13 labelled, mass
spectral fragmentation of, 263
Aziridines, ring opening of by acetonitrile,
273

Background, in mass spectra, contribution

to from metastable transitions, 423,
428
spectra, subtraction of, 94, 115
Back-reaction potentials, 392
Badlist in DENDRAL, 142, 147
Baeyer-Villiger mechanism, involvement of in
enzymatic conversion of progesterone
to testosterone, 356
Bandwidth of detection systems, in high
resolution mass spectrometers, 48
required for gas chromatography-mass
spectrometry, 108
Barbital, determination of, 332
Barbiturates, chemical ionization mass spectra
of, 201
mass spectrometric analysis of, 332
specific identification of by mass spectrometry,
333
Bar graphs, production of by gas chromatograph-
mass spectrometer-computer system, 115
Base-labile peptide bonds, 313
Baseline, electronic definition of, 69
elimination of by thresholding, 59
Beam splitters used in microdensitometers, 28
Benzaldehyde-oxygen-18 in ozonolysis of
stilbene, 271
Benzamide, formation of from ammonia. 273
Benzene, deuterated, mass spectrum of, 384
1,4-dideutero, metastable transitions in the
mass spectrum of, 439
doubly charged ions in the mass spectrum of,
446, 451
electron impact mass spectrum of, measured
at 10,000 eV energies, 204
fragmentation, map of, 441
of the molecular ion of, 435
gas chromatography-mass spectrometry of, 92
ionization of, ring opening during, 400
loss of, acetylene from the molecular ion of,
metastable peak for, 422
hydrogen atom from the molecular ion of,
metastable peaks for, 422
metastable ions in the mass spectrum of, 422
metastable transitions in the mass spectrum of,
435
noncyclic fragment ions from, 397
structure of the doubly charged molecular ion
of, 446
use of in ion bombardment methods, 212
Benzonitrile, activation of emitters for use in
field ionization by, 180
kinetic shifts in the mass spectrum of, 436
loss of HCN from the molecular ion of, 435
2,4,6-trideutero, mass spectrum of, 440
Benzophenones, cleavage of to give benzoyl ions,
396
Benzoquinone monooxime, 2,6-dimethoxy,
oxygen-18 labelled, 280
Benzoyl ion, formation of, from benzophenones,
396
from tetraphenylfuran, 385
fragmentation of, 437
N-Benzoyl peptides, mass spectra of, 315

Benzoyl radicals, formation of from tetra-
    phenylfuran, 385
Benzvalenes as intermediates during isomeri-
    zation of aromatic ions, 439
Benzyl, acetate, isobutane chemical ioniza-
    tion mass spectra of, 195, 409
  alcohol, carbon–13 labelled, mass spectrum
    of, 261, 262
  bromide-1-carbon-13, mass spectrum of, 261
  cation, formation of in peptide mass spectra,
    300
    heat of formation of, 437
    loss of acetylene from, 378
    structure of, 261, 378
  group, mass spectra of compounds contain-
    ing, 378
  phenyl ethers, mass spectra of, 395
  radical, formation of in peptide mass spectra,
    300
2-Benzyl-5-benzylidenecyclopentanone,
    photodimer of, 282
Benzyl-α-d2-chloride and alcohol, mass
    spectra of, 379
S-Benzylcysteine, -containing peptides,
    N-trifluoroacetyl derivatives of, mass
    spectra of, 304
  conversion of cysteine to in peptides, 303
N-Benzyloxycarbonyl peptides, mass spectra
    of, 315
    thermal instability of, 315
Benzyne intermediates in high temperature
    reactions, 277
Bethe's equation for ionization cross section,
    204
Bibenzyl, deuterium-labelled, mass spectrum
    of, 379
Bibenzyls, mass spectra of, 389
Bicyclic aromatic compounds, metastable
    transitions in the mass spectra of, 440
Bile, acids, measurement of biliary excretion
    of, 337
  new sterols and bile acids from, identifica-
    tion of by mass spectrometry, 341
Bimolecular processes of ions, 407
Biochemical research, use of jet separator in,
    109
Biologically important, amines, gas chromatog-
    raphy-mass spectrometry of, 94
  natural materials, identification of by
    mass spectrometry, 360
Biological systems, metals in, analysis for, 330
Biopolymers, mass spectra of, attempts to
    obtain, 211
Biosynthesis, and metabolism, distinction
    between, 221
  of steroids, use of oxygen-18 in studies of,
    221
  use of stable isotopes in studies of, 221
Biosynthetic studies, use of oxygen-18 in, 221
Biphenyls, substituted, loss of substituent
    from, 392
Bit capacity of analog-to-digital converters, 66
Bleeding of liquid phase from gas chromato-

graphic columns, 93
Blocking of magnetic tapes carrying mass
    spectral data, 61
Blood, gases dissolved in, 329
  analysis of by mass spectrometry, 328
  metals in, 331
  pH of, 329
  plasma, metals in, 330
  vessels, insertion of cannulas into for mass
    spectrometric sampling, 329
Bond, cleavage in field ionization, 188
  strengths, effect of on cleavage reactions,
    388
Bone, metals in, 331
Boronate, phenyl, derivatives of dinucleoside
    phosphates, 349
Bradikinin, amino acid sequence of, 318
Brain tissue, analysis of gases in, 330
Branches from apical nodes, 136
Broadening of metastable peaks due to energy
    release, 421
Bromine, -containing N-acyl groups in peptide
    mass spectra, 314
  isotopes, 376
  loss from 1-bromo and 1,1-dibromo-2,3-
    dimethylcyclopropanes, 402
1-Bromo-2,3-dimethylcyclopropane, loss of
    bromine from, 402
Bronsted acids in chemical ionization, 195
Buffers, data, use of in scanning of high
    resolution mass spectra, 60
Butadiene, radical cation, formation of from
    4-vinylcyclohexene radical cation, 402
  system, electrocyclic reactions of in the
    mass spectrometer, 399
1,3-Butadiene, 1,4-diphenyl, carbon-13 label-
    led, mass spectrum of, 266
n-Butane, collision-induced metastable transi-
    tions in the mass spectrum of, 429
  DENDRAL notation for, 130
  electron impact mass spectrum of, measured
    at 10,000 eV energies, 204
  fragmentation map of, 442
  linear notation for, 127
  metastable transitions in the mass spectrum
    of, 442
isoButane, chemical ionization mass spectra of
    t-amyl and benzyl acetate, 409
  DENDRAL notation for, 130
  linear notation for, 127
  and methane, comparison of as reagent gases
    for chemical ionization, 195
  as source of the t-butyl ion in chemical
    ionization mass spectrometry, 195
1-Butanol-2,2-dideutero, dehydration of, 380
1-Butene, carbon-13 labelled, mass spectrum
    of, 266
  fragmentation of, studies on using carbon-13,
    386
  loss of methyl in the mass spectrum of, 266
cis-2-Butene, metastable spectrum of, 425
isoButene, formation of from the t-butyl ion
    in chemical ionization mass spectrometry
    195

trans-2-Butene, metastable spectrum of, 425
Butethal, determination of, 332
Butter fat, lipids in, 346
sec-Butyl, -2-d-acetate, McLafferty rearrangement of, 378
-3-d-acetate, McLafferty rearrangement of, 378
Butylation of tryptophan during removal of t-butyloxycarbonyl protecting groups, 319
1-(Butylcarbamoyl)-2-benzimidazole carbamate, mammalian metabolism of, 360
t-Butylcyclohexyl, alcohols, photoionization mass spectra of, 209
mesylates, photoionization mass spectra of, 209
isoButylene, DENDRAL notation of, 139
proton affinity of, 409
Butyl esters, double hydrogen transfer reactions in, 380
t-Butyl ion, in chemical ionization mass spectra, 195
as reagent ion in chemical ionization mass spectrometry, 195
Butyl isopropyl ethers, mass spectra of, 381
t-Butyloxycarbonyl groups in peptides, side reactions during the removal of, 319
di-t-Butyl phthalate, mass spectrum of, 269
Butyl radicals in DENDRAL notation, ranking of, 132
Butyne, metal-catalyzed conversion of to hexamethylbenzene, 275

Cabbage leaf, hydroxylated fatty acids in, 346
Calculated shapes of metastable peaks, 447
Calculation of, energetics of ion formation and decomposition, 397
ionization potentials, 397
of alkanes, 397
mass spectra, with the Quasi-Equilibrium Theory, 404
from structures by DENDRAL programs, 171
metastable peak shapes, 450
reaction pathways, 398
Calibration of photographic emulsions, 19
C9 alkanes, gas chromatography-mass spectrometry of, 92
Campesterol, trimethylsilyl ether, mass spectrum of, 335
Cannulas, used in blood-gas analysis, heparinization of, 329
use of to sample gases dissolved in blood, 329
Canon of DENDRAL, order, 130
Canons, DENDRAL, for acyclic structures, 140
Capillary columns, high resolution, 106
in gas chromatography-mass spectrometry, 106

Carbalkoxy, groups, loss of from glutamic and aspartic acid residues in peptides, 298
side chains in peptides, loss of alcohol from, 298
Carbohydrates, field ionization mass spectra of, 188
Carbomethoxy group, C—terminal, as source of alkyl groups in intermolecular reactions, 301
expulsion of from S-carboxymethylcysteine residues in peptides, 303
Carbon, dioxide, partial pressure of in blood and expired gas, 328
monoxide, loss of, from benzoyl cation, 437
in field ionization mass spectroscopy, 189
from molecular ions of aromatic ethers, 257
−13, labelling in fragmentation of 1-butene, 386
use of in, biochemical problems, 354
chemistry and biochemistry, 218
studies of the fragmentation of toluene, 261
studies of mass spectra of heterocyclic systems, 263
−14, detection of by, autoradiography, 384
mass spectrometric methods, 354
mass spectrometric detection of, 220
use of in studying fate of herbicides, 358
Carbonyl, compounds, deuterium exchange into, 220
oxygen−18 exchange into, 220
group of ketones, cleavage at, 147
Carboxamide groups, loss of from glutamine and asparagine residues in peptides, 298
S—Carboxamidomethylcysteine, conversion of cysteine to in peptides, 303
Carboxyl groups, in peptides, esterification of, 312
side chain, in peptide mass spectra, 297
S—Carboxymethylcysteine, conversion of cysteine to in peptides, 303
Carboxypeptidase, use of to degrade proteins, 319
Cardiac, aglycones, trimethylsilyl ethers of, mass spectra of, 342
muscle, bovine, cholesteryl alkyl ethers in, 341
output, measurement of, 328
Cardiovascular disease, steroids in, 334
Caries formation, role of metals in, 331
Carrier gas, flow rate of jet molecular separators, 97
Cassine, chemical ionization mass spectra of, 193
Catalytic processes, in field ionization, difficulties relating to field-free systems, 186
mass spectra, 186
Catechol, attack on by molecular oxygen, 225
enzymatic cleavage of, 225
enzymatic formation of, from anthranilic acid, 355
from 2-fluorobenzoic acid, 357

Cellobiose, octamethyl, field ionization
    mass spectrum of, 188
Center of gravity of peaks in mass spectra,
    70
Centers of peaks, determination of on
    photoplates, 31
Centroids of, alkanes, 131
    peaks, calculation of, 70
        errors in the determination of, 63
        use of in the electrical recording of
            mass spectra, 44
    tree graph, 129
Ceramides, structure determination of, 348
    trimethylsilylation of, 348
Cesium in plasma, 331
Charge, density in HMO calculations, 398
    distribution in ions, calculation of in field
        ionization, 186
    as a driving force in mass spectral reactions,
        394
    exchange, in chemical ionization, 192
        mass spectra, 200
        spectra in chemical ionization source, 195
    localization, at quanidino groups in
        arginine-containing peptides, 304
        in ions, 180
            formed by field ionization, 187
            formed from peptide derivatives, 292
            in peptides containing monoamino
                dicarboxylic acids, 298
    transfer reactions in chemical ionization,
        192
Chelate ring, influence of on ionization
    potential of metal chelates, 394
Chelates, hydrolysis of, 269
    metal, site of ionization in, 394
Chemi-ionization, 207
    and chemical ionization, distinction
        between, 207
    definition of, 192
Chemical cleavage of proteins, 289
Chemical instability, theory of DENDRAL,
    143
Chemical ionization, by alkyl carbonium ion
    transfer, 192
    Bronsted acids in, 195
    by charge exchange, 192
    and chemi-ionization, distinction between,
        207
    definition of, 192
    dimensions of source for, 191
    and field ionization mass spectra, similari-
        ties in, 188
    in field ionization sources, 185
    ion current monitoring in, 192
    methane in, 190
    negative ions in, 201
    by proton transfer, 192
    sensitivity of, 203
    sources, collisions in, 194
        insulation problems in, 192
    standard mass spectrometer for, 191
Chemical ionization mass spectra, of aliphatic

    alcohols, 192
    of amino acids, 202
    of barbiturates, 201
    of benzyl acetate, 195
    of cassine, 193
    σ^+ correlations in, 409
    of cycloalkanes, 192
    of decane isomers, 189
    of dihydrocholesterol methyl ether, 193
    of drugs and drug metabolites, 200
    entropy changes in, 409
    of esters, 192
    of higher hydrocarbons, 192
    of hydrocarbons, 192, 195
    isotope labelling studies in, 194
    mechanism of fragmentation in, 199
    metastable ions in, 199
    primary and secondary events in, 199
    protonated dimers in, 409
    satellite ions in, 195
    similarities with field ionization mass spectra,
        189
    of steroidal acetates, 189
    of steroids, 201
    use to obtain activation energies and pre-
        exponential factors, 409
    variation of with temperature, 195
Chemical ionization mass spectrometry, 408
    in drug identification, 333
    of natural products, 409
    of peptides, 315
    in peptide sequencing, 202, 315
    primary events in, 198
    reviews of, 192, 408
    thermal equilibration in, 195
Chemical ionization mass spectroscopy, 190
Chemical ionization mass spectrum of,
        anagyrine, 200
    cholesteryl acetate, 201
    codeine, 200
    ephedrine, 197
    hydrogen, 192
    methane, 192
    methionine, 204
    noble gas-hydrogen mixtures, 192
    promethazine, 201
    quinine, 195
    Sandoptal*, 201
    Valium*, 195
Chess, computer as competitor in, 123
Chlordiazepoxide, metabolism of, 350
Chlorinated insecticides, field ionization mass
        spectra of, 189
Chlorine, -containing acyl groups in peptide
        mass spectra, 314
    isotopes, 376
    oxidation of thiocarbonates, 273
    -36, use of in studies of the fate of herbicides,
        358
2-Chloro-4-amino-6-ethylamino-s-triazine,
        enzymatic formation of from simazine,
        358
trans-1,4-bis(2-Chlorobenzoylaminomethyl)

* Registered.

cyclohexene as inhibitor of cholesterol biosynthesis, 341
1–Chlorobutane-3,3-dideutero, loss of hydrogen chloride from, 380
2–Chlorobutyric acid, enzymatic conversion of to 2-hydroxybutyric acid, 250
1–Chloroethane, 1,2-diphenyl, carbon−13 labelled, mass spectrum of, 265
p–Chloroethylbenzene-nitrogen mixtures, mass spectra of, 411
Chloroneb, metabolism of in soils, 359
1–Chloropentane, -3,3-dideutero, loss of hydrogen chloride from, 380
  -4,4-dideutero, loss of hydrogen chloride from, 380
4–Chloropentane, (s,r)-2-deutero-, mass spectrum of, 382
  (s,s)-2-deutero-, mass spectrum of, 382
Chlorophyll, removal of hydrocarbon side chain of, 344
Cholanoic acids, 3,7-dihydroxy, identification of in human feces, 339
Cholesta-5,7,24-trien-3-ol, identification of in lung tissue, 341
Cholestanoic acid, 3,7-dihydroxy, identification of mass spectrometry, 341
Cholest-7-en-3-beta-ol, formation of from cholest-8-en-3-beta-ol, 244
  trimethylsilyl ether of, mass spectrum of, 244
Cholest-8-en-3-beta-ol, conversion to cholest-7-en-3-beta-ol, 244
Cholesterol, 221
  biosynthesis, 240
    inhibitors of, 341
  enzymatic cleavage of the side chain of, 221
  identification of, 341
  measurement of biliary excretion of, 337
  trimethylsilyl ether, mass spectrum of, 335
Cholesteryl, acetate, chemical ionization mass spectrum of, 201
  alkyl ethers in bovine cardiac muscle, 341
  hexadecyl ether, mass spectrum of, 341
Chromatographic, peaks, broadening of in gas chromatography-mass spectrometry, 106
  dislocation of in silicon rubber membranes, 107
  resolution, loss of in gas chromatography-mass spectrometry, 106
Chromium-VI, rates of oxidation with, 268
City gas, analysis of by gas chromatography-mass spectrometry, 92
Class I mass spectral reactions, 399
Class II mass spectral reactions, 399
  Cleavage, of amide bonds in the mass spectra of peptide derivatives, 292
  a of amines, 159
  of aromatic ketones to acyl ions, 395
  of bonds in field ionization processes, 188
  of carbon-carbon bonds in, alcohols, 380
    cystine residues in peptides, 302
    lysine and ornithine residues in peptides, 304

methionine-containing peptides, 302
    pyrimidyl ornithine residues in peptides, 306
  of carbon-nitrogen bonds in, S-carboxymethylcysteine residues in peptides, 303
    lysine and ornithine residues in peptides, 304
    mass spectra of cyclic peptides and depsipeptides, 308
  of carbon-sulfur bonds in, cystine residues of peptides, 302
    methionine-containing peptides, 302
  a of ethers, 153
  a to heteroatoms in electron impact mass spectra, 196
  of peptide chains in mass spectrometry, 295
  reactions, effect on of ionizing voltage, 406
  of the ring of cyclic peptides and depsipeptides in mass spectra, 307
  of sulfur-sulfur bonds in cystine residues in peptides, 302
Clinical laboratory, use of gas chromatography-mass spectrometry in, 334
Clipping of peaks in detection systems, 50
Clocks, use of in data acquisition systems, 60
Clusters of ions formed by expansion of molecular beams, 210
CNSi as a stationary phase in gas chromatography-mass spectrometry, 95
Coating of, emitter tips with compounds to obtain field ionization mass spectra, 188
  supports for gas chromatography, 95
Codeine, chemical ionization mass spectrum of, 200
  electron impact mass spectrum of, 200
Cold trap, in-line, pressure reduction with, 92
Collagen in gelatin on photoplates, 17
Collectors, fixed, in gas analyzers, 329
Collector slit, dimensions of, 451
Collimation, according to energy of ions formed by field ionization, 185
  of light beam in microdensitometers, 27
Collision-induced, metastable ions, 408
  metastable transitions, 212, 429
    caused by an inert gas in the analyzer tube, 429
    collection of daughter ions formed in, 429
    in the mass spectrum of n-butane, 429
    pressure-dependence of, 429
    use of the determination of organic structures, 430
  reactions in ion cyclotron resonance spectroscopy, 411
Collision processes, absence of rearrangements in, 408
  energetics of, 408
Collisions, as a cause of unimolecular decomposition of ions, 429
  in cyclotron region of ion cyclotron resonance spectrometers, 191
  number of in chemical ionization sources, 194
Colorimetry in metal analysis, 330

Combination of, gas chromatography-mass spectrometry with high resolution mass spectra, 87

isotopic distributions and mass defects in determination of the formulas of ions, 377

Combined membrane-porous silver separator, 103

Commercially available, labelled compounds, 220

photoplates, 24

Communication between computer and mass spectrometer, means of, 78

Comparator, linear, use of in photoplates, 27

use of in recording of high resolution mass spectra, 59

Comparison of, mass spectra by computer, 83

radicals in DENDRAL notation, 141

Competing fragmentation processes, metastable transitions in, 433, 438

Competitive processes in, large molecules, 405

mass spectral reactions, 403

Competitive reactions, in fragmentation, introduction of by substituents, 388

of ions, 395

Complete list of structures corresponding to a formula, truncation of in DENDRAL, 142

Complex matastable peaks, 453

in the mass spectra of fluorobenzene and ethyl fluoride, 453

Complex molecules, application of the Quasi-Equilibrium Theory to, 433

Composition, of apical node in DENDRAL notation, 140

in DENDRAL notation, 140

of ions, determination of, in high resolution mass spectra, 377

from isotopic abundances, 376

identification of, 375

of radicals, handling of by DENDRAL notation, 139

Compound class, identification of by computer, 80

Compression of time scale, use of analog tape recording for, 61

Computer, analysis of high resolution and metastable ion mass spectral data in peptide sequencing, 314

assisted identification of, molecular ions, 81

specific compounds, 80

calculation of isotope distribution patterns, 376

comparing of mass spectra, 83

determination of peptide sequences from mass spectral data, 314

handling of metastable transitions data, 442

interpretation of mass spectra, 80

fragmentation maps in, 443

processing of data, 7

Computerization in gas chromatography-mass spectrometry, 93

Computers, problem-solving capabilities of, 122

reduction of cost of, 79

small, use of in recording of high resolution mass spectra, 60

use of, in calculation of mass spectra from Quasi-Equilibrium Theory, 404

in gas chromatography-mass spectrometry, 114

core storage requirements of, 116

Concave topped metastable peaks, 453

Configuration at C_5 of 3,6-diketo steroids by mass spectral methods, 342

Confusion of pairs of amino acids due to indiscriminate methylation of peptides, 316

Connectivity tables for organic molecules, 126

Consecutive, metastable transitions, 428

reactions in mass spectrometry, 403

σ Constants, for aromatic substituents, 387

correlation with ionization efficiencies, 392

Constraints, heuristic, upon DENDRAL algorithm, 167

upon structure generation in Heuristic DENDRAL program, 144

Contact telereader, use of in digitization of mass spectra, 3

Conversion, of distance to mass in photoplate data, 32

times of analog-to-digital converters, 57

Copper, use of in porous silver membrane separator, 99

Cord plasma, estrogens in, 339

Core storage, recording of mass spectra in, 60

Corn seedlings, metabolism of herbicides in, 358

Coronene, formation of from hexahelicene, 400

mass spectrum of, 400

Corrections, first order, in time-to-mass conversion routines, 74

for non-ideal behavior of magnet scanning systems, 47

σ Correlations, of appearance potentials, 390

in mass spectral processes, 388

σ Correlations in chemical ionization mass spectra, 409

Cortisone, fragmentation map of, 443

Cost of, computers, reduction of, 79

data acquisition systems, 2

high and low resolution mass spectrometry, 353

mass spectrometric systems, justification of, 364

Cotton buds, lipids in, 346

Coulomb potential in field ionization, 183

Coumarins, metastable peaks in the mass spectra of, 455

Counter-ions, abstraction of from solids, 211

Counters, use of in data acquisition systems, 60

Coupled gas chromatography-mass spectrometry, 334

Coupling, direct, of gas chromatograph to mass spectrometer, 93

Cranberries, volatile components of, 363

Critical, distance in field ionization, 184
  slope method of measuring ionization
    potentials, 390
Crossing of states in ions, 405
Cross section, for ionization, Bethe's equa-
    tion for, 204
  of ionization of molecules, 190
Cyanate, potassium, nitrogen–15 labelled,
    281
isoCyanic acid, loss of a nitrogen atom from
    the molecular ion of, 454
Cyclic, depsipeptides, fragmentation of, 309
  mass spectra of, 307
  peptides, mass spectra of, 307
  structures, DENDRAL program for genera-
    tion of, 125
Cycling of magnetic mass spectrometers in
    gas chromatography-mass spectrome-
    try, 116
Cyclization of, diphenylamine radical cation,
    402
  1,4-diphenylbutadiene, 403
  diphenylmethyl cation at high ionizing
    voltages, 402
  stilbene radical cation, 402
Cycloalkanes, chemical ionization mass
    spectra of, 192, 195
Cyclobutenedicarboxylic acid-muconic acid
    system, 400
1,2-Cyclohexanedione, condensation of with
    guanidinio groups of arginine, 305
Cyclohexanol, mass spectrum of, 382
cis-3,5-d2-cis–Cyclohexanol, loss of water
    from, 382
cis-3,5-d2-trans–Cyclohexanol, loss of water
    from, 382
Cyclohexyl chloride, mass spectrum of, 382
cis-3,5-d2-cis–Cyclohexyl chloride, loss of
    hydrogen chloride from, 382
cis-3,5-d2-trans–Cyclohexyl chloride, loss of
    hydrogen chloride from, 382
Cyclooctadiene, deuterium labelled, rear-
    rangement of to tricyclooctane, 276
trans–Cyclooctene, formation of in Hofmann
    elimination reactions, 280
Cyclopropane rings, identification of in
    kansamycolones, 363
Cyclotetradepsipeptides, natural, structure
    determination of, 309
Cysteic acid, -containing peptides, thermal
    instability of, 304
  conversion of cysteine to in peptides, 303
Cysteine, -containing peptides, desulfuriza-
    tion of, 317
  effect of on thermal stability of peptides, 315
  enzymatic conversion of to cysteine sulfinic
    acid, 356
  peptides, formation of from cystine peptides
    in mass spectra, 302
  in peptides, derivatization of for mass
    spectrometry, 303
    methylation of, 317
    specific fragmentation reactions of, 303

residues in peptides, conversion of to
    deuteroalanine residues, 318
  sulfinic acid, enzymatic formation of from
    cysteine, 356
Cystine, -containing peptide mass spectra as
    sums of cysteine peptide mass spectra,
    303
  -containing peptides, desulfurization of, 317
  in peptides, specific fragmentation reactions
    of, 302
Cytoplasmic aspartate transaminase, amino
    acid sequence of, 320
Cytosine, trimethylsilyl derivative of, mass
    spectrum of, 15

Daly detector, 424
  use of to suppress normal mass spectra, 425
Dansyl, amino acids, identification of by mass
    spectrometry, 319
  derivatives of amines, coupled gas chromatogra-
    phy-mass spectrometry of, 362
Data, acquisition, in gas chromatography-mass
    spectrometry, 114
  in scanning instruments, 57
  systems, advantages of, 2
    cost of, 2
    high speed, 44
  processing, 67
  reduction, in photoplate work, 29
    preliminary, on-line, 68
  from spark source mass spectrometers,
    automatic handling of, 331
Daughter ion, singly charged, formed from
    doubly charged parent ion, 450
Daughter ions, apparent mass of, 425
  appearance potentials of, 403
  formed in collision-induced metastable
    transitions, 429
  heat of formation of, errors in, 398
  intensity of relative to parent and metastable
    ions, 436
  loss of, at the β-slit in double focussing
    instruments, 448
  in the electric sector, 425
  relation of to parent and metastable ions, 421
DDE, enzymatic formation of from DDT, 359
  metabolism of in soils, 359
  photochemistry of, 359
DDT, metabolism of in soils, 359
Deactivation of supports for gas chromatography,
    95
Dead volume in gas chromatography-mass
    spectrometry interfaces, 107
Decane isomers, chemical ionization mass spectra
    of, 189
  field ionization mass spectra of, 189
Decomposition and formation of ions, calcula-
    tion of the energetics of, 397
Decomposition of ions, during acceleration,
    metastable transitions, caused by, 422
  in field-free regions, 421
  in field ionization, rules governing, 187
  rate constants for, 403

rates of, 389, 403
Deconvolution of, lines on photoplates, 29
  unresolved multiplets in high resolution
    mass spectra, 70
Defects, mass, and isotopic distributions in
    the determination of the formulas
    of ions, 377
Defense secretion of, arthropod, glomerin as,
    361
  larva, 362
Deflection, of ions, metastable transitions
    occurring after, 423
  metastable transitions occurring before,
    420
  of very high mass, problems with, 212, 364
  in mass spectrometers, fundamental equa-
    tions governing, 45
Defluorination, enzymatic, of monofluoaracetic
    acid, 249, 355
Defocussing, of metastable ions, possibilities
    of in peptide structure determination,
    314
  of metastables, use of in ion bombardment
    methods, 212
  techniques in double focussing instruments,
    16
Degree of apical node, 139
  in DENDRAL notation, 140
Degree-of-freedom, effect, in fragmentation
    reactions, 389
  of substituents on numbers of, 389, 396
  in a precursor ion, relationship of metastable
    ion intensity to, 408
  vibrational, importance of number of, 436
Dehydration, of alcohols in mass spectral
    fragmentations, 380
  of amide groups in the mass spectra of
    peptides, 294
  of amino acid side chains in peptide mass
    spectra, 297
  of imidazolinonylornithine-containing
    peptides, 307
  and intermolecular methylation in peptide
    mass spectra, 301
  in peptide mass spectra, influence of glycine
    upon, 294
  of peptides, thermal nature of, 294
Dehydroalanine residues, formation of from
    cysteine residues in peptide mass
    spectra, 303
24–Dehydrocholesterol, mass spectrum of, 341
Dehydroepiandrosterone, sulfate in serum, 335
  trimethylsilyl ether, mass spectrum of, 336
Dehydrohalogenation in the mass spectra of
    alkyl halides, 380
1,3–Dehydrohalogenation reactions in mass
    spectrometry, 382
11–Dehydro-17-hydroxy estradiol, identifica-
    tion of, 339
Dehydroxylation of steroids, 254
Delay time in silicon rubber membranes, 102
Dempster mass spectrometers, metastable
    transitions in, 430

use of in gas analysis, 328
DENDRAL, algorithm, Heuristic constraints
    upon, 167
  canons for acyclic structures, 140
  complete structures list in, truncation of,
    142
  enumeration, general principles of, 131
  handling of, acetylenic and olefinic carbons
    by, 137
    optical and geometrical isomerism problems
      by, 141
    radical composition by, 139
    unsaturation and heteroatoms by, 137
    unsaturation in radicals by, 139
  notation, for alkanes, 130
  notations, conversion of to structural
    formulas, 141
    derivation of from structural formulas, 141
  order, canon of, 130
  program, heuristic, 126, 144
  project, 125
  theory of chemical instability, 143
  tree, of isomers, logical permutations through,
    133
    pruning of, 143
    terminating twig of, 139
  valence restrictions in, 137
Density of energy states in ions, 393
Dental, plague, metals in, 331
  tissues, metals in, 331
Dentine, metals in, 331
Deoxyadenosine, biosynthesis of, 234
  oxygen–18 labelled, high resolution mass
    spectrum of, 234
  mass spectrum of, 235
Deoxycorticosterone, enzymatic formation of
    from progesterone, 356
11–Deoxycortisol, enzymatic formation of
    from 17-hydroxyprogesterone, 356
Depsipeptides, cyclic, mass spectra of, 307
Derivation methods of measuring ionization
    potentials, 391
Derivation of, free peptides for mass spectrome-
    try, 291
  peptides for mass spectrometry, 312
Desmosterol, mass spectrum of, 341
Desorption, field, of ions, 188
Desulfurization of peptides, 317
Detection, of metastable transitions in the
    field-free regions of mass spectrome-
    ters, 424
  of multiplets in high resolution mass spectra,
    70
  systems use in scanning high resolution mass
    spectrometers, 48
Detector for gas chromatograph, use of mass
    spectrometers as, 14
Determination of, empirical formulas from
    isotopic compositions, 376
  structures by mass spectrometry, 279
Dethiogliotoxin, high resolution mass spectrum
    of, 239
Detoxification mechanisms, in biological systems
    249

Deuterated compounds, metastable transitions in, 439
Deuteration of peptides during desulfurization, 318
Deuterium, cyanide, loss of from the molecular ion of 2,4,6-trideuterobenzonitrile, 440
  exchange into carbonyl compounds, 220
  labelled, DDT, metabolism of, 359
    methionine, use of in biosynthetic studies, 246, 357
    propanes, metastable transitions in the mass spectra of, 440
  labelling, in studies of cholesterol biosynthesis, 240
    use of in fragmentation studies, 259, 378ff
  retention of during enzymatic hydroxylation of aromatic rings, 357
  scrambling in organic ions, 439
  use of in, biochemical problems, 354, 357
    chemistry and biochemistry, 218
    organic reaction mechanism studies, 275
    peptide sequencing, 282
    studies of, mass spectral fragmentation reactions, 378
    fragmentation of toluene, 261
    study of mass spectral fragmentation mechanisms, 255
Deuteroalanine residues in peptides, formation of from cysteine residues, 318
2—Deutero-4-methyl phenol, enzymatic formation of from p—Deuterotoluene, 357
p—Deuterophenylalanine, retention of deuterium in during conversion to tyrosine, 357
2—Deuteropyridine, metastable transitions in the mass spectrum of, 439
  scrambling of the label in, 439
p-Deuterotoluene, enzymatic conversion of to 2—Deutero-4-methyl phenol, 357
Deuterotrimethylsilyl groups, use of in fragmentation mechanism studies, 258
Development of photoplates, 26
Diacylated ornithine residues in peptides, mass spectrometry of, 305
Dialkyl ethers, β-cleavage in, 204
Dialogs between spectroscopist and computer, use of in interpretational programs, 87
1,4—Diaminobutane, identification of in desert locust, 362
Diaryl carbonates, mass spectra of, 256
Diarylethylenes, fragmentation reactions of, 403
20,25—Diazacholesterol as inhibitor of cholesterol biosynthesis, 341
Diazepam, metabolism of, 350
Diazomethane, as agent for esterification of peptides, 312
  in methanol as agent for esterification of methanol-soluble peptides, 313
Dibenzofurans, loss of carbon monoxide from molecular ions of, 257

Dibenzyl, deuterium-labelled, mass spectrum of, 379
Dibenzylnitrosamine, photochemical rearrangement of, 274
Dibenzyls, mass spectra of, 389
1,1—Dibromo-2,3-dimethylcyclopropane, loss of bromine from, 402
2,6—Dibromophenol in marine acorn worm, 363
Dicarboxylic acids in shale, 346
Dichloroacetic acid, enzymatic dechlorination of, 250
3,4—Dichloroaniline, enzymatic formation of from Diuron, 358
1,1—Dichloro-2,2-bis(p-chlorophenyl)-2-deuteroethane, enzymatic formation of from deuterium-labelled DDT, 359
1,4—Dichloro-2,5-dimethoxy benzene, metabolism of in soil, 359
1,4-Dichloro-2-methoxy phenol, enzymatic formation of from 1,4-dichloro-2,5-dimethoxy benzene, 359
3,4-Dichloronitrobenzene, enzymatic formation of from Diuron, 358
3-(3,4—Dichlorophenyl-1-methyl urea, enzymatic formation of from Diuron, 358
Dictionary, ordering in DENDRAL notation, 131
  position, uniquely defined by DENDRAL notation, 133
  of structural isomers, production of by DENDRAL structure generator, 170
1,4—Dideuterobenzene, metastable transitions in the mass spectrum of, 439
Dieldrin, detection of by mass spectrometry, 334
Dielectric constants of films formed in field ionization sources, importance of, 184
Diethyl, ether, mass spectrum of, 438
  ketone, loss of ethane from, 405
  phthalate, mass spectrum of, 269
Differences between ionization and appearance potentials, 392
Differential pumping of chemical ionization mass spectrometer, 192
Differentiation, use of to detect peak maxima, 62
Diffusion, processes in gas chromatography-mass spectrometry, 93
  separators, use of hydrogen or helium in, 93
1,1—Difluoroethylene, metastable transitions in the mass spectrum of, 435
Digital, comparator, circuit, use of for peak sensing, 4
  use of in recording of high resolution mass spectra, 59
  computer, controlled analog-to-digital converters, 57
    miniac, 5
    programmable, 79
    systems, obsolescence of, 57
    use of, to focus mass spectrometer automatically, 56
    to scale mass spectral data on line, 56
  conversion, maximum permissible error in, 65

magnetic tape recording of high resolution
   mass spectra, 60
recording devices in the recording of high
   resolution mass spectra, 60
samples, number required to define a peak,
   64
smoothing of mass spectral data, 71
thresholding of mass spectra, 60
Digitization of, electron multiplier output, 8
mass spectra, 3
Digitizers, 4, 79
Digitizing, on-line, of mass spectral peaks, 63
Diglycerides, identification of from bovine
   phosphatidyl serines, 348
Dihydro cholesterol methyl ether, chemical
   ionization mass spectrum of, 193
1,2-Dihydro-1,2-dihydroxy naphthalene,
   enzymatic formation from naphthalene,
   250, 357
S−(1,2−Dihydro-2-hydroxy-1-naphthyl)
   glutathione, enzymatic formation of
   from naphthalene, 251
1,2−Dihydro naphthalene, 1,2 oxide,
   enzymatic formation of from naphthalene, 250, 357
1,4−Dihydropyridines, 4-deutero-N-alkyl-
   3-cyano, aromatization of in mass
   spectrometry, 383
Dihydrosphingosine, mass spectrum of, 227
Dihydroxy, alkanes, mass spectra of, 343
   fatty acid esters, formation of from un-
      saturated acids, 258
   pregnan-20-one isomers, separation and
      identification of by GC−MS, 110, 112
3,7−Dihydroxy cholanoic acids, identification of in human feces, 339
3,7−Dihydroxy cholestanoic acid, identification of by mass spectrometry, 341
20,22−Dihydroxycholesterol, enzymatic
   conversion of to pregnenolne acetate,
   356
20,22R−Dihydroxy cholesterol, 223
3,21−Dihydroxypregnan-20-one, metabolism
   of, 254
9,10−Dihydroxy stearic acid, formation of
   from oleic acid, 258
Diketopiperazines, formation of from free
   dipeptides, 290
3,6−Diketo steroids, determination of configuration at C5 of by mass spectral
   methods, 342
Dilation of time scale in analog tape recording
   of mass spectra, 65
Dimensions of, collector slit, 451
   emulsions used in photoplates, 17
   β-slit, 451
Dimers, of polystyrene, 211
   protonated, in field ionization mass spectra
      of alcohols, 189
2,6−Dimethoxybenzoquinone monooxime,
   oxygen−18 labelled, 280
3,4−Dimethoxy-β-phenethylamine, identification of Parkinson's disease pink spot
   as, 363

Dimethyl, acetamide, use of in exhaustive
   N-methylation of peptides, 316
cholesterol, identification of by mass
   spectrometry, 339
muconate, mass spectrum of, 226
pyrimidyl ornithine, formation of from
   arginine residues in peptides, 305
succinate, mass spectrum of, 241
sulfoxide, sodium derivatives of, use of in
   exhaustive N-methylation of peptides, 31
Dimethylamine, identification of in desert
   locust, 362
5−Dimethylaminonaphthalene-1-sulfonyl
   derivatives of amines, 362
1,2−Dimethylcyclohexene, formation of
   from pyrolysis of esters, 270
Dimethylformamide, use of as solvent in
   Raney nickel desulfurization of
   peptides, 317
O,O−Dimethylprostaglandin E1 ethyl ester,
   biosynthesis of, 224
incorporation of oxygen−18 into, 224
mass spectrum of, 224
2,4−Dinitrophenylamino acids, identification
   of by mass spectrometry, 319
2,4−Dinitorphenylarginine, methyl ester,
   mass spectrum of, 304
2,4−Dinitrophenyl derivatives of peptides,
   involatility of, 315
Dinucleoside phosphate derivatives, mass
   spectra of, 349
Dioxygenase, involvement of, in metabolism
   of o-fluorobenzoic acid, 357
in prostaglandin E1 biosynthesis, 356
Dipeptides, conversion of to diketopiperazines,
   290
free, mass spectra of, 290
isomeric leucines in, attempts to differentiate between, 296
Diphenylamine radical cation, cyclization of,
   402
1,4−Diphenylbutadiene, cyclization of, 403
1,4−Diphenyl-1,3-butadiene, carbon−13
   labelled, mass spectrum of, 266
1,2−Diphenyl-1-chloroethane, carbon−13
   labelled, mass spectrum of, 265
1,2−Diphenylcyclobutanes, fragmentation
   of, 403
1,2−Diphenylethane, carbon−13 labelled,
   mass spectrum of, 261
Diphenylmethane, deuterium-labelled, mass
   spectrum of, 379
Diphenylmethyl cation, cyclization of at high
   ionizing voltages, 402
1,3−Diphenyl propene, carbon−13 labelled,
   mass spectrum of, 266
Diphenyl sulfide, carbon−13 labelled, mass
   spectrum of, 264
Dipole moments, importance of in field
   ionization, 187
Direct sequencing of peptides by mass
   spectrometry, 291
Discharge, electrical, use of in photoionization

sources, 209

Discrimination, effect of on energy released during fragmentation, 451

effects, influence of on metastable peaks, 445

at the intermediate β-slit in double focussing instruments, 452

at the β-slit in double focussing instruments, 448

Dish-shaped metastable peaks, 444, 447

Disks, recording of mass spectra on, 60

Display system in microdensitometers, 28

Disrotatory processes in mass spectral reactions, 400

Dissociation, of electrons from atoms, field-induced, 182

field-induced, following ionization in field ionization sources, 185

statistical theory of, 393

Distance, critical, in field ionization, 184

-mass conversion on photoplates, 32

Distortion, of mass spectral peaks, 65

in peak profiles during scanning of high resolution mass spectra, 51

Distribution of internal energies of precursor ions, 388

Diuron, metabolism in corn seedlings of, 358

Docosapeptide sequenced by mass spectrometry of its methylated derivative, 316

n-Dodecane, identification of in scent glands of larvae, 362

Dopa, biosynthesis of from tyrosine, 229

Doriden*, determination of, 333

metabolites of in urine, 334

Double bonds, location of, by mass spectrometry, 258, 343

in sphinga-4, 14-dienine, 347

means of coding on teletypewriter, 142

Double hydrogen, migration processes in tyrosine peptides, 301

transfer reactions in butyl esters, 380

"Double McLafferty rearrangement" processes, 148

enol form of product ion in, 410

Doublet or singlet states, reactions occurring from, 399

Doublets, encountered in high resolution stable isotope studies, 220

resolving power required to separate, 219

Doubly-charged ions, fragmentation of, 444

given by hydrocarbons, 453

in the mass spectrum of, benzene, 451

and other aromatic hydrocarbons, 446

2-fluoropropene, 451

reactions of, 399

Driving force in mass spectral reactions, charge as, 394

Drug metabolism studies, use of accelerating voltage alternator in, 114

Drug metabolites, chemical ionization mass spectra of, 200

Drugs, chemical ionization mass spectra of, 200

identification of by chemical ionization mass spectrometry, 333

mass spectrometric determination of, 332

metabolism of, study of by mass spectrometry, 350

use of stable isotopes in studies of, 221

placental transfer of, 351

Dynamic range of recording systems, 62

Dynamic resolving power, 47, 49

dependence of sensitivity on in scanning instruments, 48

measurement of, 56

Ecdysones, mass spectrometry of, 341

Edges in topological graph expressions of structure, 128

Edman degradation, combination of with mass spectrometry in peptide sequencing, 318

use of, to remove specific amino acid residues prior to mass spectrometry, 318

in structure determination of proteins, 290

γ-Effect in isotope effects, 383

π-Effect in isotope effects, 383

Effective number of oscillators in ions, 396

Effective oscillators, number of used in Quasi-Equilibrium Theory calculations, 406

Efferent branches in DENDRAL notation, 136

Efficiency of, ionization, correlation with sigma constant, 392

radioactive sources, 205

separators, calculation of, 95

EGSS as a stationary phase in gas chromatography-mass spectrometry, 95

8,11,14-Eicosatrienoic acid, enzymatic conversion of to prostaglandin E1, 224, 356

Electrical recording of mass spectra, 43

Electric sector, failure of to transmit fragment ions, 423

loss of ions in, 425

metastable transitions in, 428

Electrocyclic reactions, antiaromatic transition states in, 399

of the butadiene system in the mass spectrometer, 399

selection rules governing stereochemistry of, 400

stereochemical course of, 399

thermal, aromatic states in, 399

Electrode preparation in analysis of trace metals, 331

Electrodes, use of for blood ph measurement, 329

Electrojet, methods of ionization, 210

use of skimmers with, 210

Electron, beam, energy in gas chromatography-mass spectrometry, 93

high energy, use of for ionization, 204

use of for ionization in mass spectroscopy, 190

bombardment, ionization, disadvantages of, 179

* Registered.

ionization by, 179
sources, disadvantages of for appearance
    potential work, 179
impact, induced dehydration processes in
    peptide mass spectra, 294
-field ionization combination source, 16
mass spectra, comparison of with photo-
    ionization mass spectra, 209
of hydrocarbons, comparison of with
    field ionization spectra, 188
of thiols, comparison with field ioniza-
    tion spectra, 188
mass spectrum of, ethylbenzene, 188
    leucine, 207
    Sandoptal*, 201
multiplier, detection of carbon—14 contain-
    ing ions with, 354
    output, digitization of, 8
    requirement for in scanning instruments, 50
removal of from metal surfaces, 181
tunneling, 182
Electronic, excitation, implication in fast mass
    spectral reactions, 399
noise, problems caused by, 65
states, isolated, fragmentations occurring
    from, 438
Electronically excited states, contribution of
    to ion reactivity, 404
reactions occurring from, 399
Electrons, transfer of from ions to molecules,
    192
Electrospraying technique for introduction of
    involatile samples to mass spectrome-
    ters, 364
Electrostatic analyzer, use of in ion bombard-
    ment methods, 212
Elemental compositions, derivation of from
    accurate masses, 74
of ions, determination of, 33
listings of, 35
Element maps, 35, 377
in peptide sequencing, 314
topographical, 39
Elimination reactions in, mass spectral frag-
    mentations, 380
hydrogen migration in, 381
    stereochemistry of, 382
    transition states in, 381
mass spectrometry, stereospecificity of, 382
1,3—Elimination reactions in mass spectrome-
    try, 382
Eluate from gas chromatograph, condensation
    and transfer to mass spectrometer of,
    92
Embden-Meyerhoff-Parnas glycolytic pathway,
    245
Emitter, in field ionization, silver as, 185
tip, temperature of in field ionization mass
    spectroscopy, 186
tips, coating of with compound to obtain
    field ionization mass spectra, 188
in field ionization, use of gold and plati-
    num as, 186
use of a-, for ionization, 204

Empirical, formula of molecular ion de-
    termination by high resolution mass
    spectrometry, 81
use of by Heuristic DENDRAL program, 14
formulas of ions, determination of, 33
from low resolution mass spectra, 376
by Heuristic DENDRAL, 163
models, correlation of fragmentation
    mechanisms with, 387
Emulphor—O, use of in gas chromatography-
    mass spectrometry, 110
Emulsions used in photoplates, 16
Enamel, dental, metals in, 331
Encoders in microdensitometer-linear com-
    parators, 29
Energetic aspects of metastable transitions,
    review of, 454
Energetics of, fragmentation processes, use
    of broad metastable peaks in studies
    on, 453
ion formation and decomposition, calcula-
    tion of, 397
mass spectral reactions, 399
Energies, in HMO calculations, 398
involved in collision processes, 408
range of in ions transmitted by electric
    sectors, 425
Energy, calculations, semi-empirical, in ion
    formation and decomposition, 397
compensation technique of measuring
    ionization potentials, 390
distribution of molecular ions, 392, 404
distributions, in parent ions, 436
division of between ion and neutral in
    fragmentation processes, 393
effect of on photographic response, 23
of electrons used for ionization, variation of,
    204
internal, distribution of in precursor ions,
    388
of fragmenting ion, variation of metastable
    peak intensities with, 439
variation of with source parameters, 439
of ions, 432
release of as translational energy during
    fragmentation reactions, 445
of ionization, effect of on mass spectra, 381
of ions, loss of by radiation, 405
relation of to metastable peak intensity,
    434
levels of molecular orbitals, effect of sub-
    stituents on, 394
range in photoionization, 209
release, broadening of metastable ions due
    to, 421
in fragmentation processes, effect on
    metastable peaks, 423
released in fragmentation, discrimination
    effects on, 451
states, in ions, density of, 393
occupied, density of, effect of substituents
    on, 389
translational, released in fragmentation
    processes, 450

* Registered.

Enhancement of molecular ion intensities by use of low filament voltages, 219

Enniatin, A, structure determination of by mass spectrometry, 310

A$_1$, structure determination of by mass spectrometry, 310

B, structure determination of by mass spectrometry, 310

B$_1$, structure determination of by mass spectrometry, 310

Enniatins, structure determination of by mass spectrometry, 310

Enol, form of product ions in a double McLafferty rearrangement, 410

and keto forms of acetone in mass spectra, 410

Enrichment of, isotopes required for mass spectral work, 219

separators, definition of, 95

Entropy, changes in chemical ionization mass spectra, 409

factors involved in rearrangement processes, 405

requirements of rearrangement processes, 188

Enzymatic, cleavage of proteins, 289

prior to mass spectrometry, 319

degradation of nucleic acids, 15

Ephedrine, chemical ionization mass spectrum of, 197

electron impact mass spectrum of, 197

Epiandrosterone, trimethylsilyl ether, mass spectrum of, 336

Equilibration, lack of in field ionization conditions, 187

processes in mass spectra, 384

Equilibrium method treatment of the relation between ground and transition states, 403

Equivalent orbital method for ion energy calculations, 397

Errors in, mass measurement during electrical recording of mass spectra, 44

peak centroid determination, 63

ESA voltage, scanning of to study metastable transitions, 427

advantages of, 427

Esperine, amino acid sequence of, 316

Ester bonds in cyclic depsipeptides, expulsion of under electron impact, 309

Esterification of, free carboxyl groups in peptides, 312

phthalic acid, 269

side chain carboxyl groups in peptides, 297

Esters, aliphatic, McLafferty rearrangements of, 383

of butyl alcohol, double hydrogen transfer reactions in, 380

chemical ionization mass spectra of, 192

field ionization mass spectra of, 189

hydroxy, field ionization mass spectra of, 189

separation of by gas chromatography, 94

17-β–Estradiol, trimethylsilyl ether, gas chromatography-mass spectrometry and quantitation of, 337

Estr-4-en-3, 17-dione, formation of in

metabolism of morethindrone, 352

Estriol, trimethylsilyl ether, mass spectrum of, 337

Estrogens, excess, excreted, identification of, 337

metabolism of, 337

in pregnancy and cord plasma, 339

Estrone, derivatives of, identification of in amniotic fluid, 339

oxygenation of in biological systems, 339

Etamycin, structure determination of by mass spectrometry, 310

Ethane, DENDRAL notation for, 130

loss of from diethyl ketone, 405

Ethanol, dideutero, formation of from hexoses, 245

field ionization mass spectrum of, 185

formation from hexoses, 245

hexadeutero, formation of from hexoses, 245

line notation for, 128

monodeutero, formation of from hexoses, 245

Penning ionization mass spectrum of compared to electron impact mass spectrum of, 208

pentadeutero, formation of from hexoses, 245

tetradeutero, formation of from hexoses, 245

Ethers, and alcohols, distinguishing between, 153

aliphatic, hydrogen rearrangements in, 380

α-cleavage of, 153

confusion of alcohols with, 153

dialkyl, β-cleavage in, 204

identification of from mass spectra by Heuristic DENDRAL programs, 146

rules for mass spectral fragmentation of, 153

structure determination of, by Heuristic DENDRAL interpretation of mass spectra, 153

from mass spectra by Heuristic DENDRAL programs, 162

success of Heuristic DENDRAL programs with, 175

unsaturated, confusion of with ketones and unsaturated alcohols, 147

N–Ethoxycarbonyl peptides, mass spectra of, 315

Ethoxy cation, various precursors of, 438

Ethoxyl cation, structure of, 409

Ethyl, acetate, identification of in desert locust, 362

acetates, deuterium labelled, mass spectra of, 245

ester of arginine, mass spectrum of, 304

fluoride, loss of hydrogen atom from the molecular ion of, 453

mass spectrum of, 453

side chain in stigmasterol, biosynthesis of, 245

undecanoate, element map of, 38

Ethylbenzene, -α-d2, mass spectrum of, 379

field ionization and electron impact mass spectra of, 188
gas chromatography-mass spectrometry of, 92
metastable peak in the mass spectrum of, 455
Ethyl-1,3-dimethylbutylamine, DENDRAL notation for, 161
Ethylene, electron impact mass spectrum of measured at 10,000 eV energies, 204
loss of from the molecular ion of, ethylbenzene, 455
hexahelicene, 400
Ethylenediamine chelates, hydrolysis of, 269
Ethylene glycol succinate as a stationary phase in gas chromatography-mass spectrometry, 95
Etiocholanolone, trimethylsilyl ether, mass spectrum of, 337
Euler's solution to the Koenigsberg bridges, 129
Evaluation function in Heuristic DENDRAL programs, 173
Even-electron ions in chemical ionization mass spectra, 195
Exchange, in molecular oxygen, 225
of the ortho protons of aromatic carboxylic acids, 440
of oxygen−18, and deuterium into carbonyl compounds during gas chromatography, 220
into ketones, 279
processes, use of for the introduction of stable isotopes into molecules, 220
Excitation energy in ions, partitioning of, 437
Excited states, of product ions, 393
study of in field ionization, 189
Exponential, dependence of mass upon time in magnet scanning, 47
functions, use of in time-to-mass conversion routines, 73
scanning, advantages of, 46
of magnetic sectors, 46
scans, relationship of peak width and resolving power in, 46
term, problems caused by in theory of mass spectral reactions, 406
Extrapolation, in mass standardization of data files, 73
linear, method of measuring ionization potentials, 390

Farnesyl pyrophosphate, -carbon−14, 242
enzymatic conversion to squalene, 241
F−60 as stationary phase in gas chromatography-mass spectrometry, 94
Fast, mass spectral reactions, implication of electronic excitation in, 399
metastables in field ionization mass spectra, 185
scanning of magnetic sector instruments for gas chromatography-mass spectrometry, 93
Fate of pesticides and insecticides, study of

by mass spectrometry, 358
Fatty acids, branched chain, identification of by mass spectrometry, 342
in shale oil, 346
hydroxylated, identification of by mass spectrometry, 346
long chain, identification of by mass spectrometry, 342
olefinic, location of the double bonds in, 343
in shale, 346
Feasibility of electrical recording of high resolution mass spectra, 44
Feedback mode, computer operation in, 2,9
of operation, 9
Fermi level of metals, 184
Field, desorption of ions, 188
emission, microscopy, 181, 182
relation of to field ionization, 181
evaporation of tungsten, 182
ion microscope, use of hydrogen and helium in, 182
Field-free, region before the electric sector, metastable transitions in, 425
regions of mass spectrometers, 420
Field-induced dissociation, of electrons from atoms, 182
following ionization in field-ionization sources, 185
Field ionization, 180, 183, 219
and chemical ionization mass spectra, similarities in, 188
critical distance in, 184
efficiency of, 180
-electron impact combination source, 16
emitter, silver as, 185
formation of ions in, 183
history of development of, 181
ions at m/e (M) and (M+1) in, 184
location of charge in ions formed by, 187
mass spectra, metastables in, 188
method of formation of intense ions in, 186
similarities with chemical ionization mass spectra, 189
mass spectra of, alcohols, 189
chlorinated insecticides, 189
decane isomers, 189
hydrocarbons, comparison with electron impact spectra, 188
hydroxy esters, 189
methyl esters, 189
nucleosides, 189
olefins, comparison with electron impact spectra, 188
simple aliphatic amines, 189
steroid acetates, 189
thiols, comparison with electron impact spectra, 188
mass spectrometry of peptides, 315
mass spectrum of ethylbenzene, 188
metastable ions in, 185
polarization forces in, 184

rearrangement processes in, 188
relation of to field emission, 181
sources, difficulty of operation of, 180
  ion-molecule reactions in, 184
  uni- and intermolecular processes in, 185
  techniques, 16
  types of ions expected in, 187
Field penetration, variations in during voltage
  scans, 426
Fields, electrical, in vicinity of metal whiskers,
  182
Filament, protection of in chemical ionization
  source, 192
Films formed in field ionization sources, 184
  ion-molecule reactions in, 184
Filtering of mass spectral signals, 71
Fixed collectors in gas analysis, 329
Fixing of photoplates, 26
Flame ionization detectors, in gas chromatogra-
  phy, 205
  use of in gas chromatography-mass spectrome-
    try, 108
Flame spectrometry of metals, 330
Flat-topped metastable peaks, 444, 446
Flavors, gas chromatography-mass spectrome-
  try analysis of, 92
Flexowriter, as an output device, 4
  use of in digitization of mass spectra, 3
Flow controllers, in gas chromatography-mass
  spectrometry, 93
  use of in temperature programming, 93
Flow rate of carrier gas in jet molecular
  separators, 97
Fluorine as a label in mass spectrometry, 384
Fluoroacetic acid, enzymatic defluorination of,
  249
Fluorobenzene, loss of hydrogen fluoride from
  the molecular ion of, 453
2–Fluorobenzoic acid, enzymatic conversion
  of to catechol and 3-fluorocatechol,
  357
  metabolism of in microorganisms, 353
p–Fluorobenzyl chloride, loss of acetylene
  and hydrogen fluoride from the
  molecular ion of, 439
  metastable transitions in the mass spectrum
    of, 439
3–Fluorocatechol, enzymatic formation of
  from 2-fluorobenzoic acid, 353, 357
Fluoroethylene, fragmentation map of, 442
  loss of hydrogen fluoride from the molecular
    ion of, 455
  metastable transitions in the mass spectrum
    of, 435
2-Fluoromuconic acid, enzymatic formation
  of from 2-fluorobenzoic acid, 353
2-Fluoropropene, doubly charged ions in the
  mass spectrum of, 451
Flutter, tape, problems caused by, 62
FM analog tape recording of mass spectra, 61
Focal plane in Mattauch-Herzog geometry, 46
Focal point in Nier-Johnson geometry, 46
Focussing, in high resolution instruments, 13

of mass spectrometer, automatic, by
  digital computer, 56
  maximization of, 47
Formaldehyde, deuterium labelled, photolysis of, 276
Formation, and decomposition of ions, calculation of the energetics of, 397
  heat of, of ions, 190
  of ions, kinetics shifts in and heats of
    formation of, 435
Formulas, of ions, identification of, 375
  of neutral fragments, determination of, 442
Fortuitin, structure determination of by mass
  spectrometry, 313
Four-membered transition states, 405
  in the McLafferty rearrangement, 378
Fourteen mass units, subdivision of mass
  spectra into groups of, 83
Fragmentation, of N-acylpeptide methyl ester
  molecular ions, 292
  of aliphatic ketones, 146
  amino acid type of, in the mass spectra of
    peptide derivatives, 293
  in aromatic systems, rate of, compared to
    scrambling, 386
  of aromatic systems, scrambling during, 385
  of benzoyl cation, 437
  of cyclic depsipeptides, 309
    effect on of hydroxy and amino acids in
      the ring, 309
  of cyclic peptides and depsipeptides under
    electron impact, 308
  different modes of, of the same ion, 435
  of 1,2-diphenylcyclobutanes, 403
  of doubly charged ions, 444
  of ethoxy cation, 436
  of 2-hexanone and 1-methylcyclobutanol to
    give the enol form of acetone, 410
  identification of atoms involved in, 377
  of ions, effect on of functional groups, 374
    mechanism of, 373
    rationalization of, 374
  in low voltage spectra, 388
  map of, benzene, 441
    n-butane, 442
    cortisone, 443
    peptide derivative, 442
    tomatidine, 443
  maps, 441
    applications of to structural problems, 442
    in computer interpretation of mass spectra,
      443
    of the fluoroethylenes, 442
    presentation of data in, 443
    redundancy check in, 442
    semi-automatic methods for determination
      of, 441
  mechanisms, correlation of empirical models
    with, 387
    use of stable isotopes in the study of, 255,
      269
  occurring from isolated electronic states, 438
  patterns, anticipation of, 377

of peptide derivatives, effect on of specific
    amino acid residues, 295
processes, competing, metastable transi-
    tions in, 433
    release of translational energy during, 446
of the pyrrolidine ring of proline in peptide
    mass spectra, 297
reactions, of alcohols and ketones, 436
    of the benzene molecular ion, 435
    degree-of-freedom effect in, 389
    effects of substituents on, 389
    importance of product ion stabilization in,
        389
    isotope effects in, 378
    multiple of ions, 437
    of primary alcohols, 436
    of secondary alcohols, 437
    of specific amino acids in peptides, 291
    of terminal diols, 437
    use of isotopes in studies of, 377
routes, effects on of bond strengths, 388
rules for peptide derivatives, 292
structures of activated complexes for, 387
of substituted, biphenyls, 392
    phenols, 392
of the N-terminal protecting group of pep-
    tide derivatives, 294
under electron impact, analogy to photo-
    chemical decomposition, 384
Fragment ions, from benzene and naphthalene,
    noncyclic nature of, 397
    failure of to be transmitted by electric sectors,
        423
    heat of formation of, 393
    identification of precursors to, 441
    of mass 4 amu less than the molecular ion,
        formation of in peptide mass spectra,
        307
    in the mass spectrum of benzene, heat of
        formation of, 436
Fragments not allowed in mass spectra, 158
Franck-Condon principle, use of in Penning
    ionization, 207
Free, dipeptides, mass spectra of, 290
    peptides, conversion of to derivatives
        suitable for mass spectrometry, 291
    mass spectra of, 290
    tripeptides, mass spectra of, 290
Frequency, of A/D conversion in gas
    chromatography-mass spectrometry, 116
factors, 395
    assignment of, 405
    dependence of rate upon, 395
    effect of substituents upon, 389
    and ionizing voltage, possible relationship,
        406
    in mass spectral reactions, 389
    in the Quasi-Equilibrium Theory, 433
    in rearrangements, best-fit values for, 405
    relation of to metastable peak intensities,
        434
response of, high resolution mass spectrome-
    ter detection systems, 48

mass spectrometer detection system, 50
Frontier electron calculaiton in the mass
    spectrum of 4-vinylcyclohexene, 402
Fructose, -1,6-diphosphate, enzymatic cleav-
    age of, 234
    -2-oxygen-18, incorporation into
        nucleosides, 234
    in pentose biosynthesis, 231
Functional groups, computer identification
    of, 84
    effect of on ion fragmentation, 374
Furan, deuterated, mass spectrum of, 384
Furanocoumarone derivatives, carbon−13
    labelled, mass spectra of, 266
Furans, fragmentation and scrambling in, 386

Gain of electron multipliers, 50
Galvonometer digitizer, 5
Game-playing programs for computers, value
    of, 123
Gangliosides, identification of by mass
    spectrometry, 348
Gas, analysis by mass spectrometry, errors in,
    329
    chromatograph, direct coupling of to mass
        spectrometers, 78
    -mass spectrometer systems, 14, 91
    chromatographic column, exchange into
        ketones of oxygen−18 and deuterium
        on, 220
    effluents, high resolution mass spectra of,
        44
    chromatography, coupled with mass
        spectrometry, 342
    of polyaminoalcohols derived from
        peptides by reduction, 291
    tritium-laden foils in, 205
    chromatography-mass spectrometry in
        peptide sequencing, 319
    of dansyl derivatives of amines, 362
    of steroid derivatives, 342
    -liquid chromatography, of derviatives of
        arginine, 305
    use of mass spectrometry in conjunction
        with, 334
Gases, analysis of by mass spectrometry, 328
    dissolved in blood, analysis of by mass
        spectrometry, 330
    methods of sampling of for mass spectrome-
        tric analysis, 329
Gastric contents, barbiturates in, 333
Gaussian, formula for ion profiles on photo-
    plates, 29
    models for mass spectral peaks, 71
    shaped metastable peaks, 421, 443
    in the mass spectrum of ethylbenzene,
        455
Gelatin, on photoplates, collagen in, 17
    use of in photoplate emulsions, 17
General patterns of fragmentation of
    peptides, 313
Generation of structures in Heuristic
    DENDRAL program, 144

Generator of isomers, DENDRAL as, 126
Geological materials, lipids in, 346
Geometrical isomers, handling of by
      DENDRAL notation, 141
Geometry of high resolution mass spectrome-
      ters, 12
Geraniol, deuterium labelled, 243
Glass used in photoplates, 16
   tolerances of, 16
Gliotoxin, biosynthesis of, 237
Gomerin, structure determination of, 361
Glucose, field ionization mass spectrum of,
      188
   glycolysis of, 245
   labelled with oxygen−18, in pentose bio-
      synthesis, 231
   -1-oxygen−18, incorporation into
      nucleosides, 234
   -2-oxygen−18, incorporation into
      nucleosides, 234
   -6-oxygen−18, incorporation into
      nucleosides, 234
Glutamic acid, in peptides, specific fragmenta-
      tion reactions of, 297
   α- and γ-peptides of, mass spectra of, 299
   residues in peptides, loss of carbalkoxy
      groups from, 298
   residues, side reactions of silver oxide with,
      316
   side reaction of sodium hydride with, 316
Glutamine, in peptides, specific fragmentation
      reactions of, 297
   residues in peptides, loss of carboxamide
      groups from, 298
N-γ-Glutamyl-2-deoxy-2-acetamido-β-d-
      glucosylamine derivatives, identifica-
      tion of, 363
Glutamyl residues in peptides, loss of ammonia
      from, 298
Glutathione in naphthalene metabolism, 251
Glutethimide, determination of, 333
Glyceraldehyde diphosphate dehydrogenase,
      245
Glycine, and alanine, confusion between fol-
      lowing peptide methylation, 316
   -arginine, lability towards base of amide
      bonds between, 313
   confusion of in peptide mass spectra, with
      aromatic and heterocyclic amino
      acids, 313
   with leucine, 313
   at the C−terminus of peptides, peculiarities
      caused by, 293, 295
   nitrogen−15 labelled, 239
   in peptides, influence of on dehydration
      processes, 294
   specific fragmentation reactions of, 295
Glycolic acid, enzymatic formation of from
      monofluoroacetic acid, 355
   oxygen−18 labelled, mass spectrum of, 249
Glycoproteins, hexosamine-aminoacid linkage
      in, 363
Glycosphingolipids, identification of by mass

spectrometry, 348
Gold, use of as an emitter tip in field ioniza-
      tion, 186
Goodlist in DENDRAL, 142, 147
Ground electronic states, implication in slow
      mass spectral reactions, 399
Ground state, relation to transition state in
      mass spectral reactions, 403
Guanidino group of arginine, condensation
      of with α- or β-dicarbonyl compounds,
      305
   in peptides, acylation of, 304
Guanosine, pentakis(trimethylsilyl), mass
      spectrum of, 231
   trimethylsilyl derivative of, mass spectrum
      of, 15
Gurney-Mott theory of latent image forma-
      tion, 21

Hair, metals in, 331
Half widths of peaks, use of in electrical
      recording of mass spectra, 44
Hall probes as mass markers, 116
Haloacetate halidohydrolase, mechanism of
      action of, 249
Halogen-containing, acids, metabolism of in
      microorganisms, 250
   acyl groups in peptide mass spectra, 314
Hamiltonian circuits, 129
Hammett, correlations, in mass spectra, 404
   limitations and complications in, 388
   in mass spectral processes, 387
   of metastable ion intensities, 389
   plots, use of in chemical ionization mass
      spectrometry, 195
   σ constants, correlation with ion intensities,
      392
Harmonic oscillators in ions, importance of
      number of, 389
Hayward line notation for organic structures,
      127
Heat of formation, of the benzyl cation, 437
   of daughter ions, errors in, 398
   of ions, 190, 393, 435
   uncertainties in, 398
Heisenberg's uncertainty principle, 182
Helium, chemical ionization mass spectra,
      protonation in, 193
   chemical ionization mass spectrum of
      dihydro cholesterol methyl ether, 193
   use, in diffusion separators, 93
   in field ion microscope, 182
Heparinization of cannulas used in blood gas
      analysis, 329
Heptadecasphing-4-enine, isolation of from
      sphingomyelin, 347
Heptafluorobutyrylpeptide methyl esters,
      mass spectra of, 294
N−Heptafluorobutyryl peptides, mass spectra
      of, 315
Heptane isomers, DENDRAL notation for, 133
1−Heptanol-5,5-dideutero, loss of water from,
      380

3–Heptanone, -2,2,4,4-d$_4$, loss of propene from the molecular ion of, 440

McLafferty rearrangement of, 440

Heptapeptide sequenced by mass spectrometry of its methylated derivative, 316

tri-n-Heptylamine, number of isomers of, 161

Heteroatomic, content, use of in computer interpretation of mass spectra, 86

plots, 39, 377

Heteroatoms, handling of by DENDRAL notation, 137

identification of, by Heuristic DENDRAL programs, 163

from mass spectra, 88

Heterocyclic, amino acid residues, influence of on dehydration processes, 294

amino acids, confusion of in peptide mass spectra with glycine, 313

in peptides, specific fragmentation reactions of, 299

compounds, mass spectra of, 263

rings in amino acid residues in peptides, N-methylation of, 316

Heuristic DENDRAL, 125, 126

structure generator, use of sub-structures in, 148

Heuristic DENDRAL program, 144

evaluation function in, 173

generation of structures by, 144

planning sub-program, 144

predictor subroutine in, 154

preliminary inference maker in, 146

structure generation in, 167

structure generation sub-program, 144

superatoms in, 159

use of, in determining structures from mass and nmr spectra, 146

high and low resolution mass spectral data in, 144

to interpret mass spectra of aliphatic ketones, 146

verification sub-program, 144, 170

Heuristic programming of computers, 123

Heuristics, definition of, 123

Hex-2-en-1-al, identification of in scent glands of larvae, 362

Hex-2-en-4-on-1-al, identification of in scent glands of larvae, 362

Hexachlorobutadiene, use of in calibration of photoplates, 19

1,2,3,4,5,6-Hexachlorocyclohexane, metabolism of in houseflies, 360

Hexadecasphing-4-enine, identification of by mass spectrometry, 347

isolation of from sphingomyelin, 347

Hexadecyl cholesteryl ether, mass spectrum of, 341

Hexahelicene, conversion of to coronene, 400

mass spectrum of, 400

metastable ion mass spectrum of, 401

racemization of optically active, 401

Hexamethylbenzene, formation of from butyne, 275

n-Hexanal, identification of in scent glands of larvae, 362

Hexane isomers in DENDRAL notation, dictionary list of, 132

1–Hexanol-4,4-dideutero, loss of water from, 380

2–Hexanone, decomposition to the enol form of acetone, 410

loss of methyl from, 408

Hexoses, conversion of to pentone phosphate, 234

enzymatic degradation of to ethanol, 245

as precursors to nucleosides, 234

2-n-Hexyl, -2-decenal-1-oxygen-18, mass spectrum of, 256

thiophene, carbon–13 labelled, mass spectrum of, 264

High energy electron beams, use of for ionization, 204

High ionizing voltage, prominence of hydrogen rearrangements at, 381

High mass range of mass spectrometers, 211

High mass region, use of to define baseline, 69

High molecular weight, compounds, difficulties in handling, 210

difficulties in mass spectrometry of, 364

mass markers, 33

mass spectrometry, jet systems in, 210

High pressure mass spectrometry, 192

High resolution, conditions, metastable peaks in, 422

mass spectra, combined with gas chromatography-low resolution mass spectra, 87

in determination of formulas of ions, 377

of gas chromatographic effluents, 44

output formats, 35

of peptide derivatives, 301

mass spectral data, arrangement of, 377

potential in structure determination, 7

processing of, 67

use of in Heuristic DENDRAL program, 144

mass spectrometers, accelerating voltage scanning of, 46

use of in gas chromatography-mass spectrometry, 109

mass spectrometric techniques to separate isobaric ions, 219

mass spectrometry, cost of compared to low resolution mass spectrometry, 353

in peptide mass spectrometry, 314

photographic techniques in, 11

in studies of fragmentation, 261

use to determine empirical formulas, 81

stable isotope studies, doublets encountered in, 220

High temperature reactions, benzyne in, 277

Histidine, containing peptides, difficulties in reduction of, 291

intermolecular alkylation reactions in, 301

in peptides, effect of on volatility, 315

specific fragmentation reactions of, 299

HMO calculations of the McLafferty rearrangement, 398

Homologous compounds, difficulties caused by in mass spectral interpretation programs, 81

Homoserine in peptides, specific fragmentation of, 297

Hospitals, use of mass spectrometers in, 330

Houseflies, metabolism in of lindane, 360

Hydrazinolysis of, arginine residues in peptides, 305
  phenylthiocarbamoyl groups, 319

Hydride, sodium, use of in exhaustive N-methylation of peptides, 316

Hydrocarbons, aromatic, chemical ionization mass spectra of, 192
  chemical ionization mass spectra of, 195
  comparison of field and electron ionization mass spectra of, 188
  doubly charged ions from, 453
  field ionization mass spectra of, 186
  higher, chemical ionization mass spectra of, 192
  loss of hydrogen from, 402
  metastable peaks in the mass spectra of, 453

Hydrogen, abstraction in elimination reactions of molecular ions, 381
  atom, loss of from the molecular ion of, benzene, metastable peak for, 422
    ethyl fluoride, 453
    methanol, metastable peak for, 423
    toluene, 428
  atoms, rearrangement of in mass spectra of ketones, 147
  chloride in methanol, use of to esterify peptides, 313
  cyanide, loss of from the molecular ions of, benzonitrile, 435
    2,4,6-trideuterobenzonitrile, 440
  field emission of, 182
    high pressure mass spectrometry of, 192
    loss or gain of in ions, errors in isotope ratios due to, 376
    use of, in diffusion separators, 93
      in field ion microscope, 182
    p-fluorobenzyl chloride, 439
    fluoroethylene, 455
  loss from hydrocarbons, 402
  migration in, glutamine and asparagine residues in peptides, 298
  fluoride, loss of from the molecular ion of, fluorobenzene, 453
  hydroxyamino acid residues in peptide mass spectra, 297
  lysine and ornithine residues in peptides, 304
  mass spectra of peptides containing pyrimidyl ornithine residues, 306
  methionine-containing peptides, 302
  peptide mass spectra, 293
  peptides containing aromatic or heterocyclic amino acids, 300
  molecule, ions formed by field emission of hydrogen gas, 182
  loss of from the molecular ion of hydrogen

sulfide, 454
  -noble gas mixtures, high pressure mass spectrometry of, 192
  rearrangements in aliphatic ethers and amines, 380
  sulfide, loss of a hydrogen molecule from the molecular ion of, 454
  transfer reactions, double, in butyl esters, 380

Hydrolysis, of chelates, mechanism of, 269
  nonspecific, of polypeptides, 291
  partial, of proteins, structure of peptides resulting from, 311

Hydroxy acids in cyclic depsipeptides, effect of on fragmentation, 309

α-Hydroxy acids in cyclic depsipeptides, 309

β-Hydroxy acids in cyclic depsipeptides, 309

Hydroxyamino acids, absence of molecular ions in spectra of peptides containing, 301
  in peptides, specific fragmentation reactions of, 297

2–Hydroxybutyric acid, enzymatic formation of from 2-chlorobutyric acid, 250

20–Hydroxycholesterol, 223

22–Hydroxycholesterol, 223

24–Hydroxycholesterol, trimethylsilyl ether mass spectrum of, 336

26–Hydroxy cholesterol, trimethylsilyl ether, mass spectrum of, 336

16–Hydroxy dehydroepiandrosterone, identification of in infant urine, 338

Hydroxy esters, field ionization mass spectra of, 189

Hydroxylase, adrenal, tyrosine, 229

Hydroxylation, enzymatic, of alkanes, 229
  of phytanic acid, 344
  of steroids by enzymatic systems, position of, 221

Hydroxyl groups, migration of during hydroxylation, 229
  in peptides, reduction of, 291

7–Hydroxy oleic acid, glycoside formation of, 230
  methyl ester, enzymatic formation of from methyl oleate, 229
  origin of hydroxyl oxygen in, 229

D–3–Hydroxy palmitic acid, identification of by mass spectrometry, 342

p–Hydroxyphenethylamine, identification of Parkinson's disease pink spot as, 363

α-Hydroxy phytanic acid, formation of from phytanic acid in liver, 344

17–Hydroxyprogesterone, 221
  enzymatic, conversion of to 11-deoxy-cortisol, 356
  formation of from progesterone, 356
  formation of from progesterone, 223

Hydroxyproline in peptides, specfiic fragmentation reactions of, 297

10D–Hydroxy stearic acid, formation of from oleic acid, 244

methyl ester, mass spectrum of, 244
13–Hydroxy stearic acid, enzymatic forma-
    tion of from linoleic acid, 343
N–Hydroxysuccimide esters of carboxylic
    acids, use of to acylate amino groups
    in peptides, 312
Hypothalamus, ovine, 15
Hypothesis, scientific, formulation of by com-
    puters, 123
Hysteresis of magnets, 47

Identification of, amino acids in peptides by
    mass spectrometry, 300
  compound class by computer, 80
  molecular ions in peptide mass spectra, 295
  specific functional groups by computer, 84
Image potential of metal surfaces, 181
Imidazole ring of histidine, alkylation of in
    mass spectrometers, 301
Imidazolinonylornithine-containing peptides,
    complications in the mass spectra of,
    307
  mass spectra of, 307
2–Imino-3-carbamoylthiazolidine, 281
Immunoglobulin, amino acid sequence in, 320
  porcine, λ-chain of, 282
  sequence of docosapeptide from, 316
Importance of mass spectrometry in peptide
    chemistry, 289
"Improbable" ions in mass spectra, 158
Impure materials, mass spectrometry of, 353
Impurities in samples, detection of in chemi-
    cal ionization mass spectra, 196
Incomplete methylation of peptides, problems
    caused by, 316
Indirect sequencing of peptides by mass
    spectrometry, 291
Indole, N-methyl, mass spectrum of, 263
  2-methyl, mass spectrum of, 263
Inductance in magnet scanning systems, 47
Induction, scientific, computer simulation of,
    174
Inert gas, use of to generate collision-induced
    metastable transitions, 429
Infant urine, "unusual" steroids in, identifi-
    cation of, 338
Infra red spectrophotometry, use of to detect
    insecticides in tissue, 334
Initial break method of measuring ionization
    potentials, 390
Insecticides, chlorinated, field ionization mass
    spectra of, 189
  fate of, use of mass spectrometry in study of,
    358
  poisoning with, 334
Instrument parameters, setting of for scanning
    of high resolution mass spectra, 55
Insulation problems in chemical ionization
    sources, 192
Integrability of photoplate, 14
Intellectual problems, solution of by computers,
    122
Intelligence, artificial, application of to

interpretation of mass spectra, 121
human and computer, comparison be-
    tween, 176
Intensest peaks in a mass spectrum, use of
    in library searching, 82
Intensities of, ions, correlation, with
    (A.P.–I.P.) values, 392
  with Hammett σ constants, 392
  effects of substituents on, 388
  metastable ions, 400
    from competing processes, 433
    dependence of on parent ion internal
        energy, 434
    measurement of, 427
    in rearrangement processes, 408
    relation of, to activation energies, 434
      to degrees of freedom in precursor
          ions, 408
      to frequency factors, 434
      to importance of fragmentation routes,
          434
    variation of with internal energy of
        fragmenting ion, 439
  molecular ions, enhancement of by use
      of low filament voltages, 220
  peaks, in field ionization mass spectra,
      absence of rules governing, 189
  in mass spectra, interpretation of, 390
Intensity, of collision-induced metastable
    ions, 408
  of ions, correlation with activation energy,
      392
  formed by field ionization, 180
  measurements of metastable peaks, ap-
      plications of, 436
  of molecular ions, preduction of, 171
Intensive Care Units, use of mass spectrome-
    ters in, 329
Interconversions of amino acid residues in
    peptides during mass spectrometry,
    313
Intermolecular, alkylation reactions in,
    histidine-containing peptides, 301
  peptide mass spectra, 295
  peptides, 313
  methylation, and dehydration in peptide
      mass spectra, 301
  of peptides, temperature dependence of,
      301
  of pyrimidyl ornithine residues in pep-
      tides, 306
  processes in field ionization sources, 185
  reactions in tryptophan-containing
      peptides, 301
  thermal methylation of imidazolinoyl-
      ornithine-containing peptides, 307
Internal, energies of ions, 432
  reference, use of in photoplate work, 14
  standards for mass marking, 32
Interpretation of, intensities of peaks in
    mass spectra, 390
  mass spectra, application, of artificial in-
      telligence to, 121

of learning machines to, 377
by computers, 80, 126
partial, by computer, 85
of peptides, 313
peptide mass spectra, simplification of, 314
scrambling experiments, 387
Introduction by exchange processes of stable
isotopes into molecules, 220
Involatile peptides, problems in mass spectrometry of, 315
Ion, beams, dimensions of in high resolution
instruments, 13
formed in photoionization, intensity of, 209
bombardment, ionization by, 212
clusters, formation of from molecular beams,
210
possible use of as mass markers, 210
current, fluctuations in, averaging out of
by photoplate integration, 16
cyclotron resonance, McLafferty and double
McLafferty rearrangements in, 410
spectrometers, mass range of, 191
spectroscopy, 191, 408, 409
review of, 409
impact on photographic plates, photographic
processes involved, 20
intensities, effects of substituents on, 388
kinetic energy spectroscopy, 427
-molecule, collisions in, chemical ionization
sources, 194
ion cyclotron resonance spectrometers, 191
reactions, 190, 408
under field ionization conditions, 185
in field ionization sources, 184
study of in ion cyclotron resonance
spectroscopy, 191
transfer of electrons, protons and alkyl
carbonium ions, 192
optical aberrations, 53
reactivities, 407
repeller voltage, effect of on relative
metastable intensities, 439
stability, effect on mass spectra, 389
statistical considerations in electrical recording of mass spectra, 44
statistics, in exponential scanning, 47
problems, caused by, 62, 65
created by in scanning of high resolution
mass spectra, 51
structure, characterization of, 437
types, 40
use of in computer interpretation of mass
spectra, 85
Ionic structure, information regarding from
metastable ion intensities, 408
Ionization, and appearance potentials, difficulties in measurement of, 390
of benzene, ring opening during, 400
spectral data relating to, 400
thermochemistry of, 400
cross section, Bethe's equation for, 204
of metastable molecules, 207
detectors in gas chromatography, 205

efficiency, correlation with signa
constant, 392
dependence upon, ionizing voltage, 392
number of valence electrons, 204
by electrojet methods, 210
by electron bombardment, 179
followed by field-induced dissociation in
field ionization sources, 185
with high energy electron beams, 204
of high molecular weight compounds, 210
by ion bombardment, 212
by laser beams, 206
with metastable molecules, 207
of molecules, cross section of, 190
Penning, 207
potential, of atoms, 183
of vinyl acetylene, 436
potentials, of alkanes, calculation of, 397
of aromatic compounds, 397
calculation of, 397
effect of, aromatic substituents on, 389
substituents on, 389
measurement of in photoionization
sources, 209
of metal chelates, 394
methods for measurement of, 390
of noble gases used in chemical ionization, 192
prediction of, 390
by protons from van de Graaff generator,
205
site of in, metal chelates, 394
methylated thioureas, 394
techniques in mass spectrometry, 179
Ionizing, efficiency, dependence on voltage,
effects of substituents on, 389
energy, dependence on of relative
metastable peak intensities, 439
effect of, on mass spectra, 381
on relative metastable intensities, 439
voltage, effect of on simple cleavages and
rearrangements, 406
and frequency factor, possible relationship, 406
Ions, from aromatic compounds, structure
of, 397
charge localization in, 180
determination of formulas of, 375
expected in field ionization, 187
field desorption of, 188
formed by, amide bond cleavage in peptides, absence of in mass spectra, 313
field ionization, intensity of, 180
monochromatization of, 185
heat of formation of, 190
heats of formation of, 435
intense, in field ionization mass spectra,
methods of formation of, 186
internal energy of, 432
kinetic shifts in the formation of, 435
of mass (M−4), significance of in peptide
mass spectra, 301
mechanism of fragmentation of, 373

at (M+14) and (M+28) in the mass spectra of pyrimidyl ornithine containing peptides, 306
residence, time of in the source, 391
   times in the source of, 420
resulting from mass spectral reactions, excited states of, 393
Iron atom, effect of on mass spectra of iron complexes, 379
Irradiation of methane-ozone mixtures, 272
Isariin, structure determination of, 311
Isariinic acid, structure determination of by mass spectrometry, 311
Isarolides A, B, and C, structure determination of by mass spectrometry, 309
Isobaric ions, difficulties in interpretation caused by, 219
   separation of by mass spectrometers, 219
Isolation of metabolites prior to mass spectrometry, difficulties with, 353
Isomeric leucines in peptides, differentiation between, 296
   difficulties of mass spectrometric identification, 296
Isomers, of a given composition, computer generation of all possible, 125
   structural, dictionary of, production of by DENDRAL structure generator, 170
   differentiation between in metastable ion spectra, 408
Isotope, abundances, determination of molecular formula from, 361
   distribution patterns, computer calculation of, 376
   effects in, McLafferty rearrangements, 383
      mass spectral, elimination reactions, 383
      fragmentation reactions, 378
   mass spectra of partly deuterated propanes, 440
   metastable transitions, 439
   enrichment requirements for low resolution work, 219
   labelling studies in chemical ionization mass spectra, 194
   ratios, errors in due to loss or gain of hydrogen, 376
Isotopes, of chlorine and bromine, 376
   natural abundance of, 376
   stable, labelling techniques for use with, 220
      use of in studies of biosynthesis, 221
   use of, in mass spectrometric studies, 354
      in studies of mass spectral fragmentations, 377
Isotopic, composition, analysis of in determination of empirical formulas, 376
   distributions and mass defects in the determination of the formulas of ions, 377
IUPAC line notation for organic structures, 127

Jet molecular separator, advantages of, 97
   beam velocity in, 98

in gas chromatography-mass spectrometry, 93, 95, 96
Jets, dimensions of in molecular separators, 97
Jet systems in mass spectrometry, 210
Juvenile hormone, structure determination of, 362

Kansamycolones, identification of, 363
Ketene, loss of from the A–ring of steroids, 223
Keto and enol forms of acetone in mass spectra, 410
Ketone groups, determination of using the mass spectra of the O-methyloximes, 343
Ketones, aliphatic, fragmentation of, 146
   interpretation of mass spectra by Heuristic DENDRAL program, 146
   McLafferty rearrangements of, 383
   structure determination of from mass spectra by Heuristic DENDRAL, 149
   success of Heuristic DENDRAL programs with, 175
   aromatic, cleavage of to give acyl ions, 395
   cleavage at carbonyl group of, 147
   confusion of with unsaturated alcohols and ethers, 147
   exchange of oxygen–18 into, 279
   fragmentation reactions in, 436
   McLafferty rearrangement in, 148
   rules for interpretation of mass spectra of, 147
   separation from alcohols of, by gas chromatography, 95
   of by gas chromatography, 94
   substructures for in Heuristic DENDRAL, 148
Kinetic aspects of metastable transitions, 431
Kinetics, of ions in a mass spectrometer source, 387
   wide range electron impact, 396
Kinetic shifts, 435
   in appearance potentials, 398
   in formation of ions, 435
   in mass spectral reactions, 391
      magnitude of, 391
   in mass spectrum of benzonitrile, 436
   measurements of, 435
   in processes giving intense metastable peaks, 436
Kodak SWR photoplates, 26
Koenigsberg bridges, Euler's solution to, 129
Koopman's theorem, 394

Label, p-fluorophenyl as in studies of scrambling, 386
Labelled, acyl groups in peptide mass spectra, 314
   compounds, commercially available, 220

Labelling, isotope, studies in chemical ionization mass spectra, 194
of molecules with stable isotopes, 220
techniques, for use with stable isotopes, 220
in mass spectrometry, 354
β-Lactams, mass spectral fragmentation of, 267
nitrogen–15 labelled, mass spectra of, 260
Lactone derived from aldosterone, 337
Lagrange polynomial, use of in time-to-mass conversion routines, 73
Lake sediment, lipids in, 346
Lanosterol, formation of, from squalene, 224
from squalene-2,3-oxide-oxygen-18, 224
identification of in brain tissue, 339
Lanosterols, identification of by mass spectrometry, 339
Large computers, use of for mass spectral data processing, 67
Larger peptides, mass spectrometry of, 292
Large organic molecules, application of simplified Quasi-Equilibrium Theory to, 404
Larvae, defense secretion of, 362
hydroxylated fatty acids in, 346
Laser-induced ionization mass spectrum of, aromatic hydrocarbons, 207
leucine, 207
Lasers, ionization by, 206
Latent image formation, Gurney–Mott theory of, 21
Mitchell theory of, 21
in photographic emulsions, 20
Latex, use of as membrane in blood gas analysis, 329
Lead, toxicology of, 331
Leading, atom as tree graph centroid, 129
bond as tree graph centroid, 129
bonds, means of coding on teletypewriter, 142
Leak, variable, use of in high resolution gas chromatography-mass spectrometry, 14
Learning machines, application of to mass spectral interpretation, 377
success of in mass spectral interpretation, 377
Leucine, confusion of in peptide mass spectra with, alanine, 313
glycine, 313
electron impact mass spectrum of, 207
and isoleucine, differentiation between in peptide mass spectra, 296
laser-induced ionization mass spectrum of, 207

in peptides, specific fragmentation reactions of, 295
and valine, confusion between following peptide methylation, 316
Libraries of mass spectra, 80
Librium*, metabolism of, 350
Lifetime of ions, dependence of degree of isotope scrambling on, 441
Light sources used in microdensitometers, 27
Limiting factors in mass spectrometry of biological molecules, 349
Lincomycin, biosynthesis of, 248
Lindane, metabolism of in house flies, 360
Linear, comparator, use of with photoplates, 27
extrapolation method of measuring ionization potentials, 390
formula, uniquely defined by DENDRAL notation, 133
measurements on photoplates, 26
notation system for organic structures, 126
regression methods for time-to-mass conversion, 73
scanning of magnetic instruments, 47
Linearity of fragment ions from benzene and naphthalene, 397
Linearization of chemical formulas, 127
Lines on photoplates, opacity of, 18
optical density of, 18
Linoleic acid, enzymatic hydroxylation of, 343
identification of in peas, 346
Linolenic acid, identification of in peas, 346
Lipidopeptides, structure determination of, 313
Lipid oxidase in soybeans, 343
Lipids, analysis of by mass spectrometry, 342
in geological specimens, 346
in preen glands of water fowl, structure of, 343
Lipopeptide from m. johnei, structure determination of by mass spectrometry, 313
Liquid phase, choice of in gas chromatography-mass spectrometry, 94
concentration of in gas chromatography-mass spectrometry, 94
LISP programming language, 144
Lithium aliminum deuteride reduction of, N-acyl peptides, 291
peptides, 290
Lithium aluminum hydride reduction of, N-acyl peptides, 291
peptides, 290

polyamino chlorides, 291
Llewellyn—Littlejohn separator, 103
Localization of charge, at guanidino groups
in arginine-containing peptides, 304
in ions, 180
formed, by field ionization, 187
from peptide derivatives, 292
Locating of, carbon—13 by nuclear magnetic
resonance spectroscopy, 354
peaks in high resolution mass spectra, 44
Logarithmic, amplification of mass spectral
signals, 61
amplifiers, 5
form, conversion of linear data to, advantages of, 67
Logical processes involved in structure
determination, 124
Long chain fatty acids, separation of by gas
chromatography, 95
Loss of, gas chromatography resolution in
gas chromatography-mass spectrometry, 106
of side chains of, hydroxyamino acids in
peptide mass spectra, 297
specific amino acids in peptide mass
spectra, 295
of water in field ionization of alcohols, 189
Low ionizing voltage, prominence of hydrogen
rearrangements at, 381
Low resolution mass spectra, determination of,
empirical formulas from, 376
peptide structure from, 314
Low resolution mass spectral data, of amines,
155
as input data, for Heuristic DENDRAL
program, 146
use of in Heuristic DENDRAL program, 144
Low voltage, spectra, fragmentation in, 388
studies on the McLafferty rearrangement, 381
Lunar landing site, absence of organic matter
in samples from, 346
Lung, measurement of gas partial pressures in,
330
tissue, metals in, 331
Lysine, acylation of side chain amino group
of, 312
in peptides, 304
carbon—13 labelled, 237
-containing peptides, difficulties of Edman
degradation with, 318
nitrogen—15 labelled, 237
and ornithine, confusion between as a
result of incomplete methylation of
peptides, 316
in peptides, specific fragmentation reactions
of, 304

Lysozyme, mass spectrum of, attempts to
obtain, 211

Mach-focussing effect in jet systems, 211
McLafferty rearrangement, 383
effect of substituents upon, 393
in field ionization mass spectra, 188
HMO calculations of, 398
ions in Heuristic DENDRAL programs, 148
isotope effect in, 383
in ketones, 148
low voltage studies on, 381
mechanism of, 378
perturbation molecular orbital treatment of,
402
plus one ions, 149
study of by ion cyclotron resonance spectroscopy, 410
of 2,2,4,4-tetradeutero-3-heptanone, 440
as a two-step process, 398
of valine residues in peptide mass spectra, 29?
Magnet, position, determination of for scanning of high resolution mass spectra,
56
positioning of in scanning instruments at
high resolving power, 55
Magnetic, deflection mass spectrometers,
fundamental equations governing, 45
field, relationship of to mass to charge ratio
and accelerating voltage, 45
sectors, exponential scanning of, 46
metastable transitions occurring in, 423
use of for high molecular weight compounds, 349
tape as an output device, 4
Magnets, large radius, possible use of in mass
spectrometers, 364
super-conducting, possible use of in mass
spectrometers, 212, 364
Manganese—VII, rates of oxidation with,
268
Manned space flight, analysis of respiratory
gases during, 330
D—Mannose, glycolysis of, 245
Markers in mass spectral fragmentation
studies, use of aryl and alkyl groups
as, 256
Mascot digitizer, 6
Mass, center of tree graph, 129
to charge ratio, relationship of to magnetic
field and accelerating voltage, 45
defects, 32
effect of on photographic response, 23
of ions, relationship to time of appearance
in magnetic scans, 73
limit, upper, of mass spectrometer, exten-

sion of, 210
markers, 79
  in gas chromatography-mass spectrometry, 116
  use of, internal standards for, 32
  ion clusters as, 210
measurement, accuracy of, 62
  average error in, 62
  errors in during electrical recording of mass spectra, 44
  range of, ion cyclotron resonance spectrometers, 191
  mass spectrometers, 211
  extension of, 210
spectra, calculation of, with Quasi-Equilibrium Theory, 403
  from structures by DENDRAL programs, 171
  of free peptides, 290
  fragments not allowed in, 158
  of peptides, 290
  containing aliphatic amino acids, 291
  interpretation of, 313
  of polyaminoalcohols derived from peptides by reduction, 291
  theoretically significant peaks in, 174
spectral, data, automatic determination of peptide structure from, 314
  high resolution, use of in Heuristic DENDRAL program, 144
  low resolution, use of in Heuristic DENDRAL program, 144
  files, abbreviation of, 83
  libraries, 80
  computer searching of, 82
  processes, Hammett correlations in, 387
  steady-state approximations in, 387
  reactions, activation energies for, 391
spectrometer, controlled by computer in feedback mode, 2
  output, automation of, 3
  use of to monitor effluent from gas chromatographs, 334
spectrometers, use of in gas chromatography-mass spectrometry combinations, 108
spectrometry, combination of with the Edman degradation in peptide sequencing, 318
  coupled with gas chromatography, 342
  use in steroid research, 342
  in peptide sequencing, early work in, 290
  reviews of, 292
  of peptides, reviews of, 290
  shortened by Edman degradation, 318
  in sequencing of peptides, 289
use of, to determine molecular weights of

cyclic peptides and depsipeptides, 309
  to identify amino acids in peptides, 300
spectrum digitizers, 3
Masses, conversion of to elemental compositions, 74
Mathematical approaches to substituent effects, 392
Mattauch-Herzog, double focussing mass spectrometer, use of with spark sources, 330
  geometry, in mass spectrometers, 12
  use of in gas chromatography-mass spectrometry, 109
  mass spectrometer, use of in respiratory gas analysis, 330
  mass spectrometers, metastable transitions in, 428
Maxima, multiple in metastable peaks, 453
  of peaks, detection of, 62
Measurement of, appearance and ionization potentials, difficulties in, 390
  appearance potentials, 391
  fragmentation maps, 441
  ionization potentials, 390
Mechanism of, fragmentation of cyclic peptides and depsipeptides, 308
  fragmentation of ions, 373
  mass spectral reactions, use of metastable ions in study of, 407
Mechanisms of fragmentation in chemical ionization mass spectra, 199
Mechanistic mass spectrometry, difficulties with, 374
Median of peaks in mass spectra, 70
Medical research, use of jet separator in, 109
Membrane, -porous silver separator, 103
  silicon rubber, 102
  used in blood gas analysis, 329
Membranes of molecule separators, working area of, 102
Memory effects in jet molecular separators, 98
Mercuric ion oxidation of propene-1-carbon-13, 275
Metabolism, and biosynthesis, distinction between, 221, 354
  and detoxification, relationship of, 249
  of drugs, use of stable isotopes in studies of, 221
  of estrone, 339
  of phytanic acid, 344
  of steroidal birth control substances, 352
  of steroids, study of with gas chromatography-mass spectrometry, 110
  studies, by mass spectrometry, 249, 350

use of stable isotopes in, 249
Metabolites of steroids, identification of
     by mass spectrometry, 338
Metal, chelates, ionization potentials of, 394
     site of ionization in, 394
   separators, silanization of, 107
   surfaces, image potential of, 181
     removal from of an electron, 181
Metallocyclobutadienes, intermediacy of in
     butyne isomerization reactions, 275
Metals, analysis for by mass spectrometry, 330
   in blood, tumors, lung tissue and bone, 331
   Fermi level of, 184
Metastable, atoms, apparatus for production of, 208
   use of to produce ions, 207
   defocussing, of ions from hexahelicene, 401
   use of in ion bombardment methods, 212
   ion, intensities, Hammett correlations of, 389
     mass spectra, calculation of with the
        Quasi-Equilibrium Theory, 404
     mass spectrum of hexahelicene, 400, 401
   ions, abundance of, 399
     in chemical ionization mass spectra, 199
     detection of by defocussing of the ion
        beam, 16
     in field ionization, 185
        mass spectra of hydroxy esters, 189
     for the loss of water from amide bonds
        in peptide mass spectra, 294
     intensities of, 396, 400, 408
     intensity of relative to parent and daughter
        ions, 436
     in the mass spectra of peptides containing
        glutamic and aspartic acids, 299
     normal, equations governing appearance
        of, 421
     in peptide mass spectra, 301
     reactivities of, 407
     relation of to parent and daughter ions, 421
     semi-automatic registration techniques for,
        314
     use of, in determining ionic structures and
        mechanisms, 407
     in peptide structure determination, 314
     in scrambling studies, 387
   mass spectrum of toluene, 426
   molecules, cross section for ionization with,
        207
   peak, intensities, dependence of on ionizing
        energy and accelerating voltage, 439
     measurement of, 427
     shape, effect of β-slit width on, 452
     shapes, calculation of, 450
        substituent effects on, 455
   peaks, applications of intensity measure-
        ments on, 436

assignment of, 422
   broad, in the mass spectra of nitrogen
        and the oxides of nitrogen, 454
   broadening of due to energy release, 421
   for competing processes, intensities of, 433
   concave topped, 453
   containing several maxima, 453
   different shapes of in different instruments,
        448
   dish shaped, 444, 447
   due to processes occurring before the
        electric sector, shapes of, 450
   effect of discrimination on, 445
   effect on of energy release during frag-
        mentation processes, 423
   factors, affecting shapes of, 421
     controlling the width of, 445
   flat-topped, 444, 446
   gaussian shaped, 443
   in high resolution mass spectra, 422
   intense, kinetic shifts in processes giving,
        436
   in the mass spectra of, coumarins, 455
     hydrocarbons, 453
        methanol and deuterium-labelled meth-
           anol, 454
     2-pyrones, 455
   in the mass spectrum of fluoroethylene, 455
   obscuring of by normal peaks, 422
   shapes of, 443, 447
   significance of, 407
   widths of, 445
processes in the mass spectrum of o-nitro-
     phenol, 423
scanning, mis-tuning of source in, 426
   range available for, 427
spectra of cis-butene-2 and trans-butene-2,
     425
suppressor, use of, 456
transition data, computer handling of, 442
transitions, 419
   appearance potential measurements in the
        study of, 435
   appearance potentials of, 436
   application of the Quasi-Equilibrium
        Theory to, 432
   applications of in organic, analytical and
        physical chemistry, 420
   caused by ions decomposing during ac-
        celeration, 422
   collision-induced, 429
   in competing fragmentation processes, 433
   in competition with each other, 438
   consecutive, 428
   in Dempster mass spectrometers, 430
   in deuterated compounds, 439

in the electric sector, 428
  of Mattauch-Herzog mass spectrome-
    ters, 429
  in the field-free region before the electric
    sector, 425
  in the field-free regions of mass spectrome-
    ters, 420
  in the first field-free region of a double
    focussing mass spectrometer, 441
  identification of precursors in, 426
  isotope effects on, 439
  kinetic aspects of, 431
  in the mass spectra of, aromatic carboxylic
    acids, 440
    benzyl derivatives, 437
    bicyclic aromatic compounds, 440
    deuterium-labelled propanes, 440
    peptides, 442
  in the mass spectrum of, benzene, 435
    benzonitrile, 435
    n-butane, 442
    deuterium-labelled phthalic acid, 440
    2-deuteropyridine, 439
    diethyl ether, 438
    1,1-difluoroethylene, 345
    p-fluorobenzyl chloride, 439
    fluoroethylene, 435
    methane, 432
    methanol, 448
    toluene, 426, 428, 429
    2,4,6-trideuterobenzonitrile, 440
  in Mattauch-Herzog mass spectrometers,
    428
  in Nier-Johnson mass spectrometers, 428
  observation of, 420
    by changing accelerating voltage–ESA
      voltage ratios, 425
    occurring, after deflection of ions, 423
      in the magnetic sector, 423
    range available in scanning of, 427
    rate constants in, 432
    study of by scanning ESA voltages, 427
    in time-of-flight mass spectrometers, 430
    use of, retarding grids in studies of, 425
    β-slits in studies of, 428
Metastables, in field ionization mass spectra,
  188
  ordinary and fast in field ionization mass
    spectra, 185
Methane, chemical ionization, 198
  mass spectra, 198
  mass spectrum of dihydro cholesterol
    methyl ether, 193
  in chemical ionization, 189, 190
  -d4, loss of from the molecular ion of
    octadeuteropropane, 440
  high pressure mass spectrometry of, 192

at high pressures, mass spectrum of, 409
  and isobutane, comparison of as reagent
    gases for chemical ionization, 195
  loss of from the molecular ion of, n-butane,
    429
  propane, 440
  metastable transitions in the mass spectrum
    of, 432
  ozonolysis of, 272
  primary ions from, 190
Methanol, deuterium labelled, metastable
    peaks in the mass spectra of, 454
  field ionization mass spectrum of, 185
  loss of, from esterified carboxylic acid side
    chains of peptides, 298
  from methyl-o-toluate, 405
  metastable transitions in the mass spectrum
    of, 448
  molecular ion of, loss of hydrogen atom
    from, metastable peak for, 423
Methionine, chemical ionization mass spectrum
    of, 204
  -containing peptides, desulfurization of, 317
  incorporation into, ubiquinone, 246
    vitamin k2, 246
  involvement in steroid methylation reac-
    tions, 342
  in peptides, specific fragmentation reactions
    of, 301
  residues in peptides, side reactions of silver
    oxide with, 316
  as source of methyl groups in lincomycin, 248
  sulfone in peptides, stability of in mass spectra,
    302
  sulfoxide in peptides, stability of in mass
    spectra, 302
  trideutero, biological alkylation by, 357
  in biosynthesis studies, 245, 246
Methoxybenzofurans, loss of carbon monoxide
    from molecular ions of, 257
2-Methoxybenzotropone, oxygen–18, carbon–13
    labelled, mass spectrum of, 257
Methoxybiphenyls, loss of carbon monoxide
    from molecular ions of, 257
7-Methoxy oleic acid, methyl ester of, mass
    spectrum of, 229
3-Methoxy tyramine, enzymatic formation
    from DOPA, 229
Methyl, acetate, identification of in desert locust,
    362
  N(o-aminophenyl)-N-(3-dimethylaminopropyl)
    anthranilate, as diuretic, metabolism
    of, 353
  cholesterol, identification of by mass spectrome-
    try, 339
  esters, of N-acylpeptides, mass spectra of, 294

field ionization mass spectra of, 294
group, elimination of from alanine in
    peptides, 295
iodide, use of in exhaustive N-methylation
    of peptides, 316
mercaptoacetate, expulsion of from
    S-carboxymethylcysteine residues in
    peptides, 303
palmitate, gas chromatography-mass
    spectrometry of, 97
    retention of by porous glass separator,
        106
radical, loss of from, 2-hexanone, 408
    molecular ions of anisoles, 390
N—Methylamino acid residues in exhaustively
    methylated peptides, 316
N—Methylated peptides, formation of by
    methylation of N-acyl peptide esters,
    316
    mass spectra of, 316
Methylated thioureas, substituent effects in,
    394
Methylation, of cysteine in peptides, 317
    exhaustive, of amide bonds in peptides, 316
    in the mass spectra of pyrimidyl ornithine-
        containing peptides, 306
    of heterocyclic rings of amino acid residues
        in peptides, 316
    of peptides, confusion between pairs of
    intermolecular, of imidazolinonylornithine-
        containing peptides, 307
        amino acids as a result of, 316
    in steroid biosynthesis, involvement of
        methionine in, 342
1—Methylcyclobutanol, decomposition of to
    the enol form of acetone, 410
Methylcyclopropylcarbinyl cation, structure
    of, 269
24—Methylene cholesterol, identification of
    by mass spectrometry, 341
Methyl (d_3) esters of N-acyl peptides, mass
    spectra of, 294
C—Methyl groups, detection of, 159
N—Methyl groups, detection of, 159
    as markers in fragmentation studies, 256
Methylhydantoins, identification of by mass
    spectrometry, 319
N—Methylindole, mass spectrum of, 263
2—Methylindole, mass spectrum of, 263
1—Methylisoquinoline, mass spectrum of, 263
Methylketo acids in shale, 346
O—Methyl oximes, mass spectra of, 342
3—Methyl-2,4-pentanedione, formation of
    from hexamethylbenzene, 276
Methylphenyl group, mass spectra of com-
    pounds containing, 378
N—Methyl-4-pyridone, nitrogen—15 labelled,
    mass spectrum of, 256
N—Methylpyrrole, mass spectrum of, 263
β—Methylstyrene oxide, hydrolysis of, 270
Methyl-o-toluate, loss of methanol from, 405
1—Mevalonate—5,5-dideutero-2-carbon-14,
    incorporation of into cholesterol, 241

Mevalonic acid, enzymatic conversion to
    squalene, 241
Mevalonic-5,5-dideutero, lactone, 242
Microdensitometer, -linear comparator,
    schematic diagram of, 29
use of with photoplates, 27
Microdensitometry of, lines on photoplates,
    26
photoplates, 27
Microorganisms, sterols from, identification
    of by mass spectrometry, 341
sterols in, 342
Microscope, field ion, use of hydrogen and
    helium in, 182
Microscopy, field emission, 182
field ion, 181
Migration of, hydrogen atoms prior to
    elimination reactions, 381
hydrogen in peptide mass spectra, 293
hydroxyl group during hydroxylation re-
    actions, 229
Miniac digital computer, 5
Mitchell theory of latent image formation, 21
Mixed acyl groups in peptide mass spectra, 314
Mixtures, analysis of by mass spectrometry, 15
of peptides, analysis of by mass spectrometry,
    303
Model compounds, use of in study of sub-
    stituent effects, 394
Models, empirical, correlation of fragmenta-
    tion mechanisms with, 387
Modification of arginine residues for mass
    spectrometry, 304
Molecular, beam techniques for introduction
    of involatile samples to mass spectrome-
    ters, 364
beams, formation of ion clusters from, 210
ion, density of states in, effect of substituents
    on, 395
identification of by computer, 81
importance of identification of, 180
ions, from N-acylpeptide methyl esters, 292
energy distribution of, 392, 404
intensity of, prediction of, 171
from pentaphenylcyclopentane diol,
    structure of, 386
in peptide mass spectra, identification of,
    295
from polystyrene, 211
from toluene, 395
orbital, highest filled, removal of electron
    from during electron impact ionization,
    394
orbitals, effect of substituents upon energy
    levels of, 394
oxygen, exchange in, 225
fate of in steroid hydroxylation, 221
separators used in gas chromatography-
    mass spectrometry, 95
weight, of cyclic peptides and depsipeptides,
    determination of by mass spectrometry,
    309
limits of in mass spectrometric work, 364

Molecules near emitter tips in field ionization, orientation of, 187
Molybdenum, role of caries formation, 331
  use of in studies of photoplate sensitivity, 23
Monamycin, structure determination of by mass spectrometry, 311
Monoamino dicarboxylic acids in peptides, specific fragmentation reactions of, 297
Monochloroacetic acid, enzymatic dechlorination of, 250
Monochromatization of ions formed in field ionization, 185
Monochromator, use of in photoionization sources, 209
Monofluoroacetic acid, enzymatic conversation-conversion of to glycolic acid, 355
  enzymatic defluorination of, 249
Monooxygenase, involvement of in metabolism of naphthalene in liver, 357
cis-cis—Muconic acid, dimethyl ester of, mass spectrum of, 226
  formed by oxidative cleavage of catechol, 225
Muconic acid-cyclobutenedicarboxylic acid system, 400
Multiple, channels of analog tape, use of in recording mass spectra, 61
  fragmentation reactions of ions, 437
  ion detection in gas chromatography-mass spectrometry, 109
Multiplets in high resolution mass spectra, detection of, 70
Multiplexers, applications of, 59
Multiplexing of various instruments into an analog-to-digital converter, 59
Multiplicity of, afferent links in DENDRAL notation, 140
  bonds, coding of on teletypewriter, 142
Multiplier gain, dependence of on various factors, 50
Multipliers, electron, gain of, 50
Multiply charged ions, formed in electrojets, 211
  metastable peaks in decomposition of, 421
Multivibrator, use in digitizers, 6
Mycoside C_{b1}, amino acid sequence of, 316

Naphthalene, 1,2-dihydro-1,2-dihydroxy, enzymatic formation of from naphthalene, 357
  metabolism of, 357
  in mammalian systems, 250
  noncyclic fragment ions from, 397
  2-phenoxy, loss of carbon monoxide from the molecular ion of, 257
Natural, abundance of isotopes, 376
  cyclotetradepsipeptides, structure determination of by mass spectrometry, 309
  products, chemical ionization mass spectrometry of, 409

Negative ions in chemical ionization, 201
Neoantimycin, structure determination of by mass spectrometry, 311
Neopentyl glycol succinate as a stationary phase in gas chromatography-mass spectrometry, 95
Neosamine—C, N,N′-diacetyl, fragmentation of, 259
Neurohypophysis, porcine, unknown peptide in, 361
Neutral, fragments, determination of the formulas of, 442
  molecules, loss of from ions in chemical ionization mass spectra, 195
Neutron activation analysis of metals, 330
Nickel, −63 as radiation source for ionization, 205
  Raney, use of in desulfurization of peptides, 317
Nier-Johnson, geometry, in mass spectrometers, 13
  use of in gas chromatography-mass spectrometry, 109
  instruments, scanning of, 46
  mass spectrometers, metastable transitions in, 428
NIH shift, 357
Nitrate, potassium, oxygen−18 labelled, 236
Nitric oxide, loss of from the molecular ion of o-nitrophenol, 446
p—Nitroaniline, mass spectral fragmentation of, 263
Nitrobenzene, conversion of to aniline in the mass spectrometer, 375
  effect of substituents on the mass spectrum of, 393
  exchange of oxygens of, 269
  loss of nitric oxide from the molecular ion of, 375
Nitrogen, atom, loss of from the molecular ion of isocyanic acid, 454
  broad metastable peaks in the mass spectrum of, 454
  oxides, broad metastable peaks in the mass spectra of, 454
Nitrogen−15, incorporation of, into 2-nitropropionic acid, 355
  into ricinine, 240, 355
  into streptomycin, 355
  use of, in biochemical problems, 237, 354, 355
  in chemistry and biochemistry, 218
  in fragmentation mechanism studies, 259
  in studies of organic reaction mechanisms, 273
Nitrones, mass spectral fragmentation of, 267
o-Nitrophenol, loss of nitric oxide from the molecular ion of, 446
  width of parent ion peak of, 423
3-Nitropropionic acid, biosynthesis of, 236, 355
  mass spectrum of, 236
Nitrosamines, deuterium and nitrogen-15 labelled, 274

Noble, gases, use of in chemical ionization, 192
gas-hydrogen mixtures, high pressure mass spectrometry of, 192
Node count in DENDRAL notation, 140
Nodes in topological graph expressions of structure, 128
Noise, acquired during transmission of mass spectral signals, 71
electronic, problems caused by, 65
peak-to-peak in mass spectra, 69
spikes in high resolution mass spectra, handling of by computer, 60
-to-signal ratio in detection systems, 50
Nominal mass, definition of, 376
Nonan-4-one, Heuristic DENDRAL interpretation of the mass spectrum of, 149
-1,1,1-trideutero, double McLafferty rearrangement in, 410
Nonapeptides, sequence determination of by mass spectrometry, 315
Norethindrone, metabolism of, 351
Norethynodrel, metabolism of, 352
Norgestrel, metabolism of, 351
studied by mass spectrometry, 340
Normal metastable transitions, pressure-dependence of, 429
Nuclear magnetic resonance, data, use of, in structure prediction, 155
in truncation of candidate structure lists, 155
spectra, use of, to identify N-methyl and C-methyl groups, 159
as input data to Heuristic DENDRAL programs, 146
Nuclear magnetic resonance spectroscopy, time averaged, 350
as supplement to mass spectrometry, 344
use of, to detect and locate carbon−13, 354
to detect, nitrogen−15, 355
oxygen−17, 355
Nucleic acid hydrolysates, 15
Nucleic acids, enzymatic degradation of, 15
Nucleosides, chemically modified, mass spectra of, 15
field ionization mass spectra of, 189
hexoses as precursors to, 234
mass spectra of, 234
oxygen−18 labelled, mass spectra of, 231
trimethylsilylation of, 15
trimethylsilyl derivatives of, mass spectra of, 232
Nucleotide derivatives, mass spectrometry of, 349
Nucleotides, mass spectrometry of, 346

Oat seed, sterols in, 341
Observation of metastable transitions, 420
Obsolescence of computer systems, 57
Octacosenoic acid in algae, identification of mass spectrometry, 346
Octadecanoic acid, esters of, mass spectra of, 230

Octadecapeptide sequenced by mass spectrometry of its methylated derivative, 316
Octamethyl cellobiose, field ionization mass spectrum of, 188
Octane isomers, DENDRAL notation for, 133
Octan-4-one, predicted mass spectrum of, 173
Oct-2-en-1-al, identification of in scent glands of larvae, 362
Oct-2-en-4-on-1-al, identification of in scent glands of larvae, 362
Odd-electron ions in chemical ionization mass spectra, 195
Off-line processing of mass spectral data, 61
Olefinic, bonds, location of by mass spectrometry, 258, 343
carbons, handling of by DENDRAL notation, 137
Olefins, comparison of field and electron ionization mass spectra of, 188
selective retention of on gas chromatography, 94
Oleic acid, in algae, identification of by mass spectrometry, 346
enzymatic hydroxylation of, 244
methyl ester, enzymatic hydroxylation of, 229
microbiological formation of from stearic acid, 244
oxidation of with osmium tetroxide-oxygen-18, 258
On-line, data acquisition, 5
systems, 8
Opacity of photoplate lines, 18
Operation of chemical ionization mass spectrometer, 203
Optical, density of photoplate lines, 18
isomers, handling of by DENDRAL notation, 141
Orange peel oil, Valencia, 259
Ordinary metastables in field ionization mass spectra, 185
Organic, molecules, connectivity tables for, 126
solvents, solubility of peptide derivatives in, 291
Orientation of molecules near emitter tips in field ionization, 187
Orifices in jet molecular separators, dimensions of, 97
Ornithine, acylation of side chain amino group of, 312
in peptides, 304
diacylated, in peptides, mass spectrometry of, 305
and lysine, confusion between as a result of incomplete methylation of peptides, 316
in peptides, specific fragmentation reactions of, 304
residues, formation of in peptides from arginine residues, 305
Oscillators, effective number of in the Quasi-Equilibrium Theory, 433
harmonic, importance of number of, 389
in ions, effective number of, 396

number of, used in Quasi-Equilibrium
  Theory calculations, 406
Oscilloscope, storage, use of to measure
  dynamic resolving power, 56
Osmium tetroxide, -oxygen-18, use of in
  location of double bonds, 258, 343
  photoionization mass spectrum of, 209
Output, data from high resolution mass
  spectra, forms of, 35
  devices for digital mass spectral data, 4
OV−1 as stationary phase in gas chromatography-mass spectrometry, 94
OV−17 as stationary phase in gas chromatography-mass spectrometry, 94
OV−25 as stationary phase in gas chromatography-mass spectrometry, 95
OV−210 as stationary phase in gas chromatography-mass spectrometry, 94
OV−225 as stationary phase in gas chromatography-mass spectrometry, 95
Overlapping of peaks in high resolution mass
  spectra, 70
Overlap populations in HMO calculations, 398
Oxidation, enzymatic, of lipids, 343
  of polyformaldehyde, 272
  of propene-1-carbon-13 by mercuric ion, 275
  reactions of chromium-VI and manganese-VII,
  rates of, 268
Oxidative cleavage of catechol, 225
Oxides of nitrogen, broad metastable peaks in
  the mass spectra of, 454
Oximes, O-methyl, mass spectra of, 342
Oxirane, protonated, as structure of ethoxy
  cation, 438
Oxygen, molecular, fate of in steroid hydroxylation, 221
  partial pressure of in blood and expired gas, 328
Oxygen−17, use of in biochemical problems, 355
Oxygen−18, exchange of into carbonyl compounds, 220
  during gas chromatography, 220
  incorporation of into, 1,2-dihydro-1,2-di-
  hydroxy naphthalene, 357
  prostaglandin E1, 356
  use of, in biochemical problems, 354, 355
  in biosynthetic studies, 221
  in chemistry and biochemistry, 218
  studies of mass spectral fragmentation
  mechanisms, 256
  in studies of organic reaction mechanisms,
  269
Oxygenase, mono, involvement of in naphthalene metabolism, 357
Oxygenation, enzymatic, of steroid hormones,
  356
Ozonolysis of, methane, 272
  stilbene, 271

Packed columns, use of in gas chromatography-mass spectrometry, 92
D-palmitic acid, 3-hydroxy, identification of
  by mass spectrometry, 342
Palmitic acid in algae, identification of by

mass spectrometry, 346
Palmityl coenzyme A, metabolism of to
  S-palmityl pantetheine, 344
Paper tape, punch, as an output device, 4
  punched, use of as a storage device for
  mass spectral data, 62
Parent ions, doubly charged, fragmentation of to give singly charged
  daughter ions, 450
  energy distribution in, 436
  intensity of relative to daughter and
  metastable ions, 436
  relation of to daughter and metastable
  ions, 421
Parkinson's disease, urinary pink spot in
  patients with, 363
Partial sequence determination of proteins,
  290
α-particles, use of for ionization, 204
Partitioning of initial excitation energy in
  ions, 437
Partition of energy between ion and neutral
  in fragmentation processes, 393
Peak, amplitude as measure of relative
  abundance, 47
  area, determination of on photoplates, 31
  as measure of relative abundance, 47
  centers, determination of on photoplates,
  31
  centroid, calculation of, 70
  intensities in field ionization mass spectra,
  absence of rules governing, 189
  matching mode, determination of static
  resolving power with, 55
  maxima, detection of, 62
  position in mass spectra, definition of, 70
  determination of, 70
  sensing devices, 4
  shapes, in field ionization mass spectra, 185
  obtained during scanning of high resolution
  mass spectra, 56
  widths, dependence upon scan speed, 46
  in exponential scans, 46
  widths, relationship to resolving power in
  exponential scans, 46
Peaks in mass spectra, gaussian model for, 71
  number of digital samples required in, 64
Peas, fatty acids in, 346
Pendant radicals in DENDRAL notation, 131
Penetration of ion in photographic emulsions,
  22
Penicillins, mass spectra of, 261
Penning ionization, 207
  Franck-Condon principle in, 207
  similarity of to Auger effect, 207
Pentachlorocyclohexene isomers, enzymatic
  formation of from lindane in houseflies, 360
n-Pentadecane, identification of in scent
  glands of larvae, 362
n-Pentadeuterobenzoyl peptides, mass spectra
  of, 315
isoPentane, DENDRAL notation for, 130

n-Pentane, DENDRAL notation for, 130
neoPentane, DENDRAL notation for, 130
   as source of the t-butyl ion in chemical
      ionization mass spectrometry, 195
Pentane-2,4-dione, 3-methyl, formation of
      from hexamethyl benzene, 276
Pentanol, field ionization mass spectrum of,
      187
   orientation of hydroxy group in with
      respect to emitter tips, 187
1-Pentanol-3,3-dideutero, loss of water from,
      380
Pentapeptides, sequence determination of by
      mass spectrometry, 315
Pentaphenyl, -cyclopentane diol, mass
      spectrum of, 386
   pyridine, specifically labelled, 385
Pentobarbital, determination of, 332
Pentose, biosynthesis of in E. Coli, 231
   phosphate, formation of from hexoses, 234
Pepsin, porcine, amino acid sequence of, 320
   structure of peptides formed by partial
      hydrolysis of, 313
Peptide, antibiotics, mass spectra of, 309
   bond cleavage during esterification with
      hydrogen chloride in methanol, 313
   bonds, base labile, 313
   derivatives, fragmentation rules for, 292
   mass spectra, dehydration of amino acid
      side chains in, 297
      effect of N- and C-terminal protecting
         groups on, 293
      facilitation of interpretation of with
         amino acid composition data, 312
      identification of molecular ions in, 295
      intermolecular alkylation reactions in, 295
      interpretation of, 313
      loss of side chains of hydroxyamino acids
         in, 297
      metastable ions in, 301
      rearrangement processes in, 293
      simplification of interpretation of, 314
   methyl esters, N-trifluoroacetyl derivatives
      of, 291
   mixtures, analysis of by mass spectrometry,
      303
   phenyl esters, mass spectra of, 294
   sequencing, by chemical ionization mass
         spectrometry, 202
      high resolution mass spectrometry in, 314
      by mass spectrometry, reviews of, 292
   structure determination, from low resolu-
      tion mass spectral data, 314
   use of metastable ions in, 314
Peptides, acylation of amino groups in, 312
   of aspartic acid, α- and β-, mass spectra of,
      299
   chemical ionization mass spectrometry of,
      315
   containing aliphatic amino acids, mass
      spectra of, 291
   containing two pyrimidyl ornithine residues,
      mass spectra of, 307

cyclic, mass spectra of, 307
   and cyclic peptides, differences in behavior
      of on electron impact, 308
   derivatization of for mass spectrometry,
      312
   determination of amino acid composition
      of by mass spectrometry, 312
   exhaustive N-methylation of the amide
      bonds in, 316
   field ionization mass spectrometry of, 315
   formed by partial hydrolysis of proteins,
      structure determination of, 311
   of glutamic acid, α- and γ-, mass spectra of,
      299
   mass spectra of, 290
   metastable transitions in the mass spectra
      of, 442
   permethylation of, 282
   photoionization mass spectrometry of, 315
   reduction of with lithium aluminum, deuteride
      290
      hydride, 290
   sequencing of, by mass spectrometry, 289
   use of deuterium in, 282
   short, gas chromatography-mass spectrome-
      try of, 319
   small, esterification of, 291
   successfully sequenced by mass spectrome-
      try, size of, 315
   thermally labile, problems in mass spectrome-
      try of, 315
   thermal stability of, effect of cysteine on,
      315
   volatility of, effect of N-acyl groups upon,
      315
      factors determining, 311
Peptidolipin−NA, structure determination of
      by mass spectrometry, 310
Perfluoroalkyl phosphonitrilates as internal
      standards, 33
Perfluorocarbon, use of as internal standard,
      32
Perfluorokerosene, base peak of, 72
   high mass region of mass spectrum of, 72
   use of as internal standard, 55
Perfluorotributylamine, use of as internal
      standard, 55
Periodate oxidation of aldosterone, 337
Permanganate, potassium, use of in locating
      double bonds by mass spectrometry,
      343
Permeability of gases through teflon, 99
Permethylation of peptides, 282
Permutations, logical, through DENDRAL
      tree of isomers, 133
Perturbation molecular orbital, theory, 399
   in mass spectral reactions, 398
   treatment of the McLafferty rearrangement,
      402
Pesticides, fate of, use of mass spectrometry
      in study of, 358
   organophosphorus, metabolism of, 360
   toxicity of, 360

Phenobarbital, determination of, 332
Phenol, -1-carbon-13, high resolution mass
    spectrum of, 263
  mass spectral fragmentation of, 262
  substituted, loss of CHO from, 392
2−Phenoxybenzotropone, oxygen−18,
    carbon−13 labelled, mass spectrum
    of, 257
2−Phenoxynaphthalene, loss of carbon
    monoxide from the molecular ion of,
    257
Phenyl, azide-1-carbon-13, mass spectral
    fragmentation of, 263
  benzyl ethers, mass spectra of, 395
  boronate derivatives of dinucleoside phos-
    phates, 349
  esters of peptides, mass spectra of, 294
  isothiocyanate, use of in Edman degrada-
    tion of proteins, 318
  substituents in aromatic systems, influence
    of on mass spectra, 386
Phenylalanine, enzymatic conversion of to
    tyrosine, 357
  hydroxylase, formation of tyrosine by from
    phenyl alanine, 357
  in peptides, specific fragmentation reactions
    of, 299
Phenylated triazines, mass spectra of, 386
9-Phenyl-9,10-dihydrophenanthrene,
    deuterium labelled, mass spectrum of,
    278
2-Phenyl glutarimide, in metabolism of
    Doriden*, 334
Phenylhydantoins, identification of by mass
    spectrometry, 319
2-Phenylisothiazole-nitrogen-15, formation
    of, 274
2-Phenyl-5-thiazolones, identification of by
    mass spectrometry, 319
Phenylthiocarbamoyl derivatives of amino
    acids in Edman degradation of
    peptides, 319
Phenylthiohydantoins from Edman degrada-
    tion, identification of by mass
    spectrometry, 318
Phosphatase, bacterial alkaline, 15
Phosphate hydroxyl groups, trimethylsilyla-
    tion of, 348
Phosphates, problems in mass spectrometry of,
    349
Phosphatidyl serines, bovine, identification
    of diglycerides from, 348
Phosphoglucose isomerase, 245
Phospholipids, function of in human bio-
    chemistry, 346
  mass spectrometry of, 346
  release into biliary ducts of, 337
Phospholypase A, structure determination of
    282
  zymogen of, sequence of heptapeptide
    from, 316
Phosphomannose isomerase, 245
Phosphonitrilates, perfluoroalkyl, as in-
    ternal standards, 33

Photochemically excited molecules, energies
    in, 398
Photochemical reactions, as analogies to mass
    spectral fragmentations, 384
  mass spectrometry in studies of, 269
Photochemistry of, DDT derivatives, 359
  prometryn, 359
Photographic, emulsions, calibration of, 19
  response curve of, 17
  plate, use of in high resolution instruments,
    13
  processes following ion impact, 20
  recording systems, 44
  response, effect of mass and energy on, 23
  techniques in high resolution mass spectrome-
    try, 11
Photoionization, 209
  advantages and disadvantages of, 209
  apparatus for, 209
  intensity of ion beams formed by, 209
  mass spectra, comparison of with electron
    impact mass spectra, 209
  of peptides, 315
  temperature effect in, 209
  range of energies in, 209
Photolysis of formaldehyde, 276
Photolytic and thermolytic processes, selection
    rules derived from, 398
Photomultiplier, use of in the Daly detector,
    424
Photoplate, data acquisition system, schematic
    diagram of, 7
  integrability of, 14
  sensitivity of, 15
  techniques, applications of in organic analysis,
    14
  use of in the analysis of small amounts of
    material, 15
  transport mechanism of, 15
Photoplates, applications of, 14
  collagen in gelatin on, 17
  commercially available, 24
  construction of, 16
  development of, 26
  fixing of, 26
  Kodak SWR, 26
  stopping of, 26
  Tech/Ops, 26
Phthalic acid, 0,0-dideutero, metastable transi-
    tions in the mass spectrum of, 440
  esterification of, 269
  ethanolysis of, 269
Phthalimide, loss of carbon dioxide from
    molecular ion of, 255
  nitrogen−15 labelled, mass spectrum of, 255
N-Phthaloyl, derivatives of peptides, in-
    volatility of, 315
  peptides, mass spectra of, 315
Phthiocerols in human tubercle bacilli, 345
Phytanic acid, enzymatic formation of from
    phytol, 344
  human metabolism of, 344
  hydroxylation of, 344
Phytenic acid, conversion of to phytanic acid

in liver, 344
conversion of to phytanic acid in liver, 344
formation from phytol, 344
isomers of, mass spectra of, 344
Phytol, occurrence in foodstuffs, 344
Phytosphingosine, N-acetyl-1,3,4-tri-0-tri-
    methylsilyl derivative of, mass
    spectrum of, 227
tetra-acetyl, origin of the oxygen atoms of,
    227, 347
Phytyl side chain of chlorophyll, enzymatic
    removal of, 344
Pink spot in urine of Parkinson's disease
    patients, 363
Placental transfer of drugs, 351
Plant, leaves, metals in, 331
sterols, identification of by mass spectrome-
    try, 341
Plants, metals in, 331
sterols in, 341
Plasma formed from methane in chemical
    ionization, 190
Platen used in microdensitometers, 28
Plate reader, data acquisition system for,
    schematic diagram of, 7
Platinum, field emission microscopy of, 183
use of as an emitter tip in field ionization,
    186
Poisoning by insecticides, 334
Polarization forces in field ionization, 184
Polarography in metal analysis, 330
Polish format in line notation, 127
compressed, 128
Pollen, sterols in, 341
Polonium—208, use of as alpha emitter in
    radiation sources, 205
Polyaminoalcohols, conversion of to poly-
    amino chlorides, 291
Polyamino chlorides, formation of from
    polyaminoalcohols, 291
intermolecular reactions of, 291
intramolecular reactions of, 291
reduction of with lithium aluminum hy-
    dride, 291
stability of, 291
derived from peptides, mass spectra of, 290
formation of from peptides by lithium
    aluminum hydride reduction, 290
gas chromatographic separation of, 291
Polyesters as stationary phases in gas
    chromatography-mass spectrometry,
    95
Polyethylene, vapor pressure of, 211
Polyformaldehyde, oxidation of, 272
Polynomials, higher term, use of in time-to-
    mass conversion routines, 73
Polypeptides, determination of the sequence
    of, 84
formation of from proteins by chemical or
    enzymatic cleavage, 289
nonspecific hydrolysis of in structure
    determination, 291
separation of, 289

Polypropylene glycol, use of in gas chromatogra-
    phy-mass spectrometry, 110
Polystyrene, mass spectrum of, 211
Pore, diameter in porous stainless steel separa-
    tors, 101
size in, porous glass separator, 98
silver membrane separators, 99
Porous, glass separator, enrichment factor in, 98
in gas chromatography-mass spectrometry,
    93, 98
pore size in, 98
silver membrane separators, enrichment factor
    of, 100
in gas chromatography-mass spectrometry, 99,
    100
gas flow rates in, 101
pore size in, 99
useful temperature range of, 101
stainless steel separators, cleaning of, 101
efficiency of, 101
enrichment factors of, 101
in gas chromatography-mass spectrometry, 101
pore size in, 101
Positioning of magnet in scanning instruments at
    high resolving power, 55
Position of, metastable peaks, 421
peaks in mass spectra, determination of, 70
ring opening in the mass spectra of cyclic
    peptides and depsipeptides, 308
Potassium, in plasma, 331
-39/potassium-41 ratios, variation of during
    active transport, 358
Potentials, of back-reactions, 392
effects of substituents upon, 389
electrical, increase in by use of metal whiskers,
    182
ionization and appearance, measurement of
    in photoionization sources, 209
Precision of analog-to-digital converters, 63
Precursor ions, distribution of internal energies
    in, 388
in metastable transitions, accuracy of meas-
    urement of mass of, 426
Precursors, of cholesterol, identification of by
    mass spectrometry, 339
of ethoxy cation, 438
of fragment ions, identification of, 441
in metastable transitions, identification of, 426
Prediction of ionization potentials, 390
Predictor, in Heuristic DENDRAL programs,
    147
subroutine in Heuristic DENDRAL program,
    154
Preen gland lipids in water fowl, structure of,
    343
Preexponential factors, extraction of from
    chemical ionization mass spectra, 409
Pregn-5-en-3,20-diol, trimethylsilyl ether, mass
    spectrum of, 337
Pregnane, diol, trimethylsilyl ether, mass
    spectrum of, 338
-3-ol-20-one, mass spectrum of, 339
-3,20-diol, mass spectrum of, 339

-3,20-diols, formation of by enzymatic dehydroxylation of sterols, 254
-3,20,21-triols, deuterium labelled, 254
identification of in human feces, 338
metabolism of, 254
17-Pregnene—C_{17},C_{20}-lyase in testicular
tissue, 223
Pregnenolone, acetate, enzymatic formation
of from 20,22-dihydroxycholesterol,
356
biosynthesis of, 223
trimethylsilyl ether, mass spectrum of, 336
Preliminary inference maker in Heuristic
DENDRAL programs, 146—148
Presentation of data in fragmentation maps,
443
Pressure, conditions of in gas chromatography-
mass spectrometry, 93
dependence on, qf collision-induced
metastable transitions, 429
of normal metastable transitions, 429
reduction, by cooling, in gas chromatography-mass spectrometry, 92
of sample, monitoring of in chemical
ionization, 192
variation of in gas chromatography-mass
spectrometry, 108
Pressures in jet molecular separators, 97
Primary, events in chemical ionization mass
spectrometry, 198, 199
ions of methane, 190
photographic processes in photoplates, 22
processes in the fragmentation of
N-acylpeptide methyl ester molecular
ions, 292
Prismanes as intermediates during the
isomerization of carbonium ions, 439
Pristane, identification of by mass spectrometry, 342
Problem-solving capabilities of computers, 122
Processing of high resolution mass spectral
data, 67
Product ions, excited states of, 393
secondary decompositions of, 388
stabilization of, 388
importance of in fragmentation reactions,
389
Product neutrals, stabilization of, 388
Profiles of, constituents of urine, 351
peaks, extraction of from raw data, 69
number of ions required for, 52
obtained by scanning of high resolution
mass spectra, 51
steroids in urine, 338
Progesterone, 221
-17-α-hydroxylase in testicular tissue, 223
biosynthesis of, 223
conversion to testosterone acetate in
cladosporium resinae, 223
enzymatic conversion of, to deoxycorticosterone, 356
to testosterone, 356
human metabolism of, 339

hydroxylation of by testicular enzymes,
356
metabolites of in urine, plasm and bile, 337
Programming, temperature, use of flow
controllers in, 93
Proline in peptides, specific fragmentation
reactions of, 297
Promethazine, chemical ionization mass
spectrum of, 201
Prometryn, photochemistry of, 359
Propane, DENDRAL notation for, 130
-d_8, loss of methane from the molecular
ion of, 440
energy distribution of molecular ions in,
392
Propanes, deuterium-labelled, metastable
transitions in the mass spectra of,
440
Propene, carbon—13 labelled, mass spectrum
of, 266
1,3-diphenyl, carbon—13 labelled, mass
spectrum of, 266
expulsion of, from leucine and isoleucine
in peptide mass spectra, 295
from valine in peptide mass spectra, 295
loss from 3-heptanone, 440
Propionaldehyde, formation of from
propene-1-carbon-13, 275
Propionic acid, 3-nitro, biosynthesis of,
236, 355
isoPropyl butyl ethers, mass spectra of, 381
Propylene, proton affinity of, 409
di-n-Propyl ketone, field ionization mass
spectrum of, 187
orientation of carbonyl group in with
respect to emitter tips, 187
Propyl radical, loss of valine side chain as
in peptide mass spectra, 295
Prostaglandin E1, biosynthesis of, 224, 356
incorporation of molecular oxygen into,
224
Prostaglandins, biosynthesis of, 345
metabolism of, 345
Protecting groups, N- and C-terminal, effect
of on peptide mass spectra, 293
for the terminal amino group of peptides,
315
Proteins, chemical and enzymatic cleavage
of, 289
degradation of by enzymes prior to mass
spectrometry, 319
enzymatic cleavage of, 289
partial hydrolysis of, structure determination of peptides resulting from, 311
sequence of amino acids in, determination
of by mass spectrometry, 289
Proton, affinities of ammonia, isobutylene
and propylene, 409
transfer, in chemical ionization, 192
during field ionization of carbohydrates,
188
Protonated dimers, equilibria involving in
chemical ionization mass spectra, 409

in field ionization mass spectra of alcohols, 189

Protonation in, field ionization mass spectra of alcohols, 189

helium chemical ionization mass spectra, 193

Protons, formed by field emission of hydrogen gas, 182

transfer of from ions to molecules, 192

from Van de Graaf generator, ionization by, 205

Pruning of DENDRAL trees, 143

Pseudo image potential, in field ionization, 183
of metals, 184

Pulmonary exchange, measurement of, 328

Pulsed systems, use of laser beams in, 206

Pumping of chemical ionization mass spectrometer, 192

Punched cards as output devices, 4

Pyranocoumarone derivatives, carbon–13 labelled, mass spectra of, 266

Pyrazines, fragmentation of, 385

Pyridine, deuterated, mass spectrum of, 384 specifically labelled, 385

Pyridines, 1,4-dihydro-4-deutero-N-alkyl-3-cyano, aromatization of in mass spectrometry, 383

Pyridyl analogs of stilbene, mass spectra of, 265

Pyrimidyl ion in the mass spectra of peptides containing pyrimidyl ornithine residues, 306

Pyrimidyl ornithine, conversion of arginine residues to in peptides, 305, 313
ornithine residues, in peptides, characteristic ions formed from, 305
thermal stability of, 305
residues at C-terminus of peptides, mass spectra of, 306

Pyrolysis in jet molecular separator, 98

2–Pyrones, metastable peaks in the mass spectra of, 455

Pyrrole, N-methyl, mass spectrum of, 263

Pyrrolidine ring of proline, fragmentation of in peptide mass spectra, 297

Pyrroline, identification of in desert locust, 362

Pyruvate kinase, 245

QF–1 as stationary phase in gas chromatography-mass spectrometry, 94

Q2 photoplates, 19

Quadratic, scan function of high resolution mass spectrometer, 73
scanning of deflection instruments, 47

Quadrupole mass spectrometers, use of in gas chromatography-mass spectrometry, 109

Quantitation in spark source mass spectrometers, accuracy of, 331

Quasi-equilibrium theory, 381, 388, 403, 407
application of, to complex molecules, 433
to complex organic molecules, 406
large organic molecules, 404
to metastable transitions, 431

continuous distribution of rate constants in, 433

effective number of oscillators in, 433

frequency factors in, 433

simplified form of, 432

use of to define mass spectral reactions, 403

Quasimolecular ions in chemical ionization mass spectra, 189

Quinazolone, identification of glomerin as, 361

Quinine, chemical ionization mass spectrum of, 195
electron impact mass spectrum of, 195

isoQuinoline, 1-methyl, mass spectrum of, 263

Racemization, rates of, studies of using oxygen–18, 269

Radiation-induced ionization of acetylene, 205
ionization by, 204
as means of loss of energy of ions, 405

Radicals, comparison of in DENDRAL notation, 141
composition of, handling of by DENDRAL notation, 139

Radioactive, carbon, mass spectrometric detection of, 220
sources, efficiency of, 205
substrates, use of in mass spectrometric studies of metabolism, 350

Radioimmunoassay of steroid hormones, 340

Ramp voltage generator, use of in studying metastable transitions, 426

Random, errors in mass measurement, 75
sampling mode of multiplexing, 58

Randomization, of carbons in aromatic ions, 439
of deuterium in, ions as a function of the lifetime of the ions, 441
organic ions, 439
in mass spectra of phenylated triazine, 386

Raney nickel, use of in desulfurization of peptides, 317

Rare earths, use of to lower work function of metals, 181

Rate, constants, calculation of from appearance potentials, errors in, 436
for decomposition of ions, 403
distribution of in the Quasi-Equilibrium Theory, 433
in field ionization mass spectra, 186
for mass spectral reactions, 403
calculation of from Quasi-Equilibrium Theory, 403
in metastable transitions, 432
for reactions of ions, 420
of mass spectral reactions, dependence of on frequency factors, 395
theory applied to mass spectral processes, 389

Rates of, decomposition of ions, 403
mass spectral reactions, parameters affecting, 389

Ratio between accelerating voltage and ESA voltage in double focussing instruments, 46
Rationalization of fragmentation of ions, 374

δ Rays, involvement of in radiation sources, 205
Reaction, mechanisms, evaluation of from substituent effects, 396
  in organic chemistry, use of stable isotopes in studies of, 268
  pathways, calculation of, 398
Reactions of, doubly charged ions, 399
  ions, calculations of, 398
Reactivities of, ions, 407
  metastable ions, 407
Reactivity of ions, control of by substituents, 394
Reagent, gases for chemical ionization, comparison of, 195
  gas in chemical ionization, compound behaving as its own, 189
  ions formed from methane in chemical ionization, 190
Real time data, acquisition, 5, 8
  processing, 8
Rearrangement, entropy factors in, 405
  of hydrogen atoms in the mass spectra of ketones, 147
  in lysine and ornithine residues in peptides, 304
  in the mass spectra of peptides containing pyrimidyl ornithine residues, 306
  processes, entropy requirements in, 188
  in field ionization, 188
  in the mass spectra of peptides, 293
  in methionine-containing peptides, 302
  in peptides containing aspartic and glutamic acids, 298
  in the side chains of hydroxyamino acids in peptide mass spectra, 297
Rearrangements, absence of in collision processes, 408
  effect on of ionizing voltage, 406
  frequency factors in, best fit values for, 405
  of hydrogen in aliphatic ethers and amines, 380
  of ions formed by electron bombardment, 180
  in peptides containing aromatic or heterocyclic amino acids, 300
Recording, devices, analog, 61
  digital, 61
  electrical, of mass spectra, 44
  photographic, of mass spectra, 44
  systems, dynamic range of, 62
Reduction of, N-acyl peptides with lithium aluminum hydride or deuteride, 291
  data from photoplates, 29
  of hydroxyl groups in peptides during sequencing, 291
  peptides with lithium aluminum, deuteride, 290

hydride, 290
  polyamino chlorides with lithium aluminum hydride, 291
Redundancy checks in fragmentation maps, 442
Reference compounds, use of in high resolution mass spectral data handling, 71
Reflecting mirrors used in microdensitometers, 27
Refsum's disease, 344
ρ-σ Relationships in mass spectral processes, 388
Relative, intensities of parent, daughter and metastable ions, 436
  metastable intensities, effect on of ion source conditions, 439
Release of translational energy in fragmentation processes, 450
Reliability of methods for measuring appearance potentials, 391
Removal of solvent in electrojets, 210
Repetitive scanning of magnetic fields in gas chromatography-mass spectrometry, 116
Reproducibility of spark source mass spectrometers, 331
Residence times of, ions in source, 391, 420
  difference between field ionization and electron impact, 188
  solutes in silicon rubber membranes, 106
Resolution of, analog-to-digital converters, 57, 63
  field emission microscopy, 183
  overlapping lines on photoplates, 29
Resolving power, dynamic, 47
  factors governing choice of, 55
  in metastable ion mass spectra, 426
  relationship to peak width in exponential scans, 46
  required for separation of selected doublets, 219
  static, 46
Resonance, steric inhibition of, effect of on mass spectra, 397
Respiratory gases, analysis of by mass spectrometry, 328
Response curve of photographic emulsions, 17
Retarding grids, use of, to detect metastable transitions in time-of-flight spectrometers, 430
  in studying metastable transitions, 425
Retro-Diels-Alder reactions of 4-vinylcyclo-hexene, 402
Reviews of, chemical ionization mass spectrometry, 192, 408
  energetic aspects of metastable transitions, 454
  ion cyclotron resonance spectroscopy, 409
  mass spectrometry of peptides, 290
  peptide sequencing of mass spectrometry, 292

Ribonucleotide reductase, deoxyadenosyl-cobalamin-dependent, 234
Ricinine, biosynthesis of, 240, 355
Ring, of cyclic peptides and depsipeptides, cleavage of under electron impact, 308
    size in cyclic depsipeptides, effect of on fragmentation, 309
RNA, transfer, 15
Rotation, internal, effect of upon mass spectral reaction rates, 389
Roundworm, hydroxylated fatty acids in, 346
Rubber membrane interface in gas chromatography-mass spectrometry, 94
Rubidium, role in caries formation, 331
Rules governing decomposition of ions in field ionization, 187

Salicylaldehyde, identification of in defense secretion of larva, 362
Sample, preparation in analysis of trace metals, 331
    pressure, monitoring of in chemical ionization, 192
        variation of in gas chromatography-mass spectrometry, 108
Sample-and-hold amplifiers, 58
    use of in scanning of high resolution mass spectra, 58
Sampling, frequencies of A/D conversion in gas chromatography-mass spectrometry, 116
    rates of analog-to-digital converters, 63
Sandoptal*, chemical ionization mass spectrum of, 201
    electron impact mass spectrum of, 201
Satellite ions in, chemical ionization mass spectra, 195
    mass spectra of pyrimidyl ornithine-containing peptides, 306
Satellite peaks due to excess translational energy in fragments, 447
Scaling, automatic, of data by digital computer, 56
Scan function, deviations from ideality of, 47
    generator, 47
Scanning, of accelerating voltage, 46
    continuously cycled, in gas chromatography-mass spectrometry, 114
    of high resolution mass spectra, 45
    data acquisition during, 57
    linear, of magnetic instruments, 47
    quadratic, of deflection instruments, 47
    speed, importance of in gas chromatography-mass spectrometry, 108
Scan speed, variation of peak width with, 46
Scent gland of larva, components of, 362
SCF—LCAO calculations in mass spectral reactions, 398
Scintillator, glass, use of in the Daly detector, 424
Scrambling, in aromatic systems, rate of, compared to fragmentation, 386

of carbons in organic ions, 439
data, information provided by on structure of ions, 387
of deuterium in, ions as a function of the lifetime of the ions, 441
    organic compounds under electron impact, 384
    organic ions, 439
experiment, various interpretations permitted by, 387
of label in 1,4-dideuterobenzene, 439
of labels in fragmentation of aromatic systems, 385
studies, use of metastable ions in, 387
use of p-fluorophenyl label in studies of, 386
Searching of mass spectral libraries by computer, 82
SE—30 as stationary phase in gas chromatography-mass spectrometry, 94
Sebacic acid, formation of from sphing—4, 14-dienine, 347
Secobarbital, determination of, 333
Secondary, decomposition of product ions, 388
    events in chemical ionization, 199
    photographic processes in photoplates, 22
Selection rules, derived from thermolytic and photolytic processes, 398
    in predicting stereochemistry of electrocyclic processes, 400
Selenium, role in caries formation, 331
Semilogarithmic plot method of measuring ionization potentials, 391
Sensing of peaks, by digital comparator circuits, 4
    devices for, 4
Sensitivity, of amino acid analyzers, 311
    of chemical ionization source used in electron impact mode, 203
    of detection systems, parameters affecting, 53
    of mass spectrometers, dependence upon, accelerating voltage, 46
        frequency response, 48
        variation of with resolving power, 51
    of spark source mass spectrometers, 331
    specks of silver sulfide in photoplates, 21
Separator, combined membrane and porous silver, 103
    jet molecular, in gas chromatography-mass spectrometry, 98
    Llewellyn-Littlejohn, 103
    porous, glass in gas chromatography-mass spectrometry, 98
    stainless steel, 101
    silicon coated silver membrane, 103
    teflon capillary, in gas chromatography-mass spectrometry, 99
    variable, conductance, 104
        slit, 104
Separators, calculation of efficiency of, 95

* Registered.

comparison of, 104
in gas chromatography-mass spectrometry,
    efficiency of, 95, 96
  enrichment of, 95, 96
  reviews of, 105
  yield of, 95, 96
Sequence of, amino acids in cyclodepsi-
    peptides, determination of by mass
    spectrometry, 310
  polypeptides, determination of, 84
Sequencing, direct, of peptides by mass
    spectrometry, 291
  indirect, of peptides by mass spectrometry,
    291
  of peptides by mass spectrometry, 289
Serine, in peptides, specific fragmentation
    reactions of, 297
  side chain of, reduction of during
    sequencing of polypeptides, 291
  and threonine, confusion between
    following peptide methylation,
    316
Serratomolide, mass spectrum of, 309
Setting up of mass spectrometer for
    scanning of high resolution mass
    spectra, 55
Shapes of metastable peaks, 443, 447
  calculations of, 447
  difference in different instruments, 448
  due to processes occurring before the
    electric sector, 450
  factors affecting, 421
Side chain of, methionine, loss of in peptide
    mass spectra, 301
  valine, loss of in peptide mass spectra,
    295
Side chains of, amino acid residues, effect
    of on fragmentation patterns of
    peptides, 295
  hydroxyamino acids, loss of in peptide
    mass spectra, 297
  hydroxy and amino acids of cyclic
    peptides and depsipeptides, expul-
    sion of, 309
Side products formed during Edman
    degradation of proteins, 318
Side reactions during acylation of peptides
    with acetic anhydride in methanol,
    312
Signals from mass spectrometers, attenua-
    tion of, 61
  logarithmic amplification of, 61
Signal-to-noise ratio in detection systems,
    50
Silanization of porous, glass separators,
    106
Signals from mass spectrometers, attenuation
    of, 61
  logarithmic amplification of, 61
Signal-to-noise ratio in detection systems, 50
Silanization of porous, glass separators, 106
  stainless steel separators, 101

of removal of an electron in electron
    impact ionization, 394
  specificity of McLafferty rearrangement,
    148
Sites of hydrogen abstraction in elimination
    reactions, 381
β–Sitosterol, trimethylsilyl ether, mass
    spectrum of, 335
Six-membered transition states, 405
  in the McLafferty rearrangement, 378
Size of peptides that have been successfully
    sequenced by mass spectrometry, 315
Skimmers, use of in electrojets, 210
Slaframine, biosynthesis of, 237
Slit, separator, 104
  widths, variation of resolving power with,
    55
β-slit, in the CEC 21–110B, 448
  dimensions of, 451
  discrimination at, 452
  narrow, use of to observe fine structure in
    metastable peaks, 453
  use of in studies of metastable transitions,
    428
  width, effect of on metastable peak shape,
    452
Slow mass spectral reactions, implication of
    ground electronic states in, 399
  methylated thioureas, 394
Silastic polymers, use of as membranes in
    blood gas analysis, 329
Silicon, use of in studies of photoplate ssensitiv-
    ity, 23
Silicon coated silver membrane separator, 103
  silicon thickness in, 103
Silicon rubber membrane, residence time of
    solutes in, 106
  separators in gas chromatography-mass
    spectrometry, 102
Silver, bromide, use of in photoplate emul-
    sions, 17
  as field ionization emitter, 185
  oxide/methyl iodide, use of in exhaustive
    N-methylation of peptides, 316
  oxide, side reactions of with various amino
    acid residues in peptides, 316
  sulfide sensitivity specks, 21
Silyl–8, use of to deactivate metal separators,
    107
Simazine, metabolism of in soil, 358
Simplified form of the Quasi-Equilibrium
    Theory, 432
Single ion, detection of, 50
Singlet or doublet states, reactions occurring
    from, 399
Singly charged ions, formation of from
    doubly charged ions, 444
Site, of ionization in, metal chelates, 394
Small computers, use of for mass spectral
    data processing, 67
Smoothing, digital, of mass spectral data,
    71

Snake venom diesterase, 15
Sodium pregnane diol glucuronidate,
    identification of by mass spectrome-
    try, 339
Soil metabolism of insecticides and pesticides,
    358
Solid state digitizer, 6
Solubility of peptide derivatives in organic
    solvents, 291
Solvent, removal of in electrojets, 210
Source for chemical ionization, dimensions
    of, 191
Sources, radiation, 204
Soybeans, volatile components of, 363
Space flight, manned, analysis of respiratory
    gases during, 330
Spark sources, use of in analysis for metals
    in biological materials, 330
Specific, compound identification by
    computer, 80
  ion detectors, use of in gas chromatography-
    mass spectrometry, 114
  patterns of fragmentation of peptides, 313
Spectral data for the ionization of benzene,
    400
Spectro-Sadic digitizer, 4
Speed of, emulsions, 18
  scans, variation of peak width with, 46
Spermine, identification of in desert locust,
    362
Sphinga-4,14-dienine, conversion of to
    sebacid acid, 347
  isolation of from sphingomyelin, 347
Sphinganine, isolation of from sphingomyelin,
    347
Sphing-4-enine, isolation of from
    sphingomyelin, 347
Sphingomyelin, analysis of, 347
  in human plasma, ceramides in, 348
  methanolysis of, 347
Sphinogosine, dihydro, mass spectrum of, 227
  isolation of from sphingomyelin, 347
  mass spectrum of, 227
Splitters, use of in gas chromatography-mass
    spectrometry, 108
Splitting of gas chromatograph output, 92
Sporidesmolide−I, mass spectrum of, 309
Sporidesmolide−III, mass spectrum of, 310
Squalene, biosynthesis of, 240, 242
  enzymatic cyclization of, 224
Squalene-2,3-oxide-oxygen-18, cyclization
    of, 224
  enzymatic conversion of to lanosterol,
    224
  synthesis of, 224
Stability, of derivatives of methionine in
    peptide mass spectra, 302
  of ions, effect of on mass spectra, 389
  of magnet scanning circuits, 47
  of polyamino chlorides, derived from
    peptides, 291
  thermal, lack of in free peptides, 290

Stabilization of, product ions, 388
  importance of in fragmentation reactions,
    389
  product neutrals, 388
Stable isotopes, use of in, biochemistry, 218
  chemistry, 218
  metabolism studies, 249
  studies of organic reaction mechanisms, 268
  study of mass spectral fragmentation
    mechanisms, 255
Standardization of mass scale in high resolu-
    tion mass spectra by computer, 72
Stanford IBM 360/67, 144
Stanford University artificial intelligence
    project, 121
Staphylomycin−S, structure determination of
    by mass spectrometry, 310
Static resolving power, 46, 49, 55
  measurement of, 55
Statistical theory of, dissociations, 393
  mass spectrometry, 186
Statistics, ion, in exponential scanning, 47
Steady-state, approximations in mass spectral
    processes, 387
  reaction hypothesis in mass spectral
    processes, 388
Stearic acid, 13-hydroxy, enzymatic formation,
    of from linoleic acid, 343
  microbiological conversion of to oleic acid,
    244
Stendomycin, amino acid sequence of, 316
Stereochemical, course of electrocyclic
    reactions, 399
  specificity in mass spectral elimination
    reactions, 382
Stereochemistry, of electrocyclic processes,
    selection rules governing, 400
  in the mass spectrum of 4-vinylcyclohexene,
    402
Stereoisomerism, handling of by DENDRAL
    notation, 141
Stereospecificity of elimination reactions in
    mass spectrometry, 382
Steric inhibition of resonance, effect of on
    mass spectra, 397
Steroid, acetates, field ionization mass
    spectra of, 189
  biosynthesis, use of oxygen−18 in studies
    of, 221
  hormones, enzymatic oxygenation of, 356
Steroidal, acetates, chemical ionization
    mass spectra of, 189
  drugs, metabolites of, 339
  metabolites, study of with gas chromatogra-
    phy-mass spectrometry, 110
Steroids, analysis of, by coupled gas
    chromatography-mass spectrometry,
    342
  by mass spectrometry, 334
  assay procedures for, based upon mass
    spectrometry, 340
  21-dehydroxylation of, 254

enzymatic hydroxylation of, 221
in infant feces, 335
trimethylsilylation of, 335
in urine, profiles of, 338
use of chemical ionization with, 201
Sterols, from animal sources, identification
    of by mass spectrometry, 341
biosynthesis of, 242
in rat feces, identification of by mass
    spectrometry, 341
Stigmasta-7,24(28)-dien-3-ol, identification of,
    341
Stigmasta-22-en-3-ol, identification of by
    mass spectrometry, 342
Stigmast-22-en-3-β-acetate, mass spectrum
    of, 245
Stigmast-7-en-3-ol, identification of, 341
Stigmast-22-en-3-β-ol, deuterium labelled,
    mass spectrum of, 245
Stigmasterol, biosynthesis of, 245
    trimethylsilyl ether, mass spectrum of,
        335
Stilbene, dihydro, carbon−13 labelled, mass
    spectrum of, 261
ozonolysis of, 271
pyridyl analogs of, mass spectra of, 265
radical cation, cyclization of, 402
Stilbenes, loss of methyl from molecular ions
    of, 265
Stopping of photoplates, 26
Storage, devices, bulk, use of in processing of
    high resolution mass spectra, 69
oscilloscope, use of to measure resolving
    power, 56
requirements of computers used in gas
    chromatography-mass spectrometry,
    116
Storing of mass spectral libraries by computer,
    82
Streptomycin, biosynthesis of, 355
Structural, formulas, conversion of to
    DENDRAL notation, 141, 142
derivation of from DENDRAL notations,
    141
isomerism, topological principles of, 143
isomers, differentiation between in
    metastable ion spectra, 408
Structure, determination, by computers,
    124
    logic of, 124
    by mass spectrometry, 279
    potential of high resolution mass
        spectrometry in, 7
    of proteins, mass spectrometry in, 290
    use of, collision-induced metastable
        transitions in, 430
        fragmentation maps in, 442
        oxygen−18 in, 279
of doubly charged molecular ion of
    benzene, 446
of ethoxy cation, 409
generation, by a Heuristic DENDRAL
    program, 144
    in Heuristic DENDRAL program, 144, 167
    constraints upon, 144

generator, in Heuristic DENDRAL program,
    146, 147
of ions, from aromatic compounds, 397
    characterization of, 437
    information provided on by scrambling
        data, 387
of molecular ions from pentaphenyl-
    cyclopentane diol, 386
Structures of, activated complexes for frag-
    mentation, 387
ions, use of metastable ions in determination
    of, 407
Studies of ion fragmentation mechanisms, 374
Styrene, gas chromatography-mass spectrome-
    try, 92
Styrene oxide, β−methyl, hydrolysis of, 270
Substituent effects, on appearance potentials,
    390, 391
on density of states in molecular ions, 395
on ion intensities, 388
on ionization potentials, 389
on the loss of NO from nitrobenzenes, 393
on the McLafferty rearrangement, 393
    mathematical approaches to, 392
on metastable peak shapes, 455
in methylated thioureas, 394
on molecular orbital energy levels, 394
upon number of degrees of freedom, 396
use of in evaluation of reaction mechanisms,
    396
Substituents, control of ion reactivities by, 394
effect of, on density of energy states in
    ions, 393
    on frequency factors, 395
    on mass spectral, fragmentation reactions,
        389
    reactions, 389
introduction of competitive reactions by,
    388
S−Substituted derivatives of cysteine, effect
    of on thermal stability of peptides, 315
Substitution on aromatic rings, rate of, 269
Subtraction of low resolution mass spectra, 219
Succinate, dimethyl, mass spectrum of, 241
Succinic anhydride, mass spectrum of, 241
Sulfanilamide, carbon−13 labelled, 262
Sulfhydryl groups, protection of in peptide
    chemistry, 303
Sulfur-sulfur bond cleavage in cystine-contain-
    ing peptides, 302
Sums of mass spectra, cystine-containing
    peptide mass spectra as, 303
Superatoms, in Heuristic DENDRAL notation,
    159
use of in DENDRAL programs, 169
validation in Heuristic DENDRAL programs,
    163
Super-conducting magnets, possible use of in
    mass spectrometers, 211, 364
Supplementary nature of data from electron
    impact and chemical ionization mass
    spectra, 202
Suppression of normal mass spectra with the
    Daly detector, 425

Surfactin, structure determination of by
mass spectrometry, 311
Symmetry, arguments of Woodward-
Hoffmann rules, 399
of peaks in high resolution mass spectra,
63
Synthesis, organic, computerized design of,
124

Tailing of peaks in field ionization mass
spectra, 185
Tandem mass spectrometers, 190, 191
Tape, flutter, problems caused by, 62
recorder, analog, for recording of mass
spectra, 61
Tartaric acid, oxygen—18 labelled, 237
Tea, identification of 3,4-dimethoxy-β-
phenethylamine in, 363
Tech/Ops photoplates, 26
Teflon, permeability of towards gases, 99
use of as membrane in blood gas analysis,
329
Teflon capillary separator, enrichment
factor of, 99
in gas chromatography-mass spectrometry,
99
variation of enrichment factor of with
temperature, 99
Teleducer, use of in digitization of mass
spectra, 3
Teletypewriter, as an output device, 4
shortcomings of in structural formula
work, 142
Temperature, dependence of chemical
ionization mass spectra on, 195
effect of, on chemical ionization mass
spectra, 409
on photoionization mass spectra, 209
programming, use of flow controllers in,
93
source, effect of on relative metastable
intensities, 439
protecting group, effect of on peptide
mass spectra, 293
N—Terminal, isoleucine in peptides, specific
ions formed from, 296
leucine in peptides, specific ions formed
from, 296
protecting group, effect of on peptide
mass spectra
C—Terminal amino acids in peptides, ions
characteristic of in peptide mass
spectra, 294
Terminating twig of DENDRAL tree, 139
Testicular tissue, steroid biosynthesis in, 223
Testosterone, 221
acetate, biosynthesis of from progesterone,
223
enzymatic formation of from progesterone,
356
Tetra-acetylphytosphingosine, origin of the
oxygen atoms of, 227
Tetra-acetyl sphingosine, isolation from
yeast, 347

1,1,3,3—Tetraalkoxypropane, reaction of
with guanidino groups of arginine
residues in peptides, 305
Tetradecasphing-4-enine, identification of by
mass spectrometry, 347
1,1,3,3—Tetramethoxypropane, derivatization of arginine in peptides with, 313
2,6,10,14—Tetramethyl-n-pentadecane, identification of by mass spectrometry, 342
Tetraphenylfuran, mass spectrum of, 385
Tetraphenylthiophene, mass spectrum of, 384
Thallium, mass spectrometric analysis of, 332
toxicity of, 332
Theoretically significant peaks in mass spectra,
174
Theoretical mass spectra, 397
derivation of with the Quasi—Equilibrium
Theory, 404
Theoretical models of ions, 397
Thermal, electrocyclic reactions, aromatic
states in, 399
equilibration in chemical ionization mass
spectrometry, 195
instability of peptides, effect of specific
amino acids upon, 315
loss of ammonia from primary amide side
chains in peptides, 298
nature of amide bond dehydration in the
mass spectra of peptides, 294
Thermally labile peptides, problems in mass
spectrometry of, 315
Thermochemical, evidence for the ionization
of benzene, 400
information derived from ion appearance
potentials, 393
Thermodynamic data, measurement of in
chemical ionization mass spectrometry,
195
Thermolytic and photolytic processes,
selection rules derived from, 398
Thin layer chromatography, use of in isolation
of drug metabolites, 351
Thiocarbonates, oxidation of by chlorine, 273
Thioethers, identification of from mass spectra
by Heuristic DENDRAL programs, 146
structure determination of from mass spectra
by Heuristic DENDRAL programs, 162
success of Heuristic DENDRAL programs
with, 175
Thiol esters, McLafferty rearrangements of,
383
Thiols, comparison of field and electron
ionization mass spectra of, 188
identification of from mass spectra by
Heuristic DENDRAL programs, 146
structure determination of from mass
spectra by Heuristic DENDRAL
programs, 162
success of Heuristic DENDRAL programs
with, 175
Thionyl chloride, in methanol as esterification
agent for peptide, 312
use of to convert polyaminoalcohols to

polyamino chlorides, 291
Thiophene, deuterated, mass spectrum of, 384
  fragmentation and scrambling in, 386
  2-n-hexyl, carbon–13 labelled, mass spectrum of, 264
  molecular ions, valence bond isomer formation in, 385
  specifically labelled, 384
Thiophenol, -1-carbon-13, high resolution mass spectrum of, 263
  mass spectral fragmentation of, 262
Thioureas, methylated, substituent effects in, 394
Threonine, in peptides, specific fragmentation reactions of, 297
  and serine, confusion between following peptide methylation, 316
Thresholding, of mass spectrometer signals, 69
  use of digital comparator for, 59
Threshold mode of operation in scanning of high resolution mass spectra, 59
Thrombogenicity, problems caused by in direct sampling of gases dissolved in blood, 329
Thymine, trimethylsilyl derivative of, mass spectrum of, 15
Thyroid stimulating hormone releasing factor, 15
Time, of appearance of a peak, relationship of to the mass of the ion, 73
  constant of scanning instruments, 48
  dilation factor, 65
  per mass unit, relationship to scan speed, 46
  scale, compression of by use of analog tape recording, 61
  sharing, use of in computers, 86
Time-averaged nuclear magnetic resonance spectroscopy, 350
Time-averaging of high resolution mass spectra, 75
Time-of-flight mass spectrometers, metastable transitions in, 430
  use of, in gas chromatography-mass spectrometry, 108
  with laser source, 206
Time-to-mass conversion techniques, 73
  in mass spectra from scanning instruments, 73
Tissues, metals in, 331
Tobacco smoke, analysis of by gas chromatography-mass spectrometry, 110
Tocopherol, gas chromatography-mass spectrometry of, 345
α–Tocopherol, isolation of from heart muscle, 342
Tocopherols, dimerization and trimerization of in vitro, 345
Toluene, carbon–13, deuterium labelled, 262
  gas chromatography-mass spectrometry of, 92

loss of hydrogen atom from the molecular ion of, 428
mass spectrum of, 429
metastable transitions in the mass spectrum of, 426, 429
molecular ions, 395
problem, 261, 378, 395, 437
  in field ionization and electron impact, 188
Toluene-α-carbon-13, mass spectrum of, 379
Toluene-α, 1-carbon-13, mass spectrum of, 379
Tomatidine, fragmentation map of, 443
Tomatoes, volatile components of, 363
Topographical element maps, 39, 377
Topological, centroid in line notation of structures, 128
  graph theory, 128
  principles of isomerism, 143
  theory of tree-graphs, 125
Total ion current, monitor, location of for gas chromatography-mass spectrometry, 108
  use of in gas chromatography-mass spectrometry, 108
  monitoring in chemical ionization, 192
Toxicity of pesticides, 360
Trace metals in biological systems, analysis for, 330
Tracers, stable isotopes as, 377
Trachea, use of in analysis of expired gases, 328
Transfer RNA, 15
Transition states, four-membered, 405
  information on provided by scrambling data, 387
  in the McLafferty rearrangement, 378
  in mass spectral elimination reactions, 381
  relation of to ground states in mass spectral reactions, 403
  six-membered, 405
Translational energy, arising from internal energy, during fragmentation reactions, 445
  released during fragmentation processes, 446
  effect of on mass spectra, 423
Transmission of, ions, by the electric sector, requirements for, 425
  in mass spectrometer, 53
  solute through silicon rubber membranes, 102
Transport, active, variation of K39/K41 ratios during, 358
  mechanism of photoplate, 15
Tree, DENDRAL, of isomers, 133
Tree graph, centroid, use of as origin in structure graph, 129
  unique center of, 129
Tree graphs, as expressions of organic structures, 129
  topological theory of, 125
  truncation of by valence restrictions, 137
TRF, structure elucidation of, 15

n-Triacontane, electron impact mass
    spectrum of, 188
Triazines, fragmentation of, 385
Tributylamine, perfluoro, use of as internal
    standard, 55
1,1,1-Trichloro-2,2-bis(p-chlorophenyl)-2-
    deuteroethane, metabolism of in
    soils, 359
Tricyclooctane, formation of from
    cyclooctadien, 276
n-Tridecane, identification of in scent
    glands of larvae, 362
2,4,6-Trideuterobenzonitrile, loss of
    hydrogen or deuterium cyanide
    from the molecular ion of, 440
Trideuteromethionine, use in biosynthetic
    studies, 245, 246
Trideuteromethyl esters of N-acyl peptides,
    mass spectra of, 294
N-Trifluoroacetyl, glycine methyl ester,
    mass spectral fragmentation of, 267
    peptide esters, gas chromatography of, 319
    peptide methyl esters, mass spectra of, 291
N-Trifluoroacetylation of peptide methyl
    esters for mass spectrometry, 291
Trifluoroacetylpeptide methyl esters, mass
    spectra of, 294
Trifluoroacetylpeptides, intensity of aldimine
    ions in, 293
    with N-terminal leucine or isoleucine, mass
        spectra of, 296
Trimethoxybiphenyl, mass spectrum of, 257
Trimethyl, tris(trideutero)methyl benzene,
    formation of, 276
Trimethylchlorosilane, use of to silanize
    porous glass separators, 106
bis-Trimethylsiloxy fatty acid esters, mass
    spectra of, 258
Trimethylsilyl, ethers as volatile derivatives,
    210
    groups, deuterated, use of in fragmentation
        studies, 258
bis-Trimethylsilylacetamide, use of in gas
    chromatography-mass spectrometry,
    106
Trimethylsilylation, of ceramides, 348
    general methods of, 342
    as a general technique for the handling of
        steroids, 335
    of nucleosides, 15
    of nucleotide derivatives, 349
    of phosphate hydroxyl groups, 348
Triose phosphate isomerase, 245
Tripeptides, free, mass spectra of, 290
Triphenylmethyl ion, loss of methyl from,
    264
Triple bonds, means of coding on teletype-
    writer, 142
Triply charged monomer ions from
    polystyrene, 211
Tris-perfluoroethyl-s-Triazine as internal
    standard, 33

Tris-perfluoroheptyl-s-Triazine as internal
    standard, 33
Tris-perfluoropropyl-s-Triazine as internal
    standard, 33
Tritium-laden, foils in gas chromatography,
    205
    tantalum as radiation source for ionization,
    205
Trityl ions, loss of methyl from, 264
Tropone, 2-methoxybenzo-, oxygen-18,
    carbon-13 labelled, mass spectrum
    of, 257
Tropylium, ion, formation of in electron
    impact mass spectra, 188
    in the mass spectra of benzyl compounds,
    378
    structure of, 261
    ions, 390
Truncation, in Heuristic programming, 125
    of isomer list in Heuristic DENDRAL
        programs, 147
    of structures list in DENDRAL, 142
    of tree graphs with valence restrictions, 137
Tryptic hydrolysis of proteins, 319
Tryptophan, butylation of during removal of
    t-butyloxycarbonyl groups, 319
    -containing peptides, intermolecular
        alkylation reactions in, 301
    identification of in peptides, 300
    incorporation of into alkaloids, 243
    in peptides, specific fragmentation
        reactions of, 299
    residues in peptides, instability of in
        acidic media, 305
    side reactions of silver oxide with, 316
Tubercle bacilli, human, phthiocerols in, 345
Tumor, kidney, metals in, 331
Tungsten, field emission microscopy of, 183
    field emission of, 182
Tuning of source after changes in accelerating
    voltage, 426
Tunneling of electrons, 182
Twig, terminating, of DENDRAL tree, 139
Types of ions, use of in computer interpreta-
    tion of mass spectra, 85
Tyrosine, -containing peptides, double
    hydrogen-migration processes in, 301
    enzymatic, formation of from phenylalanine,
    357
    oxidation of to Dopa, 229
    hydroxylase, adrenal, 229
    hydroxyl group of, migration of hydrogen
        from, 300
    in peptides, effect of on volatility, 315
    specific fragmentation reactions of, 299

Ubiquinone, incorporation of deuterium
    into from trideuteromethionine, 357
Ubiquinones, detection of by gas chromatogra-
    phy-mass spectrometry, 345
Uncertainty principle, Heisenberg's, 182
Unimolecular, decomposition of ions following
    collisions, 429

processes, in field ionization sources, 185
of ions, 407
reactions of ions, rate constants of, 420
"Unlikely" molecules, prevention of construction of in DENDRAL program, 169
Unsaturated acids, location of double bonds in, 258
Unsaturation, degree of, use of in computer interpretation of mass spectra, 86
in DENDRAL notation, 140
handling of by DENDRAL notation, 137
in radicals, handling of by DENDRAL notation, 139
"Unusual" steroids in infant urine, identification of, 338
Uranium, mass spectrometric determination of, 332
toxicology of, 332
Urea, nitrogen–15 labelled, 281
2–Ureido-2-thiazoline, formation of from 2-amino-2-thiazoline, 281
Urine, analysis for lead in, 331
drug metabolites in, 334
Uropygiols, structure and mass spectra of, 343

Valence, electrons, number of, dependence, of ionization efficiency upon, 204
restrictions in DENDRAL notation, 137
Valence bond isomers, formation of in thiophenes, 385
formed from aromatic carbonium ions, 440
postulated in fragmentation of aromatic compounds, 384
Valencia orange peel oil, 259
Valine, and leucine, confusion between following peptide methylation, 316
in peptides, specific fragmentation reactions of, 295
Valium*, chemical ionization mass spectrum of, 195
electron impact mass spectrum of, 195
metabolism of, 350
Val(6)-peptidolipin-NA, structure determination of by mass spectrometry, 310
Van de Graaff generator as proton source, 205
Van der Waals forces in Penning ionization and chemi-ionization, 207
Vaporization of high molecular weight compounds, 210
Variable, apertures, use of with silicon rubber membranes, 102
conductance separators, 104
slit separator, slit size in, 104
Variations in sensitivity with accelerating voltage, 46

Verification subroutine in Heuristic DENDRAL programs, 170
Vesicular glands, homogenate of, 224
Vibrational degrees of freedom, importance of number of, 436
Vibrationally excited states, reactions occurring from, 399
Vindoline, biosynthesis of, 242
Vinyl acetylene, as a fragment ion, appearance potential of, 436
ionization potential of, 436
4–Vinylcyclohexene, deuterium labelled, mass spectrum of, 402
radical cation, collapse of to butadiene radical cation, 402
bis-(Vinylcyclopentadienyl-alpha-d)iron, mass spectrum of, 379
Viscosic acid, amino acid sequence of, 316
Vitamin, B6, metabolic degradation of, 251
D, metabolism of, 341
25-hydroxylated derivatives of, identification of, 341
E, mass spectrum of, 342
K2, biosynthesis of, 246
incorporation of deuterium into from trideuteromethionine, 357
Volatility, lack of in free peptides, 290
limitations set by on peptide sequencing by mass spectrometry, 315
of peptides, effect of different amino acids upon, 315
factors determining, 311

Water, loss of from amide groups in the mass spectra of peptides, 294
"Whiskers," use of to increase electrical potentials, 182
Whisky, volatile components of, 363
Wide range electron impact kinetics, 396
Widths of, metastable peaks, 445
factors affecting, 445
peaks in, exponential scans, 46
mass spectrum of o-nitrophenol, 423
Wiswesser line notation for organic structures, 127
Woodward-Hoffmann rules, 399
applications of in mass spectrometry, 399
Woodward-Hoffman classification, in the cyclization of 1,4-diphenyl-butadiene, 403
of disrotatory processes, 400
in elimination reactions, 402
in mass spectral reactions, 398
Work function of metals, 181

Yield of separators, definition of, 95

Zymogen of phospholypase A, sequence of heptapeptide from, 316

*Registered.